Diversity and Commonality in Animals

The book series Diversity and Commonality in Animals publishes refereed volumes on all aspects of zoology, with a special focus on both common and unique features of biological systems for better understanding of animal biology. Originating from a common ancestor, animals share universal mechanisms, but during the process of evolution, a large variety of animals have acquired their unique morphologies and functions to adapt to the environment in the struggle for existence. Topics covered include taxonomy, behavior, developmental biology, endocrinology, neuroscience, and evolution. The series is an official publication of The Zoological Society of Japan.

More information about this series at http://www.springer.com/series/13528

Shuichi Shigeno • Yasunori Murakami •
Tadashi Nomura
Editors

Brain Evolution by Design

From Neural Origin to Cognitive Architecture

Springer

Editors
Shuichi Shigeno
Stazione Zoologica Anton Dohrn
Naples, Italy

Yasunori Murakami
Ehime University
Matsuyama, Japan

Tadashi Nomura
Kyoto Prefectural University of Medicine
Kyoto, Japan

ISSN 2509-5536 ISSN 2509-5544 (electronic)
Diversity and Commonality in Animals
ISBN 978-4-431-56467-6 ISBN 978-4-431-56469-0 (eBook)
DOI 10.1007/978-4-431-56469-0

Library of Congress Control Number: 2016962454

Printed on acid-free paper

This Springer imprint is published by Springer Nature
The registered company is Springer Japan KK
The registered company address is: Chiyoda First Bldg. East, 3-8-1 Nishi-Kanda, Chiyoda-ku, Tokyo 101-0065, Japan

Preface

The present book is a new, detailed examination to explain how elegant brains have been shaped by simple principles in evolution. Classic comparative studies have revealed great diversity of neural networks and complex behaviors in many animal groups. The recent integrative molecular, developmental, physiological, and psychological approaches, however, have revealed unexpected commonality in the basic structures and functions across animal phylogeny. The structural frameworks of the nervous systems and brains are often replicated in artificial intelligence or machines constructed by human activities, suggesting that functional similarities provide a common design for information processing systems through biophysical constraints.

The book introduces the origin of neurons with the single-cell creatures without neurons and then goes on to primordial types in invertebrates such as cnidarians, flatworms, molluscs, insects, and chordates, with a great abundance of the brains of vertebrates: fish, reptiles, birds, and mammals, including whales and humans. Recently, a number of research investigations of diverse and minor organisms have been conducted, and we need to keep up to date. Each chapter provides professional and detailed topics about brain evolution; however, this book as a whole is arranged along a simple concept to find something of common design. Also, non-organisms such as models and materials are covered to explore the designs in the origin and evolutionary processes, but they are not comprehensive. The topics are provided in a timely manner because novel techniques emerged rapidly, for example, as seen in next-generation sequencers and omics (e.g., genomics, proteomics, metabolomics, and connectomics) approaches. With the explosion of big data, the neural-related genes and molecules are now on the radar.

Importantly, now the neural networks have been taken notice of. For instance, Europe's €1 billion science and technology projects, such as the Human Brain Project, were launched in 2013 to analyze brain connectomics. The big interdisciplinary plan, the Blue Brain Project, also aims to understand the small mammalian brain. Furthermore, with the rise of recently advanced artificial neural networks, there is enthusiasm for the development of neural network models. The views of brain evolution provide an essential opportunity to generate ideas for novel neuron-

and brain-inspired computation. For that reason, this book will show the reader how to extract meaningful neural structures in nature.

For undergraduates, graduate students, and professional scientists who seek a deeper understanding, this volume demonstrates how to find the basic principles shaping brains that provided higher cognitive functions in the course of evolution. Our ambition is that the book will stimulate students, particularly young scientists, to delve into problems remaining in this discipline. Many authors were selected from young Japanese scientists, and this work is part of a series of publications of the Zoological Society of Japan.

Naples, Italy Shuichi Shigeno
Matsuyama, Japan Yasunori Murakami
Kyoto, Japan Tadashi Nomura

Contents

Part I The Origins of Neurons and Networks

1 **Physical Ethology of Unicellular Organisms** 3
Shigeru Kuroda, Seiji Takagi, Tetsu Saigusa, and Toshiyuki Nakagaki

2 **Molecular Characteristics of Neuron-like Functions in Single-Cell Organisms** 25
Shingo Maegawa

3 **Back Through Time: How Cnidarians and Basal Metazoans Shed Light on Ancient Nervous Systems** 45
Hiroshi Watanabe

Part II The Rise of Diverse Brain Types

4 **Functional Specification of a Primitive Bilaterian Brain in Planarians** 79
Takeshi Inoue

5 **The Computation and Robustness of the Mini-Cognitive Centers of Terrestrial Mollusks: An Exquisite Outcome of Brain Evolution** 101
Ryota Matsuo

6 **Insect Brains: Minute Structures Controlling Complex Behaviors** ... 123
Michiyo Kinoshita and Uwe Homberg

7 **Identifying Vertebrate Brain Prototypes in Deuterostomes** 153
Takehiro G. Kusakabe

8 **Genome and Transcriptome-Wide Research of Brain Evolution** 187
Atsushi Ogura

Part III Cognitive Systems

9 The Origin of Vertebrate Brain Centers 215
Yasunori Murakami

10 Adaptive Radiation and Vertebrate Brain Diversity: Cases of Teleosts 253
Naoyuki Yamamoto

11 Molecular Profiling Reveals Insight into Avian Brain Organization and Functional Columnar Commonalities with Mammals 273
Kazuhiro Wada, Chun-Chun Chen, and Erich D. Jarvis

12 The Neocortex and Dorsal Ventricular Ridge: Functional Convergence and Underlying Developmental Mechanisms 291
Wataru Yamashita and Tadashi Nomura

13 Molecular Investigations of the Structure and Development of the Brain of Carnivores 311
Yohei Shinmyo, Tomohisa Toda, Kosuke Masuda, Yoshio Hoshiba, Haruka Ebisu, Naoyuki Matsumoto, and Hiroshi Kawasaki

14 Evolution of the Mammalian Brain with a Focus on the Whale Olfactory Bulb 329
Takushi Kishida

15 The Evolution and Function of Sleep 343
Yu Hayashi and Chih-Yao Liu

16 Prefrontal Anatomical Architecture and Top-Down Behavioral Control in Human and Nonhuman Primates 367
Yosuke Morishima

Part IV Models and Designs

17 Organisational Principles of Connectomes: Changes During Evolution and Development 387
Roman Bauer and Marcus Kaiser

18 Muscular-Hydrostat Computers: Physical Reservoir Computing for Octopus-Inspired Soft Robots 403
Kohei Nakajima

19 Brain Evolution as an Information Flow Designer: The Ground Architecture for Biological and Artificial General Intelligence 415
Shuichi Shigeno

Part I
The Origins of Neurons and Networks

Chapter 1
Physical Ethology of Unicellular Organisms

Shigeru Kuroda, Seiji Takagi, Tetsu Saigusa, and Toshiyuki Nakagaki

Abstract In this chapter, some behaviours of unicellular organisms that appear to be smart or intelligent are reported. Two topics are the focus from two major groups of eukaryotic unicellular organisms, amoebae and ciliates: (1) anticipatory capacity of periodic environmental events in an amoeba and (2) environment-induced development of a new type of behaviour in a ciliate. A mechanism of these behaviours is discussed, based on a mechanical equation of motion. Ethology (the science of animal behaviour) of unicellular organisms is recently being studied from a physical point of view. We propose to call this kind of study {physical ethology}. Physical ethology may give us some hints about the origin of primitive intelligence.

Keywords Ciliate • Learning • Adaptability • Primitive intelligence • Membrane excitability • Ethology • Nonlinear dynamics

1.1 Introduction

The cell is the building block of all organisms, which can be the minimum set of life, because the cell is not alive if it is divided into subsystems: this implies that all essence to be common throughout the entire range of organisms must be found in a cell. So, the cell is the interface between an assembly of merely materials and the functional states of living systems.

In fact, many kinds of smart behaviour that look like a primitive form of intelligence in some sense have been reported and compared with so-called intelligent behaviours in higher animals (Jennings 1906; Bray 2009; Eisenstein 1975; Corning and Von Burg 1973; Trewavas 2003, 2005).

S. Kuroda (✉) • T. Nakagaki
Physical Ethology Laboratory, Research Institute for Electronic Science, Hokkaido University, Sapporo 001-0020, Japan
e-mail: nakagaki@es.hokudai.ac.jp

S. Takagi
Faculty of Complex and Intelligent Systems, Future University Hakodate, Hakodate, Japan

T. Saigusa
Department of Cellular Physiology, Institute of Nephrology, Niigata University, Niigata, Japan

S. Shigeno et al. (eds.), *Brain Evolution by Design*, Diversity and Commonality in Animals, DOI 10.1007/978-4-431-56469-0_1

Although the list of such behaviours is getting longer and longer, the mechanism of the smartness is being studied from a mechanical point of view. This is an attempt to understand an origin of organismic smartness in nature and is an exciting problem to be tackled by truly interdisciplinary science in this century. Here, we wish to shed light on the attempt.

Two of the impressive examples are maze solving (Nakagaki et al. 2000, 2001; Tero et al. 2006, 2007) and anticipating periodic events (Saigusa et al. 2008) by amoebae. *Physarum*, which recently has become a focus as an interesting model organism, is a huge amoeba that constructs a transport network of information and materials and mimics an anthropogenic network of public transportation. This description implies that there may be a common nature of network formation shared by man and *Physarum* (Tero et al. 2010).

The other example of smart behaviour is the environment-induced development of new types of behaviour in a ciliate (Kunita et al. 2014). In the behaviour of a ciliate, swimming activity is regulated by membrane potential (Naitoh 1990), and an equation of motion for the membrane potential is described by the reaction–diffusion equation of the Hodgkin–Huxley type, which was originally proposed for the squid neuron. Because of the similarity of excitability of membrane potential, the ciliate is often studied as a model system of neuronal activity (Eisenstein 1975; Naitoh 1990).

So far, the attempts to seek for a physical origin of smart behaviours are just at the beginning, but this line of study, which we propose to name physical ethology, is exciting and promising for the future.

In this chapter, as an introduction to the physical ethology of a cell, we describe in more detail one of the symbolic topics rather than the original report already studied: "*Physarum* anticipates periodic events" (Saigusa et al. 2008). Although a mathematical model for anticipatory behaviour was proposed and discussed, we do not repeat it here. [Please see the reference (Saigusa et al. 2008) for explanation of the model, which supplies the main body of this chapter.]

After that, we briefly mention one more topic of the swimming behaviour of ciliates because the cilia are the other main character of Protozoa as well as the amoebas. A new kind of smart behaviour in ciliates, reported recently in 2014 (Kunita et al. 2014), is described in short as a summary. The mechanism is reduced to the dynamics of membrane potential that controls the beating activity of the cilia. So, this is another typical example of physical ethology.

At the end of this chapter, some concluding remarks are given on the possible origin of intelligent behaviour that has emerged at the level of a single cell.

1.2 Anticipatory and Recall Behaviour in Response to Periodic Stimulation in the Plasmodia of *Physarum polycephalum*

1.2.1 Overview and Background

We report on adaptive responses to periodic stimulation in *Physarum polycephalum* (Mycetozoa), which exhibits primitive anticipatory and recall behaviour. The behaviour studied here involved the spontaneous periodic slowdown of migration just after a series of periodic stimulations had been experienced (anticipatory behaviour). This slowdown subsequently disappeared but then reappeared after a single stimulation (recall behaviour). The behaviour displayed was characterized by the following features. (1) The anticipatory slowdown depended on the size of the body. (2) The response varied across different parts of the body. (3) Recall behaviour appeared even in organisms that had failed to display anticipation earlier. (4) The cellular rhythm of the contraction movement was much slower during both the spontaneous slowdown (anticipation) and recall behaviour than during free locomotion; the same response was shown during the slowdown directly induced by external stimulation. The results obtained here give new insight about how this ability of anticipation and recall emerges at the cell level.

The repetition of a particular event often leads to the formation of memory and learning in organisms. For example, after mice were fed several times at regular intervals, they learned to anticipate the next feeding time (Roberts and Church 1978; Church 1978; Meck et al. 1984). When bees were given nectar at 9:00 a.m. every day, they gathered at the scheduled time even if nectar was not given (Gould and Gold 1988). These are elegant examples of anticipatory behaviour towards periodic events.

In 2008 we reported that such behaviour can be observed even in a single cell, despite the absence of any brain or nervous system (Saigusa et al. 2008). Similar behaviours could be observed in other protozoa (*Brepharisma*) and a plant (*Chara*) (Kunita et al. 2013), but the anticipatory behaviours in these species were not so clear as statistical fluctuations were large and the organisms often failed the anticipation. Thus, further characterization is needed as this finding is interesting from an evolutionary point of view and gives a hint as to the cellular origin of primitive intelligence (Ball 2008).

The anticipatory behaviour reported in the paper by (Saigusa et al. 2008) is as follows. The plasmodium of the true slime mould *Physarum polycephalum* moves rapidly under favourable conditions, but stops moving when experiencing less favourable conditions. Plasmodia exposed to unfavourable conditions (low temperature and low humidity), presented in three consecutive pulses at constant intervals, were found to reduce their locomotive speed in response to each episode. When subsequently subjected to favourable conditions, the plasmodia spontaneously reduced their locomotive speed at the point in time when the next unfavourable episode would have occurred. This finding implies that the plasmodia can anticipate impending environmental change. After this behaviour had been evoked several times in the course of favourable conditions, the locomotion of the plasmodia returned to normal; however, the anticipatory response could subsequently be induced again by a single unfavourable pulse, implying recall of the memorized periodicity.

To focus on deviation and fluctuation of response, we added new experiments and analysed the deviation by reexamining a large set of data that included not only the new data but also the previous data (Saigusa et al. 2008); this is the first point to be considered here.

The second point is dependency on the internal conditions of the cell. In general, behavioural responses to environmental conditions depend on the internal conditions of the cell, but little is known about it. So, we will test the possible factors of cellular conditions: size of body, spatial inhomogeneity in parts of a large body, and effects on rhythmic contraction of the cell.

We study further details of the anticipatory behaviour and report several new findings and insights.

1.2.2 Materials and Methods

Organisms and Culture The plasmodium of *Physarum polycephalum* (wild type) was cultured using oat flakes on an agar plate in a rectangular plastic dish (25×35 cm^2) at 26 °C and humidity of 90 % according to conventional methods.

Observation of Locomotion Velocity Under Periodic Changes in Atmospheric Temperature and Humidity The tip portion of the large cultured organism was cut out and placed at the end of a narrow rectangular area (0.5×28 cm) on the agar plate, hereafter referred as an arena, as shown in Fig. 1.1a. The wet weight of the plasmodial mass that was initially placed in the arena varied from 5 to 40 mg, depending on the particular experiment. The arena containing the specimen was kept in the dark in an incubator (KCL-10000; Eyela) in which the atmospheric humidity and temperature were controllable. The migrating organism was illuminated from below by a matrix of infrared LEDs and viewed by a CCD camera. Under the culture conditions used, the plasmodium started to move towards the other end of the arena approximately half an hour later and was allowed to move freely for 4 h. The atmospheric temperature and humidity were then changed to 23 °C and 60 % for 10 min, conditions that represent the stimulation. This stimulation was repeated three times at intervals of τ ($\tau = 30, 40, 50, 60, 70, 80, 90$ min). The position of the extending front of the migrating plasmodium was measured every 10 min; typical time courses of the tip position are shown in Fig. 1.1b.

Typical Response Analysed in This Study, Originally Reported in 2008 (Saigusa et al. 2008) A typical time course of speed and acceleration of migration as well as the time course of temperature and humidity are shown in Fig. 1.1c, as reported in the previous papers. According to the terminology used in the previous paper, the periodic stimulations were numbered from S1 to S2 and S3 (indicated by gray bars),

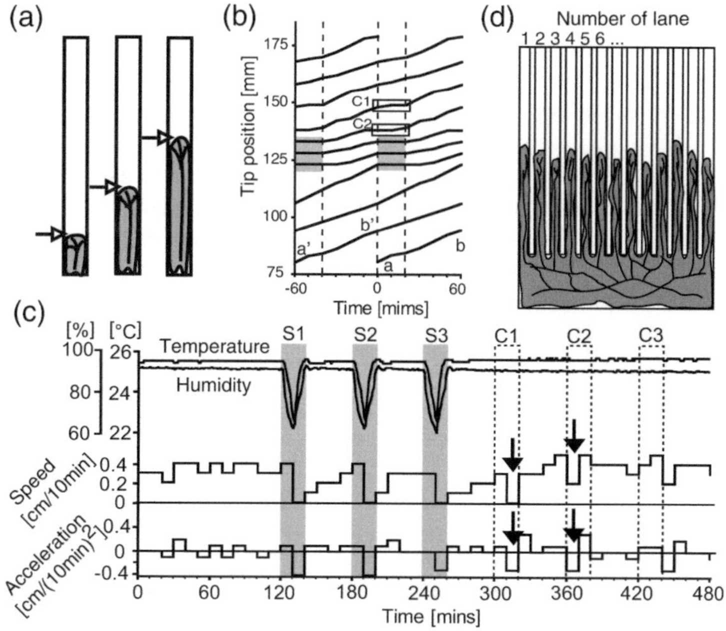

Fig. 1.1 Schematic illustration of experimental setups used to study *Physarum*. (**a**) Illustration of migration along a narrow lane from bottom to top. *Arrow* indicates the frontal tip of the organism, the position of which was used to calculate locomotion speed. Arena size was 0.5×28 cm^2. (**b**) Typical plot of tip position with respect to time. The plot has been cut into sections of 60 min that are stacked vertically; each section includes the preceding 60 min (plotted as negative time) to help recognize periodic responses. For instance, the line a–b is identical to a'–b'. The column colored *grey* indicates the times at which stimulation occurred. (**c**) Typical time course for locomotion speed in response to periodic stimulation. *Black arrows* indicate spontaneous slowdown (SPS) at times when the next stimulation would have occurred if continued periodically. The wet weight of the initially inoculated organism was 15 mg. Speed and acceleration were averaged every 10 min. Sn and Cn were timing of actual stimulation and the following periodic timing (no stimulation). This response was originally reported in the previous paper (Saigusa et al. 2008). (**d**) Setup for the 'comb' experiment in which the arena was shaped like a comb with 12 teeth into which the plasmodium could extend a pseudopod-like protrusion. The migration speed was measured along every tooth. Tooth size was 0.5×25 cm^2

and the following periodic points in time (no stimulation was applied) were C1, C2, C3, and so on (indicated by dashed rectangles). In C1 and C2 only, the locomotion speed often decreased spontaneously, so this response was called spontaneous in-phase slowdown (SPS) hereafter.

An Arena with Comb-like Shape Designed for Testing Inhomogeneity of Response Over the Whole Body Another arena in the form of a comb that consisted of many parallel slender corridors with a common base was prepared (Fig. 1.1d). The large plasmodium was placed in the base and started to extend along the many corridors while remaining as a single entity.

Spatiotemporal Observation for Rhythmic Contraction According to the conventional method (Takagi and Ueda 2008), we measured contractile activity in the plasmodium. The plasmodium was illuminated by the 20 × 20 matrix of infrared light emission diode (wave length, 920 nm) from below, and was viewed from above by the black-and-white CCD camera (spatial resolution, 480 × 640; time resolution, 1/30 s). As the gray level of the video image reflected the thickness of the plasmodium that changed periodically with accompanying to active rhythmic contraction, a darker pixel indicated a thicker part of plasmodium. The relationship between optical density on the image and the real thickness of organism was standardized by the calibration. We used the standard free software for image analysis Image J that was prepared by the National Institutes of Health (USA). A spatiotemporal pattern of oscillation periods was visualized as shown in Fig. 1.6.

1.2.3 Results

Effect of Body Size on Anticipatory Behaviour Figure 1.2 shows averaged time courses of the migration speed. The plasmodia that were initially inoculated varied in wet weight, and the difference in body size was classified into three groups: 5–9 mg ($n = 27$), 10–19 mg ($n = 68$), and 20–40 mg ($n = 21$). The initial speed before stimulation was similar for all three groups but differed after stimulation. Although the group with large body size recovered its previous migration activity after stimulation, the groups of medium and small body size did not move as fast as before stimulation. This finding implies that stimulation caused serious damage to the smaller organisms. In the group with body size from 5 to 9 mg wet weight,

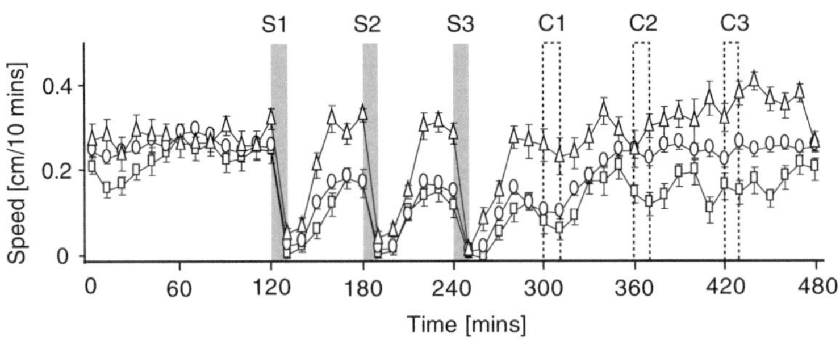

Fig. 1.2 Confirmation of anticipatory behaviour in *Physarum* obtained by measuring averaged speed of locomotion. Different body masses were initially inoculated for the three time courses shown: small (5–9 mg, $n = 27$, *lower line* with *rectangular* symbols), medium (10–9 mg, $n = 68$, *middle line* with *circular symbols*), and large (20–56 mg, $n = 21$, *upper line* with *triangular symbols*). Anticipatory behaviour is observed at C1 and C2. Mean ± standard error of mean. Statistical significance of difference was tested by pairwise t test with Bonferroni correction (a method of multiple comparison). The p value was less than 0.001 between the small and the large, and between the medium and the large, at the time points of C1 and C2, and between all pairs at C3

some plasmodia did not migrate at all after the first stimulation (data not shown). Such cases were excluded from the results shown in Fig. 1.2 and 21 experiments when finally averaged. When stronger stimulation was applied (for example, 18 °C and 50 % humidity), even larger plasmodia were damaged by a single stimulation.

The spontaneous slowdown response that was observed depended on a balance between body size and the magnitude of the environmental perturbations. Anticipatory behaviour was shown in all three groups at the times of the fourth (C1) and fifth (C2) virtual stimulations. The anticipatory response was clearer for smaller plasmodia because the ratio of the migration speeds in the two phases of slowdown and recovery was higher than for larger plasmodia. The smaller organisms demonstrated clearer anticipation but suffered more serious damage. The anticipation capacity thus depends on both internal and external conditions.

Response in Local Parts of the Body Figure 1.3 shows time courses of the migration acceleration for the comb-shaped plasmodium (see Materials and Methods for details). The lane numbers indicate different parts of the body corresponding to different 'teeth' of the comb-shaped arena. After stimulation had been applied at times S1, S2, and S3, anticipatory behaviour was expected at times C1, C2, and C3. Anticipatory behaviour was indeed observed as the overall tendency (black squares indicate deceleration). However, the expected response was not always observed in individual lanes. For example, it was absent in lanes 1–3, 7, and 10 at C1, and present in lanes, 6, 8, 9, and 11. The varying behaviour between different local parts of the body implies that the anticipatory response is subject to fluctuations and uncertainty. This comb-type experiment was repeated three times, and similar inhomogeneity of response was observed in local parts of the body.

Response to Different Periods of Stimulation Figure 1.4 shows the occurrence of spontaneous slowdown in response to various periods of stimulation in the range 30–90 min. In all cases, the periodic stimulation was applied three times (S1, S2, and S3) and the time axis was set to zero where the last stimulation (S3) was given. The first, second, and third occasions of virtual stimulation are indicated by C1, C2, and C3, respectively. At C1, the statistical occurrence of slowdown was high (10 %–50 %) for all periods of stimulation. The baseline occurrence of slowdown in the absence of stimulation was approximately 10 %. At C2, a high occurrence of slowdown

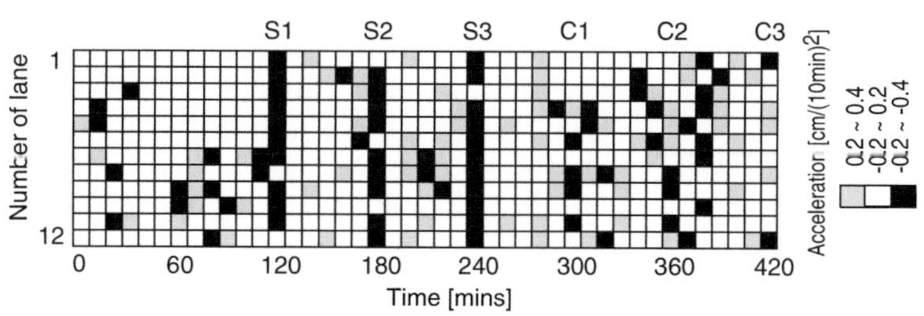

Fig. 1.3 Acceleration as a function of time for different lanes of the comb-shaped arena. The *grey squares* indicate positive acceleration and the *dark squares* indicate braking

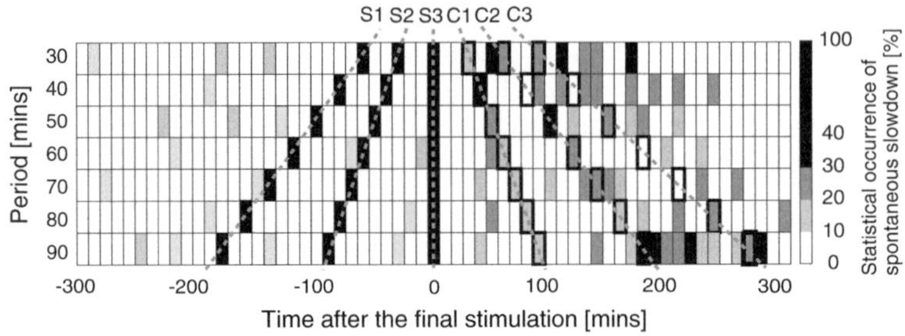

Fig. 1.4 Anticipatory response for stimulation periods ranging from 30 min to 90 min. The *grey scale* indicates the statistical occurrence of spontaneous slowdown. Zero time was defined as the time when the last stimulation S3 was applied. Spontaneous slowdown was defined to occur when acceleration ≤ -0.2 mm/(10 min). The numbers of repeated experiments for each stimulation period were 36 (30 min), 46 (40 min), 34 (50 min), 121 (60 min), 40 (70 min), 67 (80 min), and 39 (90 min). The wet weight of the organism was 10–19 mg. Data that were already used in the previous paper (Saigusa et al. 2008) were included in this analysis while we supplemented new data of 97 repeats to make this new analysis of fluctuation sufficiently reliable. We analyzed all the data, not only the previous ones but also the new data

was still observed although the timing of the response was sometimes shifted a little earlier or later. At C3, the anticipatory response was no longer significant as the timing greatly fluctuated. In each experiment for a given stimulation period, there was a degree of deviation in the time at which spontaneous slowdown was observed. The anticipatory responses were often advanced or delayed by as much as a quarter of a period, which can be seen as a characteristic of this type of behaviour. Nevertheless, there is a clear tendency towards a high probability of occurrence along line C1 in Fig. 1.4. We conclude that anticipatory spontaneous slowdown occurred for a wide range of periods from 30 min to 90 min.

Recall Behaviour With and Without Anticipatory Behaviour Figure 1.5 shows time courses of slowdown occurrence and averaged speed in response to a single stimulation that was applied 5 h after the series of three periodic stimulations. After the anticipatory response at C1, the locomotion speed quickly returned to the value that it had before the sequence of stimulations. A further single stimulation was then given at S4. In response, spontaneous slowdown was displayed at C6, which corresponds to the period of stimulation previously applied for S1–S3 (1 h). This response implies that the organism had stored a memory of the periodicity somewhere in its body that was later retrieved. This type of behaviour can be seen as a primitive form of recall. The white bars and circles in Fig. 1.5 indicate the responses of a group of organisms that showed anticipatory behaviour at C1, whereas the black bars and circles represent the group that failed to display this behaviour. Recall behaviour was observed even for the group that failed anticipatory behaviour, although the response was slightly weaker than that of the successful group. Thus, recall was not restricted to individuals that succeeded in showing anticipatory behaviour, which implies that storing a memory is a process that occurs

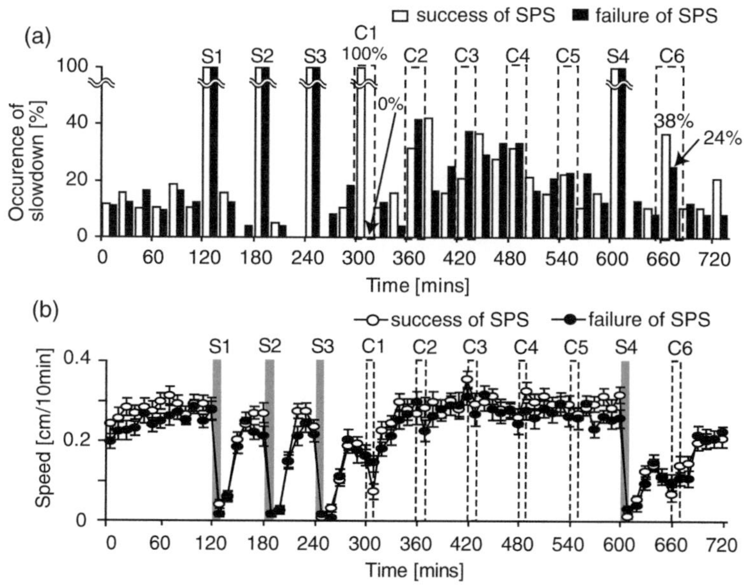

Fig. 1.5 Recall behaviour in response to stimulation S4 applied after the anticipatory response has disappeared, studied by measuring the statistical occurrence of slowdown (**a**) and averaged speed (**b**) over 40 repeats. The criterion for spontaneous slowdown was acceleration ≤ -0.2 mm/(10 min). The wet weight of the organism was 15 mg. The data of the success of SPS shown in Fig. 1.5b were the same as those previously published (Saigusa et al. 2008). The number of repeats was 40, and this number of repeats was the sum of the new data and the previous experiments (Saigusa et al. 2008). Mean $\pm$ standard error of mean. Statistical significance of difference was tested by t test. The p value was not observed to be less than 0.01 between the success and the failure groups at the time points of C1, C2–C5, and C6

in the body of both groups. Despite fluctuations in the times at which anticipation and recall took place, it is clear that *Physarum* is able to store meaningful information in the form of time memory.

Modulation of Rhythmic Contraction During Spontaneous Slowdown Figure 1.6 shows a typical time course of the rhythmic contraction that occurred when the organism showed anticipatory behaviour. The plasmodium usually displays periodic cycles of contraction and relaxation over the entire body with an average period τ of approximately 1 min, as shown by the spatiotemporal plot of the oscillation period in Fig. 1.6a. The oscillation period of the entire body became 70 s longer when stimulation was applied at S1, S2, and S3. During the spontaneous slowdown of locomotion speed at C1, slower oscillations were observed (τ 90 s), but this increase in period was smaller than that at S1, S2, and S3. Figure 1.6b, c shows the time courses of the oscillation period, averaged over the entire body, and of the locomotion speed, respectively. An increase in the period of the contraction rhythm was clearly observed at C1 in Fig. 1.6b. A minimum in the time course of the speed was also observed at C1 (Fig. 1.6c), as occurred at S1, S2, and S3. That is,

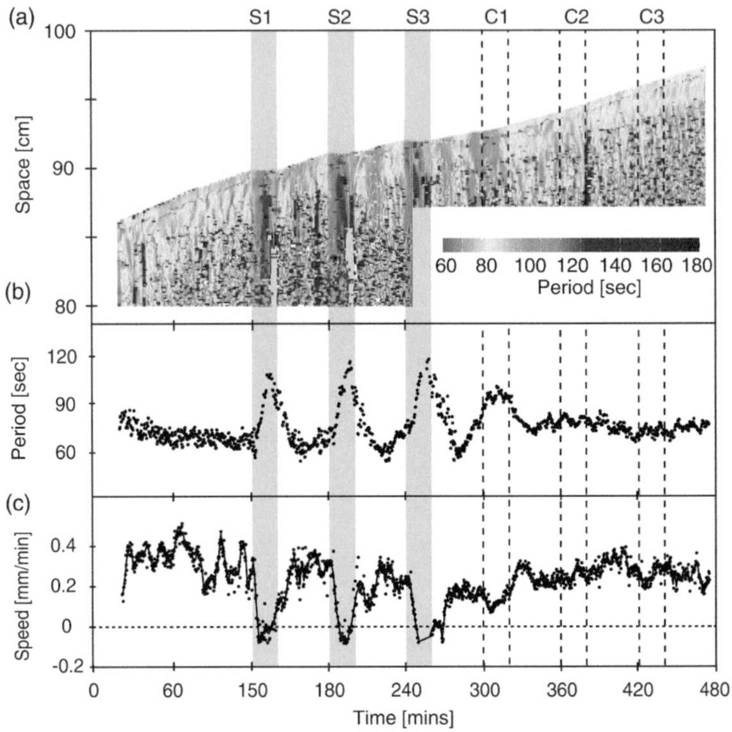

Fig. 1.6 Modulation of cellular rhythm in response to periodic stimulation. (**a**) Period of rhythmic contraction as a function of time. (**b**) Time course of contraction period, averaged over frontal part of organism (1 cm from extending tip). (**c**) Speed of locomotion with respect to time: a typical response in all three repeats. Body weight initially inoculated was 10–19 mg

the rhythmic contraction slowed down whenever the locomotion speed decreased, whether induced by external stimulation or as a result of spontaneous anticipatory behaviour. This result implies that the elementary process of anticipatory response involves not only the overall behaviour of locomotion but also the local kinetics of biochemical oscillatory reactions. Figure 1.7 shows the changes in oscillation period (Fig. 1.7a) and locomotion speed (Fig. 1.7b) associated with the recall behaviour at C6 (the experimental conditions were the same as in Fig. 1.5.). In the time course of the contraction period, a maximum similar to that observed for the anticipatory response was displayed at C6. The time course of the locomotion speed showed a maximum between S4 and C6 and a minimum at the beginning of C6. That is, a slowdown in both contraction rhythm (maximum of period) and locomotion (minimum of speed) occurred during the recall behaviour around C6. In this sense, the recall behaviour was similar to both the anticipatory behaviour and the slowdown induced by external stimulation.

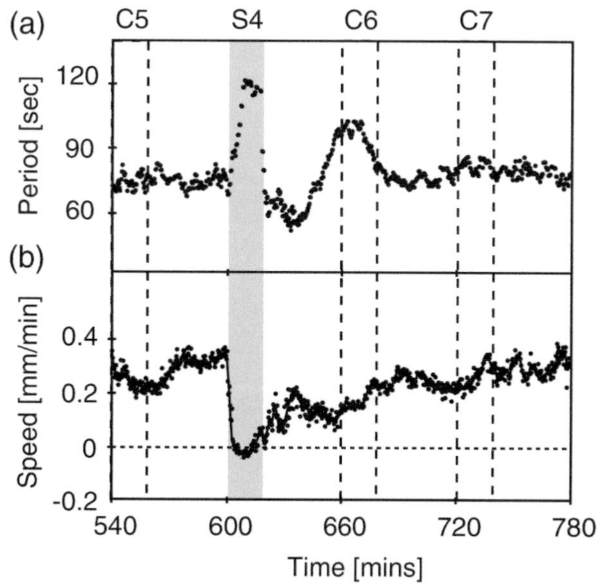

Fig. 1.7 Modulation of contraction rhythm during recall response. (**a**) Time course of contraction period after stimulation S4. (**b**) Time course of locomotion speed: a typical response in all three repeats. Body weight initially inoculated was 10–19 mg

1.2.4 Discussion

Further Examination of Data and Confirmation of Dependency of Applied frequency Although the previous paper shows just the findings of the anticipatory behaviour and, in particular, information on dependency of applied periodicity is very limited, the current paper shows a more consistent analysis of data, in which not only the new data but also the previous data of the experiment are included. Figure 1.4 gives an overall picture of anticipatory response in the parameter space that was spanned by the applied periodicity and time. The anticipatory response was observed to show a deviation of time to spontaneous slowdown around the correct time points C1 and C2, but the anticipatory response was clearly observed around C1 when we looked through a series of periodicity. The time deviation of SPS increased as the time progressed from C1 to C3. Around C3, the anticipatory response was no longer observed. When we consider justification for the physical mechanism proposed for the anticipatory behaviour, statistical distribution of the deviation may give us helpful information.

As shown in Fig. 1.5a, the plasmodia that failed the first anticipation (0 % at C1) were able to succeed in the second and third anticipation (approximately 30–40 % at C2 and C3). This is remarkable. It is interesting to check whether this response is reproduced by the previously proposed mathematical model.

Size Effect In general, it is known that the larger plasmodium is more resistive against sudden changes in environmental conditions such as temperature and humidity. For instance, although one large organism can survive against a drop in humidity although some part of the large body is necrotic, a much smaller one cannot survive against the same drop in humidity. This difference may be because loss of water vapour from the surface of the body is relatively higher in smaller organisms: the surface–volume ratio increases as the body size decreases, in principle. So, we expect that smaller organisms tend to be more sensitive to the periodic environmental change. If body size, however, is smaller than the critical size, the organism can die or try to transform to a resting stage such as the spore and sclerotium. The anticipatory behaviours observed here obey a balance between maintenance of vegetative stage (plasmodium) and morphological transformation to resting stage (spore and sclerotium). It is reasonable that a smaller plasmodium is sensitive to the periodic changes in humidity and temperature as facing to higher risk of survival.

Inhomogeneity of Response in a Whole Body We discuss the result that the anticipatory response was not homogeneous in the whole body but differed from part to part: one part might be successful while another part is not. According to the implication of the previously published model for the anticipatory response in *Physarum* (Saigusa et al. 2008), the capacity is based on the chemical kinetics of a complicated network of biochemical reactions. It is reasonable that biochemical reactions are very similar over the body but states of reaction do not always synchronize because chemical diffusion and protoplasmic streaming may not be always sufficient for maintaining perfect synchronization throughout the body. In fact, some chemicals such as ATP, Ca^{2+}, and NAD(P)H show inhomogeneous distribution in the body and moreover such inhomogeneity is actively related to the development of amoeboid behaviour in *Physarum* (Nakamura and Kamiya 1985; Ueda et al. 1990; Ueda 1993; Yoshiyama et al. 2010). So, spatial differences of anticipatory response may result from the inhomogeneity of dynamic states of reaction kinetics.

Recall with No Anticipation In the results shown in this paper, the plasmodium that failed the anticipatory response sometimes succeeded in the recall behaviour. According to the mathematical model previously proposed (Saigusa et al. 2008), the core process in kinetic motion of biochemical reactions in the plasmodium is that two kinds of phase synchronization take place: synchronization (or formation of phase cluster, in other words) between biochemical oscillators that have the same natural frequency (the same biochemical species, for instance), and synchronization between the these phase clusters that have a slightly different natural frequency. The second synchronization is related to recall behaviour whereas the first one is related to storing the memory of periodicity. As these two synchronization phenomena can be separable, we may expect that recall is possible without anticipation. It is interesting that this possibility could be examined in the conventional mathematical model for anticipatory and recall behaviours in the future.

The plasmodia that failed the first anticipation (0 % at C1; Fig. 1.5a) were able to succeed the second and third anticipation (approximately 30–40 % at C2 and C3). This is remarkable. It is interesting to check whether this response is reproduced by the previously proposed mathematical model.

Effects on Contraction Rhythms In this paper, information on intracellular states was shown for the first time: modulation of intracellular rhythm. The contraction rhythm (the typical period is 1–2 min) was modulated in response to the environmental changes and the same modulation was confirmed in the spontaneous slowdown of anticipatory and recall behaviours. Because this rhythmicity was much faster than we had considered in the mathematical modelling (applied periods were 30–90 min) in the previous paper (Saigusa et al. 2008), the information from this result does not help to examine the model justification directly. At least, we can extract the message that oscillatory activity in the plasmodium was modulated in relationship to development of anticipatory and recall behaviours. Further observation and examination of frequency modulation at a slower time scale is interesting in future studies to clarify the physical mechanism of the anticipatory and recall behaviour at cellular level.

Possible Approach to Physical Mechanism We have confirmed the previous finding (Saigusa et al. 2008) by reexamining not only new data but also the previous data together. *Physarum* exhibits anticipatory and recall behaviour in response to periodic changes in environmental conditions. Humans have a similar capacity. Our study implies that this type of capacity is not specific to a single species but rather may be found in multiple species distributed over a wide range of the phylogenetic tree. Therefore, the underlying mechanism should be considered as a common process. One approach to studying such a mechanism is to search for specific genes and proteins that are involved, but an alternative approach is the modelling of physical dynamics. According to the latter, a simple and general model can be proposed: anticipatory and recall behaviour is reproduced by the collective motion of independent oscillators with a wide range of natural frequencies. This is a possible model for the anticipatory behaviour, but another set of model equations is also proposed (Tachikawa 2010). Although the discussions on the possible physical mechanism are still ongoing, it is remarkable that higher capacities such as anticipation and recall are created by the simple and general model of differential equations. The study of how protozoa anticipate periodic events hints at the origin of time memory from an evolutionary point of view.

Evaluation Method for Memory Capacity for Complexity of Time Sequence In this report, the stimulation was applied in regularly periodic fashion. The time sequence of the stimulation could be made more complex. For instance, possible sequences are (1) alternative changes in the stimulation magnitude, strong and weak, over a regular period; (2) double-frequency stimulations such as alternating long and short stimulation periods; and (3) a chaotic sequence of stimulations. According to the ideas presented here, we may be able to design a standard test to assess the capacity of time memory throughout the phylogenetic tree that will be performed in the future.

1.3 Electric Control of Behaviour in *Paramecium*

Paramecium is a genus of popular unicellular ciliated protozoa that you can find easily in the pond near your office. Their average length and width are about 200 μm and 60 μm, respectively. Although they have a simpler body plan than ours, their behaviours are so rich that we can be glued to the microscope for long moments. They have many cilia (about 10,000) on their body surfaces, and they swim by ciliary beating that is spatiotemporally coordinated through the entire body so that they form a metachronal wave.

They can change their swimming directions by changing the directions of the effective stroke of ciliary beating. Ordinarily, in a uniform environment, they swim forwards. However, when they encounter an obstacle, they swim backwards for a while by reversing the direction of the effective stroke of ciliary beating (reversal of ciliary beating), and after swinging their anterior end, they begin to swim forward in a new direction. This behaviour is called avoiding reaction (Jennings 1906).

The relationships between the behaviour and electrophysiological property of *Paramecium* has been intensively studied for more than 50 years. During normal forward swimming, they have an internally negative resting potential that is provided by the intracellular high K^+ and low Ca^{2+} concentrations. It was known that the increase in intraciliary concentration of free calcium ion ($[Ca^{2+}]_i$) is necessary for the reversal of ciliary beating, and it is produced by depolarization of the cellular membrane (Naitoh and Eckert 1969; Naitoh and Kaneko 1972).

The appearance of the avoiding reaction behaviour is explained as follows: When the organism collides a obstacle at the front end, the mechanosensitive Ca^{2+} channel opens. The resulting depolarization that occurred around the receptors spreads immediately to the entire body including cilia. The depolarization of the ciliary membrane opens the voltage-sensitive Ca^{2+} channels on the membrane so that Ca^{2+} flows into the cilia from the outside of the body. The increase of $[Ca^{2+}]_i$ provides the reversal of ciliary beating so that the organism begins to swim backwards. The ciliary reversal is continued during $[Ca^{2+}]_i$ over a certain critical concentration (Naitoh and Kaneko 1972). The relaxation of the $[Ca^{2+}]_i$ leads it to resume the forward swimming. These data shows that an excitable cellular membrane in the protozoan acts as a receptor of the external stimulus and a transmitter between the receptor and the effector, whereas those are enacted by neural systems in eumetazoans, including human beings (Fig. 1.8).

What kinds of adaptability can be shown by *Paramecium*? If they exist, we can expect to understand them from the point of view of the membrane potential phenomena and their succeeding biochemical reaction. Actually, various behaviours in Protozoa have been reported since 100 years ago (Jennings 1906; Smith 1908; Bramstedt 1935; Applewhilte 1979). Smith (Smith 1908) found a novel behaviour of *Paramecium* in a dead-ended capillary tube. In this experiment, the organism was put into a narrow tube of which the width was smaller than its length, enough not to be able to turn by the avoiding reaction. In the beginning, the organism swims forward towards the closed end. After colliding with the tube end at its front part,

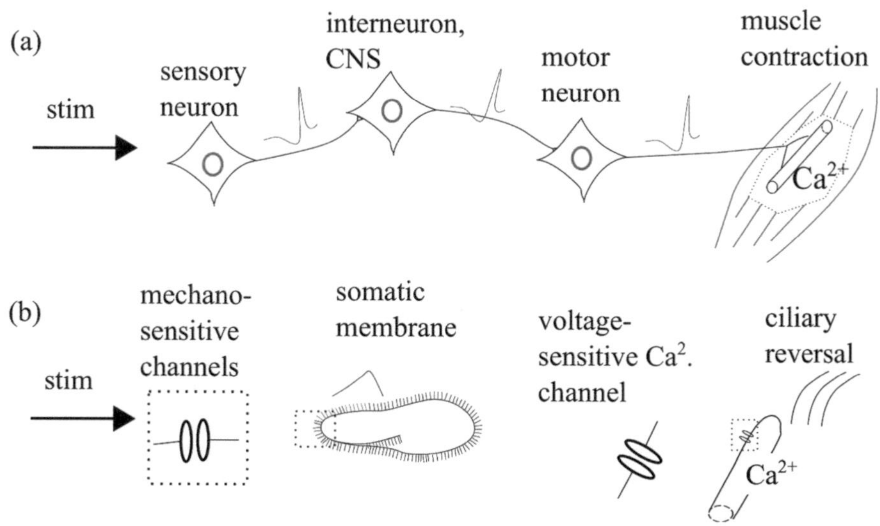

Fig. 1.8 Signal transmission by electrogenesis in (1) Eumetazoa and (2) Protozoa

it swims backward about its own length and resumes swimming forwardly; this is a part of the ordinary avoidance reaction and was repeated at least a dozen times. Finally, after some struggling, the organism succeeded to change its body direction by bending the body into a U-shape.

How do they behave when they encounter a more difficult situation? Recently we reported another novel behavior of *Paramecium* in a dead-ended capillary tube (Kunita et al. 2014). We used a narrower tube so that *Paramecium* cannot change its body direction by bending its body. The typical behaviour is as follows: After the organism was put into the tube, as in Smith's experiment, the ordinary avoiding reaction appeared at the closed end. That is, the organism approached a forward end and repeated back-and-forth swimming with a short distance (<0.5 mm). After that, for about 1 min, the distance of backward swimming gradually increased and finally reached a maximum distance (3–4 mm). It continued the long-distance backward swimming for a few minutes. We called the emergent long-distance backward swimming 'long-term backward swimming (LBS),' whereas that in the ordinary avoiding reaction is short-term backward swimming (SBS).

1.3.1 *Paramecium* Model

What is a underlying mechanism of the emergence of LBS? Naitoh (Naitoh 1990) has reported that a long-term application of outward current to the paramecium provided long-lasting backward beating of cilia even after the early inward Ca^{2+} current disappeared. Naitoh suggested the long-lasting backward swimming is caused by a small-amplitude long-lasting inward Ca^{2+} current that flows into the same Ca^{2+} channel for the early inward Ca^{2+} current. Based on such previous studies, we exploited a *Paramecium* model using a conductance-based model and attempt to demonstrate the LBS.

Ciliary Membrane Let us begin with modelling of the ciliary membrane based on a conductance-based model. The total ciliary membrane current (I_M) is the sum of the capacitive (I_c) and the ionic currents (I_i), and so:

$$I_M = I_c + I_i = C_m \frac{dE}{dt} + I_i, \tag{1.1}$$

where E is the potential difference between the between the outside and the inside of the membrane and the outward current is defined as positive current. The resting equilibrium potential (E_r) is realized during ordinary forward swimming. For simplicity of notation, we use $V = E - E_r$ instead of E in the following. It is known that the voltage-sensitive Ca^{2+} channel and also the voltage- sensitive K^+ channel are located primarily in the ciliary membrane whereas they are not found in the somatic membrane (Dunlap 1977; Machemer and Ogura 1979; Eckert and Brehm 1979). Then, the ionic current I_i is given by

$$I_i = I_{Ca} + I_K + I_L \tag{1.2}$$

$$= g_{Ca}(V,t)(V - V_{Ca}) + g_K(V,t)(V - V_K) + \overline{g}_L(V - V_L), \tag{1.3}$$

where I_{Ca} and I_K are Ca^{2+} and K^+ currents, respectively; I_L is a relatively small voltage-independent conductance of undetermined ions; and V_{Ca}, V_K, and V_L are the equilibrium potential for the respective ions. We use the following conductance formula, which was proposed by Naitoh (Naitoh 1979; Naitoh and Sugino 1984).

$$g_{Ca}(V,t) = \overline{g}_{Ca} m(V,t)^5 \left(1 - (1 - h(V,t))^5\right), \tag{1.4}$$

$$g_K(V,t) = \overline{g}_K n(V,t). \tag{1.5}$$

Here, each gate variable m, h, or $n \in [0, 1]$ develops by $dz/dt = \alpha_z(V)(1 - z) - \beta_z(V)z$, where $\alpha_z(V)$, $\beta_z(V)$ are the pair of voltage-dependent reaction rates ($z = m, h, n$) that were determined by reference to the measurement data available from (Hirano et al. 2005).[1] In the following, time, potential, current density, conductance,

[1]The expression of the functions for the voltage-dependent reaction rates (α, β) were determined by reference to the measurement data available from (Hirano et al. 2005) as follows: $\alpha_m(V) = 0.39(46.40 - V)/(\exp(0.039(46.40 - V)) - 1)$, $\beta_m(V) = 0.65 \exp(-V/15)$, $\alpha_h(V) = 0.05 \exp(-V/50)$, $\beta_h(V) = 1.0(\exp(0.032(V - 39.29)) + 1)$, $\alpha_n(V) = 0.038(58.58 - V)/(\exp((58.58 - V)/8.17) - 1)$, $\beta_n(V) = 0.10 \exp(-V/68)$. However the α_h was determined ad hoc, because its data were not available. The constant values in this paper were set as follows (Naitoh 1990; Hirano et al. 2005). $E_r = -30$, $E_{Ca} = +116$, $E_K = -41$,

capacity and concentration are given in ms, mV, μA/cm^2, mS/cm^2, μF/cm^2, and μM, respectively.

Mechanical Stimulation at the Dead-End and Depolarization A mechanical stimulation at the capillary dead end induces a depolarizing receptor potential from the mechano-sensitive channels distributed on the anterior part of the somatic membrane. This potential change in the local membrane area spreads to the entire membrane almost instantaneously (Dunlap 1977) as a result of the successive induction of the outward capacitive current passing through the non-depolarized adjacent membrane. In our model, the resultant current on the ciliary membrane is implemented as $I_M = I_{\mathrm{mech}}(>0)$ in (1) when the organism is at the capillary dead end ($x = 0$); otherwise, $I_M = 0$.

Relationship Between Intraciliary Ca^{2+} Concentration and Swimming Velocity The swimming velocity (v) against the intraciliary free Ca^{2+} concentration ([Ca^{2+}]$_{\mathrm{i}}$) has been obtained in chemically skinned cells (Naitoh and Kaneko 1972), where [Ca^{2+}]$_{\mathrm{i}}$ during normal forward swimming is 10^{-2} μM and the sign of the swimming velocity switch is near 1 μM. Figure 1.9b shows the [Ca^{2+}]$_{\mathrm{i}}$ versus v graph that is used in the simulation and was determined by reference to the experimental data in (Naitoh and Kaneko 1972), but the velocity (v) was rescaled by a factor of ten because the velocity of the normal forward-swimming specimen is expected to be 1–2 mm/s (Kunita et al. 2014).

Estimation of Intraciliary Ca^{2+} Concentration In contrast to the membrane potential, calcium ions cannot traverse between cellular body and cilia because of cytoplasm buffer action (Naitoh 1990). Extraciliary Ca^{2+} passes only through the ciliary membrane. On the other hand, the Ca^{2+} are exported to the outside by calcium pumps on the ciliary membrane. So, we assumed that [Ca^{2+}]$_{\mathrm{i}}$ is approximately determined by the contributions of Ca^{2+} channels and Ca^{2+} pumps on the ciliary membrane and is simply estimated by the following equation:

$$\frac{d[\mathrm{Ca}^{2+}]_i}{dt} = -\frac{I_{\mathrm{Ca}} - I_{\mathrm{Ca}}^0}{2F \times 10^3}\gamma_{\mathrm{sv}} - \frac{[\mathrm{Ca}^{2+}]_i - [\mathrm{Ca}]_i^0}{1 + \left([\mathrm{Ca}^{2+}]_i - [\mathrm{Ca}]_i^0/K_m\right)}\gamma_{\mathrm{pu}} \quad (1.6)$$

where F is Faraday constant [C/mol], γ_{sv} is a constant proportional to the surface-to-volume ratio of the cilium, and γ_p[ms^{-1}] is a rate constant depending on the Ca^{2+} pump performance. K_{m} [μM] is the concentration at which the pump operates at half its maximum rate, and I_{Ca}^0 and $[\mathrm{Ca}^{2+}]_i^0$ are the small constants of Ca^{i} Ca^{2+} current and [Ca^{2+}]$_{\mathrm{i}}$ during the ordinary forward swimming, respectively.

$E_L = -25, V. = E. - E_r.$ $\bar{g}_{Ca} = 0.67$, $\bar{g}_{Ca} = 1.34$, $\bar{g}_L := \left(g_{Ca}^{\infty}(0)V_{Ca} + g_K^{\infty}(0)V_K\right)/V_L$, $C_m = 1.0$. Ca K

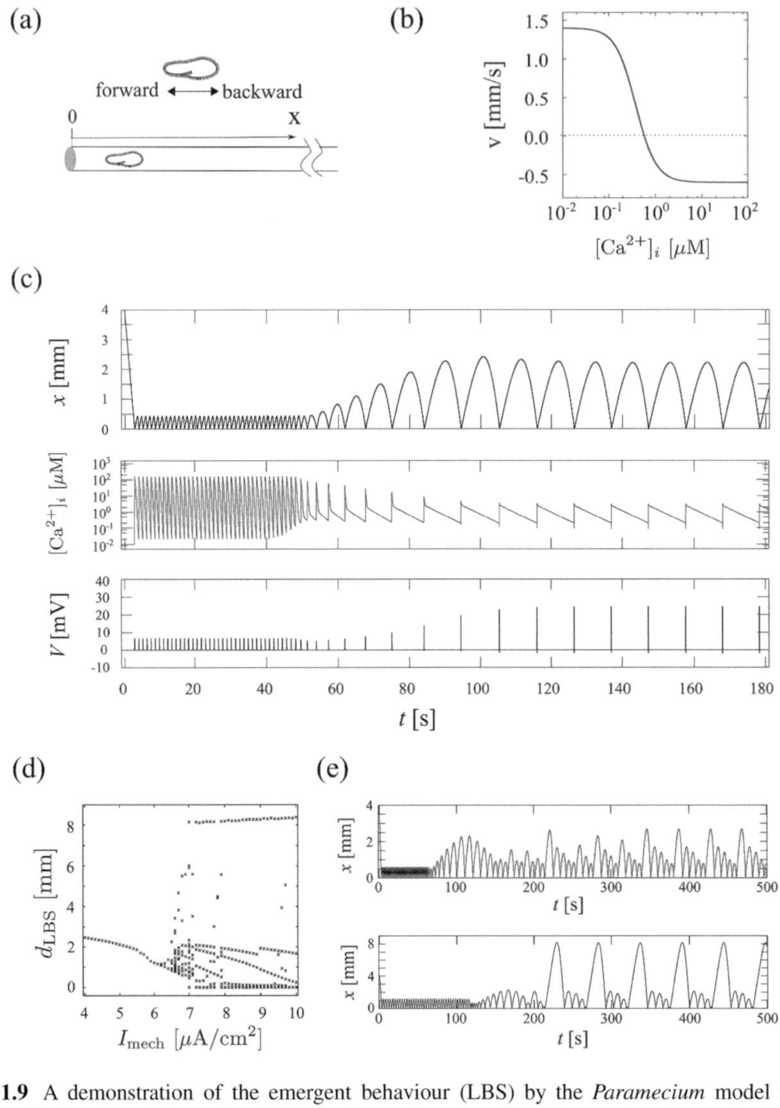

Fig. 1.9 A demonstration of the emergent behaviour (LBS) by the *Paramecium* model in the narrow dead-end capillary tube. (**a**) Experimental setting. The *Paramecium* is confined in the glass capillary tube (0.08 mm diameter, 40–50 mm length) of which the end was closed with mineral oil (Kunita et al. 2014). (**b**) Intraciliary Ca^{2+} concentration versus swimming velocity. $v = v_{-\infty} + (v_{\infty} - v_{-\infty})/(1 + b[\mathrm{Ca}]^{-a})$ (mm/s) where $v_{-\infty} = 1.40$, $v_{\infty} =$ i -0.60, $a = 2.0$, $b = 2.91 \times 10^{-7}$. (**c**) Time courses of the position of *Paramecium* (*upper panel*), membrane potential (*middle panel*), and intraciliary Ca^{2+} consecration. The *vertical lines* through the three panels indicate the timings when the organism reached the dead end ($x = 0$) with forward swimming. $I_{mech} = 5$, $K_m = 1 \times 10^2$, $\gamma_{pu} = 1.0 \times 10^{-2}$, and $\gamma_{sv} = 10 \times$ area/vol, where area/vol $= 2.0 \times 10^8$ (cm^2/l) is the surface:volume ratio of a cilium (Machemer 1974). (**d**) A bifurcation diagram of the backward distance (d_{LBS}) during the LBS period depending on the sensitivity to mechanical stimulation (I_{mech}). (**e**) Time courses of the position of the *Paramecium* model with $I_{mech} = 6.6$ (*upper panel*) and 7.7 (*lower panel*). In all simulations, the time constant of the slow-inactivated Ca^{2+} channel was set to be 2000 times that of the ordinary channel

Slow Ca^{2+} Current Induced by Repetitive Stimulation The small-amplitude, long-lasting inward Ca^{2+} currents mentioned by Naitoh (1990) are assumed to be recruited by repetitive collisions at the forward end of the capillary tube. This is implemented in our model in that an ordinary Ca^{2+} channel on the ciliary membrane is modified to a slow inactivated one with a probability δp every time an action potential happens, although it becomes normal again exponentially over time.[2]

Simulation Results Figure 1.9c shows an example of the time course of the position and the ciliary electrophysiological states of the *Paramecium* model. The organism showed LBS after SBS ($t<50$s) via a gradual increase of the backward swimming distance (upper panel of Fig. 1.9c). The shift from SBS to LBS was caused by the interaction between the organism and environment via the collision-induced recruitment of the slow-inactivated ciliary Ca^{2+} channel. Because the temporal position and the electrochemical states of the organism are determined by both the internal dynamics (Eqs. 1.1–1.6) and the history of the collision with the tube end, the model behaviour can be more diverse than that derived from only the internal dynamics. Our model includes several parameters of which values may vary through the specimen or environmental uncertain factors. I_{mech} is one of such parameters and it relates to the sensitivity to mechanical stimulations. Figure 1.9d shows dependence of backward distances after emergence of LBS on I_{mech}. As an increase of I_{mech}, various qualitative changes occurred, including doubling of the backward distances, sudden occurrence of a new long backward distance, intervening of SBS-like behaviour, and so on. Such a bifurcation diagram suggests the behaviour during LBS can vary from simple to complex and vice versa dependent on specimens and environmental details. Actually, several variants of the LBS behaviour were observed in our experiments. However, the quantitative analysis of such aspects are still not developed. Further investigation is left for future studies.

1.4 Comparative Remarks in Single-Celled Organisms and Higher Organisms

Although the capacity of memory and learning is generally assumed to be limited to higher organisms, a number of studies have suggested that these attributes can also be displayed by unicellular organisms (Bray 2009; Eisenstein 1975; Corning and Von Burg 1973; Reid et al. 2012, 2013; Hinkle and Wood 1994) as well as by nonliving systems such as spin echo and electronic circuits (Pershin et al. 2009;

[2] In the simulation, the δp assumed to be proportional to $n_s(n-n_s)/n^2$ so that n_s grows and saturates in a sigmoidal manner, where n is the total number of the Ca^{2+} channels and n_s is the number of the modified channels. The update equation for $p=n_s/n$ is given as $p(t)=(p(t_i)+k_p p(t_i)(1-p(t_i)))\exp(-(t-t_i)/\tau_p)+p_\varepsilon$, where t_i is the most recent collision time, $0<k_p\leq 1$ is a growth rate, τ_p is a relaxation time constant, and p_ε is a sufficiently small positive constant.

Chung and Choe 2009). This perception is supported by the fact that unicellular organisms have survived successfully for billions of years. It is thus reasonable to expect a primitive form of memory and learning in the most elementary living systems. The study of the extent to which this capacity is possessed by unicellular organisms, and the manner by which it is realized, is interesting with respect to the evolutionary origin of intelligence.

Acknowledgments This research was supported by JSPS KAKENHI grant no. 26310202, and by a Grant-in-Aid for Scientific Research on Innovative Area 'Fluctuation and Structure' (no. 25103006) and 'Cross talk between moving cells and microenvironment' (no. 25111726) from Mext Japan, and by Strategic Japanese-Swedish Research Cooperative Program, Japan Science and Technology Agency (JST).

References

Applewhite P (1979) Learning in protozoa. In: Biochemistry and physiology of protozoa, vol 1. Academic Press, New York, pp 341–353

Ball P (2008) Cellular memory hints at the origins of intelligence. Nature (Lond) 451:358

Bramstedt F (1935) Dressurversuche mit paramecium caudatum und stylonychia mytilus. Z Vergl Physiol 22:490–516

Bray D (2009) Wetware: a computer in every living cell. Yale University Press, New Haven

Chung JR, Choe Y (2009) Emergence of memory-like behavior in reactive agents using external markers. In: 21st IEEE international conference on tools with artificial intelligence, pp 404–408

Church RM (1978) The internal clock. In: Hulse SH, Fowler H, Honing WK (eds) Cognitive processes in animal behavior. Erlbaum, Hillsdale

Corning WC, Von Burg R (1973) Protozoa. In: Corning WC, Dyal JA, Willows AOD (eds) Invertebrate learning. Plenum Press, New York, pp 49–122

Dunlap K (1977) Localization of calcium channels in paramecium caudatum. J Physiol 271:119–133

Eckert R, Brehm P (1979) Ionic mechanisms of excitation in *Paramecium*. Annu Rev Biophys Bioeng 8:353–383

Eisenstein EM (1975) Aneural systems and neurobiology: a point of view. In: Eisenstein EM (ed) Aneural organisms in neurobiology. Plenum Press, New York, pp 1–4

Gould JL, Gold CG (1988) The honey bee (Scientific American Library, vol 186). Freeman, San Francisco

Hinkle DJ, Wood DC (1994) Is tube-escape learning by protozoa associative learning. Behav Neurosci 108:94–99

Hirano A, Tsuji T, Takiguchi N, Ohtake H (2005) Modeling of the membrane potential change of paramecium for mechanical stimuli. Trans SICE 41:351–357

Jennings H (1906) Behavior of the lower organisms. Indiana University Press, Bloomington

Kunita I, Sato S, Saigusa T, Nakagaki T (2013) Ethological response to periodic stimulation in *Chara* and *Brepharisma*. In: Natural computing and beyond, Proceedings in Information and Communications Technology (PICT), vol 6. Springer, New York, p 3–3

Kunita I, Kuroda S, Ohki K, Nakagaki T (2014) Attempts to retreat from a dead-ended long capillary by backward swimming in *Paramecium*. Front Microbiol 5:270. doi:10.3389/fmicb.2014.0027

Machemer H (1974) Frequency and directional responses of cilia to membrane potential changes in *Paramecium*. J Comp Physiol 92:293–316

Machemer H, Ogura A (1979) Ionic conductances of membranes in ciliated and deciliated paramecium. J Physiol 296:49–60
Meck WH, Church RM, Olton DS (1984) Hippocampus, time, and memory. Behav Neurosci 98:3–22
Naitoh Y (1979) Membrane currents in voltage clamped *Paramecium* and their relations to ciliary motion. Acta Protozool 18:1–6
Naitoh Y (1990) Behavior of unicellular animals (in Japanese). University of Tokyo Press, Tokyo
Naitoh Y, Eckert R (1969) Ionic mechanisms controlling behavioral responses of *Paramecium* to mechanical stimulation. Science 164:964–965
Naitoh Y, Kaneko H (1972) Reactivated triton-extracted models of *Paramecium*: modification of ciliary movement by calcium ions. Science 176:523–524
Naitoh Y, Sugino K (1984) Ciliary movement and its control in *Paramecium*. J Protozool 3:31–40
Nakagaki T, Yamada H, Tóth A (2000) Maze-solving by an amoeboid organism. Nature (Lond) 407:470
Nakagaki T, Yamada H, Tóth A (2001) Path finding by tube morphogenesis in an amoeboid organism. Biophys Chem 92:47–52
Nakamura S, Kamiya N (1985) Regional difference in oscillatory characteristics of *Physarum plasmodium* as revealed by surface pH. Cell Struct Funct 10:173–176
Pershin YV, La Fontaine S, Di Ventra M (2009) Memristive model of amoeba learning. Phys Rev E 80:021926
Reid CR, Latty T, Dussutour A, Beekman M (2012) Slime mold uses an externalized spatial memory to navigate in complex environments. Proc Natl Acad Sci USA 109(43):17490–17494
Reid CR, Beekman M, Latty T, Dussutour A (2013) Amoeboid organism uses extracellular secretions to make smart foraging decisions. Behav Ecol 24(4):812–818
Roberts S, Church RM (1978) Control of an internal clock. J Exp Psychol Anim Behav Process 4:318–337
Saigusa T, Tero A, Nakagaki T, Kuramoto Y (2008) Amoebae anticipate periodic events. Phys Rev Lett 100:018101
Smith S (1908) The limits of educability in *Paramecium*. J Comp Neurol Psychol 18:499–510
Tachikawa M (2010) A mathematical model for period-memorizing behavior in *Physarum plasmodium*. J Theor Biol 263:449–454
Takagi S, Ueda T (2008) Emergence and transitions of dynamic patterns of thickness oscillation of the plasmodium of the true slime mold *Physarum polycephalum*. Physica D 237:420–427
Tero A, Kobayashi R, Nakagaki T (2006) Physarum solver: a biologically inspired method for road-network navigation. Physica A 363:115
Tero A, Kobayashi R, Nakagaki T (2007) Mathematical model for adaptive transport network in path finding by true slime mold. J Theor Biol 244:553–564
Tero A, Takagi T, Saigusa T, Ito K, Bebber DP, Fricker MD, Yumiki Y, Kobayashi R, Nakagaki T (2010) Rules for biologically-inspired adaptive network design. Science 327:439–442
Trewavas A (2003) Aspects of plant intelligence. Ann Bot 92:1–20
Trewavas A (2005) Green plants as intelligent organisms. Trends Plant Sci 10:413–419
Ueda T (1993) Pattern dynamics in cellular perception. Phase Transit 45:93–104
Ueda T, Nakagaki T, Yamada T (1990) Dynamic organization of ATP and birefringent fibrils during free locomotion and galvanotaxis in the plasmodium of *Physarum polycephalum*. J Cell Biol 110:1097–1102
Yoshiyama S, Ishigami M, Nakamura A, Kohama K (2010) Calcium wave for cytoplasmic streaming of *Physarum polycephalum*. Cell Biol Int 34:35–40

Chapter 2
Molecular Characteristics of Neuron-like Functions in Single-Cell Organisms

Shingo Maegawa

Abstract Single-cell organisms can respond to stimuli from the environment, including chemical and tactile stimuli, to survive and propagate. Thus, single-cell organisms appear to show neuron-like functions. In this review, I investigate neuron-like functions in bacteria (*Escherichia coli*) and ciliates (*Paramecium* and *Tetrahymena*). In *E. coli*, six chemotaxis-specific (*che*) genes have been identified as critical in the ability of organisms to react to stimuli from the environment. The *che* genes encode signaling molecules to transmit information from receptors to motor proteins that regulate some *E. coli* behavior. Thus, the Che proteins are thought to form a "central processing unit (CPU)"-like complex in *E. coli*. The eukaryotic single-cell organisms *Paramecium* and *Tetrahymena* have also been employed for understanding the molecular mechanisms underlying ciliate behaviors. Paramecia uptake calcium ions and show membrane excitation when they receive a repulsive stimulus, similar to neurons. In addition, the calcium ions function as the second messengers through calmodulin activity and regulate the concentration of cAMP in cilia. Increment changes in cAMP concentration in the cilia result in changes in their beating pattern, which alters the behaviors of paramecia. Moreover, our recent results indicate that the neurotransmitter serotonin is involved in physical functions in *Tetrahymena thermophila*. These results indicate clearly that bacteria and ciliates are equipped with neuron-like functions. The discussion addresses whether single-cell organisms have intelligence, emotion, and mind.

Keywords Bacteria • Ciliates • Behavior • Signal processing • Central processing unit • Learning

---

S. Maegawa, Ph.D. (✉)
Department of Intelligence Science and Technology, Graduate School of Informatics, Kyoto University, Kyoto, Japan

Kyoto University, Yoshida Honmachi, Sakyo, Kyoto 606-8501, Japan
e-mail: shingo@i.kyoto-u.ac.jp

S. Shigeno et al. (eds.), *Brain Evolution by Design*, Diversity and Commonality in Animals, DOI 10.1007/978-4-431-56469-0_2

2.1 What is a "Neuron-like" Function?

In vertebrates, neurons are essential for processing sensory information from the environment to regulate physiological conditions and behaviors. Accordingly, a "neuron-like" function is the ability to process information from the environment and to evoke a specific response. Organisms live in many types of environments. Some of these environments are unstable, such as deserts or deep-sea environments, and may be severe. However, unique organisms are able to inhabit these environments and survive. Organisms in any environment must sense, recognize, and adjust to dynamic environmental changes to survive and propagate. However, many questions remain as to the evolution of organism abilities to respond to the environment.

Single-cell organisms can regulate physiological conditions and their behaviors in response to stimuli from the environment (Fig. 2.1), even without a cellular network such as the central nervous system found in vertebrates. In fact, several studies have shown that single-cell organisms such as bacteria, ciliates, and slime molds can sense and adapt to the environment and display learning and memory. Here, I especially introduce about the molecular mechanisms of neuron-like function in prokaryotes (*Escherichia coli*) and eukaryotes (*Paramecium* and *Tetrahymena*), including molecular mechanisms of the transduction of environmental signals.

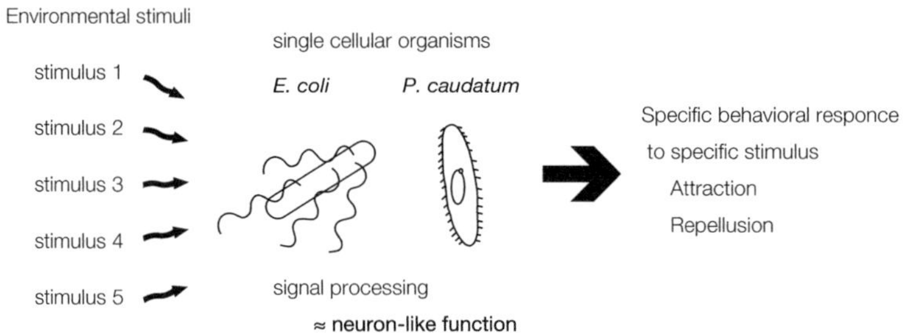

Fig. 2.1 Neuron-like function in single-celled organisms. Single-celled organisms such as *Escherichia coli* and *Paramecium caudatum* can receive external stimuli from the environment and demonstrate specific behavioral responses (attraction or repulsion) to a specific stimulus. In this chapter, the signal-processing pathway starting from external stimuli to specific behavioral response is defined as a neuron-like function of single-cell organisms

2.2 Neuron-like Functions in Prokaryotes

2.2.1 Swimming Behaviors in Bacteria

Bacteria swim freely in freshwater or saltwater to obtain nutrients. Observations of bacterial swimming patterns indicate the presence of switching mechanisms for smooth swimming and a tumbling behavior when they receive no specific stimuli (Berg 1975; Berg and Brown 1972). After the tumbling behavior, bacteria usually swim smoothly in a random orientation (Berg 1971; Berg and Brown 1972). The tumbling behavior is important to change the direction of swimming in bacteria (Fig. 2.2). This random reorientation from each tumbling is thought to enable bacteria to explore the surrounding environment in a random pattern (Berg 1971; Berg and Brown 1972).

In the late nineteenth century, German researchers described that bacteria swim toward external stimuli and environments that are optimal for survival and away from environments that are dangerous (Koshland 1980). Evidence indicates that bacteria can receive external stimuli and display specific behaviors to specific stimuli (Koshland 1980). In addition, bacteria can regulate the tumbling behavior. For example, bacteria suppress the tumbling behavior when they approach a preferable environment. In contrast, bacteria may enter a dangerous environment if the tumbling behavior is not evoked (Koshland 1980; Macnab and Koshland 1972). Thus, the regulation of the frequency of the tumbling behavior is sufficient to bias the random walk of bacteria and enable swimming toward preferable environments or escaping from danger.

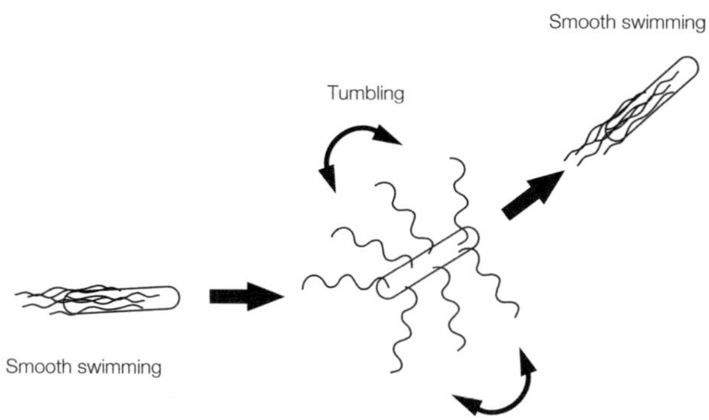

Fig. 2.2 Two modes of swimming of *Escherichia coli*. In a bacterium demonstrating smooth swimming, the flagella rotate counterclockwise and are fastened together. With time, the bacterium shows tumbling behavior by a stochastic process to randomly choose a new direction of swimming. Importantly, the flagella move clockwise during tumbling. After tumbling, a bacterium swims smoothly again. Thus, bacterial swimming is regulated by the direction of rotation of the flagella

2.2.2 Sensory and Motor Systems in Bacteria

Several genetic studies have been conducted to understand the mechanisms underlying the chemotactic response in bacteria. The results revealed that bacteria possess receptors for external stimuli and a motor apparatus for regulating swimming behavior. Five chemoreceptors that capture external stimuli, such as amino acids, carbohydrates, and oxygen, were identified (Bibikov et al. 1997; Boyd et al. 1983; Kondoh et al. 1979; Manson et al. 1986; Reader et al. 1979). These receptors are components of the bacterial "sensory" system, similar to human eyes, ears, etc.

The bacterial motor apparatus is the flagella on the bacteria cell surface (Depamphi and Adler 1971; Iino 1969; Macnab and Koshland 1974). Observation of the swimming pattern of *E. coli* demonstrates that flagella are held close when bacteria swim smoothly to one direction, but are spread wide apart when bacteria undergo the tumbling behavior (Fig. 2.2) (Macnab and Koshland 1974). In addition, flagella move in a counterclockwise (CCW) rotation similar to a propeller when a bacterium swims smoothly. In contrast, when evoking the tumbling behavior, flagella turn to move clockwise (CW), which results in disruption of the bundle of the flagella (Fig. 2.2) (Larsen et al. 1974). Taken together, these results demonstrate that bacteria sense external stimuli with specific receptors and regulate the rotation of their flagella, which results in swimming forward or changing direction randomly. These observations imply the presence of central processing units (CPUs) that bridge the receptors and flagella.

2.2.3 Exploring the CPU in E. coli

A computer CPU integrates all information from inputs and returns processed results as outputs to peripheral devices such as displays and speakers. The CPU is often compared to the human brain because of the similarity in functions. In terms of processing and integrating all information from the environment and returning specific outputs, bacteria can be considered to have a CPU because they can process sensory information and return outputs as behaviors (Falke et al. 1997; Parkinson 2004).

To explore the molecular components of bacterial CPUs, Warrick and coworkers isolated *Salmonella* mutants that exhibited no specific response to any chemical (Warrick et al. 1977). One group of mutants exhibited continuous tumbling in chemical gradients. In contrast, another group of mutants showed only smooth swimming in chemical gradients. The mutant genes were identified as three groups of genes forming signal transduction pathways in bacteria. The first group of genes encodes chemical receptors, as expected. These receptors bind to specific ligands, which indicates that the genes function in sensing the environment, similar to human peripheral sensory neurons. The second group of genes encodes the proteins required for the formation of flagella and the motor apparatus mentioned in the previous section. These proteins are clearly essential for locomotion through bacterial swimming, a muscle-like function in bacteria. The last group of genes encodes the proteins that bridge the bacterial sensory system to the bacterial locomotive apparatus, the bacterial CPU.

2.2.4 Che Proteins Are Components of the CPU in *E. coli*

Six genes were identified from the mutant screening as encoding components of the CPU in *E. coli* (Fig. 2.3; Table 2.1) (Bourret and Stock 2002; Hazelbauer et al. 2008; Warrick et al. 1977). The genes are required for transmitting information from receptors to flagella to regulate attractive or repellent behaviors in bacteria. Thus, the genes were called chemotaxis-specific (*che*) genes. One of the genes encodes the CheA protein (Smith and Parkinson 1980). CheA is a histidine protein kinase that transfers phosphoryl groups from ATP to the histidine residues of target proteins (Hess et al. 1988a). CheA is activated by autophosphorylation, and then transfers the phosphoryl groups onto another *che* gene product, CheY (Hess et al. 1988b; Wylie et al. 1988). Phosphorylated CheY interacts with flagellar motor proteins to enhance the probability of CW rotation (Kuo and Koshland 1987). As already mentioned, CW rotation results in tumbling and changing the direction of swimming. When bacteria receive attractants, CheA is not activated, resulting in the inhibition of CheY. Thus, bacteria can suppress the tumbling behavior. The regulation of CheA and CheY activity underlies the molecular mechanism of the bacterial CPU.

An additional *che* gene product is CheW, which associates receptor complexes (Borkovich et al. 1989; Borkovich and Simon 1990; Ninfa et al. 1991). CheW integrates receptor activity to regulate CheA activation. The enzyme CheZ is a protein phosphatase for CheY (Hess et al. 1987). The activity of CheZ is important to inhibit CheY activity and facilitates short-period tumbling in bacteria.

The remaining two enzymes, CheB and CheR, are regulators for receptors. Both enzymes function to modify glutamate residues in chemoreceptors. CheR is an *S*-adenisylmethionine-dependent methyltransferase that converts the carboxy groups of specific glutamate residues into uncharged methyl esters (Springer and Koshland 1977). In contrast, CheB is an esterase, reverting the glutamate modified by CheR (Stock and Koshland 1978). Modification of specific glutamate residues affects the affinity of receptors to their ligands (Levit and Stock 2002; Okumura et al. 1998; Sourjik and Berg 2002). Methylated receptors show lower affinity to their ligands. These mechanisms seem to be important to adjust the sensitivity of receptors. In addition, CheB is activated by phosphorylated CheA (Hess et al. 1988b; Lupas and Stock 1989; Stewart et al. 1990), which indicates that the activation of signal transduction starting from CheA removes the methyl group from the glutamate residues of receptors, resulting in an increment change of affinity of receptors to

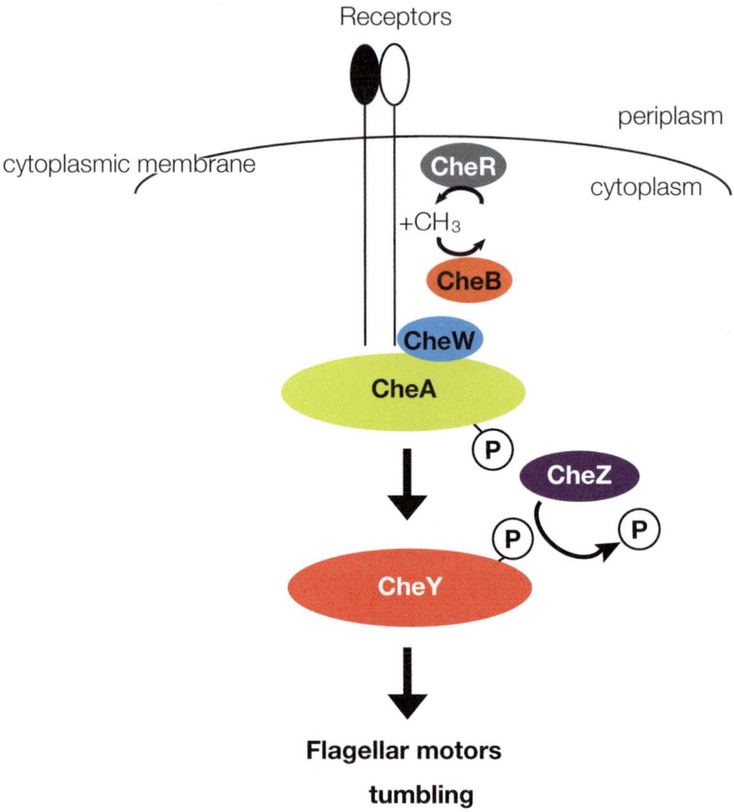

Fig. 2.3 The bacterial "CPU" is composed of Che proteins. A complex of CheA and CheW interacts with receptors in the cytoplasm. When CheA is phosphorylated and activated, the activated CheA transfers the phosphoryl group onto CheY. Phosphorylated CheY can regulate flagellar motors to increase the probability of the clockwise rotation of flagella (i.e., induce the tumbling behavior). CheZ is a phosphatase for CheY, indicating that CheZ can inhibit CheY activity and change the swimming mode from tumbling to smooth swimming. CheB and CheR regulate receptor sensitivity through the addition or removal of methyl groups to the glutamate residues of receptors. In addition, CheB is activated by CheA, indicating a feedback loop. When bacteria sense attractants, CheA activation is inhibited. Then, CheY is also inhibited, causing the inhibition of tumbling. Suppression of the tumbling behavior results in long smooth swimming to approach the center of the attractants

ligands. Thus, there is a feedback regulation of the sensitivity of receptors, which represents the molecular mechanism of adaptation in bacteria. Taken all these results together, bacteria appear to have network-type feedback loops to regulate their behavior.

Table 2.1 Summary for gene products regulating signal transduction in bacteria

Gene product	Functions	References
CheA	Histidine protein kinase	Smith and Parkinson (1980)
	Integration of all information from receptors	Hess et al. (1988a)
CheY	Regulator of flagella motors	Hess et al. (1988b)
		Wylie et al. (1988)
CheW	Regulator of CheA activity	Borkovich et al. (1989)
		Ninfa et al. (1991)
CheZ	Protein phosphatase	Hess et al. (1987)
	Inhibitor of CheY	
Che R	Methyltransferase	Springer and Koshland (1977)
	Modifier of receptors	Stock and Koshland (1978)
Che B	Esterase	Stock and Koshland (1978)
	Modifier of receptors	Lupas and Stock (1989)

2.2.5 Is the Bacterial CPU Common to Eukaryotes?

Is this mechanism underlying bacterial signaling pathways conserved in eukaryotes? To answer this question, genes homologous to the *che* genes in prokaryotes should be identified in eukaryotes. However, there are only a couple of studies examining homologous *che* genes in eukaryotes (Chang et al. 1993; Ota and Varshavsky 1993). Genes encoding the histidine protein kinase, such as *cheA*, have been isolated from yeast (Ota and Varshavsky 1993) and plants (Chang et al. 1993). Thus, similar pathways in eukaryotes may be conserved, but the pathways may not have important roles in eukaryotic cells because no apparent homologous genes were found based on database search in humans, mice, zebrafish, and *Drosophila*. In fact, serine and threonine protein kinases are major kinases in signal transduction in eukaryotic cells, suggesting that serine and threonine protein kinases became major players during evolution (Kennelly and Potts 1996). However, the basic logic for the regulation of sensitivity through networks based on feedback loops appears to have been conserved during evolution.

2.3 Neuron-like Functions in Eukaryotes

2.3.1 What Happens in Eukaryotic Single-Cell Organisms?

Eukaryotic single-cell organisms can sense their environment and evoke specific responses, similar to bacteria. From the late nineteenth century, ciliates such as *Paramecium* sp., *Tetrahymena* sp., and *Euplotes* sp. have been used to investigate behaviors (Vanhouten 1994).

Paramecia have been used as a eukaryotic model organism to investigate the molecular mechanisms regulating their behavior, as well as to understand essential biological processes such as gene expression (Allen and Gibson 1972; Kung et al. 1975; Sommerville and McTavish 1982). Large numbers of paramecia can be grown, which allows them to be a good model system. In addition, induction and isolation of mutants can be accomplished within one laboratory. Thus, many genes responsible for the regulation of behaviors in *Paramecium* have been identified to better understand the molecular mechanisms of its "neuron-like" function.

2.3.2 Swimming Behaviors and Chemotaxis in Paramecium

Paramecia swim smoothly until they sense a physical obstacle by touch or repellants (Fig. 2.4) (Nakatani 1970; Vanhouten 1978). Once paramecia receive the information of touch or repellents, they show backward swimming and waving of the anterior part of the cell (Naitoh and Eckert 1969a, b). Paramecia choose their orientation randomly, and then start smooth swimming.

Paramecia can sense external stimuli and respond with specific behaviors. Paramecia are, in general, attracted to bacterial metabolites such as acetate, folate, glutamate, and biotin (Bell et al. 1998; Preston and Vanhouten 1987; Vanhouten 1994; Yang et al. 1997). In contrast, GTP and quinidine act as repellents to paramecia (Clark et al. 1993, 1997; Preston et al. 1990).

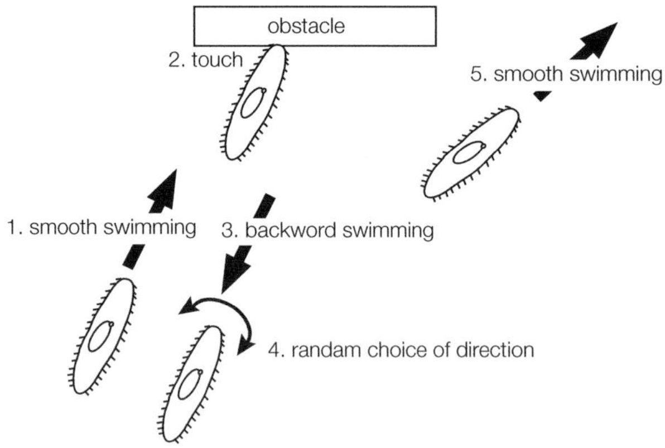

Fig. 2.4 Two modes of swimming in paramecia A paramecium swims smoothly to one direction. The paramecium shows backward swimming when it senses an obstacle or repellant. The paramecium then shows waving of cilia to choose randomly a new direction of swimming. Thus, a paramecium can maneuver around an obstacle or avoid high concentrations of repellants. A paramecium suppresses direction change and increases swimming speed when it senses an attractant. Please note that bacteria and paramecia have similar strategies for swimming behavior

Interesting observations have been reported on the effects of acetic acid on paramecia (Bell et al. 2007; Johnson 1929). Acetic acid is an attractant for paramecia at low concentrations. However, highly concentrated acetic acid has a repulsive effect on paramecia. Thus, paramecia can become trapped within a ring-shaped area surrounding the highest concentration of acetic acid when they swim in a gradient of acetic acid. A bubble of carbon dioxide has a similar effect. These phenomena can be explained as follows. First, paramecia can sense a low concentration of acetic acid as an attractant. As they approach an area of a higher concentration of acetic acid, the escape behavior is induced. After the escape, paramecia sense that the concentration of acetic acid is decreasing. Thus, paramecia can become trapped in an area between low acetic acid concentration as an attractant and high acetic acid concentration as a repellant. These observations clearly indicate that specific chemicals evoke specific behaviors in *Paramecium*, suggesting that paramecia have signal-processing mechanisms.

Tetrahymena is also attracted to proteins, peptides, and amino acids (Leick and Hellunglarsen 1992). Moreover, one report demonstrated that a neurotransmitter, β-endorphin, was an attractant to *Tetrahymena* (Oneill et al. 1988). The effects of neurotransmitters on ciliates are discussed later.

2.3.3 Regulation of Behaviors During Chemotaxis in Paramecia

As already mentioned, there are two characteristic behaviors in chemotaxis of paramecia: approach and escape. Several reports demonstrated that the swimming speed of paramecia increases when they approach an attractant, and that the escape behavior including backward swimming is suppressed during approach (Eckert et al. 1972; Hemmersbachkrause et al. 1992; Vanhouten et al. 1982). In contrast, the swimming speed of paramecia decreases when they sense a repellent, and then they show backward swimming and changes in swimming direction. This regulation resembles the mechanisms in bacteria just described. Differences between *Paramecium* and *E. coli* are observed in the regulation of swimming speed and backward swimming in paramecia before reorientation. Both *Paramecium* and *E. coli* seem to regulate the probability of reorientation to a corresponding external stimulus.

2.3.4 Sensory System and Motor Apparatus

Naitoh and Eckert (1969a, b) demonstrated that paramecia reverse the ciliary beat as an avoidance behavior when stimulated by touching with a needle in the anterior part, and that *Paramecium* showed an increment of ciliary beat as an attractive behavior when it was stimulated in the posterior part. These results suggest that different areas of the *Paramecium* body evoke different behaviors to the same stimulus. Each body part may be responsible for each different stimulus, similar to the human body and brain. In addition to the touch response, paramecia have specific receptors for attractants or repellents. These receptors are essential components for the sensory system in paramecia. The localization of receptors is also important for the regulation of the behaviors already described.

Cilia, which are the motor apparatus, consist of several proteins. The detailed structure and beating mechanisms of motile cilia have been reviewed in detail previously (Doughty and Dryl 1981; Wiederhold 1976). Microtubules and dynein are important for active beating. In fact, $NiCl_2$, an inhibitor of dynein ATPase, can inhibit ciliary beating in paramecia (Larsen and Satir 1991; Naitoh 1966). Naitoh and coworkers demonstrated that paramecia extracted with Triton X-100 can swim in the presence of ATP and that reversal of ciliary beating can be induced by adding Ca^{2+} and ATP (Naitoh and Kaneko 1972). The results indicate that the cilia can be regulated by ATP as an energy source and Ca^{2+}, suggesting that paramecia can regulate behavior by controlling the concentration of Ca^{2+}. Of course, the potassium ion also has a crucial role in ciliary beating (Preston and Vanhouten 1987).

2.3.5 The Molecular Mechanisms Bridging Receptors and Cilia in Paramecium

The molecular mechanisms underlying signal transduction from receptors to cilia are not fully understood. However, evidence suggests that the signal transduction pathway in paramecia is similar, in part, to those in neurons.

Paramecia used to be referred to as a swimming neuron because they show dynamic change in the electrical characteristics of the cell membrane during chemotaxis (Eckert et al. 1972; Naitoh et al. 1972; Vanhouten 1979). Paramecia show depolarization when they are stimulated with repellents, decreasing attractants, or simple physical stimuli such as touch (Naitoh 1966; Naitoh and Eckert 1973). Paramecia swim backward when they face stimuli that can evoke the depolarization. After the backward swimming, they change direction randomly, and swim away from the stimulus. In contrast, paramecia show hyperpolarization when they detect attractants, resulting in an increment change of swimming speed and the suppression of backward swimming (Preston and Vanhouten 1987; Van Houten et al. 2000). Interestingly, the mechanisms of membrane potential and firing in paramecia are essentially the same as those in neurons (Adoutte et al. 1981; Naitoh et al. 1972; Nakaoka and Ooi 1985; Oka et al. 1986).

Receptors for cAMP and glutamate have been identified as attractants and characterized (Bell et al. 1998; Ramoino et al. 2014; Vanhouten et al. 1991). Both these receptors bind specifically to the ligand and seem to release calcium and potassium ions from paramecia through calcium and potassium ion pumps (Adoutte et al. 1981; Doughty and Dryl 1981). Thus, attractants such as cAMP and glutamate hyperpolarize membranes through these mechanisms. In contrast, paramecia change the permeability of the cell membrane for Ca^{2+}, resulting in depolarization (Naitoh and Kaneko 1972; Nakaoka and Ooi 1985). The Ca^{2+}-dependent action potential controls the angle and frequency of ciliary beating (Machemer 1976).

The structures of receptors in paramecia are similar to those in vertebrates. The isolated receptor for glutamate is closely related to the human umami receptor, a G-protein-coupled receptor (Van Houten et al. 2000). In addition, activities downstream of the receptors are closely related to signal transduction pathways in vertebrates. The hyperpolarization of the cell membrane in paramecia triggers increment changes of the intracellular, especially intraciliary, concentration of cAMP (Hennessey et al. 1985; Schultz et al. 1992). Taken together, it is likely that activation of the receptor for glutamate results in activation of adenylate cyclase to synthesize cAMP. This cAMP increases the frequency of ciliary beating (Bonini and Nelson 1988).

The next indication for the similarity of paramecia to neurons is the identification of calmodulin in cilia (Maihle et al. 1981) and that guanylate cyclase is activated by Ca^{2+}-calmodulin (Klumpp et al. 1983). In addition, endogenous kinases activated by cAMP have been identified (Van Houten et al. 2000). Thus, Ca^{2+} and cAMP seem to function as second messengers to regulate ciliary beating. However, the molecular mechanisms that enable Ca^{2+} to induce reversal of ciliary beating remain unclear.

Taken together, these findings indicated that paramecia have signal-processing pathways that are closely related to those observed in neurons in vertebrates and that calcium ions and cAMP are important mediators of signaling from receptors to cilia.

2.3.6 Memory and Learning in Ciliates

Memory and learning are important functions of the human brain. Is there an equivalent to our memory and learning in paramecia? To answer this question, several investigations have been carried out and reported. Jensen (Jensen 1957a, b) reported the training of paramecia with a platinum needle covered with bacteria. After several training sessions, paramecia approached a bare platinum needle without bacteria; these observations provided evidence for classical conditioned learning in paramecia. In addition, two papers reported similar results from experiments for trial-and-error learning in paramecia (French 1940, Hanzel and Rucker 1972). These studies suggest that paramecia can learn and hold memory for a short period. However, several criticisms have been raised concerning their experimental designs and interpretation. Recently, Armus et al. (2006) trained paramecia with cathodes (attractant, unconditioned stimulus) and light (conditioned stimulus). Their results demonstrate that paramecia are attracted to a lighted area without the cathode after six repeated training sessions, suggesting that paramecia may be able to learn and maintain memory. However, the molecular mechanisms involved in these processes remain unclear. Further studies are required for elucidating the mechanism of learning in *Paramecium*.

2.3.7 Neurotransmitters and Hormones in Ciliates

Glutamate is a well-known neurotransmitter in the brain. Glutamate also has important functions as an attractant in paramecia. In addition, γ-animobutyric acid (GABA) affects swimming behavior in paramecia (Bucci et al. 2005). Bucci et al. (2005) demonstrated that pharmacological treatments of paramecia with agonist or antagonist of the $GABA_A$ receptor alter the swimming behavior. Moreover, paramecia release GABA into their environment by neuronal-like exocytosis (Ramoino et al. 2010). The reason for GABA release by paramecia is unclear. Taken together, these results show that paramecia can respond to the neurotransmitters glutamate and GABA.

Tetrahymena can react physiologically to serotonin, histamine, and insulin (Csaba and Lantos 1973, 1975). Interestingly, *Tetrahymena* synthesizes adrenocorticotropic hormone (ACTH), β-endorphin, serotonin, and triiodothyronine (T_3) in response to stress from formaldehyde and high temperatures, suggesting that *Tetrahymena* may show a general response to the stressors (Csaba and Pallinger 2008). Thus, *Paramecium* and *Tetrahymena* may utilize neurotransmitters and hormones to regulate their behavior and adaptation in general.

2.3.8 Serotonin Is Involved in Physiological Functions in Tetrahymena

As mentioned, serotonin was detected in *Tetrahymena* by an antibody against serotonin even under normal growth conditions (Csaba and Pallinger 2008). Binding sites for serotonin are located on the surface of the cell membrane of *Tetrahymena* (Csaba and Lantos 1973), suggesting that *Tetrahymena* has receptor(s) for serotonin. Serotonin is an important neurotransmitter in the human brain that is synthesized in the raphe and may be related to depression (Stockmeier 1997). In addition, whole-genome sequencing revealed that *Tetrahymena* has a gene encoding tryptophan hydroxylase (TPH), which is a rate-limiting enzyme for serotonin biosynthesis (Eisen et al. 2006). These findings suggest that serotonin is produced in paramecia and *Tetrahymena* and that the synthesized serotonin has biological and physiological functions in ciliates.

Specific signals corresponding to serotonin have been observed in *Tetrahymena* when starved (Fig. 2.5c, d). A specific inhibitor of TPH, *para*-chlorophenylalanine (PCPA), inhibits serotonin synthesis (Fig. 2.5e, f). These results demonstrate that

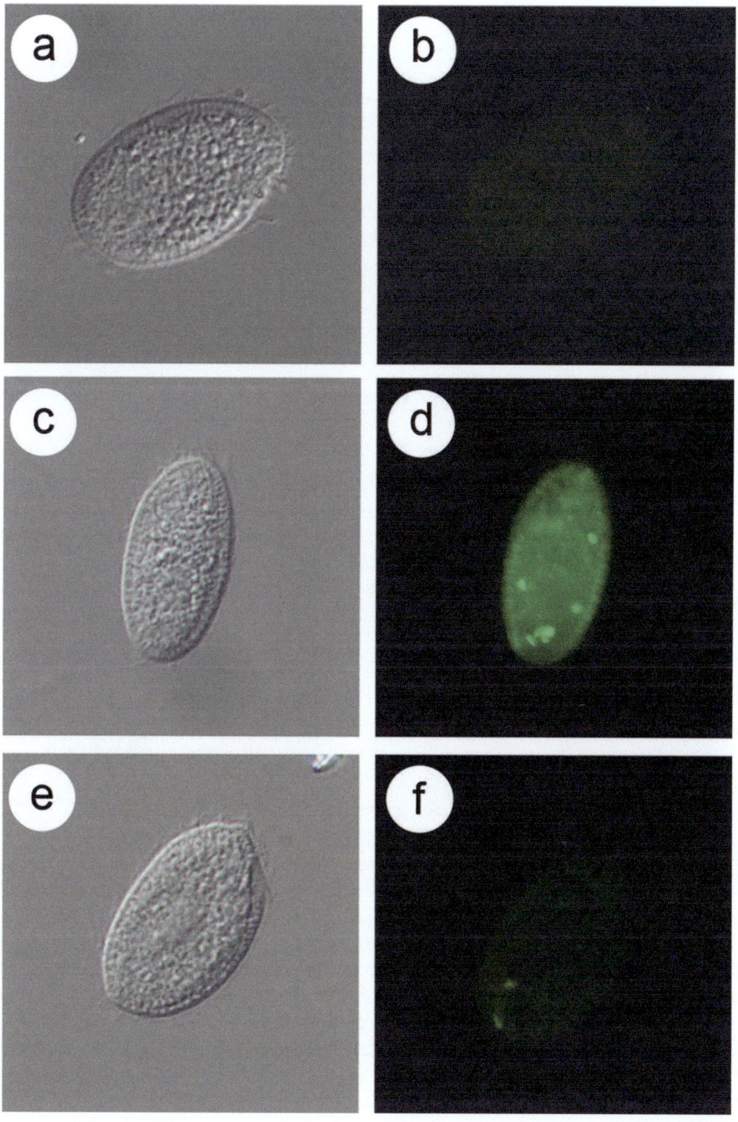

Fig. 2.5 Serotonin is synthesized in *Tetrahymena thermophila*Immunological staining of serotonin in a *Tetrahymena* cell. Starvation results in serotonin synthesis (**c**, **d**). The presence of 4-chlorophenylacetic acid (PCPA, an inhibitor of tryptophan hydroxylase) inhibits the synthesis of serotonin (**e**, **f**). No significant staining was observed without anti-serotonin antibody (**a**, **b**). **a**, **c**, **e** Differential interference contrast (DIC) images. **b**, **d**, **f** Fluorescent images

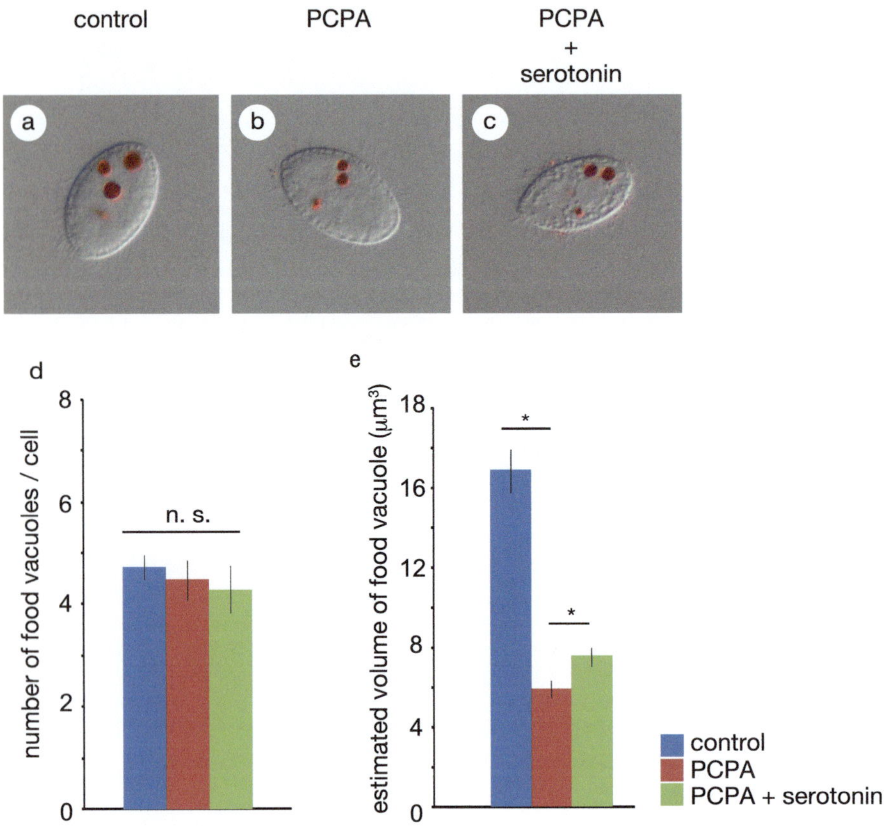

Fig. 2.6 Inhibition of serotonin synthesis causes smaller food vesicles in *Tetrahymena thermophilaTetrahymena* show four or five food vesicles (*red* vesicles in the panel) when starved (**a**). During starvation, PCPA was applied to inhibit serotonin synthesis. The treatment of PCPA did not affect the number of food vesicles, but reduced the volume of food vesicles (**b**, **d**, **e**). Supplementation of serotonin partially rescued the phenotype (**c**, **d**, **e**). Food vesicles were visualized with *red* ink. The average of the number of food vesicles (indicated in **d**) is ($n = 10$). Volumes of food vesicles were estimated by the diameter of the food vesicles in the images, and the average of estimated volumes of food vesicles is shown in (**e**) ($n = 10$). n.s. not significant. *; $p < 0.05$ ($n = 10$)

TPH is responsible for serotonin synthesis in *Tetrahymena*. In addition, exposure of *Tetrahymena* to a specific inhibitor of tryptophan hydroxylase results in a smaller food vesicle (Fig. 2.6b, e) and a slower initiation of mating behaviors compared with nontreated controls (Fig. 2.7d, e, f), even if the inhibitor did not affect the number of food vesicles (Fig. 2.6d) and meiosis (Fig. 2.7). Supplementation of serotonin can rescue the volume of the food vesicle (Fig. 2.6c, e) and the mating behavior (Fig. 2.7). These results indicate that tryptophan hydroxylase in *Tetrahymena* functions to produce serotonin, and that synthesized serotonin is required for food uptake and initiation of conjugation. Human patients suffering from depression

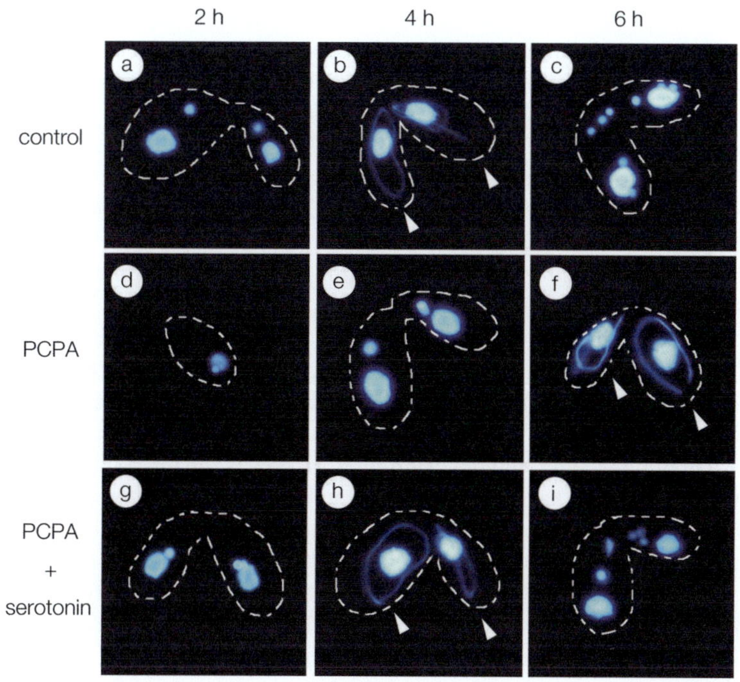

Fig. 2.7 Inhibition of serotonin synthesis also affects timing of conjugation *Tetrahymena* usually start conjugation when they are exposed to starvation and they find conjugation partners having different conjugation types. After 2 h, two cells showed conjugation in the control group (**a**). The conjugated cells showed elongated small nuclei (*white arrowheads* in **b**), indicating that meiosis started at 4 h after starvation. Several small nuclei were observed in the conjugated cells at 6 h after starvation. In contrast to the control group, PCPA treatment results in a delay of conjugation (**d**, **e**, **f**). Importantly, the conjugation processes, such as elongated small nuclei, were detected (*white arrowheads* in **f**). Supplementation of serotonin with PCPA completely suppressed the delay (**g**, **h**, **i**). Nuclei were visualized with Hoechst 33342. The *dashed white line* in all panels indicates the outline of cells

show less social activity, less appetite, and less sexual desire (Blundell 1992; Stockmeier 1997). If the phenotype observed in *Tetrahymena* is similar to human depression, serotonin may have evolutionally conserved functions for survival of organisms. Thus, serotonin might affect the "mind" in ciliates.

2.4 Are Neuron-like Functions in Single-Cell Organisms an Indication of Emotion or Mind?

In the late nineteenth century, Herbert Spencer Jennings began analyzing the behavior of protozoa. He sought to understand emotion and/or mind in animals, as revealed in his paper entitled "The psychology of a protozoan" (Jennings 1899).

Of course, emotion and mind cannot be measured and examined directly with biological methods in animals because we cannot communicate and interview the animals. Nevertheless, emotion and mind in bacteria and ciliates can be considered philosophically.

As described previously, bacteria and ciliates can respond to external signals and make "decisions" to approach or escape the signals. If the function of emotion and mind is to respond to external signals and to determine behavior, bacteria and ciliates seem to have well-established emotion and mind. Based on these criteria for the function of emotion and mind, bacteria, ciliates, yeast, invertebrates, vertebrates, and plants should possess emotion and mind.

If bacteria and ciliates merely show a reflex to external stimuli, then it is difficult to explain human behaviors. Human behaviors can be considered as complex reflexes if emotion and mind are not considered. Of course, an estimation of emotion and mind in humans can be attained through questioning, and we can measure emotion and mind by interviewing examinees. However, the mechanisms of emotion and mind in the brain are too difficult to understand only within one scientific field. We can only fully reveal the mechanisms of emotion and mind in the brain by aligning expertise and knowledge from the fields of biology, neurophysiology, psychology, computer science, and informatics. Thus, in the future, emotion and mind may be understood in humans and other living organisms, including bacteria and ciliates.

References

Adoutte A, Ling KY, Forte M, Ramanathan R, Nelson D, Kung C (1981) Ionic channels of *Paramecium*: from genetics and electrophysiology to biochemistry. J Physiol 77:1145–1159

Allen S, Gibson I (1972) Genome amplification and gene expression in ciliate macronucleus. Biochem Genet 6:293–313

Armus HL, Montgomery AR, Jellison JL (2006) Discrimination learning in paramecia (P. caudatum). Psychol Rec 56:489–498

Bell WE, Karstens W, Sun Y, Van Houten JL (1998) Biotin chemoresponse in *Paramecium*. J Comp Physiol A Sens Neural Behav Physiol 183:361–366

Bell WE, Preston RR, Yano J, Van Houten JL (2007) Genetic dissection of attractant-induced conductances in *Paramecium*. J Exp Biol 210:357–365

Berg HC (1971) How to track bacteria. Rev Sci Instrum 42:868–871

Berg HC (1975) Chemotaxis in bacteria. Annu Rev Biophys Bioeng 4:119–136

Berg HC, Brown DA (1972) Chemotaxis in *Escherichia coli* analyzed by 3-dimensional tracking. Nature (Lond) 239:500–504

Bibikov SI, Biran R, Rudd KE, Parkinson JS (1997) A signal transducer for aerotaxis in *Escherichia coli*. J Bacteriol 179:4075–4079

Blundell JE (1992) Serotonin and the biology of feeding. Am J Clin Nutr 55:155–159

Bonini NM, Nelson DL (1988) Differential regulation of *Paramecium* ciliary motility by cAMP and cGMP. J Cell Biol 106:1615–1623

Borkovich KA, Simon MI (1990) The dynamics of protein-phosphorylation in bacterial chemotaxis. Cell 63:1339–1348

Borkovich KA, Kaplan N, Hess JF, Simon MI (1989) Transmembrane signal transduction in bacterial chemotaxis involves ligand-dependent activation of phosphate group transfer. Proc Natl Acad Sci USA 86:1208–1212
Bourret RB, Stock AM (2002) Molecular information processing: lessons from bacterial chemotaxis. J Biol Chem 277:9625–9628
Boyd A, Kendall K, Simon MI (1983) Structure of the serine chemoreceptor in *Escherichia coli*. Nature (Lond) 301:623–626
Bucci G, Ramoino P, Diaspro A, Usai C (2005) A role for GABA(A) receptors in the modulation of *Paramecium* swimming behavior. Neurosci Lett 386:179–183
Chang C, Kwok SF, Bleecker AB, Meyerowitz EM (1993) *Arabidopsis* ethylene-response gene ETR1: similarity of product to 2-component regulators. Science 262:539–544
Clark KD, Hennessey TM, Nelson DL (1993) External GTP alters the motility and elicits an oscillating membrane depolarization in *Paramecium tetraurelia*. Proc Natl Acad Sci USA 90:3782–3786
Clark KD, Hennessey TM, Nelson DL, Preston RR (1997) Extracellular GTP causes membrane-potential oscillations through the parallel activation of Mg^{2+} and Na^{+} currents in *Paramecium tetraurelia*. J Membr Biol 157:159–167
Csaba G, Lantos T (1973) Effect of hormones on Protozoa: studies on phagocytotic effect of histamine, 5-hydroxytryptamine and indoleacetic acid in *Tetrahymena pyriformis*. Cytobiologie 7:361–365
Csaba G, Lantos T (1975) Effect of insulin on glucose uptake of protozoa. Experientia (Basel) 31:1097–1098
Csaba G, Pallinger E (2008) A general response to stressors by the unicellular *Tetrahymena*: effect of stress on the hormone levels. Cell Biochem Funct 26:797–800
Depamphi ML, Adler J (1971) Attachment of flagellar basal bodies to cell envelope: specific attachment to outer, lipopolysaccharide membrane and cytoplasmic membrane. J Bacteriol 105:396–407
Doughty MJ, Dryl S (1981) Control of ciliary activity in *Paramecium*: an analysis of chemosensory transduction in a eukaryotic unicellular organism. Prog Neurobiol 16:1–115
Eckert R, Friedman K, Naitoh Y (1972) Sensory mechanisms in *Paramecium*. 1. Two components of electric response to mechanical stimulation of anterior surface. J Exp Biol 56:683
Eisen JA, Coyne RS, Wu M, Wu D, Thiagarajan M, Wortman JR, Badger JH, Ren Q, Amedeo P, Jones KM et al (2006) Macronuclear genome sequence of the ciliate *Tetrahymena thermophila*, a model eukaryote. PLoS Biol 4:1620–1642
Falke JJ, Bass RB, Butler SL, Chervitz SA, Danielson MA (1997) The two-component signaling pathway of bacterial chemotaxis: a molecular view of signal transduction by receptors, kinases, and adaptation enzymes. Annu Rev Cell Dev Biol 13:457–512
French JW (1940) Trial and error learning in *Paramecium*. J Exp Psychol 26:609–613
Hanzel TE, Rucker WB (1972) Trial and error learning in *Paramecium*: replication. Behav Biol 7:873
Hazelbauer GL, Falke JJ, Parkinson JS (2008) Bacterial chemoreceptors: high-performance signaling in networked arrays. Trends Biochem Sci 33:9–19
Hemmersbachkrause R, Briegleb W, Hader DP (1992) Swimming behavior of paramecium: first results with the low-speed centrifuge microscope (NIZEMI). Life Sciences and Space Research, vol xxiv (1). Gravitat Biol 12:113–116
Hennessey T, Machemer H, Nelson DL (1985) Injected cyclic-AMP increases ciliary beat frequency in conjunction with membrane hyperpolarization. Eur J Cell Biol 36:153–156
Hess JF, Oosawa K, Matsumura P, Simon MI (1987) Protein phosphorylation is involved in bacterial chemotaxis. Proc Natl Acad Sci USA 84:7609–7613
Hess JF, Bourret RB, Simon MI (1988a) Histidine phosphorylation and phosphoryl group transfer in bacterial chemotaxis. Nature (Lond) 336:139–143
Hess JF, Oosawa K, Kaplan N, Simon MI (1988b) Phosphorylation of three proteins in the signaling pathway of bacterial chemotaxis. Cell 53:79–87
Iino T (1969) Genetics and chemistry of bacterial flagella. Bacteriol Rev 33:454

Jennings HS (1899) The psychology of a protozoan. Am J Psychol 10:503–515
Jensen DD (1957a) Experiments on learning in *Paramecia*. Science 125:191–192
Jensen DD (1957b) More on learning in *Paramecia*. Science 126:1341–1342
Johnson WH (1929) The reactions of *Paramecium* to solutions of known hydrogen ion concentration. Biol Bull 57:199–224
Kennelly PJ, Potts M (1996) Fancy meeting you here! A fresh look at "prokaryotic" protein phosphorylation. J Bacteriol 178:4759–4764
Klumpp S, Kleefeld G, Schultz JE (1983) Calcium calmodulin-regulated guanylate cyclase of the excitable ciliary membrane from *Paramecium*: dissociation of calmodulin by $La\text{-}3^{+}$: calmodulin specificity and properties of the reconstituted guanylate cyclase. J Biol Chem 258:2455–2459
Kondoh H, Ball CB, Adler J (1979) Identification of a methyl-accepting chemotaxis protein for the ribose and galactose chemoreceptors of *Escherichia coli*. Proc Natl Acad Sci USA 76:260–264
Koshland DE (1980) Bacterial chemotaxis in relation to neurobiology. Annu Rev Neurosci 3:43–75
Kung C, Chang SY, Satow Y, Vanhouten J, Hansma H (1975) Genetic dissection of behavior in *Paramecium*. Science 188:898–904
Kuo SC, Koshland DE (1987) Roles of CHeY and CHeZ gene products in controlling flagellar rotation in bacterial chemotaxis of *Escherichia coli*. J Bacteriol 169:1307–1314
Larsen J, Satir P (1991) Analysis of Ni^{2+}-induced arrest of *Paramecium* axonemes. J Cell Sci 99:33–40
Larsen SH, Reader RW, Kort EN, Tso WW, Adler J (1974) Change in direction of flagellar rotation is basis of chemotactic response in *Escherichia coli*. Nature (Lond) 249:74–77
Leick V, Hellunglarsen P (1992) Chemosensory behavior of *Tetrahymena*. Bioessays 14:61–66
Levit MN, Stock JB (2002) Receptor methylation controls the magnitude of stimulus–response coupling in bacterial chemotaxis. J Biol Chem 277:36760–36765
Lupas A, Stock J (1989) Phosphorylation of an N-terminal regulatory domain activates the CHeB methylesterase in bacterial chemotaxis. J Biol Chem 264:17337–17342
Machemer H (1976) Interactions of membrane-potential and cations in regulation of ciliary activity in *Paramecium*. J Exp Biol 65:427–447
Macnab RM, Koshland DE (1972) Gradient-sensing mechanism in bacterial chemotaxis. Proc Natl Acad Sci USA 69:2509
Macnab R, Koshland DE (1974) Bacterial motility and chemotaxis: light-induced tumbling response and visualization of individual flagella. J Mol Biol 84:399–406
Maihle NJ, Dedman JR, Means AR, Chafouleas JG, Satir BH (1981) Presence and indirect immunofluorescent localization of calmodulin in *Paramecium tetraurelia*. J Cell Biol 89:695–699
Manson MD, Blank V, Brade G, Higgins CF (1986) Peptide chemotaxis in *Escherichia coli* involves the tap signal transducer and the dipeptide permease. Nature (Lond) 321:253–256
Naitoh Y (1966) Reversal response elicited in nonbeating cilia of *Paramecium* by membrane depolarization. Science 154:660
Naitoh Y, Eckert R (1969a) Ciliary orientation: controlled by cell membrane or by intracellular fibrils. Science 166:1633
Naitoh Y, Eckert R (1969b) Ionic mechanisms controlling behavioral responses of *Paramecium* to mechanical stimulation. Science 164:963
Naitoh Y, Eckert R (1973) Sensory mechanisms in *Paramecium*. 2. Ionic basis of hyperpolarizing mechanoreceptor potential. J Exp Biol 59:53–65
Naitoh Y, Kaneko H (1972) Reactivated Triton-extracted models of *Paramecium*: modification of ciliary movement by calcium ions. Science 176:523
Naitoh Y, Eckert R, Friedman K (1972) Regenerative calcium response in *Paramecium*. J Exp Biol 56:667
Nakaoka Y, Ooi H (1985) Regulation of ciliary reversal in Triton-extracted *Paramecium* by calcium and cyclic adenosine monophosphate. J Cell Sci 77:185–195
Nakatani I (1970) Effects of various chemicals on the behavior of *Paramecium caudatum*. J Fac Sci Hokkaido Univ Ser VI Zool 17:401–410

Ninfa EG, Stock A, Mowbray S, Stock J (1991) Reconstitution of the bacterial chemotaxis signal transduction system from purified components. J Biol Chem 266:9764–9770

Oka T, Nakaoka Y, Oosawa F (1986) Changes in membrane potential during adaptation to external potassium ions in *Paramecium caudatum*. J Exp Biol 126:111–117

Okumura H, Nishiyama SI, Sasaki A, Homma M, Kawagishi I (1998) Chemotactic adaptation is altered by changes in the carboxy-terminal sequence conserved among the major methyl-accepting chemoreceptors. J Bacteriol 180:1862–1868

O'Neill JB, Pert CB, Ruff MR, Smith CC, Higgins WJ, Zipser B (1988) Identification and characterization of the opiate receptor in the ciliated protozoan, *Tetrahymena*. Brain Res 450:303–315

Ota IM, Varshavsky A (1993) A yeast protein similar to bacterial two-component regulators. Science 262:566–569

Parkinson JS (2004) Signal amplification in bacterial chemotaxis through receptor teamwork. ASM News 70:575

Preston RR, Vanhouten JL (1987) Chemoreception in *Paramecium tetraurelia*: acetate and folate-induced membrane hyperpolarization. J Comp Physiol A Sens Neural Behav Physiol 160:525–535

Preston RR, Saimi Y, Kung C (1990) Evidence for two K^+ currents activated upon hyperpolarization of *Paramecium tetraurelia*. J Membr Biol 115:41–50

Ramoino P, Milanese M, Candiani S, Diaspro A, Fato M, Usai C, Bonanno G (2010) Gamma-amino butyric acid (GABA) release in the ciliated protozoon *Paramecium* occurs by neuronal-like exocytosis. J Exp Biol 213:1251–1258

Ramoino P, Candiani S, Pittaluga AM, Usai C, Gallus L, Ferrando S, Milanese M, Faimali M, Bonanno G (2014) Pharmacological characterization of NMDA-like receptors in the single-celled organism *Paramecium primaurelia*. J Exp Biol 217:463–471

Reader RW, Tso WW, Springer MS, Goy MF, Adler J (1979) Pleiotropic aspartate taxis and serine taxis mutants of *Escherichia coli*. J Gen Microbiol 111:363–374

Schultz JE, Klumpp S, Benz R, Schurhoffgoeters WJC, Schmid A (1992) Regulation of adenylyl cyclase from *Paramecium* by an intrinsic potassium conductance. Science 255:600–603

Smith RA, Parkinson JS (1980) Overlapping genes at the CHeA locus of *Escherichia coli*. Proc Natl Acad Sci USA Biol Sci 77:5370–5374

Sommerville J, McTavish C (1982) The effect of temperature change on gene expression in *Paramecium primaurelia*. Biochim Biophys Acta 698:158–166

Sourjik V, Berg HC (2002) Receptor sensitivity in bacterial chemotaxis. Proc Natl Acad Sci USA 99:123–127

Springer WR, Koshland DE (1977) Identification of a protein methyltransferase as CheR gene product in bacterial sensing system. Proc Natl Acad Sci USA 74:533–537

Stewart RC, Roth AF, Dahlquist FW (1990) Mutations that affect control of the methylesterase activity of CheB, a component of the chemotaxis adaptation system in *Escherichia coli*. J Bacteriol 172:3388–3399

Stock JB, Koshland DE (1978) Protein methylesterase involved in bacterial sensing. Proc Natl Acad Sci USA 75:3659–3663

Stockmeier CA (1997) Neurobiology of serotonin in depression and suicide. Neurobiol Suicide Bench Clin 836:220–232

Van Houten JL, Yang WQ, Bergeron A (2000) Chemosensory signal transduction in *Paramecium*. J Nutr 130:946S–949S

Vanhouten J (1978) Two mechanisms of chemotaxis in *Paramecium*. J Comp Physiol 127:167–174

Vanhouten J (1979) Membrane potential changes during chemokinesis in *Paramecium*. Science 204:1100–1103

Vanhouten J (1994) Chemosensory transduction in eukaryotic microorganisms: trends for neuroscience. Trends Neurosci 17:62–71

Vanhouten J, Martel E, Kasch T (1982) Kinetic analysis of chemokinesis of *Paramecium*. J Protozool 29:226–230

Vanhouten JL, Cote BL, Zhang J, Baez J, Gagnon ML (1991) Studies of the cyclic adenosine-monophosphate chemoreceptor of *Paramecium*. J Membr Biol 119:15–24

Warrick HM, Taylor BL, Koshland DE (1977) Chemotactic mechanism of *Salmonella typhimurium*: preliminary mapping and characterization of mutants. J Bacteriol 130:223–231

Wiederhold ML (1976) Mechanosensory transduction in sensory and motile cilia. Annu Rev Biophys Bioeng 5:39–62

Wylie D, Stock A, Wong CY, Stock J (1988) Sensory transduction in bacterial chemotaxis involves phosphotransfer between ChE proteins. Biochem Biophys Res Commun 151:891–896

Yang WQ, Braun C, Plattner H, Purvee J, VanHouten JL (1997) Cyclic nucleotides in glutamate chemosensory signal transduction of *Paramecium*. J Cell Sci 110:2567–2572

Chapter 3
Back Through Time: How Cnidarians and Basal Metazoans Shed Light on Ancient Nervous Systems

Hiroshi Watanabe

Abstract The origin of neurons and the evolution of the central nervous system (CNS) are not well understood. The physiological nature of primitive neurons has not been elucidated, and whether the CNSs of extant bilaterians originated with an array of nerve nets or with a primordial neuronal aggregation is unknown. The nervous systems of cnidarians, the closest sister branch to bilaterians, manifest similarities to bilaterian nervous systems, including developmental mechanisms and cellular features. For example, the cnidarian neurons are electrically excitable, communicating with other neurons or muscles via chemical synapses, and forming diffuse neural networks with significant condensations along the main body axis.

Recent genomic and gene expression data from cnidarians and other basal metazoans have provided hints to reconstruct the evolutionary history of neurons and the CNS. Genes involved in neuronal physiological functions are conserved among bilaterians, cnidarians, and even sponges. The latter possess sensory cells, but not neurons, providing insights into the origin of neurons. Accumulating evidence shows that cnidarians develop a neural condensation, a "semi-centralized nervous system (semiCNS)," composed of multiple neuronal cell types. Although the development and function of cnidarian nervous systems, especially the semiCNS, remain largely unexplored, numerous molecular signatures shared by cnidarians and bilaterians help us to understand early processes of neural centralization.

Keywords cnidarians • ctenophores • placozoans • sponges • Evolution • Protoneuron • Nervous system

H. Watanabe (✉)
Evolutionary Neurobiology Unit, Okinawa Institute of Science and Technology Graduate University, 1919-1 Tancha, Onna-son, Kunigami, Okinawa 904-0495, Japan
e-mail: hiroshi.watanabe@oist.jp

S. Shigeno et al. (eds.), *Brain Evolution by Design*, Diversity and Commonality in Animals, DOI 10.1007/978-4-431-56469-0_3

3.1 Introduction

The anatomically and functionally organized network of a nervous system serves the operational center for animal behaviors. Regionalized condensations of neurons, including the brain of bilaterians, have essential roles for cognitive functions in which neurons process information about ambient stimuli and sometimes store it as individual experiences.

At a very early stage of animal evolution, neurons may have originated as unspecialized cells with sensory, neurosecretory, and contractile functions. These ancestral, multifunctional cells became segregated into distinct cell types with either specific sensory, neuronal, or contractile function (Mackie 1970). Neurons formed extended cellular processes, or neurites, connecting to a specific neighboring cells via synapses. This "neural" system seems to have evolved for a rapid and specific signal transmission from sensory cells to a certain specific cell clusters such as a contractile units of muscles and a ciliomotor systems. In contrast to cell–cell communication mediated by undirected diffusion of signaling chemicals, the directed and restricted mode of synaptic communication between connected neurons allows animals to execute coordinated body movements in response to specific environmental contexts.

The origin of the nervous system is one of the most exciting questions in biology. There has long been interest in the use of basal metazoans, animal lineages that diverged early in animal evolution, including poriferans, placozoans, ctenophores, and cnidarians (Fig. 3.1)—to understand the early evolutionary processes of animal-specific traits such as the nervous system. In recent years, thanks to sequencing of the basal metazoan genomes, evolutionary biologists have made spectacular advances in unveiling primitive neuronal components. Recent findings in the basal metazoans have also raised several important questions, including whether a nervous system arose only once, or multiple times, and whether neural condensations in bilaterian and cnidarian branches reflect a homologous ancestral nature or a paraphyletic neural characteristics. Answers to these questions are pivotal in reconstructing the molecular and cellular features of the nervous systems that existed in ancestral metazoans.

In this chapter, I first provide an overview of genetic repertoires of "neural" components found in basal metazoan genomes and anticipate the genetic and cellular natures of primordial neurons. I then focus on molecular and anatomical features and on physiological functions of the nervous systems in extant cnidarians. Finally, I discuss the nature of primordial neural assemblies that may have been present before divergence of the Cnidaria and Bilateria.

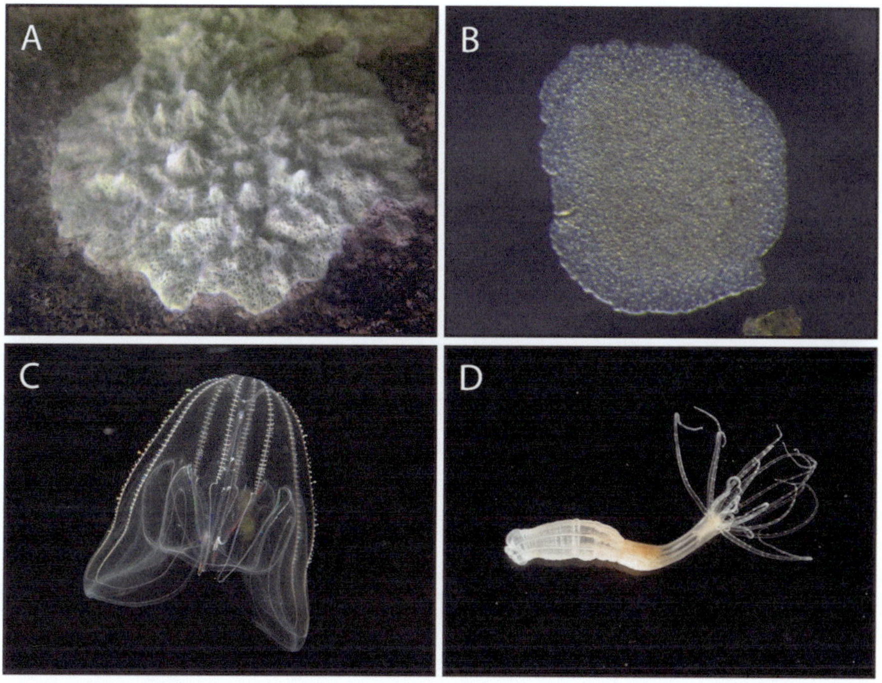

Fig. 3.1 Images of basal metazoans. (**a**) *Ephydatia fluviatilis* (*Porifera*). (**b**) *Trichoplax adhaerens* (Placozoa). (**c**) *Bolinopsis mikado* (Ctenophora). (**d**) *Nematostella vectensis* (Cnidaria). [Photographs courtesy of Dr. Noriko Funayama (*E. fluviatilis*), Dr. Hiroaki Nakano (*T. adhaerens*), and Ms. Noriko Ishikawa (*B. mikado*)]

3.2 Neural Gene Repertoires in Basal Metazoans

3.2.1 Poriferans

The phylum Porifera (sponges) comprises basal metazoans that do not possess bona fide neurons. Transcriptomic and genomic data from all four classes of poriferans (Hexactinellida, Demospongiae, Homoscleromorpha, and Calcarea) revealed that this basal metazoan lineage possesses surprisingly complex gene components believed to have been involved in the development and function of nervous systems (Table 3.1) (Simionato et al. 2007; Riesgo et al. 2014). Genomic analyses of *Amphimedon queenslandica* (Demospongia) have identified poriferan homologues for bilaterian neural genes such as *SoxB*, *Lhx*, and proneural basic helix-loop-helix (bHLH) transcription factors, *Elav*/*Musashi*-like RNA-binding protein (RBP) genes, and Notch signaling molecules (Richards et al. 2008; Larroux et al. 2008; Srivastava et al. 2010a; Fortunato et al. 2012; Richards and Degnan 2012). *AmqbHLH1*, a bHLH transcription factor gene that seems to belong to the atonal-related protein (Arp) superfamily, is expressed in globular cells of parenchymella larvae of *A.*

Table 3.1 Molecular and cellular features of sensory cells and neurons of metazoan animals

		Porifera	Placozoa	Ctenophora	Cnidaria	Bilateria
Neurosecretory cells		+	+	+	+	+
Neurons		–	–	+	+	+
Transcription factor genes	Proneural bHLH (expr./func.)	+ (sensory cells/n.d.)	+ (n.d./n.d.)	+ (n.d./n.d.)	+ (sensory cells and neurons/differentiation)	+ (sensory cells and neurons/differentiation)
	SoxB (expr./func.)	+ (sensory cells/n.d.)	+ (n.d./n.d.)	+ (neurons/n.d.)	+ (neurons/differentiation)	+ (neurons/differentiation)
Neural synapses	Electric synapse (gap junctions)	–	–	–	+[a]	+
	Synapses	–	–	+	+	+
Peptides	Conserved neuropeptides	–[b]	+	–[c]	+	+
	Peptide-gated ion channels	+	+	+	+	+
Chemical transmitters	Chemicals	Glu, GABA, Gly NA, AD, 5-HT[d]	n.d.[e]	Glu, GABA	Glu, GABA, Gly, DA, NA, AD, 5-HT, Ach	Glu, GABA, Gly, DA, NA, AD, 5-HT, Ach
	Genes	GAD, AAAH, AADC, DBH, AChE	GAD, AAAH, AADC, DBH, AChE	GAD, AAAH, AChE	GAD, AAAH, AADC, DBH, ChAT, AChE	GAD, PH, TH, TpH, AADC, DBH, AANAT, HIOMT, ChAT, AChE
	Neural function	–	–	+	+	+

[a]Gap junctions between small number of neurons have been found only in hydrozoans, but not the other cnidarian classes

[b]Although no neuropeptides have been identified in poriferan genomes, cnidarian LWamide neuropeptide treatment has been found to trigger settlement of poriferan larvae (Whalan et al. 2012)

[c]Although none of the evolutionary conserved neuropeptides (e.g., RFamide and LWamide) have so far been identified from ctenophore genomes, neuronal FMRFamide immunoreactivity has been observed (Jager et al. 2011)

[d]Chemical transmitters known to modulate the contracting behavior of poriferan species

[e]Trichoplax genome encodes genes involved in synthesis and vesicular transport of several chemical transmitters (NA, AD) (Srivastava et al. 2008)

queenslandica (Fig. 3.2). The globular cells, putative sensory cells in sponges, are located in the outer epithelium along the larval primary axis. Richards and colleagues found that the *AmqbHLH1* gene is coexpressed with genes for the Notch/Delta signaling pathway during cellular differentiation (Richards et al. 2008; Richards and Degnan 2012). In addition to the globular cells, the *A. queenslandica* larvae bear distinct sensory cells in anterior and posterior regions. In the anterior region, sensory flask cells develop in the outer epithelium to regulate larval metamorphosis (Nakanishi et al. 2015). A ciliated pigmented ring of photo-sensitive sensory cells develops at the posterior end of the larvae (Leys and Degnan 2001). Although genes specific to flask cells have not yet been found, pigmented ring cells express a number of genes that are involved in bilaterian sensory cell development, for example, proneural *Achaete-Scute homolog* (*Ash*), *SoxB2*, *Pax2/5/8*, and *Lhx* (Fig. 3.2) (Larroux et al. 2006; Richards et al. 2008; Srivastava et al. 2010b; Degnan et al. 2015). Clear homologues of *SoxB* genes have been identified in the two demosponge species, *A. queenslandica* (Larroux et al. 2006, 2008) and *Ephydatia muelleri*, and in a calcareous sponge, *Sycon ciliatum* (Fortunato et al. 2012). In *S. ciliatum* and *A. queenslandica*, the *SoxB* genes are expressed in cruciform cells (putative sensory cells or their precursors) during larval development (Fortunato et al. 2012). Cruciform cells also express *Elav* and *Musashi*, as well as *Pax* and *Six* transcription factors involved in formation of bilaterian eyes and other sensory organs (Fortunato et al. 2014).

Phylogenetic analyses of poriferan genes have shown to be rich in molecular components involved in formation of the postsynaptic density (PSD) (Sakarya et al. 2007; Alié and Manuel 2010; Srivastava et al. 2010a; Riesgo et al. 2014). Genomes of all four poriferan classes contain PSD genes, with little variation among species, suggesting that these genes existed in the common poriferan ancestor. It should be noted that poriferan homologues for PSD components *Dlg*, *Homer*, *Grip*, *Cript*, and *Gkap* are coexpressed dominantly or exclusively in the sensory globular cells of *Amphimedon parenchymella* larvae (Sakarya et al. 2007). Concurrent expression of multiple postsynaptic gene homologues may support the existence of a macromolecular complex (Sakarya et al. 2007; Emes et al. 2008; Ryan and Grant 2009). The existence of PSD genes in poriferan genomes, however, does not necessarily connote the appearance of functional PSD in the common poriferan ancestor, because a significant number of PSD genes have also been identified even in unicellular organisms such as choanoflagellates (Alié and Manuel 2010; Burkhardt et al. 2014; Burkhardt 2015). Indeed, no clear morphological feature consistent with a PSD or a synapse has been observed in sponges.

These findings suggest that the genetic mechanisms giving rise to both sensory cells and neurons have a deep evolutionary root (Fig. 3.3). However, genetic and signaling mechanisms regulating early commitment and later differentiation of the poriferan sensory cells still remain largely unknown. Additionally, functional and molecular dissection of the PSD protein complex in the poriferan sensory cells will

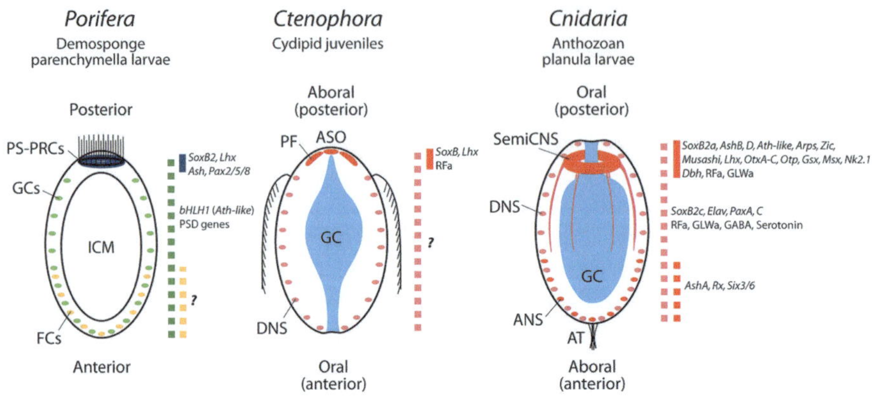

Fig. 3.2 The larval body plans of basal metazoans and expression patterns of neural markers. Regionalized expression of neural marker genes along the primary body axis of three basal metazoan larvae. In Porifera, the *blue line* indicates neurogenic gene expression in photosensitive pigmented ring cells. The *dashed green* and *dashed yellow lines* denote sensory globular cells and flask cells, respectively. The *red lines* in Ctenophora and Cnidaria indicate posterior neural aggregations from their diffuse nervous systems. The *dashed pink line* in Cnidaria shows pervasive expression of neural marker genes. Ctenophores possess a diffuse nervous system, whereas no expression of neural marker genes has been shown. The *dashed red line* in Cnidaria indicates expression of neural markers for aboral nervous system. *ANS* aboral nervous system, *ASO* apical sensory organ, *AT* apical tuft, *DNS* diffuse nervous system, *GC* gastric cavity, *FCs* flask cells, *GCs* globular cells, *semiCNS* semi-centralized nervous system, *PF* polar field, *PS-PRCs* photo-sensitive pigment ring cells

help us to explain how the "post"synaptic proteins are implicated in function of the sensory (usually "pre"synaptic) cells.

3.2.2 *Placozoans*

The phylum Placozoa contains at least 19 groups of disc-shaped marine invertebrates, a 1- to 2-mm in diameter. They are simple metazoans with two epithelial layers and some cell types interspersed in between, but they lack neurons. Recent studies on *Trichoplax* have described six somatic cell types comprising ciliated dorsal and ventral epithelial cells, lipophil cells, fiber cells, crystal cells, and gland cells (Smith et al. 2014).

Homologues of neurogenic *SoxB*, *Lhx*, and group A bHLH transcription factors related to *Ash* or *Arp* and to most components of synaptic vesicles and PSD, including synaptotagmin, have been identified in the genome of the placozoan, Trichoplax adhaerens (Fig. 3.3) (Srivastava et al. 2008; Gyoja 2014). Although the physiological functions and genetic signatures of placozoan cell types are largely unknown, immunostaining using anti-FMRFamide antibody revealed that this simple animal lacking neurons deploys RFamidergic neurosecretory cells in the

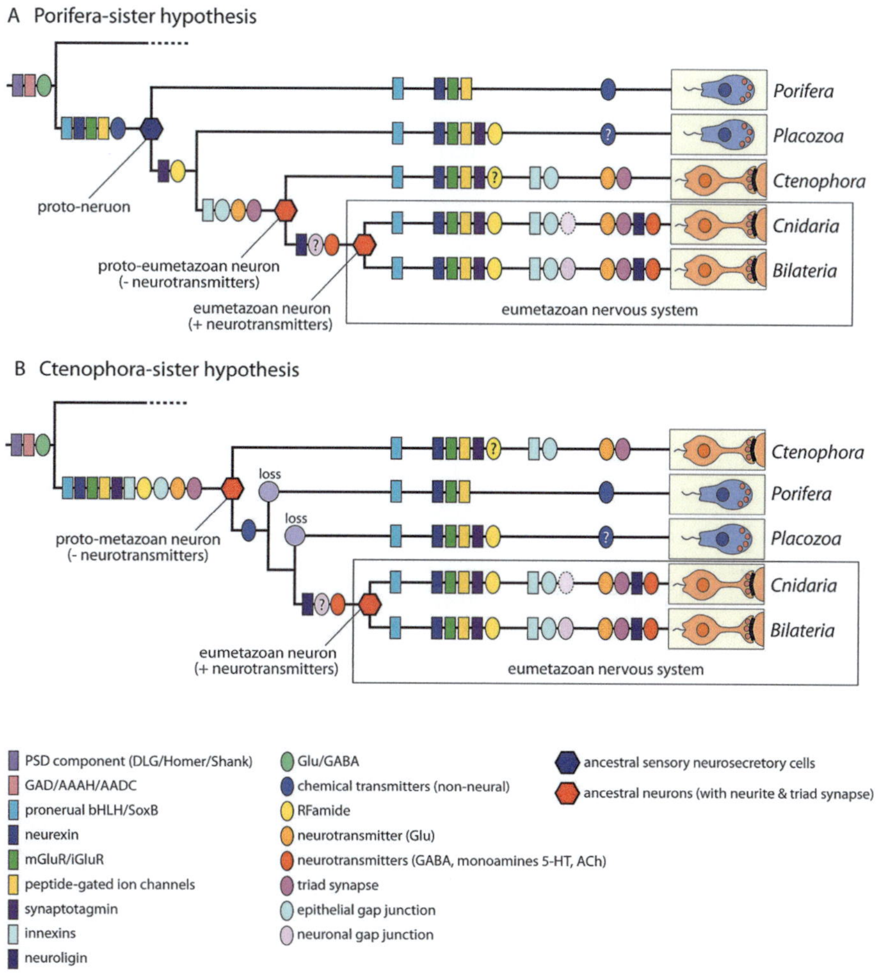

Fig. 3.3 Two scenarios of neuronal evolution. Key genetic and physiological innovations underlying neural organization in metazoan evolution. *Rectangles* and *ellipses* indicate acquisition of selected gene families and physiological and cellular properties, respectively. The presence of neurosecretory cells and neurons with neurites and synapses are shown to the *right*, respectively. The emergence of postulated ancestral cell types is shown in hexagons. (**a**) In the conventional metazoan tree, the "porifera-sister hypothesis," "proto-neurons" may have existed in the common metazoan ancestor. This postulates that ancestral cells may have resembled the neurosecretory cells lacking neurites and synapses that are seen in modern poriferans and placozoans. Genes for glutamatergic and peptidergic systems may already have been deployed for cell–cell communications in the common metazoan ancestor. After the common ancestor of Ctenophora/Cnidaria/Bilateria branched off, gap junctions (innexins), neurites, and synapses evolved. (**b**) In the ctenophora-sister hypothesis, one assumption is that gap junctions and these neuronal characteristics were deployed in the common metazoan ancestor. In this scenario, glutamatergic and probably peptidergic proto-metazoan nervous systems, gap junctions, neurites, and synapses have been lost in poriferan and placozoan lineages

marginal body region (Schuchert 1993a). Smith and colleagues have demonstrated that in the ciliated gland cells, an FMRFamide-like neuropeptide is coexpressed with synaptic vesicle proteins such as syntaxin, SNAP-25, and synapsin, suggesting neurosecretory functions of this cell type (Smith et al. 2014). Expression and function of the neurogenic transcriptional factors in the gland cells remain to be explored.

3.2.3 Ctenophores

Ctenophores, collectively known to as comb jellies, are a group of neuron-bearing marine invertebrates with controversial ancestry. They are thought to be genetically less complex, because ctenophore genomes only have a few Wnt and homeobox genes and apparently do not seem to encode any of the micro-RNA homologues that have been identified to date in cnidarians and bilaterians (Fortunato et al. 2015; Maxwell et al. 2012). Recent phylogenomic analyses and comparisons of genetic repertoires suggest that this orphan animal lineage is a sister group to all other extant metazoans (the ctenophora-sister hypothesis) (Fig. 3.3b) (Dunn et al. 2008; Ryan et al. 2013; Moroz et al. 2014; Whelan et al. 2015a). Other phylogenetic analyses, however, proposed that ctenophores and cnidarians form a clade with bilaterians in the *Eumetazoa* (animals with nerve and muscle cells) (Fig. 3.3a) (Philippe et al. 2009, 2011; Pick et al. 2010). With increasing transcriptomic and genomic data from basal metazoan species, systematic errors that can cause mis-positioning of basal metazoan taxa are now under careful scrutiny (Pisani et al. 2015; Whelan et al. 2015a, 2015b). Nevertheless, understanding the ctenophoran nervous system at the molecular and cellular level is essential to reconstruct the ancestral nervous systems.

Ctenophores have nerve nets with mesogleal fibers and tentacular nerves. These nerve nets exhibit numerous condensations associated with the apical sensory organs/polar fields and tentacle bulbs (Harbison 1985). Ctenophore genomes, as well as poriferan genomes, contain homologues of neurogenic transcription factors including *Lhx*, *bHLH*, *Six*, and *SoxB*, as well as neural RBP genes, *Elav* and *Musashi*, that are involved in early neural development of bilaterians. Poriferans and ctenophores have some genes for axon guidance molecules, including semaphorin, plexin, and an ephrin receptor, while others, such as netrin and Unc-5, are absent from the genomes of both phyla (Srivastava et al. 2010a; Ryan et al. 2013; Moroz et al. 2014). Most of the genes involved in the formation of bilaterian PSDs have been identified in both ctenophoran and poriferan genomes, but they lack certain genes, such as *Erbin* and *Neuroligin* (Srivastava et al. 2010a; Riesgo et al. 2014; Ryan et al. 2013; Moroz et al. 2014). These comparative genomic data indicate that certain axon guidance molecules and scaffolding proteins were absent in the common ancestor of these basal metazoans.

Glutamate appears to be the best transmitter candidate for ctenophoran neuromuscular transmission (Table 3.1) (Moroz et al. 2014). Ctenophoran genomes possess a gene related to glutamate decarboxylase (GAD) that synthesizes γ-aminobutyric acid (GABA). There is no clear genetic evidence supporting synthesis of any other bilaterian neurotransmitters. Immunohistochemical and biochemical analyses using *Pleurobrachia bachei* have failed to detect conventional neurotransmitters [e.g., monoamines and acetylcholine (ACh)] (Hay-Schmidt 2000; Moroz et al. 2014). While these data could be interpreted to mean that the ctenophoran nervous system evolved independently (Moroz et al. 2014), a substantial set of neuronal marker genes in the nervous systems of ctenophores and cnidarians/bilaterians imply a common evolutionary origin (Watanabe et al. 2014a; Marlow and Arendt 2014; Jékely et al. 2015). For example, *SoxB* and *Lhx* transcription factors and RFamide-like neuropeptides are expressed in neurons in the apical region of ctenophores (Fig. 3.2) (Jager et al. 2008, 2011; Simmons et al. 2012). In accordance with this hypothesis, ctenophores, cnidarians, and bilaterians, but not poriferans or placozoans, have neuronal synapses (Fig. 3.3) (Hernandez-Nicaise 1973).

The lack of unambiguous evidence for neurotransmitter use by ctenophores could argue for an independent origin of the ctenophoran nervous system; however, nonneural chemical transmitter localization and functions are observed among cnidarians and bilaterians. And in poriferans, many of the transmitters are involved in modulation of contractile behavior (see following). Thus, the absence of chemical neurotransmission may not be a reliable basis for falsifying the homology of nervous systems in early branching metazoans.

3.2.4 Protoneurons: An Ancestral Neurosecretory Cells?

Many primary ciliated larvae of marine invertebrates possess the apical sensory neurosecretory cells expressing RFamide and Wamide neuropeptides (Lacalli 1983; Dickinson and Croll 2003; Nielsen 2005; Tessmar-Raible et al. 2007; Byrne et al. 2007; Conzelmann et al. 2011, 2013; Conzelmann and Jékely 2012). Sensory neurosecretory cells have repeatedly been considered in different evolutionary contexts as ancestral neuronal cells or "protoneurons" (Vígh and Vígh-Teichmann 1982; Tessmar-Raible et al. 2007; Sakarya et al. 2007; Richards et al. 2008; Jékely et al. 2015). Since vertebrate and invertebrate deuterostomes possess an assembly of ciliated sensory neurosecretory cells contacting the cerebrospinal fluid in the lumen of the CNS (vertebrates) or ambient seawater (invertebrates), ciliated sensory neurosecretory cells have been regarded as a phylogenetically old neurosecretory cell type, the "protoneuron" (Vígh and Vígh-Teichmann 1982; Vigh et al. 2004). Detailed molecular and cellular dissections of the developing apical neurosecretory cell cluster of annelid larvae have suggested that the sensory neurosecretory cells already existed in the common bilaterian ancestor (Tessmar-Raible et al. 2007; Conzelmann et al. 2013; Tosches and Arendt 2013; Marlow et al. 2014; Nielsen 2015). RFamidergic sensory neurosecretory cells have also been identified in cnidarian planula larvae (Plickert 1989; Leitz and Lay 1995; Gajewski et al. 1996). Although peptidergic neurophysiological features of the ctenophoran nervous system remain highly understudied, a rich neural gene repertoire, including peptide-gated ion channels (PGICs) in poriferan, placozoan, and ctenophoran genomes indicate that the conceptual protoneuron might be traced back to the sensory neurosecretory cell types that may have existed in the common metazoan ancestor (Fig. 3.3) (Smith et al. 2014; Jékely et al. 2015). This idea is supported by the fact that the RFamide neuropeptides are expressed exclusively in cnidarian nervous systems and placozoan neurosecretory gland cells. Since conventional chemical neurotransmitters don't seem to be neuron-specific in cnidarians, it would be useful to identify neuropeptides in neurons and sensory cells in sponges and ctenophores.

3.3 Cnidarian Nervous Systems

The Cnidaria is a large and successful phylum containing more than 9,000 species, and in phylogenetic terms, it represents the closest sister group to all bilaterians. The Cnidaria is divided into two major lineages: the Anthozoa (corals, sea anemones, and sea pens) and the Medusozoa, consisting of four classes: Hydrozoa (hydras and marine hydrozoans), Cubozoa (box jellyfish), Scyphozoa (true jellyfish), and Staurozoa (stalked jellyfish). Cnidarians usually have a life cycle containing a polyp stage and a medusa (jellyfish) stage (Bridge et al. 1992, 1995; Odorico and Miller 1997; Schuchert 1993b; Collins 2002; Collins et al. 2006). The Anthozoa are thought to retain ancestral characteristics, including 1) polyps that never metamorphose into medusae, 2) few derived genomic features, and conserved mitochondorial genome structure (circular in anthozoans and bilaterians, but linear in medusozoans), 3) nematocysts (the cnidarian-specific stinging apparatus) that are less elaborate and diversified than those of medusozoans (Pantin 1966; Willmer 1990; Bridge et al. 1995; Medina et al. 2001; Collins 2002; Dunn et al. 2008).

The privileged phylogenetic position of the Cnidaria as the closest sister group to the Bilateria has made these animals one of the most useful models for deciphering the genetic basis for the early evolution of sophisticated nervous systems, such as the central nervous system (CNS). In addition to the phylogenetic position of the Cnidaria, transcriptomic and genomic data from various cnidarian species, including the anthozoans *Nematostella vectensis* and *Aiptasia* sp. (sea anemones), *Acropora digitifera* and *Acropora millepora* (stony corals), and the hydrozoans *Hydra magnipapillata, Cladonema pacificum*, and *Clytia hemisphaerica*, have made the Cnidaria even more useful for comparative and functional molecular studies of the evolution of neurodevelopmental events (Kortschak et al. 2003; Technau et al. 2005; Putnam et al. 2007; Shinzato et al. 2011; Watanabe et al. 2014b; Baumgarten et al. 2015).

Comparative genomic analyses of neural genes using *Nematostella vectensis* (Anthozoa) and *Hydra magnipapillata* (Hydrozoa) have revealed extensive conservation of the neural gene repertoire, not only between these distantly related cnidarian species, but also among other metazoan lineages. In general, cnidarian genomes possess a nearly complete set of homologous genes involved in bilaterian neurotransmission and neurodevelopmental processes, including neurogenesis, neuronal specification, and neural network formation (Fig. 3.3; Table 3.1) (for more detail, see Watanabe et al. 2009; Galliot et al. 2009; Galliot and Quiquand 2011).

Chemical synapses, in which signal transmission is effected by neuropeptides and small chemicals (acetylcholine, monoamines, and amino acids), are essential for both slow and fast transmission. Neuropeptides are related to peptide hormones and are often involved in slow transmission in bilaterian nervous systems. Among cnidarians, neuronal cells, mainly ganglionic neurons, have been shown to possess synapses with lucent-core and dense-core vesicles (Davis et al. 1968; Kinnamon and Westall 1981; Westfall and Kinnamon 1978, 1984; Kass-Simon and Pierobon 2007; Pierobon 2012). Small chemical neurotransmitters are generally packaged in small, lucent-core synaptic vesicles that are clustered in presynaptic terminals. Neuropeptides are generally stored in large, dense-cored vesicles that are distributed at presynaptic terminals and also along neuronal processes (for review, see Pierobon 2012; Fujisawa and Hayakawa 2012). These observations suggest that neurotransmission and neuromodulation mediated by both chemical transmitters and neuropeptides are involved in cnidarian neurotransmission.

3.3.1 Peptidergic Nervous Systems

Neuropeptides serve essential and pleiotropic neural functions in both cnidarians and bilaterians. The cnidarian nervous system is rich in neuropeptides, including the short amidated neuropeptides, RFamide and LWamide (Fujisawa 2008; Watanabe et al. 2009). These neuropeptides emerged before the Cnidaria/Bilateria evolutionary split (Jékely 2013). Among cnidarians, RFamide has been studied most extensively. The first sign of *Rfamide* expression can be detected in some cells of embryonic epithelium at the blastula stage (Marlow et al. 2009; Richards and Rentzsch 2014). Here, it contributes to formation of a loose plexus of the neurons at the base of the ectoderm. At the planula stage, RFamide-positive neurons are well developed around the anterior (aboral) region of hydrozoan and scyphozoan larvae (Fig. 3.4) (Martin 1988, 1992; Plickert 1989; Leitz and Lay 1995; Gajewski et al. 1996; Gröger and Schmid 2001; Seipel et al. 2004; Nakanishi et al. 2008; Marlow et al. 2009). In the more basal cnidarian group, sea anemones, the RFamidergic neurons form a neural cluster at the oral region in addition to the nerve net (Fig. 3.4) (Marlow et al. 2009; Watanabe et al. 2014a). Oral and pharyngeal development of the RFamidergic neurons at polyp and medusae stages has widely been observed among cnidarian phyla (Grimmelikhuijzen 1985; Plickert 1989; Grimmelikhuijzen et al. 1988, 1991; Koizumi et al. 1992; Mackie and Stell 1984; Mackie et al. 1985; Anderson et al. 2004). It probably constitutes a plesiomorphic neural character of cnidarians. The oral arrangemet of the RFamidergic neurons in cnidarian polyps can be anatomically divided into a hypostomal accumulation of neuronal perikaria and ring-shaped neurite bundles (nerve ring) that formed between the hypostome

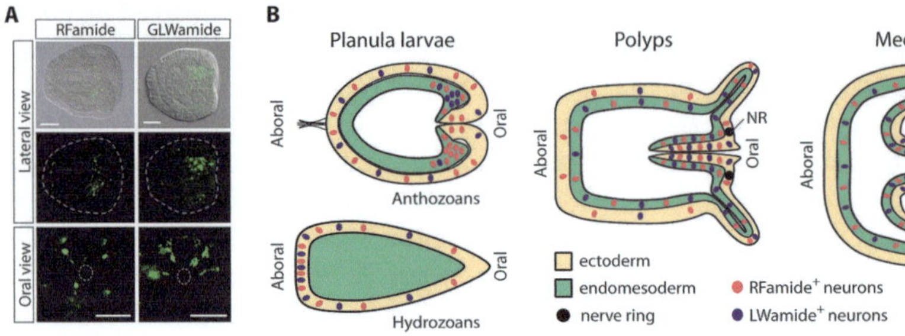

Fig. 3.4 Semicentralization of the cnidarian oral/blastoporal nervous system. (**a**) Lateral and oral views of neural subsystems expressing RFamide or GLWamide neuropeptides of *Nematostella* planula larvae. Note that these neural populations are localized mainly around the pharynx. GLWamidergic neurons exhibit an additional bias along the secondary axis, which is orthogonal to the primary oral–aboral axis. *Dotted white circles* in *lower panels* indicate the blastopore. *Bars* in *upper panels* (lateral view) 100 μm; *bars* in *lower panel*s (oral view) 50 μm. (**b**) Schematic views of regionalization of RFamidergic (RFamide$^+$) and LWamidergic (LWamide$^+$) neurons in the cnidarian life stages. Blastoporal condensation of these neurons develops in the planula larva stage of *Nematostella* (Anthozoa). The sensory function of the apical organ/tuft is not clear yet. Among medusozoans such as hydrozoans, neurons are developed predominantly in the aboral (or apical) region of the planula larvae. During the polyp and medusa stages, the oral/pharyngeal neural condensations become more conspicuous and nerve rings develop around the mouth opening (polyp) or at tentacle bases (medusa). *NR* nerve ring, *M* manubrium, *INR* inner nerve ring, *ONR* outer nerve ring

and tentacle bases (Mackie and Stell 1984; Mackie et al. 1985; Grimmelikhuijzen et al. 1988, 1991; Koizumi et al. 2014). These distinct neural architectures found in polyps could be related to a neuronal condensation at the manubrium (pharynx) and the nerve ring at the bell margin in medusae (Fig. 3.4). Neural expression of LWamide neuropeptide has also been demonstrated in a wide variety of cnidarian species (Leitz and Lay 1995; Schmich et al. 1998a; Mitgutsch et al. 1999; Takahashi et al. 2003; Watanabe et al. 2014a). In addition to the RFamidergic nervous system, LWamidergic neurons develop in the oral/pharyngeal neural condensation, and during the medusa stage, the nerve ring around the bell margin (Fig. 3.4) (Schmich et al. 1998a; Koizumi et al. 2014; Watanabe et al. 2014a). These expression patterns suggest that RFamide and LWamide have critical functions in the oral nervous system.

Activity of peptidergic neurons is required for various coordinated behaviors in cnidarians. Myoactivities of cnidarian neuropeptides have been well characterized in polyps. The RFamide neuropeptide family is believed to regulate the myoactivity and modulatory role of muscle contraction (McFarlane et al. 1987, 1991; Shimizu and Fujisawa 2003; Fujisawa 2008). A *Hydra* LWamide neuropeptide (Hym-248) is reported to directly induce relaxation of myoepithelial cells in the body column and tentacles (Takahashi et al. 2003). In addition to the myoactivity control, the RFamidergic and LWamidergic neurons have pivotal roles in medusozoans in positive and negative phototactic behavior (Katsukura et al. 2004; Plickert and

Schneider 2004). The cnidarian nervous system forms synaptic connections with both epitheliomuscular cells and neighboring neurons (Westfall et al. 1971, 2002; Westfall 1973, 1987; Westfall and Kinnamon 1978, 1984; Kinnamon and Westfall 1982). Dense-core, neuropeptide-containing vesicles accumulate in the presynaptic area in nerve terminals and at *en passant* synapses (Koizumi et al. 1989; Westfall and Grimmelikhuijzen 1993; Westfall et al. 1995), indicating that localized release of neuropeptides at synapses serves directed and restricted signal transmission. This idea finds support in the unidirectional neural conduction and partial contraction of tentacles in *Hydra* (Rushforth and Hofman 1972; Shimizu 2002). These findings, taken together, indicate that in cnidarians, the structural organization of the nervous system is indispensable for neural regulation of behavior during all life stages.

3.3.2 Classical Chemical Neurotransmitters

Although the cnidarian nervous system is rich in neuropeptides, accumulating physiological, biochemical, and ultrastructural evidence suggest that classical neurotransmitters and neuromodulaters are also involved in neural function (Pierobon 2012). Cnidarian genomes possess a number of genes showing significant similarity to bilaterian genes for synthesis and degradation of classical chemical neurotransmitters, including acetylcholine (ACh), monoamines, GABA, and nitric oxide (NO). Homology searches of bilaterian genes for the chemical neurotransmitter receptors and transporters have identified a large number of putative cnidarian counterparts (Watanabe et al. 2009; Anctil 2009; Marlow et al. 2009; Chapman et al. 2010; Oren et al. 2014). Since many of these genes probably existed in the common poriferan and cnidarian ancestor, they seem to have deep evolutionary roots with non-neural functions. However, physiological and immunohistological data imply that small chemical transmitters and modulators play roles in the cnidarian neural functions.

Glutamate is the most common excitatory neurotransmitter among bilaterians. The *Nematostella* genome has a large number of genes for metabotropic and ionotropic glutamate receptors (mGluR and iGluR, respectively) (Anctil 2009). Among four classes of the iGluRs, including NMDA, AMPA, Delta, and kainate receptors, *Nematostella* genes show greater similarity to the NMDA and AMPA classes. Immunohistochemical studies indicate that the cnidarians have several neuronal and sensory cell populations expressing iGluR-like proteins (Kass-Simon and Scappaticci 2004). Glutamate accumulates in processes of these neural cells (Martin 2004; Delgado et al. 2010). Glutamate and its agonists, NMDA, AMPA, and kainite, appear to have an excitatory functions in control of epitheliomuscular cell contraction and discharge of the nematocysts (Kass-Simon et al. 2003; Scappaticci et al. 2004; Scappaticci and Kass-Simon 2008). Other excitatory neurotransmitters, such as serotonin, dopamine, and adrenaline, as well as neuromodulatory molecules, including NO and carbon monoxide (CO), are involved in cnidarian neural functions (Kass-Simon and Pierobon 2007; Pierobon 2012). Cnidarians possess homologues to bilaterian genes such as choline acetyltransferase (ChAT), acetylcholinesterase

(AChE), and nicotinic ACh receptor subunits (Anctil 2009). ACh induces musculoepithelial contraction in cnidarians (Lentz and Barrnett 1961; Scappaticci and Kass-Simon 2008).

The *Nematostella* genome includes receptors for inhibitory transmitters, GABA and glycine. Sequences of GABAb and glycine receptors show a higher degree of similarity to bilaterian receptors than do GABAa sequences, which are more distantly related to their bilaterian counterparts and which lack certain amino acid residues important for GABA binding (Anctil 2009). Immunohistochemical analyses for GABA have demonstrated the neuronal localization of this transmitter in sea anemones and cubozoan medusae (Martin 2004; Marlow et al. 2009; Delgado et al. 2010), implying a role in cnidarian neural function. This idea is supported by data from the sea fan, *Eunicella cavolini* (Anthozoa), in which GABA signaling molecules such as GAD, the enzyme that produces GABA, the vesicular GABA transporter (VGAT), and the GABAb receptor are expressed in neuronal cells (Girosi et al. 2007). Glycine and taurine, a glycine receptor agonist, also localize in cnidarian neurons. A taurine-like immunoactivity in the ganglia and sensory neurons was reported from anthozoan and scyphozoan species (Carlberg et al. 1995; Anctil and Minh 1997; Nakanishi et al. 2008). Neurophysiological functions of glycine and taurine have been analyzed in *Hydra*, and both chemicals are involved in the peristaltic contraction of the epitheliomuscular cells and in the chemosensory response (Pierobon et al. 2001; Ruggieri et al. 2004; Kass-Simon and Pierobon 2007). An interesting finding is that taurine can function as an excitatory neuromuscular transmitter in cnidarians (Carlberg et al. 1995; Anctil and Minh 1997). Taken together, it appears that these classical neurotransmitters had already been used in the nervous system before the Cnidaria/Bilateria cladogenesis (Fig. 3.3; Table 3.1).

3.3.3 Nonneural Functions of Classical Transmitters

While a large variety of ultrastructural, biochemical, and neurophysiological data from cnidarians indicates that the conventional neurotransmitters and neuromodulators are involved in neural control of cnidarian behavior, nonneural localization and functions of these chemicals have also been reported (for review, see Kass-Simon and Pierobon 2007). Glutamate, for instance, has been detected in nonneural cells in the oral/pharyngeal region of sea anemones (Anctil and Carette 1994; Oren et al. 2014). Expression patterns of genes for GAD and AChE suggest that GABA and ACh can also be metabolized in nonneuronal epithelial cells (Denker et al. 2008; Takahashi and Hamaue 2010; Oren et al. 2014). Nonneuronal metabolism of ACh is likely because the gene encoding the choline transporter (ChT) was expressed in epithelium of *Hydra* polyps that were depleted of all neural cells (Chapman et al. 2010).

Interestingly, molecular evidence suggests that receptors for glutamate and GABA were present before plants and animals diverged (Lam et al. 1998; Moroz

2001). Glutamate and GABA act as important chemical messengers in organisms lacking neurons, such as poriferans, plants, and even unicellular protists, in which these molecules are involved in contraction, feeding, sensory systems, and development (Lam et al. 1998; Moroz 2001; Bouche et al. 2003; Davenport 2002; Elliott and Leys 2010). Clearly, glutamine- and GABA-based transmission systems predated the development of the nervous systems (Fig. 3.3) (Parker 1910; Jones 1962; Pavans de Ceccatty 1974a, 1974b, 1979; Mackie 1970, 1979, 1990; Nickel 2004). In poriferans, some classical transmitters control contraction behaviors regulated by contractile cells, such as myocytes/actinocytes (Boury-Esnault and Rützler 1997) and/or pinacocytes (Nickel et al. 2011). mGluRs and iGluRs have been identified in poriferan genomes (Perovic et al. 1999; Srivastava et al. 2010a; Riesgo et al. 2014; Burkhardt et al. 2014). Glutamate treatment of adult poriferans induced contraction and propagation of a stereotypical behavior, inflating and deflating the canal system (Ellwanger et al. 2007; Elliott and Leys 2010). Recent phylogenetic studies have demonstrated that poriferans have genes involved in synthesis of monoamines (dopamine, adrenaline, and serotonin) and ACh, that include GAD, DOPA decarboxylase, tryptophan hydroxylase (TpH), and dopamine β-hydroxylase (DBH) (Srivastava et al. 2010a; Riesgo et al. 2014). Consistent with the existence of the GAD and TpH/DBH genes, GABA and serotonin have been detected in poriferans, where they stimulate and modulate contractions (Ellwanger and Nickel 2006; Ellwanger et al. 2007; Ramoino et al. 2007, 2011; Elliott and Leys 2010).

Genomic and transcriptomic analyses have tentatively identified genes for the synthesis, degradation, and transport of classical transmitters in basal metazoans; however, sequence data for these enzymes are not sufficiently conclusive to confirm their substrates and catabolites, thereby decreasing the reliability of transmitter prediction, especially in lower metazoans. For example, in cnidarian genomes, genes related to bilaterian choline/carnitine acetyltransferase subfamiles have been found. However, detailed analyses of the amino acids involved in substrate discrimination have obscured whether the catalytic efficiency of these cnidarian enzymes favors choline more than carnitine as a substrate (Chapman et al. 2010). One should exercise great caution when considering genes of the aromatic amino acid hydroxylase (AAAH) family, including TH and TpH, and of the methyltransferase family, including phenylethanolamine *N*-methyltransferase (PNMT). Since a large genetic repertoire for chemical neurotransmitter sysnthesis exists in poriferans, more detailed sequence analysis of these basal metazoan homologues and biochemical examination of their real catabolites will help us to understand the ancient means of intercellular communication and the evolutionary history of neurotransmitters.

3.3.4 Electrical Synapses and Gap Junctions

In addition to chemical neurotransmission, there are electrical synapses in hydromedusae that fire synchronously, serve pacemaker functions and coordinate contraction of the epitheliomuscular cells (Campbell et al. 1976; Passano and McCullough

1963; Shimizu and Fujisawa 2003; Takaku et al. 2014). Electric synapses are specialized and physically connected transcellular channels enabling rapid bidirectional communication between two neighboring cells. In *Hydra,* gap junctions form between ectodermal cells and between endodermal cells, and mediate electrical coupling of these epitheliomuscular cells (Hand and Gobel 1972; Wood 1977, 1979; Fraser et al. 1987). In medusae, striated muscle-like cells, forming the innermost layer of the bell that are not directly connected to neurons, are coordinated so that contraction occurs by epithelial conduction through gap junctions (Mackie 1990). Electron microscopic studies have shown that in hydrozoans certain neurons are connected with gap junctions (Westfall et al. 1980; Takaku et al. 2014).

Gap junctions are formed by two unrelated gene families, innexins in protostomes and connexins in deuterostomes (Phelan 2005; Scemes et al. 2007). Although connexins do not exist in the cnidarian genomes sequenced to date, 17 and 8 genes encoding innexins have been identified in the genomes of the hydrozoans, *Hydra* and *Clytia* (Chapman et al. 2010), respectively. In *Hydra*, innexin-1 is expressed in ectodermal epithelial cells, suggesting that the innexin-1 is a component of the ectodermal gap junctions (Alexopoulos et al. 2004; Chapman et al. 2010). Recently, characterization of innexin-2 has shown that it forms neural gap junctions and coordinates spontaneous contraction of the body column (Takaku et al. 2014). In hydromedusae, expansion of innexin genes has probably been accompanied by synchronous epithelial conduction via gap junctions (Mackie and Passano 1968; Spencer 1974; Anderson 1980).

In contrast to hydrozoans, anthozoans and scyphozoans do not seem to have gap junctions and no empirical evidence has so far suggested any neuronal electrical coupling in their nervous systems (Mackie et al. 1984; Mackie 1990). Only one innexin/pannexin-like gene is present in the *Nematostella* genome, and it is absent in the other anthozoan, *Acropora digitifera* (Shinzato et al. 2011), suggesting that the innexin genes in hydromedusae have undergone independent expansions after their divergence from other cnidarian lineages. The existence of electrical synapses between neurons is therefore not a common feature in Cnidaria. Although poriferans and placozoans do not have proteins that form gap junctions, the ctenophore genomes encode multiple innexin genes (Moroz et al. 2014). These gene products probably form many gap junctions in ciliated cells of the apical organ and in endodermal cells of the comb plates (Satterlie and Case 1978). This suggests that gap junctions were established between epithelial cells, but not neurons, before the Cnidaria and Ctenophora diverged.

3.4 Anatomical and Physiological Features of the Cnidarian Nervous System

The Cnidaria are often used in textbooks to show a primitive nerve net, a neuronal network connected by a mesh of neurite processes. In the freshwater polyp, *Hydra* (Hydrozoa), which shows the simplest body plan and nervous system among cnidarian polyps, neurons are connected to other neurons and to the epitheliomuscular cells, forming a diffuse neural meshwork throughout the animal body (Westfall et al. 2002). The nerve net is composed of sensory cells and interneuronal ganglion nerve cells. Sensory cells have elongated cell bodies with a ciliary cone at the apical end. Ganglion cells extend neurites at the basal end of epithelial cells (Davis et al. 1968; Kinnamon and Westall 1981; Westfall and Kinnamon 1978; 1984). Immunohistochemical studies of neuropeptides have demonstrated that the nerve plexus is composed of neuronal subtypes expressing distinct neuropeptides that are distributed in a polarized way with respect to the body axis (Koizumi et al. 2004).

In addtion to the diffuse nervous system, cnidarians also possess regionally restricted and condensed nervous systems (Figs. 3.2, 3.4). A cluster of neurosecretory cells with sensory functions develops on the aboral (apical) side, mainly among lecithotrophic (yolk-feeding) larvae. In the oral region, a neuronal accumulation with organized neurite fasciculations, called a "nerve ring" or "oral nervous system," has been observed in planktotrophic (plankton-feeding) larvae and polyps (Koizumi 2007; Koizumi et al. 2014; Watanabe et al. 2009; Marlow et al. 2009; Layden et al. 2012). Although anthozoans bear neuronal condensation at the oral region (mouth and pharynx), ring-shaped neural architecture is observed more clearly in hydrozoans at the base of the oral tip (hypostome) of polyps, as in *Hydra oligactis*; it is most evident at the bell margin of medusae. In cubozoan medusae, the nerve ring connects to the elaborate visual sensory system at the tentacle bulbs (Mackie 1990; Koizumi et al. 2014).

3.4.1 Aboral Nervous Systems and Apical Sensory Organs

The apical sensory organ, an anterior cluster of ciliated sensory neurosecretory cells (Richter et al. 2010), has widely been observed in the ciliated larvae of marine invertebrates (Lacalli 1983; Nielsen 2005; Tessmar-Raible et al. 2007; Conzelmann et al. 2011; Dickinson and Croll 2003; Byrne et al. 2007). In primary larvae of marine invertebrates, the apical organ comprises sensory cells with neurosecretory characters, and is assumed to help control of larval swimming behavior and metamorphosis (Chia and Bickell 1978; Chia and Koss 1979; Hadfield et al. 2000; Conzelmann and Jékely 2012; Conzelmann et al. 2013). In annelids, an apical sensory cell cluster comprises neurosecretory cell types expressing several neuropeptides, including RFamide and Wamide (Conzelmann et al. 2011, 2013; Conzelmann and Jékely 2012). These neuropeptides regulate swimming depth and settlement of pelagic larvae (Conzelmann et al. 2011, 2013). Similarly, in cnidarians, the aboral ectoderm of planula larvae harbors RFamidergic and LWamidergic sensory cells (Plickert 1989; Leitz and Lay 1995; Gajewski et al. 1996). These ciliated neuropeptide-positive cells are thought to have a sensory neurosecretory function. Although sensory functions of the apical tuft/organ formed at the aboral pole of the planula larvae remain obscure, RFamidergic and LWamidergic sensory neurons in the aboral half in various cnidarian larvae appears to be involved in the metamorphosis of free-swimming planula larvae into benthic polyps (Chia and Bickell 1978; Chia and Koss 1979; Leitz et al. 1994; Gajewski et al. 1996; Takahashi et al. 1997; Schmich et al. 1998b; Iwao et al. 2002; Hatta and Iwao 2003; Katsukura et al. 2003, 2004; Erwin and Szmant 2010; Takahashi and Hatta 2011). Since settlement and metamorphosis of cnidarian pelagic larvae are induced by marine biofilms (Müller 1969; Morse and Morse 1991; Leitz and Wagner 1992), neuropeptide-expressing sensory cells are probably implicated in perception of environmental signals from suitable sites on the benthos. The aboral neurosecretory system of cnidarian larvae appears also to allow orientation toward light for coordinating the diurnal cycle of migration. RFamide and LWamide peptides are involved in control of the creeping behavior of planulae toward a light source (positive phototaxis), as the phototaxis of the planulae was drastically suppressed or promoted by exogenous RFamide peptide or LWamide peptide, respectively (Katsukura et al. 2004; Plickert and Schneider 2004).

3.4.2 Oral/Pharyngeal Nervous Systems

Immunohistochemical analyses of neuropeptide-expressing neurons have shown that, in addition to their aborally localized neurosecretory cells, cnidarians develop elaborate nerve structures in the oral region. Because the basic cnidarian body plan shows radial symmetry with a single mouth opening, the oral nerve plexus sometimes shows an annular architecture called 'nerve ring' (Figs. 3.2, 3.4) (Grimmelikhuijzen and Spencer 1984; Grimmelikhuijzen 1985; Koizumi et al. 1992; Mackie and Meech 2000; Yi-Chan et al. 2001; Mackie 2004; Garm et al. 2006, 2007; Satterlie 2011). The highest level of morphological and physiological elaboration of the nerve ring has been observed especially at the bell margin of medusae (Koizumi et al. 2014). A nerve ring comprising neuronal subsets with distinct neurophysiological functions was found in most medusae investigated so far. The nerve ring of *Aglantha digitale* (Hydrozoa), for example, has been divided into at least seven subsystems with distinct physiological properties (Mackie and Meech 1995a, 1995b, 2000; Mackie 2004). Communication among the subsystems allows complex behavioral control, including swim contractions of the medusae (Mackie and Meech 1995b). Several medusae bear a nerve ring connected to a sophisticated eye-bearing sensory complex at the base of the tentacles (Singla 1974; Yamamoto and Yoshida 1980; Singla and Weber 1982; Laska and Hündgen 1982; Nilsson et al. 2005). Visually guided behavioral patterns are observed in these cnidarian classes, and especially in cubomedusae, these patterns are quite complex (Hartwick 1991; Hamner et al. 1995; Matsumoto 1995). Ring-like neurite bundles and a neuronal condensation at the oral side have been reported from polyps and even from planula larval stages (Matsuno and Kageyama 1984; Grimmelikhuijzen 1985; Koizumi 2007; Marlow et al. 2009; Watanabe et al. 2014a). Immunostaining of neuropeptides and gene expression analyses of neurogenic genes demonstrate that neuronal differentiation starts in the blastula epithelium and progressively increases on the oral side during larval developmental stages. This oral neurogenic domain develops into a semi-centralized nervous system (semiCNS) on the oral side of the planula larvae and primary polyps that comprises several subsystems (Fig. 3.4). The cnidarian oral nervous system is therefore regarded as a considerable degree of neuronal condensation and has been regarded as the beginning of the bilaterian CNS (Holland 2003; Davis et al. 1968; Koizumi 2007; Tosches and Arendt 2013; Holland et al. 2013; Marlow et al. 2014; Watanabe et al. 2014a; Nielsen 2015). More detailed comparisons of the bilaterian CNS and the cnidarian semiCNS are needed, however.

3.5 Development of Cnidarian Nervous Systems

Cnidarian homologues for the proneural bHLH genes, *Ash* and *Arp*, as well as *SoxB* genes, have been identified in a wide range of cnidarian species and are expressed in the neural cell progenitors (Grens et al. 1995; Müller et al. 2003; Hayakawa et al. 2004; Lindgens et al. 2004; Seipel et al. 2004; Magie et al. 2005; Simionato et al. 2007; Layden et al. 2012; Watanabe et al. 2014a). In *Nematostella*, a series of gene function analyses of these neurogenic transcription factors and *Elav1* have unveiled genetic mechanisms essential for differentiation of ectodermal and endodermal neurons. The first sign of neurogenesis in *Nematostella* is the salt-and-pepper-like expression of the neurogenic transcription factors *NvSoxB2c* (also called *NvSoxB2* or *NvSoxB(2)*), *NvAth-like (NvArp3)*, and *NvAshA*, and various neural markers, including *NvElav1* and *Rfamide*, in blastula epithelium (Magie et al. 2005; Marlow et al. 2009; Layden et al. 2012; Nakanishi et al. 2012; Richards and Rentzsch 2014; 2015). The patterns of *NvSoxB2c* and *NvAshA* expression suggest that the early embryonic epithelium of *Nematostella* has the potential to generate various neuronal cell types that form a diffuse nerve net (Fig. 3.2) (Magie et al. 2005; Layden et al. 2012). Gene function analyses indicate that they are required for the development of *NvElav1*-positive neuronal populations in the endoderm and *NvRfamide*-positive cells in the lateral ectoderm (Nakanishi et al. 2012). During gastrulation, the expression of transcription factors, such as *NvRx* and *NvAshA*, is localized mainly in the aboral half of the embryos, indicating that these genes may be involved in development of the sensory cells/neurons in the aboral region. The genetic mechanism responsible for development of the aboral nervous system is still unclear. *Six3* is a homeodomain transcription factor with a central role in the development of anterior sensory and neural structures in bilaterians (Steinmetz et al. 2010). In *Nematostella*, the *NvSix3/6* gene is expressed in the aboral region where the aboral sensory cells and neurons develop, suggesting that the *NvSix3/6* has an evolutionarily conserved role in demarcating the anterior neurosensory region, both in Cnidaria and Bilateria (Sinigaglia et al. 2013). Although an inhibition of the *NvSix3/6* gene in *Nematostella* embryos did not have a significant effect on *NvRfamide* expression, its possible involvement in expression of aboral neural genes *NvRx* and *NvAshA* has not been explored.

In addition to their aborally-biased nervous systems, the oral region of *Nematostella* larvae expresses a number of neurogenic genes and markers (Fig. 3.2) (Magie et al. 2005; Shinzato et al. 2008; Marlow et al. 2009; Nakanishi et al. 2012; Layden et al. 2012; Watanabe et al. 2014a). The blastopore region of the gastrulae develops into a prominent oral neurogenic domain, showing dominant or exclusive expression of *NvSoxB2* homologues, *NvAsh* and *NvArp* homologues, and *NvMusashi* and *NvRfamide*. Early neurogenic markers for, and later differentiation of, neuropeptide-expressing neurons in the oral nervous system are severely reduced or absent in the embryos injected with morpholino antisense oligonucleotides against early oral neurogenic transcription factors such as *NvSoxB2a, NvAshB*, and *NvAth-like/NvArp3* (Watanabe et al. 2014a), which suggests that the blastoporal side of the early embryos has a distinct neurogenic capacity. It has been shown that the development of the oral nervous system is dependent on β-catenin and bone morphogenetic protein (BMP) signaling pathways. The function of these signaling activities in development of the cnidarian oral nervous system is highly reminiscent of their functional patterning of the bilaterian the CNS (Watanabe et al. 2014a).

3.6 Outlook

Accumulating molecular and cellular evidence has led many researchers to propose that the Urbilateria, the last common bilaterian ancestor, had a condensed nervous system with more or less specific anatomical and/or physiological features (Hirth et al. 2003; De Robertis 2008; Tomer et al. 2010; Strausfeld 2010; Strausfeld and Hirth 2013; Bailly et al. 2013; Holland et al. 2013), whereas others favor the idea that the bilaterian CNS started with a diffuse nerve net (Gerhart et al. 2005; Pani et al. 2012; Arendt et al. 2016). Since the Cnidaria is the closest sister group to bilaterians, it is important to decipher molecular and cellular features of diffuse and regionalized components of the cnidarian nervous system. Expression analyses of CNS genes in *Nematostella* suggest that the aboral sensory system and the oral nervous system of cnidarians are related to the anterior part of the bilaterian brain, including sensory organs, and the posterior part of the brain with trunk nervous systems, respectively (Tosches and Arendt 2013; Marlow et al. 2014; Arendt et al. 2016). Recent gene function data suggest that rudimentary centralization of the oral nervous system might be an antecedent characteristic of the bilaterian CNS (Watanabe et al. 2014a). However, more detailed functional analyses of CNS genes in the Cnidaria are required to better explain how the first step in nervous system centralization may have been accomplished. Studies on nervous systems of cnidarians and ctenophores, as well as the sensory systems of poriferans and placozoans, are also strongly needed to reconstruct the early evolution of the nervous system.

Acknowledgments I am grateful to Thomas W. Holstein and Osamu Koizumi for critical and helpful comments on the manuscript, and to Noriko Funayama, Hiroaki Nakano, and Noriko Ishikawa for providing the beautiful photographs of basal metazoans. I am deeply indebted to

Annika Guse, Elizabeth A. Hambleton, and Steven Douglas Aird for correcting my English. I apologize to all researchers whose work could not be cited owing to space constraints. I, alone am responsible for any errors or controversial views.

References

Alexopoulos H, Böttger A, Fischer S et al (2004) Evolution of gap junctions: the missing link? Curr Biol 14:R879–R880

Alié A, Manuel M (2010) The backbone of the post-synaptic density originated in a unicellular ancestor of choanoflagellates and metazoans. BMC Evol Biol 10:34. doi:10.1186/1471-2148-10-34

Anctil M (2009) Chemical transmission in the sea anemone *Nematostella vectensis*: a genomic perspective. Comp Biochem Physiol Part D Genomics Proteomics 4:268–289. doi:10.1016/j.cbd.2009.07.001

Anctil M, Carette JP (1994) Glutamate immunoreactivity in non-neuronal cells of the sea anemone *Metridium senile*. Biol Bull 187:48–54. doi:10.2307/1542164

Anctil M, Minh CN (1997) Neuronal and nonneuronal taurine-like immunoreactivity in the sea pansy, *Renilla koellikeri* (Cnidaria, Anthozoa). Cell Tissue Res 288:127–134

Anderson PA (1980) Epithelial conduction: its properties and function. Prog Neurobiol 15:161–203

Anderson PA, Thompson LF, Moneypenny CG (2004) Evidence for a common pattern of peptidergic innervation of cnidocytes. Biol Bull 207:141–146. doi:10.2307/1543588

Arendt D, Denes AS, Jékely G et al (2008) The evolution of nervous system centralization. Philos Trans R Soc Lond B Biol Sci 363:1523–1528. doi:10.1098/rstb.2007.2242

Arendt D, Tosches MA, Marlow H (2016) From nerve net to nerve ring, nerve cord and brain-evolution of the nervous system. Nat Rev Neurosci 17:61–72. doi:10.1038/nrn.2015.15

Bailly X, Reichert H, Hartenstein V (2013) The urbilaterian brain revisited: novel insights into old questions from new flatworm clades. Dev Genes Evol 223:149–157. doi:10.1007/s00427-012-0423-7

Baumgarten S, Simakov O, Esherick LY et al (2015) The genome of *Aiptasia*, a sea anemone model for coral symbiosis. Proc Natl Acad Sci USA 112:11893–11898. doi:10.1073/pnas.1513318112

Bouche N, Lacombe B, Fromm H (2003) GABA signaling: a conserved and ubiquitous mechanism. Trends Cell Biol 13:607–610

Boury-Esnault N, Rützler K (1997) Thesaurus of sponge morphology. Smithson Contrib Zool 596:1–55. doi:10.5479/si.00810282.596

Bridge DC, Cunningham W, Schierwater B et al (1992) Class-level relationships in the phylum Cnidaria: evidence from mitochondrial genome structure. Proc Natl Acad Sci USA 89:8750–8753

Bridge D, Cunningham CW, DeSalle R et al (1995) Class-level relationships in the phylum Cnidaria: molecular and morphological evidence. Mol Biol Evol 12:679–689

Burkhardt P (2015) The origin and evolution of synaptic proteins – choanoflagellates lead the way. J Exp Biol 218:506–514. doi:10.1242/jeb.1102

Burkhardt P, Grønborg M, McDonald K et al (2014) Evolutionary insights into premetazoan functions of the neuronal protein homer. Mol Biol Evol 31:2342–2355. doi:10.1093/molbev/msu178

Byrne M, Nakajima Y, Chee FC, Burke RD (2007) Apical organs in echinoderm larvae: insights into larval evolution in the Ambulacraria. Evol Dev 9:432–445. doi:10.1111/j.1525-142X.2007.00189.x

Campbell RD, Josephson RK, Schwab WE, Rushforth NB (1976) Excitability of nerve-free hydra. Nature (Lond) 262:388–390. doi:10.1038/262388a0

Carlberg M, Alfredsson K, Nielson SO et al (1995) Taurine-like immunoreactivity in the motor nerve net of the jellyfish *Cyanea capillata*. Biol Bull 188:78–82. doi:10.2307/1542069

Chapman JA, Kirkness EF, Simakov O et al (2010) The dynamic genome of *Hydra*. Nature (Lond) 464:592–596. doi:10.1038/nature08830

Chia FS, Bickell L (1978) Mechanisms of larval settlement and the induction of settlement and metamorphosis: a review. In: Chia FS, Rice ME (eds) Settlement and metamorphosis of marine invertebrate larvae. Elsevier, New York, pp 1–12

Chia FS, Koss R (1979) Fine structural studies of the nervous system and the apical organ in the planula larva of the sea anemone *Anthopleura elegantissima*. J Morphol 160:275–298

Collins AG (2002) Phylogeny of medusozoa and the evolution of cnidarian life cycles. J Evol Biol 15:418–431

Collins AG, Marques AC, Jankowski T et al (2006) Medusozoan phylogeny and character evolution clarified by new large and small subunit rDNA data and an assessment of the utility of phylogenetic mixture models. Syst Biol 55:97–115. doi:10.1080/10635150500433615

Conzelmann M, Jékely G (2012) Antibodies against conserved amidated neuropeptide epitopes enrich the comparative neurobiology toolbox. Evo Devo 3:23. doi:10.1186/2041-9139-3-23

Conzelmann M, Offenburger SL, Asadulina A et al (2011) Neuropeptides regulate swimming depth of *Platynereis* larvae. Proc Natl Acad Sci USA 108:E1174–E1183. doi:10.1073/pnas.1109085108

Conzelmann M, Williams EA, Krug K et al (2013) The neuropeptide complement of the marine annelid *Platynereis dumerilii*. BMC Genomics 14:906. doi:10.1186/1471-2164-14-906

Davenport R (2002) Glutamate receptors in plants. Ann Bot 90:549–557. doi:10.1093/aob/mcf228

Davis LE, Burnett AL, Haynes JF (1968) Histological and ultrastructural study of the muscular and nervous system in *Hydra*. 2. Nervous system. J Exp Zool 167:295–331. doi:10.1002/jez.1401670304

De Robertis EM (2008) The molecular ancestry of segmentation mechanisms. Proc Natl Acad Sci USA 105:16411–16412. doi:10.1073/pnas.0808774105

Degnan DM, Adamska M, Richards GR et al (2015) Sponges. In: Wanninger A (ed) Evolutionary developmental biology of invertebrates, vol 1. Springer , New York, pp 65–106

Delgado LM, Couve E, Schmachtenberg O (2010) GABA and glutamate immunoreactivity in tentacles of the sea anemone *Phymactis papillosa* (LESSON 1830). J Morphol 271:845–852. doi:10.1002/jmor.10838

Denker E, Chatonnet A, Rabet N (2008) Acetylcholinesterase activity in *Clytia hemisphaerica* (Cnidaria). Chem Biol Interact 175:125–128. doi:10.1016/j.cbi.2008.03.004

Dickinson AJG, Croll RP (2003) Development of the larval nervous system of the gastropod *Ilyanassa obsoleta*. J Comp Neurol 466:197–218. doi:10.1002/cne.10863

Dunn CW, Hejnol A, Matus DQ et al (2008) Broad phylogenomic sampling improves resolution of the animal tree of life. Nature (Lond) 452:745–749. doi:10.1038/nature06614

Elliott GR, Leys SP (2010) Evidence for glutamate, GABA and NO in coordinating behaviour in the sponge, *Ephydatia muelleri* (Demospongiae, Spongillidae). J Exp Biol 213:2310–2321. doi:10.1242/jeb.039859

Ellwanger K, Nickel M (2006) Neuroactive substances specifically modulate rhythmic body contractions in the nerveless metazoon *Tethya wilhelma* (Demospongiae, Porifera). Front Zool 3:7. doi:10.1186/1742-9994-3-7

Ellwanger K, Eich A, Nickel M (2007) GABA and glutamate specifically induce contractions in the sponge *Tethya wilhelma*. J Comp Physiol A Neuroethol Sens Neural Behav Physiol 193:1–11. doi:10.1007/s00359-006-0165-y

Emes RD, Pocklington AJ, Anderson CN et al (2008) Evolutionary expansion and anatomical specialization of synapse proteome complexity. Nat Neurosci 11:799–806. doi:10.1038/nn.2135

Erwin PM, Szmant AM (2010) Settlement induction of *Acropora palmata* planulae by a GLW-amide neuropeptide. Coral Reefs 29:929–939. doi:10.1007/s00338-010-0634-1

Fortunato SA, Adamski M, Bergum B et al (2012) Genome-wide analysis of the sox family in the calcareous sponge *Sycon ciliatum*: multiple genes with unique expression patterns. Evo Devo 3:14. doi:10.1186/2041-9139-3-14

Fortunato SA, Leininger S, Adamska M (2014) Evolution of the Pax-Six-Eya-Dach network: the calcisponge case study. Evodevo 5:23. doi:10.1186/2041-9139-5-23

Fortunato SA, Adamski M, Adamska M (2015) Comparative analyses of developmental transcription factor repertoires in sponges reveal unexpected complexity of the earliest animals. Mar Genomics 24:121–129. doi:10.1016/j.margen.2015.07.008

Fraser SE, Green CR, Bode HR et al (1987) Selective disruption of gap junctional communication interferes with a patterning process in hydra. Science 237:49–55. doi:10.1126/science.3037697

Fujisawa T (2008) Hydra peptide project 1993–2007. Dev Growth Differ 50:S257–S268. doi:10.1111/j.1440-169X.2008.00997.x

Fujisawa T, Hayakawa E (2012) Peptide signaling in *Hydra*. Int J Dev Biol 56:543–550. doi:10.1387/ijdb.113477tf

Gajewski M, Leitz T, Schlossherr J et al (1996) LWamides from cnidaria constitute a novel family of neuropeptides with morphogenetic activity. Dev Genes Evol 205:232–242. doi:10.1007/BF00365801

Galliot B, Quiquand M (2011) A two-step process in the emergence of neurogenesis. Eur J Neurosci 34:847–862. doi:10.1111/j.1460-9568.2011.07829.x

Galliot B, Quiquand M, Ghila L et al (2009) Origins of neurogenesis, a cnidarian view. Dev Biol 332:2–24. doi:10.1016/j.ydbio.2009.05.563

Garm A, Ekström P, Boudes M et al (2006) Rhopalia are integrated parts of the central nervous system in box jellyfish. Cell Tissue Res 325:333–343. doi:10.1007/s00441-005-0134-8

Garm A, Poussart Y, Parkefelt L et al (2007) The ring nerve of the box jellyfish *Tripedalia cystophora*. Cell Tissue Res 329:147–157. doi:10.1007/s00441-007-0393-7

Gerhart J, Lowe C, Kirschner M (2005) Hemichordates and the origin of chordates. Curr Opin Genet Dev 15:461–467. doi:10.1016/j.gde.2005.06.004

Girosi L, Ferrando S, Beltrame F et al (2007) Gamma-aminobutyric acid and related molecules in the sea fan *Eunicella cavolini* (Cnidaria: Octocorallia): a biochemical and immunohistochemical approach. Cell Tissue Res 329:187–196. doi:10.1007/s00441-007-0408-4

Grens A, Mason E, Marsh JL et al (1995) Evolutionary conservation of a cell fate specification gene: the hydra achaete-scute homolog has proneural activity in *Drosophila*. Development (Camb) 121:4027–4035

Grimmelikhuijzen CJ (1985) Antisera to the sequence Arg-Phe-amide visualize neuronal centralization in hydroid polyps. Cell Tissue Res 241:171–182. doi:10.1007/BF00214639

Grimmelikhuijzen CJ, Spencer AN (1984) FMRFamide immunoreactivity in the nervous system of the medusa *Polyorchis penicillatus*. J Comp Neurol 230:361–371. doi:10.1002/cne.902300305

Grimmelikhuijzen CJ, Hahn M, Rinehart KL et al (1988) Isolation of pyroGlu-Leu-Leu-Gly-Gly-Arg-Phe-NH2 (Pol-RFamide), a novel neuropeptide from hydromedusae. Brain Res 475:198–203. doi:10.1016/0006-8993(88)90219-3

Grimmelikhuijzen CJ, Graff D, Koizumi O et al (1991) Neuropeptides in coelenterates: a review. Hydrobiologia 216(217):555–563. doi:10.1007/978-94-011-3240-4_78

Gröger H, Schmid V (2001) Larval development in Cnidaria: a connection to Bilateria? Genesis 29:110–114. doi:10.1002/gene.1013

Gyoja F (2014) A genome-wide survey of bHLH transcription factors in the placozoan *Trichoplax adhaerens* reveals the ancient repertoire of this gene family in metazoan. Gene (Amst) 542:29–37. doi:10.1016/j.gene.2014.03.024

Hadfield MG, Meleshkevitch EA, Boudko DY (2000) The apical sensory organ of a gastropod veliger is a receptor for settlement cues. Biol Bull 198:67–76. doi:10.2307/1542804

Hamner WM, Jones MS, Hamner PP (1995) Swimming, feeding, circulation and vision in the Australian box jellyfish, *Chironex fleckeri* (Cnidaria, Cubozoa). Mar Freshw Res 46:985–990. doi:10.1071/MF9950985

Hand AR, Gobel S (1972) The structural organization of the septate and gap junctions of *Hydra*. J Cell Biol 52:397–408

Harbison GR (1985) On the classification and evolution of the Ctenophora. In: Conway Morris S, George JD, Gibson R, Platt HM (eds) The origins and relationships of lower invertebrates. Oxford University Press, Oxford, pp 78–100

Hartwick RF (1991) Observations on the anatomy, behaviour, reproduction and life cycle of the cubozoan *Carybdea sivickisi*. Hydrobiologia 216(217):171–179. doi:10.1007/BF00026459

Hatta M, Iwao K (2003) Metamorphosis induction and its possible application to coral seedlings production. In: Saxena N (ed) Recent advances in marine science and technology, 2002. Japan International Science and Technology Federation, Akasaka, pp 465–470

Hayakawa E, Fujisawa C, Fujisawa T (2004) Involvement of hydra achaete-scute gene CnASH in the differentiation pathway of sensory neurons in the tentacles. Dev Genes Evol 214:486–492. doi:10.1007/s00427-004-0430-4

Hay-Schmidt A (2000) The evolution of the serotonergic nervous system. Proc R Soc Lond B Biol Sci 267:1071–1079

Hernandez-Nicaise ML (1973) The nervous system of ctenophores. III. Ultrastructure of synapses. J Neurocytol 2:249–263

Hirth F, Kammermeier L, Frei E et al (2003) An urbilaterian origin of the tripartite brain: developmental genetic insights from *Drosophila*. Development (Camb) 130:2365–2373

Holland ND (2003) Early central nervous system evolution: an era of skin brains? Nat Rev Neurosci 4:617–627. doi:10.1038/nrn1175

Holland LZ, Carvalho JE, Escriva H et al (2013) Evolution of bilaterian central nervous systems: a single origin? Evodevo 4:27. doi:10.1186/2041-9139-4-27

Iwao K, Fujisawa T, Hatta M (2002) A cnidarian neuropeptide of the GLWamide family induces metamorphosis of reef-building corals in the genus *Acropora*. Coral Reefs 21:127–129. doi:10.1007/s00338-002-0219-8

Jager M, Queinnec E, Chiori R et al (2008) Insights into the early evolution of SOX genes from expression analyses in a ctenophore. J Exp Zool B Mol Dev Evol 310:650–667. doi:10.1002/jez.b.21244

Jager M, Chiori R, Alié A et al (2011) New insights on ctenophore neural anatomy: immunofluorescence study in *Pleurobrachia pileus* (Müller, 1776). J Exp Zool B Mol Dev Evol 316B:171–187. doi:10.1002/jez.b.21386

Jékely G (2013) Global view of the evolution and diversity of metazoan neuropeptide signaling. Proc Natl Acad Sci U S A 110:8702–8707. doi:10.1073/pnas.1221833110

Jékely G, Papsand J, Nielsen C (2015) The phylogenetic position of ctenophores and the origin(s) of nervous systems. Evodevo 6:1. doi:10.1186/2041-9139-6-1

Jones WC (1962) Is there a nervous system in sponges? Biol Rev 37:1–50. doi:10.1111/j.1469-185X.1962.tb01602.x

Kass-Simon G, Scappaticci AA (2004) Glutamatergic and GABAnergic control in the tentacle effector systems of *Hydra vulgaris*. Hydrobiologia 530/531:67–71

Kass-Simon G, Pierobon P (2007) Cnidarian chemical neurotransmission, an updated overview. Comp Biochem Physiol A146:9–25. doi:10.1016/j.cbpa.2006.09.008

Kass-Simon G, Pannaccione A, Pierobon P (2003) GABA and glutamate receptors are involved in modulating pacemaker activity in hydra. Comp Biochem Physiol A Mol Integr Physiol 136:329–342. doi:10.1016/S1095-6433(03)00168-5

Katsukura Y, David CN, Grimmelikhuijzen CJ et al (2003) Inhibition of metamorphosis by RFamide neuropeptides in planula larvae of *Hydractinia echinata*. Dev Genes Evol 213:579–586. doi:10.1007/s00427-003-0361-5

Katsukura Y, Ando H, David CN et al (2004) Control of planula migration by LWamide and RFamide neuropeptides in *Hydractinia echinata*. J Exp Biol 207:1803–1810. doi:10.1242/jeb.00974

Kinnamon JC, Westall JA (1981) A three-dimensional serial reconstruction of neuronal distributions in the hypostome of *Hydra*. J Morphol 168:321–329. doi:10.1002/jmor.1051680308

Kinnamon JC, Westfall JA (1982) Types of neurons and synaptic connections at hypostome–tentacle junctions in *Hydra*. J Morphol 173:119–128. doi:10.1002/jmor.1051730110

Koizumi O (2007) Nerve ring of the hypostome in hydra: is it an origin of the central nervous system of bilaterian animals? Brain Behav Evol 69:151–159. doi:10.1159/000095204

Koizumi O, Wilson JD, Grimmelikhuijzen CJ et al (1989) Ultrastructural localization of RFamide-like peptides in neuronal dense-cored vesicles in the peduncle of *Hydra*. J Exp Zool 249:17–22. doi:10.1002/jez.1402490105

Koizumi O, Itazawa M, Mizumoto H et al (1992) Nerve ring of the hypostome in hydra. I. Its structure, development, and maintenance. J Comp Neurol 326:7–21. doi:10.1002/cne.903260103

Koizumi O, Sato N, Goto C (2004) Chemical anatomy of hydra nervous system using antibodies against hydra neuropeptides: a review. Hydrobiologia 530(531):41–47. doi:10.1007/s10750-004-2636-x

Koizumi O, Hamada S, Minobe S et al (2014) The nerve ring in cnidarians: its presence and structure in hydrozoan medusae. Zoology (Jena) 118:79–88. doi:10.1016/j.zool.2014.10.001

Kortschak RD, Samuel G, Saint R et al (2003) EST analysis of the cnidarian *Acropora millepora* reveals extensive gene loss and rapid sequence divergence in the model invertebrates. Curr Biol 13:2190–2195. doi:10.1016/S0960-9822(03)00872-8

Lacalli TC (1983) The brain and central nervous system of Müller's larva. Can J Zool 61:39–51. doi:10.1139/z83-004

Lam HM, Chiu J, Hsieh M et al (1998) Glutamate-receptor genes in plants. Nature (Lond) 396:125–126. doi:10.1038/24066

Larroux C, Fahey B, Liubicich D et al (2006) Developmental expression of transcription factor genes in a demosponge: insights into the origin of metazoan multicellularity. Evol Dev 8:150–173. doi:10.1111/j.1525-142X.2006.00086.x

Larroux C, Luke GN, Koopman P et al (2008) Genesis and expansion of metazoan transcription factor gene classes. Mol Biol Evol 25:980–996. doi:10.1093/molbev/msn047

Laska G, Hündgen M (1982) Morphologie und Ultrastruktur der Lichtsinnesorgane von *Tripedalia cystophora* Conant (Cnidaria, Cubozoa). Zool Jb Anat 108:107–123

Layden MJ, Boekhout M, Martindale MQ (2012) *Nematostella vectensis* achaete-scute homolog NvashA regulates embryonic ectodermal neurogenesis and represents an ancient component of the metazoan neural specification pathway. Development (Camb) 139:1013–1022. doi:10.1242/dev.073221

Leitz T, Lay M (1995) Metamorphosin A is a neuropeptide. Dev Genes Evol 204:276–279. doi:10.1007/BF00208495

Leitz T, Wagner T (1992) The marine bacterium *Alteromonas espejiana* induces metamorphosis of the hydroid *Hydractinia echinata*. Mar Biol 115:173–178. doi:10.1007/BF00346332

Leitz T, Morand K, Mann M (1994) Metamorphosin A: a novel peptide controlling development of the lower metazoan *Hydractinia echinata* (Coelenterata, Hydrozoa). Dev Biol 163:440–446. doi:10.1006/dbio.1994.1160

Lentz TL, Barrnett RJ (1961) Enzyme histochemistry of *Hydra*. J Exp Zool 147:125–149. doi:10.1002/jez.1401470204

Leys SP, Degnan BM (2001) Cytological basis of photoresponsive behavior in a sponge larva. Biol Bull 201:323–338. doi:10.2307/1543611

Lindgens D, Holstein TW, Technau U (2004) Hyzic, the *Hydra* homolog of the zic/odd-paired gene, is involved in the early specification of the sensory nematocytes. Development (Camb) 131:191–201

Mackie GO (1970) Neuroid conduction and the evolution of conducting tissues. Q Rev Biol 45:19–332

Mackie GO (1979) Is there a conduction system in sponges? In: Boury-Esnault N, Levi C (eds) Biologie des Spongiaires. CNRS, Paris, pp 145–151

Mackie GO (1990) The elementary nervous system revisited. Am Zool 30:907–920

Mackie GO (2004) Central neural circuitry in the jellyfish *Aglantha*: a model "simple nervous system". NeuroSignals 13:5–19. doi:10.1159/000076155

Mackie GO, Meech RW (1995a) Central circuitry in the jellyfish *Aglantha digitale*. I. The relay system. J Exp Biol 198:2261–2270

Mackie GO, Meech RW (1995b) Central circuitry in the jellyfish *Aglantha digitale*. II. The ring giant and carrier systems. J Exp Biol 198:2271–2278

Mackie GO, Meech RW (2000) Central circuitry in the jellyfish *Aglantha digitale*. III. The rootlet and pacemaker systems. J Exp Biol 203:1797–1807

Mackie GO, Passano LM (1968) Epithelial conduction in hydromedusae. J Gen Physiol 52:600–621

Mackie GO, Stell WK (1984) FMRF-amide-like immunoreactivity in the neurons of medusae. Am Zool 24:36A

Mackie GO, Anderson PAV, Singla CL (1984) Apparent absence of gap junctions in two classes of *Cnidaria*. Biol Bull 167:120–123

Mackie GO, Singla CL, Stell WK (1985) Distribution of nerve elements showing FMRF-amide-like immunoreactivity in hydromedusae. Acta Zool 66:199–210

Magie CR, Pang K, Martindale MQ (2005) Genomic inventory and expression of Sox and Fox genes in the cnidarian *Nematostella vectensis*. Dev Genes Evol 215:618–630. doi:10.1007/s00427-005-0022-y

Marlow H, Arendt D (2014) Evolution: ctenophore genomes and the origin of neurons. Curr Biol 24:R757–R761. doi:10.1016/j.cub.2014.06.057

Marlow HQ, Srivastava M, Matus DQ et al (2009) Anatomy and development of the nervous system of *Nematostella vectensis*, an anthozoan cnidarian. Dev Neurobiol 69:235–254. doi:10.1002/dneu.20698

Marlow H, Tosches MA, Tomer R et al (2014) Larval body patterning and apical organs are conserved in animal evolution. BMC Biol 12:7. doi:10.1186/1741-7007-12-7

Martin VJ (1988) Development of nerve cells in hydrozoan planulae: II. Examination of sensory cell differentiation using electron microscopy and immunocytochemistry. Biol Bull 175:65–78. doi:10.2307/1541957

Martin VJ (1992) Characterization of a RFamide-positive subset of ganglionic cells in the hydrozoan planular nerve net. Cell Tissue Res 269:431–438

Martin VJ (2004) Photoreceptors of cubozoan jellyfish. Hydrobiologia 530(531):135–144. doi:10.1007/s10750-004-2674-4

Matsumoto GI (1995) Observations on the anatomy and behaviour of the cubozoan *Carybdea rastonii* Haacke. Mar Freshw Behav Physiol 26:139–148

Matsuno T, Kageyama T (1984) The nervous system in the hypostome of *Pelmatohydra robusta*: the presence of a circumhypostomal nerve ring in the epidermis. J Morphol 182:153–168. doi:10.1002/jmor.1051820204

Maxwell EK, Ryan JF, Schnitzler CE et al (2012) MicroRNAs and essential components of the microRNA processing machinery are not encoded in the genome of the ctenophore *Mnemiopsis leidyi*. BMC Genomics 13:714. doi:10.1186/1471-2164-13-714

McFarlane ID, Graff D, Grimmelikhuijzen CJ (1987) Excitatory actions of Antho-RFamide, an anthozoan neuropeptide, on muscles and conducting systems in the sea anemone *Calliactis parasitica*. J Exp Biol 133:157–168

McFarlane ID, Anderson PA, Grimmelikhuijzen CJ (1991) Effects of three anthozoan neuropeptides, Antho-RWamide I, Antho-RWamide II and Antho-RFamide, on slow muscles from sea anemones. J Exp Biol 156:419–431

Medina M, Collins AG, Silberman JD et al (2001) Evaluating hypotheses of basal animal phylogeny using complete sequences of large and small subunit rRNA. Proc Natl Acad Sci USA 98:9707–9712. doi:10.1073/pnas.171316998

Mitgutsch C, Hauser F, Grimmelikhuijzen CJ (1999) Expression and developmental regulation of the hydra-RFamide and hydra-LWamide preprohormone genes in *Hydra*: evidence for transient phases of head formation. Dev Biol 207:189–203. doi:10.1006/dbio.1998.9150

Moroz L (2001) Gaseous transmission across time and species. Am Zool 41:304–320. doi:10.1093/icb/41.2.304

Moroz LL, Kocot KM, Citarella MR et al (2014) The ctenophore genome and the evolutionary origins of neural systems. Nature (Lond) 510:109–114. doi:10.1038/nature13400

Morse DE, Morse ANC (1991) Enzymatic characterization of the morphogen recognized by *Agaricia humilis* (scleractinian coral) larvae. Biol Bull 181:104–122. doi:10.2307/1542493

Müller WA (1969) Auslösung der Metamorphose durch Bakterien bei den Larven von *Hydractinia echinata*. Zool Jb Anat 86:84–95

Müller P, Seipel K, Yanze N et al (2003) Evolutionary aspects of developmentally regulated helix-loop-helix transcription factors in striated muscle of jellyfish. Dev Biol 255:216–229. doi:10.1016/S0012-1606(02)00091-X

Nakanishi N, Yuan D, Jacobs DK et al (2008) Early development, pattern, and reorganization of the planula nervous system in *Aurelia* (Cnidaria, Scyphozoa). Dev Genes Evol 218:511–524. doi:10.1007/s00427-008-0239-7

Nakanishi N, Renfer E, Technau U et al (2012) Nervous systems of the sea anemone *Nematostella vectensis* are generated by ectoderm and endoderm and shaped by distinct mechanisms. Development (Camb) 139:347–357. doi:10.1242/dev.071902

Nakanishi N, Stoupin D, Degnan SM et al (2015) Sensory flask cells in sponge larvae regulate metamorphosis via calcium signaling. Integr Comp Biol (pii: icv014). doi: 10.1093/icb/icv014

Nickel M (2004) Kinetics and rhythm of body contractions in the sponge *Tethya wilhelma* (Porifera: Demospongiae). J Exp Biol 207:4515–4524. doi:10.1242/jeb.01289

Nickel M, Scheer C, Hammel JU et al (2011) The contractile sponge epithelium sensu lato–body contraction of the demosponge *Tethya wilhelma* is mediated by the pinacoderm. J Exp Biol 214:1692–1698. doi:10.1242/jeb.049148

Nielsen C (2005) Larval and adult brains. Evol Dev 7:483–489. doi:10.1111/j.1525-142X.2005.05051.x

Nielsen C (2015) Larval nervous systems: true larval and precocious adult. J Exp Biol 218:629–636. doi:10.1242/jeb.109603

Nilsson DE, Coates MM, Gislén L et al (2005) Advanced optics in a jellyfish eye. Nature (Lond) 435:201–205. doi:10.1038/nature03484

Odorico DM, Miller DJ (1997) Internal and external relationships of the Cnidaria: implications of primary and predicted secondary structure of the 5′-end of the 23S-like rDNA. Proc Biol Sci 264:77–82. doi:10.1098/rspb.1997.0011

Oren M, Brikner I, Appelbaum L et al (2014) Fast neurotransmission related genes are expressed in non nervous endoderm in the sea anemone *Nematostella vectensis*. PLoS One 9:e93832. doi:10.1371/journal.pone.0093832

Pani AM, Mullarkey EE, Aronowicz J et al (2012) Ancient deuterostome origins of vertebrate brain signalling centres. Nature (Lond) 483:289–294. doi:10.1038/nature10838

Pantin CFA (1966) Homology, analogy and chemical identity in the Cnidaria. In: Rees WJ (ed) The Cnidaria and their evolution. Academic Press, New York, pp 1–15

Parker GH (1910) The reactions of sponges with a consideration of the origin of the nervous system. J Exp Zool 8:765–805. doi:10.1002/jez.1400080102

Passano LM, McCullough CB (1963) Pacemaker hierarchies controlling the behaviour of hydras. Nature (Lond) 199:1174–1175. doi:10.1038/1991174a0

Pavans de Ceccatty M (1974a) Coordination in sponges. The foundations of integration. Am Zool 14:895–903

Pavans de Ceccatty M (1974b) The origin of the integrative systems: a change in view derived from research on coelenterates and sponges. Perspect Biol Med 17:379–390

Pavans de Ceccatty M (1979) Cell correlations and integration in sponges. Colloq Int CNRS, Biologie des Spongiaires 291:123–135

Perovic S, Krasko A, Prokic I et al (1999) Origin of neuronal-like receptors in metazoa: cloning of a metabotropic glutamate/GABA-like receptor from the marine sponge *Geodia cydonium*. Cell Tissue Res 296:395–404. doi:10.1007/s004410051299

Phelan P (2005) Innexins: members of an evolutionarily conserved family of gap-junction proteins. Biochim Biophys Acta 1711:225–245. doi:10.1016/j.bbamem.2004.10.004

Philippe H, Derelle R, Lopez P et al (2009) Phylogenomics revives traditional views on deep animal relationships. Curr Biol 19:706–712. doi:10.1016/j.cub.2009.02.052

Philippe H, Brinkmann H, Lavrov DV et al (2011) Resolving difficult phylogenetic questions: why more sequences are not enough. PLoS Biol 9:e1000602. doi:10.1371/journal.pbio.1000602

Pick KS, Philippe H, Schreiber F et al (2010) Improved phylogenomic taxon sampling noticeably affects nonbilaterian relationships. Mol Biol Evol 27:1983–1987. doi:10.1093/molbev/msq089

Pierobon P (2012) Coordinated modulation of cellular signaling through ligand-gated ion channels in *Hydra vulgaris* (Cnidaria, Hydrozoa). Int J Dev Biol 56:551–565. doi:10.1387/ijdb.113464pp

Pierobon P, Minei R, Porcu P et al (2001) Putative glycine receptors in *Hydra*: a biochemical and behavioural study. Eur J Neurosci 14:1659–1666. doi:10.1046/j.0953-816x.2001.01792

Pisani D, Pett W, Dohrmann M et al (2015) Genomic data do not support comb jellies as the sister group to all other animals. Proc Natl Acad Sci U S A 112:15402–15407. doi:10.1073/pnas.1518127112

Plickert G (1989) Proportion-altering factor (PAF) stimulates nerve cell formation in *Hydractinia echinata*. Cell Differ Dev 26:19–27. doi:10.1016/0922-3371(89)90780-6

Plickert G, Schneider B (2004) Neuropeptides and photic behavior in *Cnidaria*. Hydrobiologia 530(531):49–57. doi:10.1007/s10750-004-2689-x

Putnam NH, Srivastava M, Hellsten U et al (2007) Sea anemone genome reveals ancestral eumetazoan gene repertoire and genomic organization. Science 317:86–94. doi:10.1126/science.1139158

Ramoino P, Gallus L, Paluzzi S et al (2007) The GABAergic-like system in the marine demosponge *Chondrilla nucula*. Microsc Res Tech 70:944–951. doi:10.1002/jemt.20499

Ramoino P, Ledda FD, Ferrando S et al (2011) Metabotropic γ-aminobutyric acid (GABAB) receptors modulate feeding behavior in the calcisponge *Leucandra aspera*. J Exp Zool A Ecol Genet Physiol 315:132–140. doi:10.1002/jez.657

Richards GS, Degnan BM (2012) The expression of delta ligands in the sponge *Amphimedon queenslandica* suggests an ancient role for notch signaling in metazoan development. Evodevo 3:15. doi:10.1186/2041-9139-3-15

Richards GS, Rentzsch F (2014) Transgenic analysis of a SoxB gene reveals neural progenitor cells in the cnidarian *Nematostella vectensis*. Development (Camb) 141:4681–4689. doi:10.1242/dev.112029

Richards GS, Rentzsch F (2015) Regulation of Nematostella neural progenitors by SoxB, Notch and bHLH genes. Development 142:3332–3342. doi:10.1242/dev.123745

Richards GS, Simionato E, Perron M et al (2008) Sponge genes provide new insight into the evolutionary origin of the neurogenic circuit. Curr Biol 18:1156–1161. doi:10.1016/j.cub.2008.06.074

Richter S, Loesel R, Purschke G et al (2010) Invertebrate neurophylogeny: suggested terms and definitions for a neuroanatomical glossary. Front Zool 7:29. doi:10.1186/1742-9994-7-29

Riesgo A, Farrar N, Windsor PJ et al (2014) The analysis of eight transcriptomes from all poriferan classes reveals surprising genetic complexity in sponges. Mol Biol Evol 31:1102–1120. doi:10.1093/molbev/msu057

Ruggieri RD, Pierobon P, Kass-Simon G (2004) Pacemaker activity in hydra is modulated by glycine receptor ligands. Comp Biochem Physiol A Mol Integr Physiol 138:193–202. doi:10.1016/j.cbpb.2004.03.015

Rushforth NB, Hofman F (1972) Behavioral and electrophysiological studies of *Hydra*. III. Components of feeding behavior. Biol Bull 142:110–131. doi:10.2307/1540250

Ryan TJ, Grant SGN (2009) The origin and evolution of synapses. Nat Rev Neurosci 10:701–712. doi:10.1038/nrn2717

Ryan JF, Pang K, Schnitzler CE et al (2013) The genome of the ctenophore *Mnemiopsis leidyi* and its implications for cell type evolution. Science 342:1242592. doi:10.1126/science

Sakarya O, Armstrong KA, Adamska M et al (2007) A post-synaptic scaffold at the origin of the animal kingdom. PLoS One 2:e506. doi:10.1371/journal.pone.0000506

Satterlie RA (2011) Do jellyfish have central nervous systems? J Exp Biol 214:1215–1223. doi:10.1242/jeb.043687

Satterlie RA, Case JF (1978) Gap junctions suggest epithelial conduction within the comb plates of the ctenophore *Pleurobrachia bachei*. Cell Tissue Res 193:87–91

Scappaticci AA, Kass-Simon G (2008) NMDA and GABA B receptors are involved in controlling nematocyst discharge in hydra. Comp Biochem Physiol A Mol Integr Physiol 150:415–422. doi:10.1016/j.cbpa.2008.04.606

Scappaticci AA, Jacques R, Carroll JE et al (2004) Immunocytochemical evidence for an NMDA1 receptor subunit in dissociated cells of *Hydra vulgaris*. Cell Tissue Res 316:263–270

Scemes E, Suadicani SO, Dahl G et al (2007) Connexin and pannexin mediated cell–cell communication. Neuron Glia Biol 3:199–208. doi:10.1017/S1740925X08000069

Schmich J, Rudolf R, Trepel S et al (1998a) Immunohistochemical studies of GLWamides in Cnidaria. Cell Tissue Res 294:169–177. doi:10.1007/s004410051167

Schmich J, Trepel S, Leitz T (1998b) The role of GLWamides in metamorphosis of *Hydractinia echinata*. Dev Genes Evol 208:267–273. doi:10.1007/s004270050181

Schuchert P (1993a) *Trichoplax adhaerens* (phylum Placozoa) has cells that react with antibodies against the neuropeptide RFamide. Acta Zool 74:115–117. doi:10.1111/j.1463-6395.1993.tb01227.x

Schuchert P (1993b) Phylogenetic analysis of the Cnidaria. J Zool Syst Evol Res 31:161–173. doi:10.1111/j.1439-0469.1993.tb00187.x

Seipel K, Yanze N, Schmid V (2004) Developmental and evolutionary aspects of the basic helix-loop-helix transcription factors Atonal-like 1 and Achaete-scute homolog 2 in the jellyfish. Dev Biol 269:331–345. doi:10.1016/j.ydbio.2004.01.035

Shimizu H (2002) Feeding and wounding responses in *Hydra* suggest functional and structural polarization of the tentacle nervous system. Comp Biochem Physiol 131:669–674

Shimizu H, Fujisawa T (2003) Peduncle of *Hydra* and the heart of higher organisms share a common ancestral origin. Genesis 36:182–186. doi:10.1002/gene.10213

Shinzato C, Iguchi A, Hayward DC et al (2008) Sox genes in the coral *Acropora millepora*: divergent expression patterns reflect differences in developmental mechanisms within the *Anthozoa*. BMC Evol Biol 8:311. doi:10.1186/1471-2148-8-311

Shinzato C, Shoguchi E, Kawashima T et al (2011) Using the *Acropora digitifera* genome to understand coral responses to environmental change. Nature (Lond) 476:320–323. doi:10.1038/nature10249

Simionato E, Ledent V, Richards G et al (2007) Origin and diversification of the basic helix-loop-helix gene family in metazoans: insights from comparative genomics. BMC Evol Biol 7:33. doi:10.1186/1471-2148-7-33

Simmons DK, Pang K, Martindale MQ (2012) Lim homeobox genes in the ctenophore *Mnemiopsis leidyi*: the evolution of neural cell type specification. EvoDevo 3:2. doi:10.1186/2041-9139-3-2

Singla CL (1974) Ocelli of hydromedusae. Cell Tissue Res 149:413–429

Singla CL, Weber C (1982) Fine structure of the ocelli of *Polyorchis penicillatus* (Hydrozoa: Anthomedusae) and their connection with the nerve ring. Zoomorphology 99:117–129. doi:10.1007/BF00310304

Sinigaglia C, Busengdal H, Leclère L et al (2013) The bilaterian head patterning gene six3/6 controls aboral domain development in a cnidarian. PLoS Biol 11:e1001488. doi:10.1371/journal.pbio.1001488

Smith CL, Varoqueaux F, Kittelmann M et al (2014) Novel cell types, neurosecretory cells, and body plan of the early-diverging metazoan *Trichoplax adhaerens*. Curr Biol 24:1565–1572. doi:10.1016/j.cub.2014.05.046

Spencer AN (1974) Non-nervous conduction in invertebrates and embryos. Am Zool 14:917–929

Srivastava M, Begovic E, Chapmann J et al (2008) The *Trichoplax* genome and the nature of placozoans. Nature (Lond) 454:955–960. doi:10.1038/nature07191

Srivastava M, Simakov O, Chapman J et al (2010a) The *Amphimedon queenslandica* genome and the evolution of animal complexity. Nature (Lond) 466:720–726. doi:10.1038/nature09201

Srivastava M, Larroux C, Lu D et al (2010b) Early evolution of the LIM homeobox gene family. BMC Biol 8:4. doi:10.1186/1741-7007-8-4

Steinmetz PR, Urbach R, Posnien N et al (2010) Six3 demarcates the anterior-most developing brain region in bilaterian animals. Evodevo 1:14. doi:10.1186/2041-9139-1-14

Strausfeld NJ (2010) Brain homology: Dohrn of a new era? Brain Behav Evol 76:165–167. doi:10.1159/000322150

Strausfeld NJ, Hirth F (2013) Deep homology of arthropod central complex and vertebrate basal ganglia. Science 340:157–161. doi:10.1126/science.1231828

Takahashi T, Hamaue N (2010) Molecular characterization of hydra acetylcholinesterase and its catalytic activity. FEBS Lett 584:511–516. doi:10.1016/j.febslet.2009.11.081

Takahashi T, Hatta M (2011) The importance of GLWamide neuropeptides in cnidarian development and physiology. J Amino Acids 2011:424501. doi:10.4061/2011/424501

Takahashi T, Muneoka Y, Lohmann J et al (1997) Systematic isolation of peptide signal molecules regulating development in hydra: LWamide and PW families. Proc Natl Acad Sci USA 94:1241–1246

Takahashi T, Kobayakawa Y, Muneoka Y et al (2003) Identification of a new member of the GLWamide peptide family: physiological activity and cellular localization in cnidarian polyps. Comp Biochem Physiol 135:309–324. doi:10.1016/S1096-4959(03)00088-5

Takaku Y, Hwang JS, Wolf A et al (2014) Innexin gap junctions in nerve cells coordinate spontaneous contractile behavior in *Hydra* polyps. Sci Rep 4:3573. doi:10.1038/srep03573

Technau U, Rudd S, Maxwell P et al (2005) Maintenance of ancestral complexity and non-metazoan genes in two basal cnidarians. Trends Genet 21:633–639

Tessmar-Raible K, Raible F, Christodoulou F et al (2007) Conserved sensory-neurosecretory cell types in annelid and fish forebrain: insights into hypothalamus evolution. Cell 129:1389–1400. doi:10.1016/j.cell.2007.04.041

Tomer R, Denes AS, Tessmar-Raible K et al (2010) Profiling by image registration reveals common origin of annelid mushroom bodies and vertebrate pallium. Cell 142:800–809. doi:10.1016/j.cell.2010.07.043

Tosches MA, Arendt D (2013) The bilaterian forebrain: an evolutionary chimaera. Curr Opin Neurobiol 23:1080–1089. doi:10.1016/j.conb.2013.09.005

Vígh B, Vígh-Teichmann I (1982) The cerebrospinal fluid-contacting neurosecretory cell: a protoneuron. In: Farner DS, Lederis K (eds) Molecules cells systems. Plenum Press, New York, pp 458–460

Vigh B, Manzano e Silva MJ, Frank CL et al (2004) The system of cerebrospinal fluid-contacting neurons. Its supposed role in the nonsynaptic signal transmission of the brain. Histol Histopathol 19:607–628

Watanabe H, Fujisawa T, Holstein TW (2009) Cnidarians and the evolutionary origin of the nervous system. Dev Growth Differ 51:167–183. doi:10.1111/j.1440-169X.2009.01103.x

Watanabe H, Kuhn A, Fushiki M et al (2014a) Sequential actions of β-catenin and Bmp pattern the oral nerve net in *Nematostella vectensis*. Nat Commun 5:5536. doi:10.1038/ncomms6536

Watanabe H, Schmidt HA, Kuhn A et al (2014b) Nodal signalling determines biradial asymmetry in *Hydra*. Nature (Lond) 515:112–115. doi:10.1038/nature13666

Westfall JA (1973) Ultrastructural evidence for a granule-containing sensory-motor-interneuron in *Hydra littoralis*. J Ultrastruct Res 42:268–282. doi:10.1016/S0022-5320(73)90055-5

Westfall JA (1987) Ultrastructure of invertebrate synapses. In: Ali MA (ed) Nervous systems in invertebrates. Plenum Press, New York, pp 3–28

Westfall JA, Grimmelikhuijzen CJ (1993) Antho-RFamide immunoreactivity in neuronal synaptic and nonsynaptic vesicles of sea anemones. Biol Bull 185:109–114. doi:10.2307/1542134

Westfall JA, Kinnamon JC (1978) A second sensory-motor-interneuron with neurosecretory granules in *Hydra*. J Neurocytol 7:365–379

Westfall JA, Kinnamon JC (1984) Perioral synaptic connections and their possible role in the feeding behavior of *Hydra*. Tissue Cell 16:355–365. doi:10.1016/0040-8166(84)90055-7

Westfall JA, Yamataka S, Enos PD (1971) Ultrastructural evidence of polarized synapses in the nerve net of *Hydra*. J Cell Biol 51:318–323

Westfall JA, Kinnamon JC, Sims DE (1980) Neuro-epitheliomuscular cell and neuro-neuronal gap junctions in *Hydra*. J Neurocytol 9:725–732

Westfall JA, Sayyar KL, Elliott CF et al (1995) Ultrastructural localization of antho-RWamides I and II at neuromuscular synapses in the gastrodermis and oral sphincter muscle of the sea anemone *Calliactis parasitica*. Biol Bull 189:280–287. doi:10.2307/1542145

Westfall JA, Elliott CF, Carlin RW (2002) Ultrastructural evidence for two-cell and three-cell neural pathways in the tentacle epidermis of the sea anemone *Aiptasia pallida*. J Morphol 251:83–92. doi:10.1002/jmor.1075

Whalan S, Webster NS, Negri AP (2012) Crustose coralline algae and a cnidarian neuropeptide trigger larval settlement in two coral reef sponges. PLoS One 7:e30386. doi:10.1371/journal.pone.0030386

Whelan NV, Kocot KM, Moroz LL et al (2015a) Error, signal, and the placement of Ctenophora, sister to all other animals. Proc Natl Acad Sci USA 112:5773–5778. doi:10.1073/pnas.1503453112

Whelan NV, Kocot KM, Halanych KM (2015b) Employing phylogenomics to resolve the relationships among cnidarians, ctenophores, sponges, placozoans, and bilaterians. Integr Comp Biol pii: icv037. doi:10.1093/icb/icv037

Willmer P (1990) Invertebrate relationships: patterns in animal evolution. Cambridge University Press, Cambridge, MA

Wood RL (1977) The cell junctions of hydra as viewed by freeze-fracture replication. J Ultrastruct Res 58:299–315

Wood RL (1979) The fine structure of the hypostome and mouth of hydra. II. Transmission electron microscopy. Cell Tissue Res 199:319–338

Yamamoto M, Yoshida M (1980) Fine structure of ocelli of an anthomedusan, *Nemiopsis dofleini*, with special reference to synaptic organization. Zoomorphology 96:169–181

Yi-Chan JL, Gallin WJ, Spencer AN (2001) The anatomy of the nervous system of the hydrozoan jellyfish, *Polyorchis penicillatus*, as revealed by a monoclonal antibody. Invertebr Neurosci 4:65–75. doi:10.1007/s101580100008

Part II
The Rise of Diverse Brain Types

Chapter 4
Functional Specification of a Primitive Bilaterian Brain in Planarians

Takeshi Inoue

Abstract The mammalian brain is remarkably complex in function as well as morphology, and therefore understanding the original evolutionary function and primitive principles of brain function is an important and classical challenge. The planarian is an emerging new model of the nervous system as well as regenerative biology because it belongs to an evolutionarily early group possessing a central nervous system (CNS), including a brain with simple architecture. Recently, it has been revealed that the planarian brain is divided into several functional and structural domains as defined by the discrete expression of homeobox genes and a surprisingly complex set of genes, and is composed of several types of neurons distinguished by their neurotransmitters and neural modulators, which are conserved with those used by vertebrates. Furthermore, the planarian brain functions as an information-processing center to produce distinct behavioral traits in response to a variety of signals arising from the external environment. This chapter summarizes recent insights into cellular and molecular mechanisms that regulate planarian brain formation and function. The relative simplicity of the planarian brain, combined with its molecular accessibility, planarian complex behavioral traits, and advances in planarian stem cell biology associated with the planarian extraordinary regenerative capacity, provide a unique opportunity to unravel molecular and cellular mechanisms underlying fundamental brain function and brain evolution.

Keywords Planarian • Platyhelminthes • Central nervous system (CNS) • RNAi • Behavior • Stem cell • Neoblast • Regeneration

---

T. Inoue (✉)
Department of Biophysics, Graduate School of Science, Kyoto University, Kitashirakawa-Oiwake, Sakyo-ku, Kyoto 606-8502, Japan

Department of Life Science, Faculty of Science, Gakushuin University, Mejiro1-5-1, Toshima-ku, Tokyo 171-8588, Japan
e-mail: takeshi.inoue@gakushuin.ac.jp

S. Shigeno et al. (eds.), *Brain Evolution by Design*, Diversity and Commonality in Animals, DOI 10.1007/978-4-431-56469-0_4

4.1 What Is a Planarian?

Freshwater planarians are free-living members (class Turbellaria) of the phylum Platyhelminthes, that is, the flatworms. The Platyhelminthes are located in an important position with respect to the evolution of metazoans (Hyman 1940; Egger et al. 2009). They are widely recognized as being among the simplest organisms possessing three tissue layers (triploblasts), bilateral symmetry, cephalization, and complex organ systems. Among the metazoans, planarians are representatives of the order Tricladida, a designation that is based on the three main branches that comprise their digestive system. Planarians have characteristic organs along the anteroposterior axis, such as a pair of eyes and auricles, and a brain with simple architecture in the anterior head (Fig. 4.1a). A pharynx in the middle portion of the body acts as a mouth and anus, and an intestinal tract instead of a blood vascular system spreads and supplies nutrients throughout the body (Fig. 4.1a, b).

One of the most notable characteristics of planarians is their high regenerative ability (Fig. 4.1c). They can regenerate whole animals, including a functional brain, from tiny fragments from almost any part of their bodies after amputation (Agata and Watanabe 1999; Newmark and Sanchez Alvarado 2002; Agata et al. 2003; Reddien and Sanchez Alvarado 2004; Umesono and Agata 2009; Umesono et al. 2011; King and Newmark 2012; Elliott and Sanchez Alvarado 2013; Rink 2013), an ability that has have been the focus of intense study by many biologists for almost 250 years. An early worker on planarians described them as "almost immortal under the edge of the knife" (Dalyell 1814). The first description of planarian regeneration was published in 1767 (Pallas et al. 1767). Later, Charles Darwin, famous as the author of *The Origin of Species*, was also interested in the regenerative ability of planarians (Darwin 1844), and Thomas Hunt Morgan, known as the originator of *Drosophila* genetics, showed that planarians can regenerate completely when cut in half transversely or longitudinally, as can fragments that are derived from cutting the worm into small pieces (Morgan 1898). The robust regenerative abilities of planarians are based on a population of pluripotent stem cells called neoblasts, which are the only mitotic somatic cells in adults and are distributed throughout the body in planarians (Fig. 4.1d) (Baguna and Slack 1981; Baguna et al. 1989; Sanchez Alvarado 2007; Shibata et al. 2010; Reddien 2013). Therefore, many biologists have been deeply interested in the molecular and cellular mechanisms underlying the nature of the neoblasts (Shibata et al. 1999; Newmark and Sanchez Alvarado 2000; Reddien et al. 2005a, b; Guo et al. 2006; Yoshida-Kashikawa et al. 2007; Hayashi et al. 2010; Rouhana et al. 2010; Tasaki et al. 2011b, a; Shibata et al. 2012; Wagner et al. 2012; Rouhana et al. 2014; Scimone et al. 2014; van Wolfswinkel et al. 2014; Shibata et al. 2016), the anteroposterior polarity (Petersen and Reddien 2008; Gurley et al. 2008; Iglesias et al. 2008; Rink et al. 2009; Petersen and Reddien 2009; Yazawa et al. 2009; Adell et al. 2009; Petersen and Reddien 2011; Umesono et al. 2013; Liu et al. 2013; Sikes and Newmark 2013; Roberts-Galbraith and Newmark 2013; Reuter et al. 2015; Lander and Petersen 2016; Scimone et al. 2016), brain regeneration (Cebria et al. 2002c; Kobayashi et al. 2007; Felix and Aboobaker 2010;

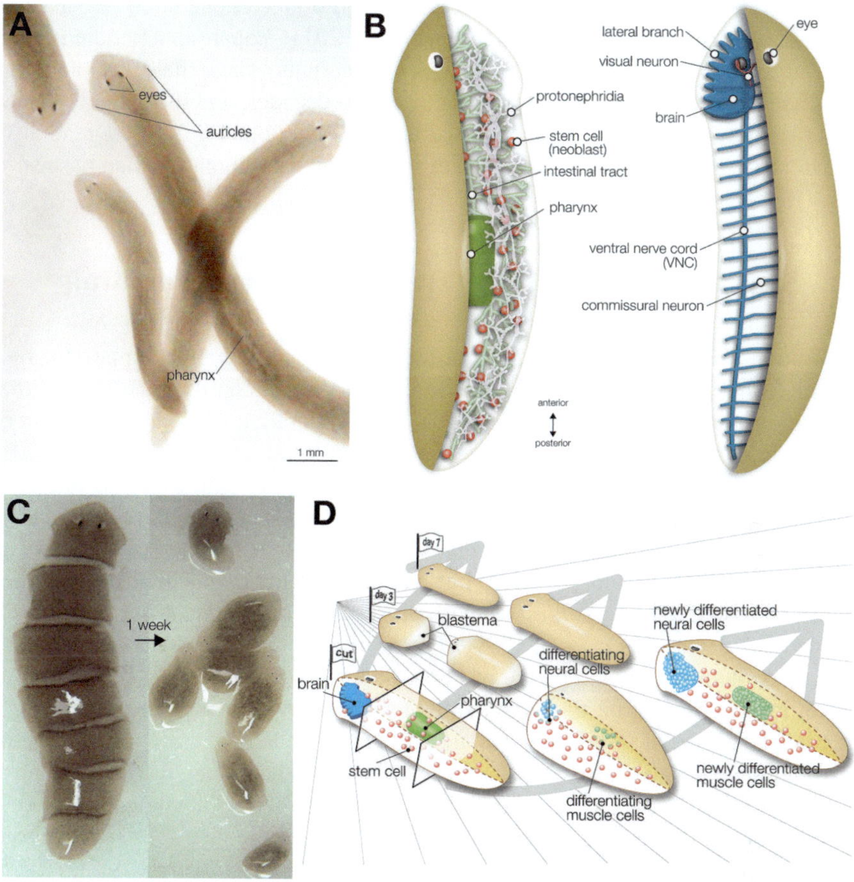

Fig. 4.1 Body structure and regenerative ability of the planarian *Dugesia japonica*. (**a**) Planarian *Dugesia japonica*. A pair of eyes and auricles are observed in the dorsal side of the head region. A pharynx is present in the central portion of the body and acts as a mouth and anus. (**b**) Schematic illustration of main organs of planarian. Intestinal tract occupies almost the entire mesenchymal space and delivers nutrition instead of a blood circulatory system. Stem cells (neoblasts, 30 % of total cells) are distributed throughout the entire mesenchymal space of the body with the exception of the pharyngeal region and the region anterior to the eyes. Planarian CNS is composed of a bilobed brain with nine lateral branches on each outer side in the anterior region of the ventral side and a pair of ventral nerve cords (VNCs) along the body. The brain is located on the dorsal side, in contrast to the VNCs. (**c**) The planarian has a robust capacity for regeneration. Even a tiny fragment cut from any part of the body can regenerate an entire individual planarian with the same anteroposterior and dorsoventral body patterning within 1 week. (**d**) Planarian regenerative ability is strictly dependent on their neoblasts. During regeneration, the neoblasts give rise to all types of cells, somatic cells such as neurons in the brain and muscle cells in the pharynx, and also germline cells

Cowles et al. 2013, 2014; Fraguas et al. 2014), and organ regeneration (Zayas et al. 2010; Lapan and Reddien 2011; Scimone et al. 2011; Lapan and Reddien 2012; Chen et al. 2013; Barberan et al. 2016), in planarians. These fields of planarian biology have advanced remarkably in the past two decades, and the insights gained from planarian studies in vivo show possible new directions for future regenerative medicine, such as reactivation of dormant stem cells in adult mammalian tissues.

4.2 Structural and Cellular Aspects of the Planarian Brain

In addition to planarian regenerative ability, they possess another important biological feature; that is, planarians belong to an evolutionarily early group that acquired a central nervous system (CNS). The planarian CNS is composed of two morphologically distinct structures: a bilobed brain, composed of about 2.0–3.0 × 10^4 neurons in an adult planarian of about 8 mm in length, with nine branches on each outer side in the anterior region of the animal, and two longitudinal ventral nerve cords (VNCs) along the body (Fig. 4.1b). These two structures had long been considered to be a united structure, but recent anatomical and molecular studies clearly indicated that the brain is actually a structure independent from the VNCs (Agata et al. 1998; Tazaki et al. 1999). The brain is composed of a cortex of nerve cells in its outer region and a core of nerve fibers in its inner region (Fig. 4.2a). A pair of eyes is located on the dorsal side at the level of the third lateral branch of the brain (Agata et al. 1998; Sakai et al. 2000), and the axons of the visual neurons form an optic chiasma and directly project to the dorsomedial region (visual center) in the brain (Fig. 4.2b) (Agata et al. 1998; Sakai et al. 2000; Okamoto et al. 2005). On the other hand, axons of chemosensory neurons in the lateral branch neurons project to the ventrolateral region of the brain (Fig. 4.2b) (Okamoto et al. 2005; Inoue et al. 2015). These morphological features of the brain structure suggested that external stimuli sensed by various organs, and the information thus acquired by sensory neurons, might be accumulated inside the brain, and then processed and integrated to transduce the signals into the activity of motor neurons.

In spite of the marked differences between vertebrates and invertebrates, the planarian CNS exhibits many morphological features similar to the vertebrate CNS, such as multipolar neurons and dendrites possessing structures resembling dendritic spines (Sarnat and Netsky 1985). In *Drosophila* and vertebrates, it has been shown that the expression of *Drosophila* cephalic gap genes *empty spiracles* (*ems*) and *orthodenticle* (*otd*), or their respective vertebrate homologues *Emx1/2* and *Otx1/2*, specifies the brain region during development (Kammermeier and Reichert 2001). Similarly, molecular analysis has revealed that the planarian *Dugesia japonica* orthologues of *otd/Otx*-related homeobox genes *DjotxA* and *DjotxB*, and *orthopedia* (*otp*)-related homeobox gene *Djotp*, which are considered to be common regulatory genes involved in regional organization of the brain, are expressed specifically in distinct regions of the brain (Fig. 4.2c) (Umesono et al. 1997, 1999), and that the planarian brain is divided into several functional and structural domains as

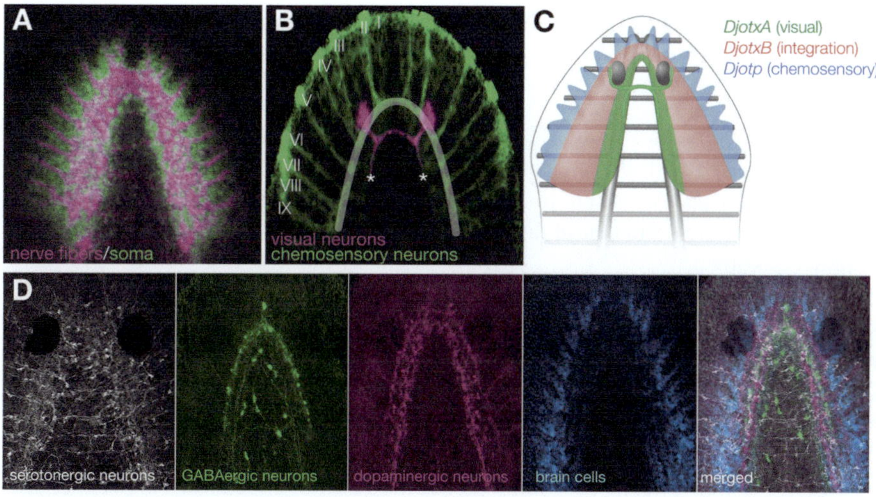

Fig. 4.2 The brain structure of the planarian. (**a**) The planarian brain visualized by in situ hybridization for *Djsyt* mRNA in *green* (somas) and visualized by immunostaining against DjSYT protein in *magenta* (nerve fibers). (**b**) Lateral branches of the planarian brain visualized with G-protein β-subunit (*green*) immunostaining and visual neurons visualized with arrestin (*magenta*) immunostaining. Axons of visual neurons form the optic chiasm and project to the medial region of the brain indicated by *asterisks*. Chemosensory neurons are located in nine lateral brain branches whose dendrites elongate toward the outer region of the head (I–IX) and whose axons project to the lateral region of the brain, indicated by a diaphanous *white* arc. (**c**) Schematic illustration of the discrete expression pattern of the brain-specific homeobox genes. These expression domains represent functional regions in the brain. Photosensory and chemosensory neurons are present in the *DjotxA*- and *Djotp*-expressing domains, respectively. (**d**) Neural subtypes in the brain showing serotonergic neurons (*white*), GABAergic neurons (*green*), dopaminergic neurons (*magenta*), and other neural cells (*cyan*). The planarian brain is more sophisticated than previously thought

defined by the discrete expression of many other neuron-specific genes (Cebria et al. 2002b, c; Mineta et al. 2003). Photosensory and chemosensory neurons are present in the *DjotxA*- and *Djotp*-expressing domains, respectively (Fig. 4.2c). Furthermore, the planarian brain is composed of many distinct neuronal populations, such as dopaminergic, serotonergic, octopaminergic, cholinergic, γ-aminobutyric acid (GABA)ergic, and glutamatergic neurons, which form distinct neuronal networks in the brain (Fig. 4.2d) (Nishimura et al. 2007a, b, 2008a, b, 2010). In addition to these neurotransmitters of monoaminergic and aminergic neurons, many types of neuropeptides may also function in the planarian CNS (Collins et al. 2010). Interestingly, the genes encoding rate-limiting enzymes of these neurotransmitters are expressed in discrete neurons (Fig. 4.2d). These observations suggest that the relatively simple planarian brain is nevertheless well organized and underpinned by a surprisingly complex set of regulatory genetic events.

4.3 Ongoing Search for Neural Stem Cells and Glial Cells in Planarians

In both mammals and *Drosophila*, neurons arise from neural stem cells and neural progenitor cells (neuroblasts) committed to the neuronal fate and possessing the potential to divide. Although all planarian neurons are known to be differentiated from neoblasts, how the neural fate and neural subtypes in a planarian are specified remains unknown.

Musashi encodes an evolutionarily highly conserved RNA-binding protein known to be expressed in the neural lineage, including neural stem cells and neural progenitor cells, across the animal kingdom (Okano et al. 2002). Although three *musashi* family genes (*DjmlgA*, *DjmlgB*, *DjmlgC*) have been identified in planarians, all of them are predominantly expressed in terminally differentiated neurons, not in proliferating cells, including putative neural stem cells (Higuchi et al. 2008). Furthermore, neither single nor combinatorial knockdown experiments of these genes by RNAi affected the gross number of neurons during brain regeneration (Higuchi et al. 2008).

In contrast, *sox* genes that encode transcription factor proteins containing the SRY (sex-determining region Y) box have important functions in the maintenance of mammalian neural stem cells. Although five *sox* family genes (*Smed-soxP1*–*Smed-soxP5*) were identified in the planarian, all these genes were expressed in *piwi*-expressing neoblasts, and RNAi experiments clearly demonstrated that *Smed-soxP1* functions in the regulation of self-renewal of planarian neoblasts (which can give rise to all differentiated cells), not in specific progenitors with neural cell fate (Wagner et al. 2012; Wenemoser et al. 2012). Rather, it is expected that there is an evolutionary association between the functions of *Smed-soxP1* in planarians and those of *Sox* genes, such as *Sox2*, in other organisms, including in ES cells. Functions of the other *sox* genes have not yet been detected in planarian neoblasts or neural stem cells.

However, more recently four other *sox* genes categorized in the SoxB group were obtained from another species of planarian, *Schmidtea polychroa* (*Spol-soxB1-1*, *Spol-soxB1-2*, *Spol-soxB2-1*, *Spol-soxB2-2*), and expression analysis during embryogenesis suggested that the *Spol-soxB1-1* and *Spol-soxB1-2* genes are expressed in putative neural progenitor cells and *Spol-soxB2-1* and *Spol-soxB2-2* genes are expressed in neural subpopulations of the central and peripheral nervous systems (Monjo and Romero 2015). Furthermore, it was found that several conserved transcription factors associated with the neural lineage are expressed in a small population of neoblasts during head regeneration (Cowles et al. 2013; Scimone et al. 2014; Cowles et al. 2014). These observations suggest that the gene regulatory network for maintenance of neural stem cells and neural specification is at least partially conserved across metazoans. Although it was reported that *Smed-soxB* promotes differentiation of an anterior subset of visual neurons (Lapan and Reddien 2012), the possible functions of *soxB* genes for neurogenesis and maintenance of neural stem cells in planarian remain obscure. Further functional analysis should reveal the molecular and genetic system for neurogenesis in planarians in the near future.

Glial cells in the brain are the other important cells differentiated from neural stem cells in vertebrates as well as invertebrates (Homem and Knoblich 2012; Kohwi and Doe 2013). Fine structural analysis using electron microscopic studies have suggested that there are glial-like cells in planarians (Morita and Best 1966). In *Drosophila, glial cells missing* (*gcm*) encodes an evolutionarily conserved transcription factor that functions for lineage specification by promoting glial cell fate during embryogenesis (Cattenoz and Giangrande 2013). Although *gcm*-related gene in *D. japonica* (*Djgcm*) shows high conservation with *Drosophila gcm*, *Djgcm* is not expressed in the neural lineage in either intact or regenerating animals (Umesono and Agata 2009). Therefore, the existence and possible features of glial cells in planarians remain obscure.

4.4 Neural Pathways in the Brain Regulating Behaviors in Planarians

Planarians display various behaviors in response to external stimuli. Since very early planarian studies, the ecology and behavior of planarians have been examined (Dalyell 1814; Moseley 1874; Parker and Burnett 1900). Raymond Pearl reported precise observations of planarian movements, including reactions to food, to chemicals, and to mechanical stimuli, and behaviors such as electrotaxis, thigmotaxis, rheotaxis, hydrotaxis, thermotaxis, and phototaxis (Pearl 1903). Since then, phototaxis, mediated by the eyes acting as light-sensing organs, has been the main focus of attention because of its association with morphologically well-characterized organs (Walter 1907; Taliaferro 1920). In addition, the auricles are thought to be important for chemoreception (Cole and Allison 1930; Pigon et al. 1974; Ferrero and Bedini 1989), and thickened ciliated cells connected to neural cells of the branches of the brain exist in the epithelium of the auricle (MacRae 1967; Okamoto et al. 2005; Inoue et al. 2015).

Recently, tractable and quantitative assays have been established for planarian behaviors such as phototaxis, thermotaxis, chemotaxis, and thigmotaxis/kinesis (Fig. 4.3a), and therefore these behaviors can now be quantified in detail for parameters such as velocity, distance, orientation, location, and time (Inoue et al. 2004; Takano et al. 2007; Blackiston et al. 2010; Talbot et al. 2014; Inoue et al. 2014, 2015). These behavioral analyses combined with the RNAi technique (Sanchez Alvarado and Newmark 1999; Rouhana et al. 2013) of neural-related genes have enabled us to gain some understanding of the molecular mechanisms that regulate neural function and behaviors. Studies on *netrin* of both *Dugesia japonica* and *Schmidtea mediterranea* and their receptor homologues (*DjnetB*, *Djunc5A*, *Smed-netrin2*, *Smed-netR*) and planarian *slit* and its receptor *robo* homologues (*Djslit*, *DjroboA*, *Smed-slit*, *Smed-roboA*) revealed that wiring and proper remodeling of the

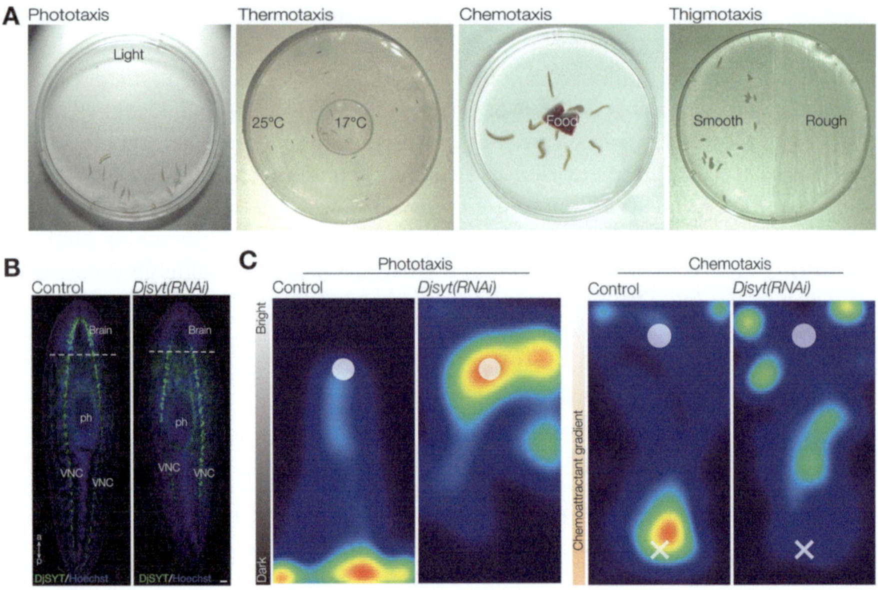

Fig. 4.3 Planarian behaviors are regulated by neural activity in the brain. (**a**) Characteristic planarian behaviors. Planarians show phototaxis, chemotaxis, thermotaxis, and thigmotaxis/kinesis in response to their light, food, temperature, and the surface under them, respectively. (**b**) Ready-knock experiment in planarian brain cells. Immunohistochemical detection of DjSYT protein, which is involved in synaptic transmission, in brain neurons of normal head-regenerated animals 7 days after decapitation (Control). Severe reduction of DjSYT protein was observed in the newly regenerated brain cells (*Djsyt(RNAi)*), although the brain was regenerated normally, as indicated by Hoechst staining in *blue*. However, strong signals of the DjSYT protein were still detected in the preexisting VNCs in the trunk region. (**c**) Heat map view of phototactic and chemotactic behavior of control and *Djsyt(RNAi)* planarians in the field exposed to light or chemoattractant concentration-gradient field. *Circle* indicates start area, and *cross* indicates the highest concentration of chemoattractant. *Djsyt(RNAi)* planarians showed random movement, whereas control planarians showed evasive behavior away from the light in the phototaxis assay and moved to and stayed in the region having the highest concentration of chemoattractant in the chemoattractant-gradient field. Neural activity in the brain is required for planarian behaviors

neural network are essential for its neural function, and these guidance molecules are required for formation of the optic chiasma, and for formation of the connection between visual neurons and the brain, and the connection between the VNCs essential for proper photosensing and locomotion (Cebria and Newmark 2005; Cebria et al. 2007; Cebria and Newmark 2007; Yamamoto and Agata 2011). Planarian homologues of immunoglobulin superfamily cell adhesion molecule (*DjCAM*) similarly function in the fasciculation of axons essential for locomotion in response to external stimuli (Fusaoka et al. 2006). In addition, synaptic recycling regulated by clathrin, adapter protein-2 (AP-2), and dynamin was shown to be required for planarian movement (Inoue et al. 2007; Talbot et al. 2014).

Furthermore, conditional gene knockdown was achieved by utilizing the combination of RNAi and the regenerative ability of planarians in a procedure named "Regeneration-dependent conditional gene knockdown (Readyknock)" (Fig. 4.3b) (Takano et al. 2007). Readyknock experiments provided clear molecular evidence that the planarian brain acts as an information-processing center to produce planarian behaviors in response to environmental stimuli, such as phototaxis, chemotaxis, thermotaxis, and thigmotaxis/kinesis, as shown by inhibition of brain neural activity by Readyknock of neuron-specific genes *Djsnap-25* and of *Djsyt*, encoding a planarian synaptosome-associated protein of 25 kDa that is a component of SNARE (soluble *N*-ethylmaleimide-sensitive fusion protein attachment protein receptor) complex, and planarian synaptotagmin, respectively, which are effector proteins for synaptic transition (Fig. 4.3c) (Takano et al. 2007; Inoue et al. 2014, 2015).

As planarian genes encoding rate-limiting enzymes, such as glutamic acid decarboxylase (GAD), tyrosine hydroxylase (TH), and tryptophan hydroxylase (TPH), for the biosynthesis of neurotransmitters such as GABA, dopamine, and serotonin, respectively, are expressed in discrete cells (Fig. 4.2c), interference with one gene leads to inhibition of the corresponding neural function. Analyses of these genes revealed neuronal subtypes and neural pathways involved in sensing external stimuli and related characteristic behaviors (Fig. 4.4). For example, GABAergic neurons

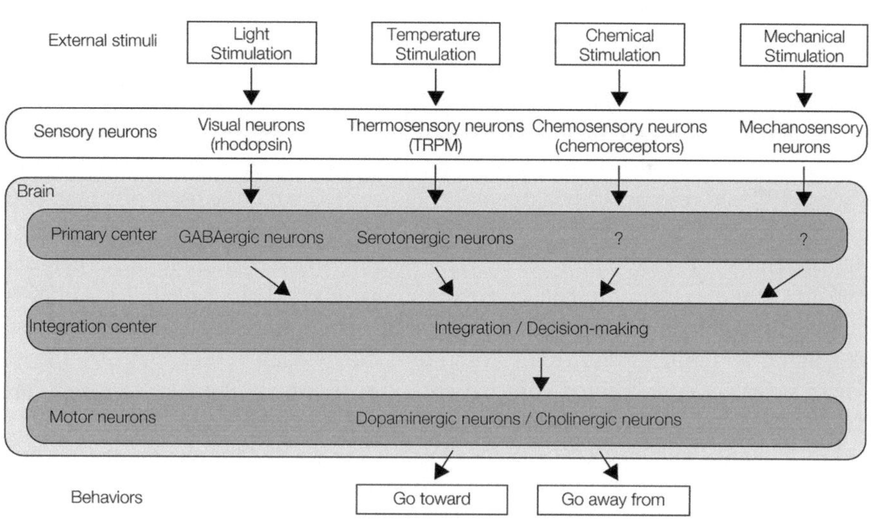

Fig. 4.4 Neural networks controlling planarian behaviors. Planarians receive various signals from the environment, through independent types of sensory neurons such as visual neurons, thermosensory neurons, and chemosensory neurons. Signals received by sensory neurons are processed by neurons in the brain, such as GABAergic neurons and serotonergic neurons. Thereafter, the various signals are integrated via certain neural networks in the brain to decide a planarian behavioral strategy. Subsequently, the planarian behaves suitably in response to its environmental conditions by controlling its motor neurons

distributed in the visual center in the brain are involved in processing the signals received from visual neurons to enable proper phototaxis (Nishimura et al. 2008b). Although serotonergic neurons are distributed throughout the body, including in the brain and VNCs (Nishimura et al. 2007b), studies on brain-specific *DjTPH(RNAi)* using the Readyknock method demonstrated that serotonergic neurons in the brain are required for processing of thermosensing signals from TRPM family gene-expressing thermosensing neurons to elicit cryophilic thermotaxis (Inoue et al. 2014). Planarian behavior is regulated in both cilia- and muscle-dependent manners (Rompolas et al. 2010), and dopaminergic neurons coordinate muscle-mediated locomotion (Nishimura et al. 2007a), whereas trunk serotonergic neurons regulate cilia-mediated locomotion (Currie and Pearson 2013), and cholinergic neurons regulate the motor function of body wall muscles (Nishimura et al. 2010).

4.5 Higher Brain Function in Planarians

In the 1950s, when it was generally thought that invertebrates do not have higher brain function, James V. McConnell showed planarian capability of learning and memory using classical conditioning with light and electrical shock (Fig. 4.5a) (Thompson and McConnell 1955). In that experiment, when a planarian was trained by repeatedly being given both electrical shock and light stimulation, planarians began to react to the light stimulus alone. Utilizing the planarian high regenerative ability, McConnell also showed that regenerated planarians that had been trained before amputation retained memory. This phenomenon was analyzed from various angles, and finally McConnell reported a "memory transfer" paradigm based on the results of "cannibalism" studies, in which a naïve planarian showed the acquisition of a conditioned response when the planarian was fed the body parts of a planarian that had learned a classical conditioning task (McConnell 1962). As a result, McConnell hypothesized that "RNA" is a memory-transferring molecule that can transfer memories from trained to untrained organisms (McConnell 1962, 1964). Although many researchers tried to confirm this phenomenon of memory transfer, they failed to replicate the result. Therefore, McConnell's successful studies of planarian learning, combined with the confusing issue of "memory transfer," are still controversial, and the phenomenon has not been verified since then.

The recently widely used methods of RNA-mediated gene expression modification show some striking parallels to the nearly forgotten method of "memory transfer through RNA" proposed by McConnell, and recent studies show that synaptic plasticity affected by local protein synthesis in neuronal dendritic spines is involved in learning and memory (Greenspan 2003; Bramham and Messaoudi 2005; Bramham and Wells 2007; Costa-Mattioli et al. 2009; Sossin and Lacaille 2010), but it seems most likely that the effects of McConnell's RNA were not the result of transfer of memory itself. Future studies using tractable quantitative behavioral, molecular, and cellular techniques to analyze the neural pathways of

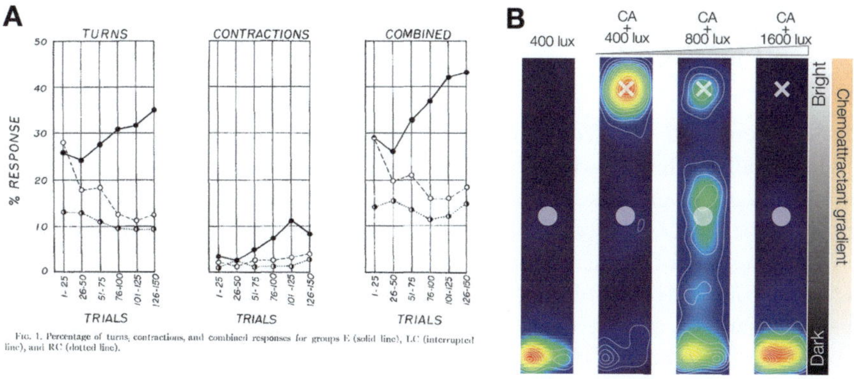

Fig. 4.5 Planarian higher brain function. (**a**) Classical conditioning experiments (reproduced from Thompson and McConnell 1955). If electrical shock is repeatedly given to a planarian after exposure to light, the planarian begins to react to the light stimulus alone. The graph shows the proportion of turns, contractions, and the combination of turns and contractions in planarians exposed only to light as related to the number of light-shock training trials. (**b**) Integrative behavior in planarian in response to simultaneous stimulation which light and chemoattractant. A planarian was placed in the center (indicated by a *circle*). Chemoattractant (CA) was dropped into the position indicated by a *cross*. When the chamber was illuminated from one side with 400 lux, the planarians immediately moved away from the light. When chemoattractant was added and the chamber was then exposed to 400 lux of light planarians showed chemotaxis but not negative phototaxis, even though they had been exposed to sufficient light to show phototaxis if light had been the sole stimulus. When planarians in a chamber containing chemoattractant were exposed to 800 lux of light, some planarians showed negative phototaxis, whereas others showed chemotaxis. When the planarians in the chamber were exposed to 1600 lux of light plus chemoattractant, they showed negative phototaxis and also showed a statistically significant failure to show chemotaxis. This planarians do not show a simple, direct response to a stimulus, but rather integrate external stimuli and then behave suitably in response to the overall conditions

planarian brain functions should be able to resolve the debate about the issue of McConnell's observations on inheritance of memory.

Recently, the first molecular and neuronal evidence that planarian shows decision-making behavior depending on the stimulus conditions by integrative brain function was reported (Figs. 4.4, 4.5) (Inoue et al. 2015). In that experiment, when planarians were exposed to light and chemoattractant simultaneously, the planarians showed chemoattractive behavior regardless of the direction of the light, although this intensity of light when presented by itself was sufficient to induce evasive behavior away from the light. In contrast, when the light intensity was increased and the same dose of chemoattractant was presented, planarians showed negative phototaxis behavior instead of chemoattractive behavior. These results demonstrated that planarians decide on a behavioral strategy from among several possible candidate behavioral strategies. Furthermore, it was shown that planarians that had specifically lost neural activity in the brain as a result of Readyknock of synaptotagmin in the brain lost such decision-making behavior (Inoue et al. 2015).

4.6 Evolutionarily Early Binocular Visual System in Planarians

The ontogeny and architecture of the eyes of vertebrates and invertebrates are fundamentally different, which has caused discussion about whether the eyes of vertebrates and invertebrates trace back to a common eye precursor or whether they are of independent evolutionary origin. However, recent studies showing similarities in the expression of transcription factors and other developmental molecular cues indicate a common origin of light-sensitive systems in all animals. It was also shown that planarian eyes share genetic similarity with vertebrate eyes, and planarian eyes express *otx*, *sine oculis*, *six1/2*, *eyes-absent*, and many other genes involved in phototransduction (Umesono et al. 1999; Pineda et al. 2000; Mannini et al. 2004; Lapan and Reddien 2012). These observations suggest that photoreceptor cells were presumably present in the Urbilateria, which may have used transcription factors for eye formation and GPCRs for phototransduction (Arendt 2003; Gehring 2004; Nilsson 2009).

Planarian eyes are composed of only two cell types: pigment cells are arranged into a semicircular eyecup, and visual bipolar neurons are composed of cell bodies, axons, and dendrites, with their cell bodies located outside the eyecup along the anteroposterior axis (Fig. 4.6a) (Carpenter et al. 1974; Asano et al. 1998). Dendrites from the visual neurons are enclosed by the pigment eyecup to form a rhabdomeric structure (Hesse 1897; Press 1959; Kishida 1967), and the axons of the visual

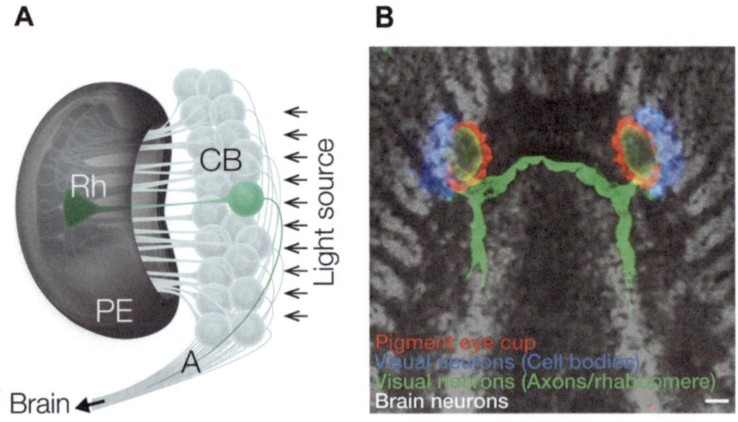

Fig. 4.6 Binocular visual system in planarian. (**a**) Schematic drawing of architecture of planarian eye. *PE* pigment eyecup, *CB* cell bodies, *Rh* rhabdomere, *A* axons. One representative visual neuron is indicated in *green*. (**b**) Planarian eyes are composed of a pigment eyecup and visual neurons. Cell bodies of visual neurons were visualized in *blue* by in situ hybridization for expression of the *rhodopsin* gene, axons, and cell bodies of visual neurons were visualized in *green* by immunostaining of arrestin, pigment eyecups were visualized in *red* by immunostaining of TPH, and brain neurons were visualized in *white* by Hoechst staining

neurons project directly to the brain (Agata et al. 1998; Sakai et al. 2000; Okamoto et al. 2005). The planarian eye is non-dioptric and cannot focus light to form an image, and instead its visual neurons might sense the light direction from one side only as a result of shielding by its pigment cells against the light from the opposite direction. Interestingly, although the architecture of eyes is diverse among organisms (Gehring 2004), animals as evolutionarily early as cnidarians, with pigment-cup ocelli, had acquired both pigment and photosensory capacity in a single cell (Nordstrom et al. 2003), and even some dinoflagellates possess photoreceptor and pigment-cup structures in a single cell (Greuet 1965; Gehring 2014), indicating that detection of the direction of light by shielding against light coming from the opposite side with pigment, rather than detection of light intensity, might have been the evolutionarily fundamental and initial adaptive function of photosensing.

For determining the direction of light, an animal can compare the light intensity perceived during its own movements in different directions. Although the binocular visual system now has important roles in stereoscopic vision or providing a wide field of view, a binocular visual system may have initially been evolutionarily acquired to sense the light coming from a wider range orientations than would be possible with a single eye with pigment. Planarians possess the primitive bilaterally symmetric eyes with simple architecture and display striking negative phototaxis through neural wiring between the eyes and the brain with hemidecussation (Fig. 4.6b) (Walter 1907; Taliaferro 1920; Agata et al. 1998; Inoue et al. 2004; Takano et al. 2007; Gehring 2014), retaining this symmetry even in the case of a multiocular planarian species (Kuchiiwa et al. 1991). These findings suggested that the acquisition of a binocular visual system processing multiple inputs in the planarian might be correlated with much higher throughputs in the brain to enable robust recognition of the direction of a light source.

4.7 Evolutionary Implications of *ndk* Function

The evolution of the CNS remains a controversial issue in biology. Platyhelminthes have been considered the most basal animals that acquired a CNS, although molecular phylogenetic analysis and synapomorphies of stem cell-related characters produce conflicting results regarding the position of the Acoela in the tree of life (Egger et al. 2009). Therefore, planarians may provide unique opportunities for investigating the evolutionary origin of the CNS from a diffuse nervous system. To track the evolutionary transition from a diffuse nervous system to the CNS in bilaterians, this section focuses on the role of *nou-darake* (*ndk*; meaning "brains everywhere" in Japanese), a gene specifically expressed in the head regions including the brain cells. *ndk(RNAi)* caused ectopic brain formation in the planarian trunk region (Fig. 4.7) (Cebria et al. 2002a). Interestingly, the ectopic brain was gradually expanded to posterior regions. *ndk* encodes a putative fibroblast growth factor (FGF) receptor-like protein lacking a tyrosine kinase domain in the intracellular region, and the ectopic brain formation by *ndk(RNAi)* was suppressed by combined

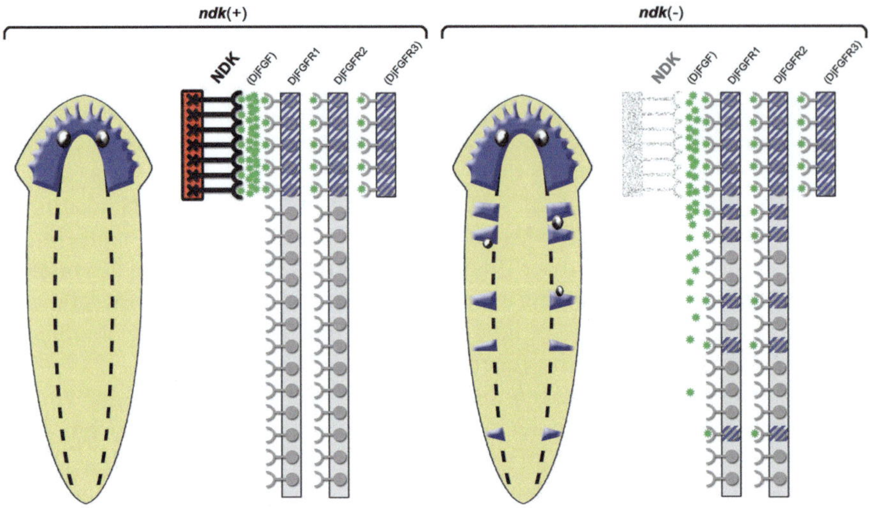

Fig. 4.7 Capture model of *ndk* function. A predicted brain-formation activator (*green*) can stimulate the differentiation of brain neurons, but may be captured by NDK and consequently not able to diffuse outside the head region. Excess brain activators are trapped by NDK, but they can diffuse to the posterior region of the body in nou-darake RNAi planarians. However, in *ndk(RNAi)/FGFR1(RNAi)/FGFR2(RNAi)* triple-knockdown planarians, ectopic brains were not formed in the trunk region because of the lack of expression of FGFR in the neoblasts. In this model, the existence of a third FGFR molecule (DjFGFR3), which has not yet been identified, must be postulated for this capture model to form a brain in the head region

inhibition of two planarian FGF receptor genes (*DjFGFR1* and *DjFGFR2*), which are expressed in planarian neoblasts (Ogawa et al. 1998, 2002), indicating that NDK negatively regulates a brain-inducing circuit based on FGF signaling outside the head region in a non-cell-autonomous manner (Fig. 4.7). Indeed, ERK activation in the anterior region regulates head formation during regeneration in planarians, and *ndk* expression acts as an outcome of an ERK signaling in the differentiating head blastema (Umesono et al. 2013). These observations give rise to the following hypothesis about NDK function in brain formation in the head region: NDK may regulate the diffusion range of brain activators (FGF or FGF-like molecules) from a putative source in the head region to the rest of the body through direct interaction with brain activators (Fig. 4.7). It may be able to bind an FGF-like brain activator molecule, but not stimulate differentiation of the brain neurons. Loss of function of *ndk* would allow these factors to travel to more posterior regions, and thus activate FGF receptors outside the head region to trigger ectopic brain formation.

Interestingly, *ndk* is evolutionarily conserved with vertebrates (Trueb 2011), and a *Xenopus* orthologue of *ndk* (*XFGFRL1/Xndk*) is expressed in the anterior region at the late gastrula stage and dramatically increased at the early neurula stage in an anterior mesendodermal region. Importantly, *XFGFRL1* binds FGFs to antagonize FGF signaling in *Xenopus* embryos (Steinberg et al. 2010), and ectopic expression of planarian *ndk* mRNA inhibited *Xenopus* gastrulation by interfering with *Xbra* expression (Cebria et al. 2002a). It is tempting to speculate that regulation of FGF signaling via an evolutionarily conserved NDK function might have a pivotal role in centralizing neurons in certain regions (Umesono and Agata 2009). Further studies are required to unravel the molecular mechanisms of formation of the CNS, including the brain.

4.8 Conclusions and Future Prospects

It is hypothesized that brain evolution may have been accompanied by the evolutionary emergence of stem cell systems, including neural stem cells that could give rise to a variety of neural cell types (Agata et al. 2006; Umesono and Agata 2009). Planarians have well-characterized somatic stem cells that are their only mitotic cells and which give rise to all cell types during regeneration, and therefore knowledge about the stem cell system of planarians might enable us to understand brain morphogenesis as well as brain evolution in vivo. Recently, combinatory analyses of quantitative planarian behavioral assays and molecular techniques such as RNAi have provided evidence about the mechanisms of brain formation and brain function in planarians. Further functional analysis of the planarian brain will provide new insights into the fundamental principles of the evolution of centralization of the nervous system and brain function, including higher brain function.

Acknowledgments I thank Dr. Kiyokazu Agata, Dr. Yoshihiko Umesono, Dr. Kaneyasu Nishimura, Mr. Taiga Yamashita, Mr. Hajime Hoshino, Mr. Yoshitaro Akiyama, Dr. Seira Shimoyama and Dr. Norito Shibata for collaboration in the planarian brain project, and Elizabeth Nakajima for critical reading of the manuscript.

References

Adell T, Salo E, Boutros M, Bartscherer K (2009) Smed-Evi/Wntless is required for beta-catenin-dependent and -independent processes during planarian regeneration. Development 136(6):905–910. doi:10.1242/dev.033761

Agata K, Watanabe K (1999) Molecular and cellular aspects of planarian regeneration. Semin Cell Dev Biol 10(4):377–383. doi:10.1006/scdb.1999.0324

Agata K, Soejima Y, Kato K, Kobayashi C, Umesono Y, Watanabe K (1998) Structure of the planarian central nervous system (CNS) revealed by neuronal cell markers. Zool Sci 15(3):433–440. doi:10.2108/zsj.15.433

Agata K, Tanaka T, Kobayashi C, Kato K, Saitoh Y (2003) Intercalary regeneration in planarians. Dev Dyn 226(2):308–316. doi:10.1002/dvdy.10249

Agata K, Nakajima E, Funayama N, Shibata N, Saito Y, Umesono Y (2006) Two different evolutionary origins of stem cell systems and their molecular basis. Semin Cell Dev Biol 17(4):503–509. doi:10.1016/j.semcdb.2006.05.004

Arendt D (2003) Evolution of eyes and photoreceptor cell types. Int J Dev Biol 47(7–8):563–571

Asano Y, Nakamura S, Ishida S, Azuma K, Shinozawa T (1998) Rhodopsin-like proteins in planarian eye and auricle: detection and functional analysis. J Exp Biol 201(9):1263–1271
Baguna J, Slack JMW (1981) Planarian neoblasts. Nature 290(5801):14–15
Baguna J, Salo E, Auladell C (1989) Regeneration and pattern formation in planarians. III. Evidence that neoblasts are totipotent stem cells and the source of blastema cells. Development 107(1):77–86
Barberan S, Fraguas S, Cebria F (2016) The EGFR signaling pathway controls gut progenitor differentiation during planarian regeneration and homeostasis. Development 143(12):2089–2102. doi:10.1242/dev.131995
Blackiston D, Shomrat T, Nicolas CL, Granata C, Levin M (2010) A second-generation device for automated training and quantitative behavior analyses of molecularly-tractable model organisms. PLoS One 5(12), e14370. doi:10.1371/journal.pone.0014370
Bramham CR, Messaoudi E (2005) BDNF function in adult synaptic plasticity: the synaptic consolidation hypothesis. Prog Neurobiol 76(2):99–125. doi:10.1016/j.pneurobio.2005.06.003
Bramham CR, Wells DG (2007) Dendritic mRNA: transport, translation and function. Nat Rev Neurosci 8(10):776–789. doi:10.1038/nrn2150
Carpenter KS, Morita M, Best JB (1974) Ultrastructure of the photoreceptor of the planarian *Dugesia dorotocephala.* I. Normal eye. Cell Tissue Res 148(2):143–158. doi:10.1007/BF00224579
Cattenoz PB, Giangrande A (2013) Lineage specification in the fly nervous system and evolutionary implications. Cell Cycle 12(17):2753–2759. doi:10.4161/cc.25918
Cebria F, Newmark PA (2005) Planarian homologs of netrin and netrin receptor are required for proper regeneration of the central nervous system and the maintenance of nervous system architecture. Development 132(16):3691–3703. doi:10.1242/dev.01941
Cebria F, Newmark PA (2007) Morphogenesis defects are associated with abnormal nervous system regeneration following roboA RNAi in planarians. Development 134(5):833–837. doi:10.1242/dev.02794
Cebria F, Kobayashi C, Umesono Y, Nakazawa M, Mineta K, Ikeo K, Gojobori T, Itoh M, Taira M, Sanchez Alvarado A, Agata K (2002a) FGFR-related gene nou-darake restricts brain tissues to the head region of planarians. Nature 419(6907):620–624. doi:10.1038/nature01042
Cebria F, Kudome T, Nakazawa M, Mineta K, Ikeo K, Gojobori T, Agata K (2002b) The expression of neural-specific genes reveals the structural and molecular complexity of the planarian central nervous system. Mech Dev 116(1-2):199–204
Cebria F, Nakazawa M, Mineta K, Ikeo K, Gojobori T, Agata K (2002c) Dissecting planarian central nervous system regeneration by the expression of neural-specific genes. Dev Growth Differ 44(2):135–146
Cebria F, Guo T, Jopek J, Newmark PA (2007) Regeneration and maintenance of the planarian midline is regulated by a slit orthologue. Dev Biol 307(2):394–406. doi:10.1016/j.ydbio.2007.05.006
Chen CC, Wang IE, Reddien PW (2013) pbx is required for pole and eye regeneration in planarians. Development 140(4):719–729. doi:10.1242/dev.083741
Cole WH, Allison JB (1930) Chemical stimulation by alcohols in the barnacle, the frog and *Planaria.* J Gen Physiol 14(1):71–86
Collins JJ 3rd, Hou X, Romanova EV, Lambrus BG, Miller CM, Saberi A, Sweedler JV, Newmark PA (2010) Genome-wide analyses reveal a role for peptide hormones in planarian germline development. PLoS Biol 8(10), e1000509. doi:10.1371/journal.pbio.1000509
Costa-Mattioli M, Sossin WS, Klann E, Sonenberg N (2009) Translational control of long-lasting synaptic plasticity and memory. Neuron 61(1):10–26. doi:10.1016/j.neuron.2008.10.055
Cowles MW, Brown DD, Nisperos SV, Stanley BN, Pearson BJ, Zayas RM (2013) Genome-wide analysis of the bHLH gene family in planarians identifies factors required for adult neurogenesis and neuronal regeneration. Development 140(23):4691–4702. doi:10.1242/dev.098616
Cowles MW, Omuro KC, Stanley BN, Quintanilla CG, Zayas RM (2014) COE loss-of-function analysis reveals a genetic program underlying maintenance and regeneration of the nervous system in planarians. PLoS Genet 10(10), e1004746. doi:10.1371/journal.pgen.1004746

Currie KW, Pearson BJ (2013) Transcription factors lhx1/5-1 and pitx are required for the maintenance and regeneration of serotonergic neurons in planarians. Development 140(17):3577–3588. doi:10.1242/dev.098590

Dalyell JG (1814) Observations on some interesting phenomena in animal physiology, exhibited by several species of planariae, illustrated by colored figures of living animals. Museum of Comparative Zoology–Biodiversity Heritage Library digitization project. Archibald Constable, Edinburgh. doi:10.5962/bhl.title.10135

Darwin C (1844) Brief descriptions of several terrestrial Planariæ, and of some remarkable marine species, with an account of their habits. J Nat Hist 14(91):241–251

Egger B, Steinke D, Tarui H, De Mulder K, Arendt D, Borgonie G, Funayama N, Gschwentner R, Hartenstein V, Hobmayer B, Hooge M, Hrouda M, Ishida S, Kobayashi C, Kuales G, Nishimura O, Pfister D, Rieger R, Salvenmoser W, Smith J, Technau U, Tyler S, Agata K, Salzburger W, Ladurner P (2009) To be or not to be a flatworm: the acoel controversy. PLoS One 4(5):e5502. doi:10.1371/journal.pone.0005502

Elliott SA, Sanchez Alvarado A (2013) The history and enduring contributions of planarians to the study of animal regeneration. Wiley Interdiscip Rev Dev Biol 2(3):301–326. doi:10.1002/wdev.82

Felix DA, Aboobaker AA (2010) The TALE class homeobox gene Smed-prep defines the anterior compartment for head regeneration. PLoS Genet 6(4):e1000915. doi:10.1371/journal.pgen.1000915

Ferrero EA, Bedini C (1989) Chemoreception in Turbellaria. Exp Biol 48(3):141–148

Fraguas S, Barberan S, Iglesias M, Rodriguez-Esteban G, Cebria F (2014) egr-4, a target of EGFR signaling, is required for the formation of the brain primordia and head regeneration in planarians. Development 141(9):1835–1847. doi:10.1242/dev.101345

Fusaoka E, Inoue T, Mineta K, Agata K, Takeuchi K (2006) Structure and function of primitive immunoglobulin superfamily neural cell adhesion molecules: a lesson from studies on planarian. Genes Cells 11(5):541–555. doi:10.1111/j.1365-2443.2006.00962.x

Gehring WJ (2004) Historical perspective on the development and evolution of eyes and photoreceptors. Int J Dev Biol 48(8-9):707–717. doi:10.1387/ijdb.041900wg

Gehring WJ (2014) The evolution of vision. Wiley Interdiscip Rev Dev Biol 3(1):1–40. doi:10.1002/wdev.96

Greenspan RJ (2003) RNA and memory: from feeding to localization. Curr Biol 13(4):R126–R127

Greuet C (1965) Structure fine de lócelle *d'Erythropsis pavillardi* Hertwig, ptéridinien Warnowidae Lindemann. CR Acad Sci Paris 261:1904–1907

Guo T, Peters AH, Newmark PA (2006) A Bruno-like gene is required for stem cell maintenance in planarians. Dev Cell 11(2):159–169. doi:10.1016/j.devcel.2006.06.004

Gurley KA, Rink JC, Sanchez Alvarado A (2008) Beta-catenin defines head versus tail identity during planarian regeneration and homeostasis. Science 319(5861):323–327. doi:10.1126/science.1150029

Hayashi T, Shibata N, Okumura R, Kudome T, Nishimura O, Tarui H, Agata K (2010) Single-cell gene profiling of planarian stem cells using fluorescent activated cell sorting and its "index sorting" function for stem cell research. Dev Growth Differ 52(1):131–144. doi:10.1111/j.1440-169X.2009.01157.x

Hesse R (1897) Untersuchungen über die organe der lichtempfindung bei niedern thieren.II. Die Augen der Plathelminthen, insonderheit der Tricladen Turbellarien. Zeit Wiss Zool 62:527–582

Higuchi S, Hayashi T, Tarui H, Nishimura O, Nishimura K, Shibata N, Sakamoto H, Agata K (2008) Expression and functional analysis of musashi-like genes in planarian CNS regeneration. Mech Dev 125(7):631–645. doi:10.1016/j.mod.2008.03.002

Homem CC, Knoblich JA (2012) *Drosophila* neuroblasts: a model for stem cell biology. Development 139(23):4297–4310. doi:10.1242/dev.080515

Hyman LH (1940) Native and introduced land planarians in the United States. Science 92(2379):105–106. doi:10.1126/science.92.2379.105

Iglesias M, Gomez-Skarmeta JL, Salo E, Adell T (2008) Silencing of Smed-betacatenin1 generates radial-like hypercephalized planarians. Development 135(7):1215–1221. doi:10.1242/dev.020289

Inoue T, Kumamoto H, Okamoto K, Umesono Y, Sakai M, Sanchez Alvarado A, Agata K (2004) Morphological and functional recovery of the planarian photosensing system during head regeneration. Zool Sci 21(3):275–283. doi:10.2108/zsj.21.275

Inoue T, Hayashi T, Takechi K, Agata K (2007) Clathrin-mediated endocytic signals are required for the regeneration of, as well as homeostasis in, the planarian CNS. Development 134(9):1679–1689. doi:10.1242/dev.02835

Inoue T, Yamashita T, Agata K (2014) Thermosensory signaling by TRPM is processed by brain serotonergic neurons to produce planarian thermotaxis. J Neurosci 34(47):15701–15714. doi:10.1523/JNEUROSCI.5379-13.2014

Inoue T, Hoshino H, Yamashita T, Shimoyama S, Agata K (2015) Planarian shows decision-making behavior in response to multiple stimuli by integrative brain function. Zool Lett 1(1):7. doi:10.1186/s40851-014-0010-z

Kammermeier L, Reichert H (2001) Common developmental genetic mechanisms for patterning invertebrate and vertebrate brains. Brain Res Bull 55(6):675–682

King RS, Newmark PA (2012) The cell biology of regeneration. J Cell Biol 196(5):553–562. doi:10.1083/jcb.201105099

Kishida Y (1967) Electron microscopic studies on the planarian eye. I. Fine structures of the normal eye. Sci Rep Kanazawa Univ 12(1):75–110

Kobayashi C, Saito Y, Ogawa K, Agata K (2007) Wnt signaling is required for anteroposterior patterning of the planarian brain. Dev Biol 306(2):714–724. doi:10.1016/j.ydbio.2007.04.010

Kohwi M, Doe CQ (2013) Temporal fate specification and neural progenitor competence during development. Nat Rev Neurosci 14(12):823–838

Kuchiiwa T, Kuchiiwa S, Teshirogi W (1991) Comparative morphological studies on the visual systems in a binocular and a multi-ocular species of freshwater planarian. Hydrobiologia 227:241–249

Lander R, Petersen CP (2016) Wnt, Ptk7, and FGFRL expression gradients control trunk positional identity in planarian regeneration. Elife 5:e12850. doi:10.7554/eLife.12850

Lapan SW, Reddien PW (2011) dlx and sp6-9 control optic cup regeneration in a prototypic eye. PLoS Genet 7(8):e1002226. doi:10.1371/journal.pgen.1002226

Lapan SW, Reddien PW (2012) Transcriptome analysis of the planarian eye identifies ovo as a specific regulator of eye regeneration. Cell Rep 2(2):294–307. doi:10.1016/j.celrep.2012.06.018

Liu SY, Selck C, Friedrich B, Lutz R, Vila-Farre M, Dahl A, Brandl H, Lakshmanaperumal N, Henry I, Rink JC (2013) Reactivating head regrowth in a regeneration-deficient planarian species. Nature 500(7460):81–84. doi:10.1038/nature12414

MacRae EK (1967) The fine structure of sensory receptor processes in the auricular epithelium of the planarian, *Dugesia tigrina*. Z Zellforsch Mikrosk Anat 82(4):479–494

Mannini L, Rossi L, Deri P, Gremigni V, Salvetti A, Salo E, Batistoni R (2004) Djeyes absent (Djeya) controls prototypic planarian eye regeneration by cooperating with the transcription factor Djsix-1. Dev Biol 269(2):346–359. doi:10.1016/j.ydbio.2004.01.042

McConnell JV (1962) Memory transfer through cannibalism in planarians. J Neuropsychiatry 3(suppl 1):S42

McConnell JV (1964) On the turning of worms: a reply to James and Halas. Psychol Rec 14:13–20

Mineta K, Nakazawa M, Cebria F, Ikeo K, Agata K, Gojobori T (2003) Origin and evolutionary process of the CNS elucidated by comparative genomics analysis of planarian ESTs. Proc Natl Acad Sci USA 100(13):7666–7671. doi:10.1073/pnas.1332513100

Monjo F, Romero R (2015) Embryonic development of the nervous system in the planarian *Schmidtea polychroa*. Dev Biol 397(2):305–319. doi:10.1016/j.ydbio.2014.10.021

Morgan TH (1898) Experimental studies of the regeneration of *Planaria maculata*. Arch Entwickl Org 7(2-3):364–397. doi:10.1007/bf02161491

Morita M, Best JB (1966) Electron microscopic studies of *Planaria*. 3. Some observations on the fine structure of planarian nervous tissue. J Exp Zool 161(3):391–411. doi:10.1002/jez.1401610307

Moseley HN (1874) On the anatomy and histology of land planarians of Ceylon; with account for their habits, and a description of two new species, and with notes on the anatomy of some European aquatic species. Philos Trans R Soc Lond 10:105–171

Newmark PA, Sanchez Alvarado A (2000) Bromodeoxyuridine specifically labels the regenerative stem cells of planarians. Dev Biol 220(2):142–153. doi:10.1006/dbio.2000.9645

Newmark PA, Sanchez Alvarado A (2002) Not your father's planarian: a classic model enters the era of functional genomics. Nat Rev Genet 3(3):210–219. doi:10.1038/nrg759

Nilsson DE (2009) The evolution of eyes and visually guided behaviour. Philos Trans R Soc Lond B Biol Sci 364(1531):2833–2847. doi:10.1098/rstb.2009.0083

Nishimura K, Kitamura Y, Inoue T, Umesono Y, Sano S, Yoshimoto K, Inden M, Takata K, Taniguchi T, Shimohama S, Agata K (2007a) Reconstruction of dopaminergic neural network and locomotion function in planarian regenerates. Dev Neurobiol 67(8):1059–1078. doi:10.1002/dneu.20377

Nishimura K, Kitamura Y, Inoue T, Umesono Y, Yoshimoto K, Takeuchi K, Taniguchi T, Agata K (2007b) Identification and distribution of tryptophan hydroxylase (TPH)-positive neurons in the planarian *Dugesia japonica*. Neurosci Res 59(1):101–106. doi:10.1016/j.neures.2007.05.014

Nishimura K, Kitamura Y, Inoue T, Umesono Y, Yoshimoto K, Taniguchi T, Agata K (2008a) Characterization of tyramine beta-hydroxylase in planarian *Dugesia japonica*: cloning and expression. Neurochem Int 53(6-8):184–192. doi:10.1016/j.neuint.2008.09.006

Nishimura K, Kitamura Y, Umesono Y, Takeuchi K, Takata K, Taniguchi T, Agata K (2008b) Identification of glutamic acid decarboxylase gene and distribution of GABAergic nervous system in the planarian *Dugesia japonica*. Neuroscience 153(4):1103–1114. doi:10.1016/j.neuroscience.2008.03.026

Nishimura K, Kitamura Y, Taniguchi T, Agata K (2010) Analysis of motor function modulated by cholinergic neurons in planarian *Dugesia japonica*. Neuroscience 168(1):18–30. doi:10.1016/j.neuroscience.2010.03.038

Nordstrom K, Wallen R, Seymour J, Nilsson D (2003) A simple visual system without neurons in jellyfish larvae. Proc Biol Sci 270(1531):2349–2354. doi:10.1098/rspb.2003.2504

Ogawa K, Wakayama A, Kunisada T, Orii H, Watanabe K, Agata K (1998) Identification of a receptor tyrosine kinase involved in germ cell differentiation in planarians. Biochem Biophys Res Commun 248(1):204–209. doi:10.1006/bbrc.1998.8915

Ogawa K, Kobayashi C, Hayashi T, Orii H, Watanabe K, Agata K (2002) Planarian fibroblast growth factor receptor homologs expressed in stem cells and cephalic ganglions. Dev Growth Differ 44(3):191–204

Okamoto K, Takeuchi K, Agata K (2005) Neural projections in planarian brain revealed by fluorescent dye tracing. Zool Sci 22(5):535–546. doi:10.2108/zsj.22.535

Okano H, Imai T, Okabe M (2002) Musashi: a translational regulator of cell fate. J Cell Sci 115(pt 7):1355–1359

Pallas PS, Dwight J, Gurney JH, Lea I, Richmond CW, Tucker MB, Philadelphia. AoNSo, filius. CFVe, Library of Congress (1767) Spicilegia zoologica: quibus novae imprimis et obscurae animalium species iconibus, descriptionibus atque commentariis illustrantur/cura P.S. Pallas, vol fasc. 1–10. Prostant apud Gottl. August. Lange, Berolini. doi:10.5962/bhl.title.39832

Parker GH, Burnett FL (1900) The reactions of planarians with and without eyes to light. Am J Physiol 4:373–385

Pearl R (1903) The movements and reactions of fresh-water planarians: a study in animal behaviour. Q J Microsc Sci 46(184):509–714

Petersen CP, Reddien PW (2008) Smed-betacatenin-1 is required for anteroposterior blastema polarity in planarian regeneration. Science 319(5861):327–330. doi:10.1126/science.1149943

Petersen CP, Reddien PW (2009) Wnt signaling and the polarity of the primary body axis. Cell 139(6):1056–1068. doi:10.1016/j.cell.2009.11.035

Petersen CP, Reddien PW (2011) Polarized notum activation at wounds inhibits Wnt function to promote planarian head regeneration. Science 332(6031):852–855. doi:10.1126/science.1202143

Pigon A, Morita M, Best JB (1974) Cephalic mechanism for social control of fissioning in planarians. II. Localization and identification of the receptors by electron micrographic and ablation studies. J Neurobiol 5(5):443–462. doi:10.1002/neu.480050506

Pineda D, Gonzalez J, Callaerts P, Ikeo K, Gehring WJ, Salo E (2000) Searching for the prototypic eye genetic network: sine oculis is essential for eye regeneration in planarians. Proc Natl Acad Sci USA 97(9):4525–4529. doi:10.1073/pnas.97.9.4525

Press N (1959) Electron microscope study of the distal portion of a planarian retinular cell. Biol Bull 117(3):511–517. doi:10.2307/1538864

Reddien PW (2013) Specialized progenitors and regeneration. Development 140(5):951–957. doi:10.1242/dev.080499

Reddien PW, Sanchez Alvarado A (2004) Fundamentals of planarian regeneration. Annu Rev Cell Dev Biol 20:725–757. doi:10.1146/annurev.cellbio.20.010403.095114

Reddien PW, Bermange AL, Murfitt KJ, Jennings JR, Sanchez Alvarado A (2005a) Identification of genes needed for regeneration, stem cell function, and tissue homeostasis by systematic gene perturbation in *Planaria*. Dev Cell 8(5):635–649. doi:10.1016/j.devcel.2005.02.014

Reddien PW, Oviedo NJ, Jennings JR, Jenkin JC, Sanchez Alvarado A (2005b) SMEDWI-2 is a PIWI-like protein that regulates planarian stem cells. Science 310(5752):1327–1330. doi:10.1126/science.1116110

Reuter H, Marz M, Vogg MC, Eccles D, Grifol-Boldu L, Wehner D, Owlarn S, Adell T, Weidinger G, Bartscherer K (2015) Beta-catenin-dependent control of positional information along the AP body axis in planarians involves a teashirt family member. Cell Rep 10(2):253–265. doi:10.1016/j.celrep.2014.12.018

Rink JC (2013) Stem cell systems and regeneration in *Planaria*. Dev Genes Evol 223(1-2):67–84. doi:10.1007/s00427-012-0426-4

Rink JC, Gurley KA, Elliott SA, Sanchez Alvarado A (2009) Planarian Hh signaling regulates regeneration polarity and links Hh pathway evolution to cilia. Science 326(5958):1406–1410. doi:10.1126/science.1178712

Roberts-Galbraith RH, Newmark PA (2013) Follistatin antagonizes activin signaling and acts with notum to direct planarian head regeneration. Proc Natl Acad Sci USA 110(4):1363–1368. doi:10.1073/pnas.1214053110

Rompolas P, Patel-King RS, King SM (2010) An outer arm dynein conformational switch is required for metachronal synchrony of motile cilia in *Planaria*. Mol Biol Cell 21(21):3669–3679. doi:10.1091/mbc.E10-04-0373

Rouhana L, Shibata N, Nishimura O, Agata K (2010) Different requirements for conserved post-transcriptional regulators in planarian regeneration and stem cell maintenance. Dev Biol 341(2):429–443. doi:10.1016/j.ydbio.2010.02.037

Rouhana L, Weiss JA, Forsthoefel DJ, Lee H, King RS, Inoue T, Shibata N, Agata K, Newmark PA (2013) RNA interference by feeding in vitro-synthesized double-stranded RNA to planarians: methodology and dynamics. Dev Dyn 242(6):718–730. doi:10.1002/dvdy.23950

Rouhana L, Weiss JA, King RS, Newmark PA (2014) PIWI homologs mediate Histone H4 mRNA localization to planarian chromatoid bodies. Development 141(13):2592–2601. doi:10.1242/dev.101618

Sakai F, Agata K, Orii H, Watanabe K (2000) Organization and regeneration ability of spontaneous supernumerary eyes in planarians: eye regeneration field and pathway selection by optic nerves. Zool Sci 17(3):375–381. doi:10.2108/jsz.17.375

Sanchez Alvarado A (2007) Stem cells and the planarian *Schmidtea mediterranea*. C R Biol 330(6-7):498–503. doi:10.1016/j.crvi.2007.05.005

Sanchez Alvarado A, Newmark PA (1999) Double-stranded RNA specifically disrupts gene expression during planarian regeneration. Proc Natl Acad Sci USA 96(9):5049–5054

Sarnat HB, Netsky MG (1985) The brain of the planarian as the ancestor of the human brain. Can J Neurol Sci 12(4):296–302

Scimone ML, Srivastava M, Bell GW, Reddien PW (2011) A regulatory program for excretory system regeneration in planarians. Development 138(20):4387–4398. doi:10.1242/dev.068098

Scimone ML, Kravarik KM, Lapan SW, Reddien PW (2014) Neoblast specialization in regeneration of the planarian *Schmidtea mediterranea*. Stem Cell Rep 3(2):339–352. doi:10.1016/j.stemcr.2014.06.001

Scimone ML, Cote LE, Rogers T, Reddien PW (2016) Two FGFRL-Wnt circuits organize the planarian anteroposterior axis. Elife 5:e12845. doi:10.7554/eLife.12845

Shibata N, Umesono Y, Orii H, Sakurai T, Watanabe K, Agata K (1999) Expression of vasa(vas)-related genes in germline cells and totipotent somatic stem cells of planarians. Dev Biol 206(1):73–87. doi:10.1006/dbio.1998.9130

Shibata N, Rouhana L, Agata K (2010) Cellular and molecular dissection of pluripotent adult somatic stem cells in planarians. Dev Growth Differ 52(1):27–41. doi:10.1111/j.1440-169X.2009.01155.x

Shibata N, Hayashi T, Fukumura R, Fujii J, Kudome-Takamatsu T, Nishimura O, Sano S, Son F, Suzuki N, Araki R, Abe M, Agata K (2012) Comprehensive gene expression analyses in pluripotent stem cells of a planarian, *Dugesia japonica*. Int J Dev Biol 56(1-3):93–102. doi:10.1387/ijdb.113434ns

Shibata N, Kashima M, Ishiko T, Nishimura O, Rouhana L, Misaki K, Yonemura S, Saito K, Siomi H, Siomi MC, Agata K (2016) Inheritance of a nuclear PIWI from pluripotent stem cells by somatic descendants ensures differentiation by silencing transposons in planarian. Dev Cell 37(3):226–237. doi:10.1016/j.devcel.2016.04.009

Sikes JM, Newmark PA (2013) Restoration of anterior regeneration in a planarian with limited regenerative ability. Nature 500(7460):77–80. doi:10.1038/nature12403

Sossin WS, Lacaille JC (2010) Mechanisms of translational regulation in synaptic plasticity. Curr Opin Neurobiol 20(4):450–456. doi:10.1016/j.conb.2010.03.011

Steinberg F, Zhuang L, Beyeler M, Kalin RE, Mullis PE, Brandli AW, Trueb B (2010) The FGFRL1 receptor is shed from cell membranes, binds fibroblast growth factors (FGFs), and antagonizes FGF signaling in *Xenopus* embryos. J Biol Chem 285(3):2193–2202. doi:10.1074/jbc.M109.058248

Takano T, Pulvers JN, Inoue T, Tarui H, Sakamoto H, Agata K, Umesono Y (2007) Regeneration-dependent conditional gene knockdown (Readyknock) in planarian: demonstration of requirement for Djsnap-25 expression in the brain for negative phototactic behavior. Dev Growth Differ 49(5):383–394. doi:10.1111/j.1440-169X.2007.00936.x

Talbot JA, Currie KW, Pearson BJ, Collins EM (2014) Smed-dynA-1 is a planarian nervous system specific dynamin 1 homolog required for normal locomotion. Biol Open 3(7):627–634. doi:10.1242/bio.20147583

Taliaferro WH (1920) Reactions to light in *Planaria maculata*, with special reference to the function and structure of the eyes. J Exp Zool 31(1):59–116. doi:10.1002/jez.1400310103

Tasaki J, Shibata N, Nishimura O, Itomi K, Tabata Y, Son F, Suzuki N, Araki R, Abe M, Agata K, Umesono Y (2011a) ERK signaling controls blastema cell differentiation during planarian regeneration. Development 138(12):2417–2427. doi:10.1242/dev.060764

Tasaki J, Shibata N, Sakurai T, Agata K, Umesono Y (2011b) Role of c-Jun N-terminal kinase activation in blastema formation during planarian regeneration. Dev Growth Differ 53(3):389–400. doi:10.1111/j.1440-169X.2011.01254.x

Tazaki A, Gaudieri S, Ikeo K, Gojobori T, Watanabe K, Agata K (1999) Neural network in planarian revealed by an antibody against planarian synaptotagmin homologue. Biochem Biophys Res Commun 260(2):426–432. doi:10.1006/bbrc.1999.0933

Thompson R, McConnell JV (1955) Classical conditioning in the planarian, *Dugesia dorotocephala*. J Comp Physiol Psychol 48(1):65–68

Trueb B (2011) Biology of FGFRL1, the fifth fibroblast growth factor receptor. Cell Mol Life Sci 68(6):951–964. doi:10.1007/s00018-010-0576-3

Umesono Y, Agata K (2009) Evolution and regeneration of the planarian central nervous system. Dev Growth Differ 51(3):185–195. doi:10.1111/j.1440-169X.2009.01099.x

Umesono Y, Watanabe K, Agata K (1997) A planarian orthopedia homolog is specifically expressed in the branch region of both the mature and regenerating brain. Dev Growth Differ 39(6):723–727

Umesono Y, Watanabe K, Agata K (1999) Distinct structural domains in the planarian brain defined by the expression of evolutionarily conserved homeobox genes. Dev Genes Evol 209(1):31–39. doi:10.1007/s004270050224

Umesono Y, Tasaki J, Nishimura K, Inoue T, Agata K (2011) Regeneration in an evolutionarily primitive brain: the planarian *Dugesia japonica* model. Eur J Neurosci 34(6):863–869. doi:10.1111/j.1460-9568.2011.07819.x

Umesono Y, Tasaki J, Nishimura Y, Hrouda M, Kawaguchi E, Yazawa S, Nishimura O, Hosoda K, Inoue T, Agata K (2013) The molecular logic for planarian regeneration along the anterior-posterior axis. Nature 500(7460):73–76. doi:10.1038/nature12359

van Wolfswinkel JC, Wagner DE, Reddien PW (2014) Single-cell analysis reveals functionally distinct classes within the planarian stem cell compartment. Cell Stem Cell 15(3):326–339. doi:10.1016/j.stem.2014.06.007

Wagner DE, Ho JJ, Reddien PW (2012) Genetic regulators of a pluripotent adult stem cell system in planarians identified by RNAi and clonal analysis. Cell Stem Cell 10(3):299–311. doi:10.1016/j.stem.2012.01.016

Walter HE (1907) The reactions of planarians to light. J Exp Zool 5(1-3):35–162. doi:10.1002/Jez.1400050104

Wenemoser D, Lapan SW, Wilkinson AW, Bell GW, Reddien PW (2012) A molecular wound response program associated with regeneration initiation in planarians. Genes Dev 26(9):988–1002. doi:10.1101/gad.187377.112

Yamamoto H, Agata K (2011) Optic chiasm formation in planarian I: cooperative netrin- and robo-mediated signals are required for the early stage of optic chiasm formation. Dev Growth Differ 53(3):300–311. doi:10.1111/j.1440-169X.2010.01234.x

Yazawa S, Umesono Y, Hayashi T, Tarui H, Agata K (2009) Planarian Hedgehog/Patched establishes anterior-posterior polarity by regulating Wnt signaling. Proc Natl Acad Sci USA 106(52):22329–22334. doi:10.1073/pnas.0907464106

Yoshida-Kashikawa M, Shibata N, Takechi K, Agata K (2007) DjCBC-1, a conserved DEAD box RNA helicase of the RCK/p54/Me31B family, is a component of RNA–protein complexes in planarian stem cells and neurons. Dev Dyn 236(12):3436–3450. doi:10.1002/dvdy.21375

Zayas RM, Cebria F, Guo T, Feng J, Newmark PA (2010) The use of lectins as markers for differentiated secretory cells in planarians. Dev Dyn 239(11):2888–2897. doi:10.1002/dvdy.22427

Chapter 5
The Computation and Robustness of the Mini-Cognitive Centers of Terrestrial Mollusks: An Exquisite Outcome of Brain Evolution

Ryota Matsuo

Abstract Terrestrial mollusks (slugs and snails) have brains with highly evolved and intriguing characteristics. These characteristics are comparable, or far superior, to those of the mammalian brain. The first ability is the capacity to acquire sophisticated olfactory memories, which probably allows them to survive in their terrestrial environments. A specialized structure in the brain, called the procerebrum, functions as a higher olfactory center, and enables the slugs and snails to accomplish complex olfactory tasks. The second characteristic is the presence of many polyploid neurons in the brain, although this is not a specific feature of the terrestrial slugs or snails but prevails widely in gastropod animal species. Polyploidy is observed primarily in motor and endocrine neurons innervating peripheral organs that continue to grow in the adult. Polyploidization seems to occur to meet the increasing demand for macromolecules in these neurons to innervate and control the growing peripheral organs. The third characteristic is an ability to regenerate an injured part of the central nervous system spontaneously. One clear example is the spontaneous regeneration of an amputated tentacle: the regenerated tentacle is equipped with all the elements of the original one. The procerebrum also exhibits structural and functional regeneration following injury, which is accomplished by enhancing ongoing neurogenesis in the adult.

Keywords *Limax* • Procerebrum • Olfactory learning • Endoreplication • Polyploidy • Neurogenesis • Regeneration

---

R. Matsuo (✉)
Fokuoka Women's University, Fukuoka, Japan
e-mail: matsuor@fwu.ac.jp

S. Shigeno et al. (eds.), *Brain Evolution by Design*, Diversity and Commonality in Animals, DOI 10.1007/978-4-431-56469-0_5

5.1 Introduction

The brain of the gastropod mollusks has long served as a model system for the study of synaptic plasticity, learning/memory, and nerve regeneration. One of the advantages in the use of gastropods is the feasibility of investigating neuronal properties in in vitro preparations. Another advantage is the large size of the cell bodies of the neurons, which allows various artificial manipulations for studying the functions of genes and proteins. Many researchers, therefore, have been engaged in the use of the sea hare *Aplysia*, which has as few as several tens of thousands of large neurons in its central nervous system (CNS) (Bailey et al. 1996).

Beside the simplicity of the brains of the marine gastropods, terrestrial pulmonate gastropod species, such as the slug *Limax* and the snails *Helix* and *Achatina*, have a very sophisticated brain, composed of more than hundreds of thousands of neurons. These animals have successfully adapted their way of life to the terrestrial environment, possibly in part because of their higher intellectual capacity as well as their well-developed olfaction. Their brains also show neuronal DNA polyploidy and exquisite regenerative capability, both of which have never been realized in the human brain.

In this chapter, I first review what has been demonstrated during the study of olfactory function and learning capability in *Limax* and *Helix* (Sect. 5.2), followed by an overview of the structure and function of their brains, focusing on the tentacles and higher olfactory center, the procerebrum (Sect. 5.3). DNA polyploidization, an unusual property intrinsic to the gastropod brain, is then introduced (Sect. 5.4). Finally, I describe and discuss the regenerative ability of the tentacles and the brain following physical injury (Sect. 5.5).

5.2 Olfactory Learning in *Limax*

5.2.1 Olfactory Associative Learning

Olfaction is an important sensory modality for the terrestrial slugs and snails. They rely primarily on olfaction to sense their surroundings in their daily lives. In 1975, Gelperin reported that the terrestrial slug *Limax* can acquire odor-aversion memory if it is exposed to CO_2 gas immediately following the intake of some specific food (e.g., mushrooms) (Gelperin 1975). The slugs avoided this food thereafter when they detected its odor, whereas they did not avoid the food odor that was not paired with the aversive stimulus (CO_2). Therefore, the slugs distinguish the learned odor from the other and form an odor-specific aversive memory. Following this report, various modified protocols have been developed for the behavioral study of odor-aversion learning in *Limax* and *Helix* (Sahley et al. 1981a; Maximova and Balaban 1984; Nakaya et al. 2001). In some of these, a single paired presentation of the CS (odor) and US (aversive stimulus) was sufficient to establish a long-term memory (i.e., a protein synthesis-dependent memory phase) (Matsuo et al. 2002). In most of the studies, the formation of odor-aversion memory was assessed by an increase in the latency until reaching the odor source.

Slugs can also acquire olfactory appetitive association learning by receiving a paired presentation of some odor and an attractive stimulus, or by simply letting the slugs eat some novel food. As a result of the appetitive learning, the behavior of the slugs was changed in the face of the conditioned odor. The memory formation was judged by an increased preference for the conditioned odor over the other, as assessed by the relative increase of time spent over the odor source (Sahley et al. 1990), the reduced latency until reaching the odor source (Yabumoto et al. 2008), or an increase in the probability of exhibiting eating behavior (Sekiguchi et al. 2010).

It has been reported that the slugs can acquire more difficult learning tasks. They understand the logical relationship between stimuli presented sequentially (Sahley et al. 1981b). In the second-order conditioning, a potato odor (CS2) is presented in combination with a carrot odor (CS1) that was aversively conditioned in advance by paired presentation with a bitter taste of quinidine sulfate (US). As a result, the slugs come to avoid not only the carrot odor but also the potato odor, suggesting that they could relate CS2 to the US through CS1. In "blocking", the slugs were conditioned with the carrot odor (CS1) by paired presentation with quinidine taste in phase 1. Then, the carrot odor and potato odor (CS2) were presented with quinidine taste in phase 2. However, this does not lead the slugs to avoid the potato odor, indicating that the CS1 is sufficient to predict the occurrence of the US and this blocked the further conditioning to the CS2. Such a blocking phenomenon was initially reported in rats by Kamin (1969). The slugs, therefore, can perform logically complex tasks sufficiently well and are as smart as mammals.

5.2.2 *Memory Consolidation and Reconsolidation*

The olfactory memory of the slugs is consolidated in a protein synthesis-dependent manner, as in most other animals. If a protein synthesis inhibitor anisomycin is systemically injected into the body cavity 30 min before the conditioning, memory retention does not persist for more than 2 days. At 24 h after the conditioning, however, they still retain and retrieve the aversive memory (Matsuo et al. 2002). When the extent of the protein synthesis inhibition is low, the memory is lost at an even later time point (3 or more days after the conditioning (Matsuo et al. 2002; Yasui et al. 2004; Suenaga and Matsuo 2016). Based on these observations, the long-term memory can be considered the phase later than 2 days after memory formation in slug; this is in clear contrast to the results obtained in other animal species wherein the protein synthesis inhibitors impair memory retention at 24 h after conditioning (Flood et al. 1972; Bull et al. 1976; Montarolo et al. 1986; Tully et al. 1994; Pedreira et al. 1996). There is currently no explanation for this peculiarity of the slug. Future investigations should focus on how the slugs retain the memory for more than 24 h without or with only a low amount of newly synthesized protein in the brain following the conditioning.

In addition to memory consolidation, memory reconsolidation has received the attention of neuroscientists, especially since the discovery of the "reconsolidation failure" by protein synthesis inhibitors administered into the amygdala immediately following the memory retrieval in rats (Nader et al. 2000). Nader et al. (2000) demonstrated that the protein synthesis inhibition in the amygdala immediately following the retrieval of the tone-dependent fear memory abolishes that memory even if it already has been fully consolidated, whereas Milekic and Alberini (2002) have reported that the effectiveness of this manipulation gradually declines in rats as time passes after the initial conditioning in inhibitory avoidance tasks. There has been extensive discussion of this discrepancy ever since these reports were published (Dudai and Eisenberg 2004; Lee 2009; Nader and Einarsson 2010).

On the other hand, the phenomenon of memory reconsolidation failure itself had been reported long before the discovery by Nader et al. (2000). In these classical studies, memory reconsolidation failure was caused by memory retrieval in combination with an electroconvulsive shock (Misanin et al. 1968), rapid cooling (Mactutus et al. 1979), or a systemic injection of anisomycin (Judge and Quartermain 1982), instead of infusion of a protein synthesis inhibitor into the specific brain area (Nader et al. 2000; Debiec et al. 2002). In the terrestrial slug *Limax*, rapid cooling immediately after the retrieval of an odor-aversion memory abolishes the memory; this effect is temporally graded according to the increase in the interval between the initial conditioning and the memory retrieval/cooling (Yamada et al. 1992; Sekiguchi et al. 1997), similar to a report by Milekic and Alberini (2002). These facts indicate that slugs and mammals share common neural underpinnings in memory reconsolidation. Although the cooling (hypothermia) and protein synthesis inhibition do not seem to target the identical molecular steps leading to the memory stabilization (Fulton et al. 2008; see also the discussion in Matsuo 2015), the temporal change of memory states will be more thoroughly investigated using *Limax* because of the greater ease of handling and feasibility of using large sample sizes amenable to statistical analysis.

5.3 Structure and Function of the Cognitive Center of *Limax*

5.3.1 Structure of the Brain

The brain of terrestrial slugs and snails is located in the head and circumscribes the esophagus (therefore, the brain is sometimes referred to as "circumesophageal ganglia"), a feature common to all gastropod mollusks. In comparison with other molluscan animals, however, the brains of the terrestrial slugs and snails contain far larger numbers of neurons within the brain (>10^5 neurons). This difference is mostly ascribable to the presence of the olfactory center, called the procerebrum (PC). Because terrestrial slugs and snails depend on olfaction to monitor their surrounding environment, a large number of neurons are devoted to the processing of olfactory information. In fact, approximately half the total neurons in the brain are found in the PC (Chase 2000).

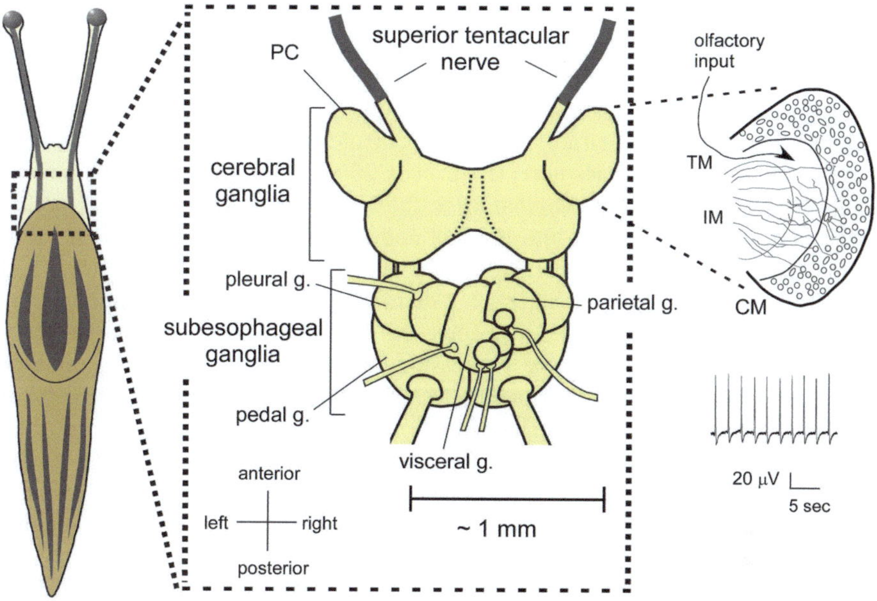

Fig. 5.1 Structure of the brain of the terrestrial slug *Limax. Right panel* is a cartoon of the procerebrum (PC); below it is a typical local field potential (LFP) oscillation recorded on the surface of the PC. Olfactory input comes into the TM layer of the PC. The *small circles* in the CM are the nonbursting (NB) neurons and the *ovals* are the bursting (B) neurons. For simplicity, an inferior tentacular nerve is not drawn. *TM* terminal mass, *CM* cell mass, *IM* internal mass

The PC, which is located on both lateral sides of the cerebral ganglia (Fig. 5.1), is ontogenetically different from the rest of the cerebral ganglion in that it derives from a distinct ectodermal invagination during embryonic development (Chase 2000). The PC consists of three structurally distinct layers: the cell mass (CM), terminal mass (TM), and internal mass (IM) layers. In the CM layer, a large number of cell bodies of small interneurons are found in dense arrangements. The TM layer and IM layer are under the cell mass layer and consist of nerve fibers derived from the small neurons in the CM layer. This layered structure is reminiscent of the cortical brain structure of vertebrates. In the TM layer, the interneurons of the PC receive olfactory inputs from the tentacles. Although the structural features have been investigated to a lesser extent for the IM layer, this layer seems to be involved in the output to the other parts of the brain as well as participating in reciprocal interactions among the PC neurons (Ratté and Chase 1997, 2000).

Although the PC neurons are seemingly a homogeneous neuronal population, there are known to be at least two distinct types of neurons that can be distinguished based on their electrophysiological properties. The majority are nonbursting (NB) neurons, and the minority are termed bursting (B) neurons. The B neurons comprise less than 10 % of the total number of PC neurons and exhibit periodic, spontaneous bursting activity. On the other hand, the NB neurons (>90 %) do not exhibit periodic bursts, but instead show infrequent firings, possibly caused by inputs from the olfactory nerves (Kleinfeld et al. 1994; Watanabe et al. 1998). The B neurons release glutamate periodically (0.7 Hz) and suppress the activity of the NB neurons around them (Watanabe et al. 1999; Matsuo et al. 2009). Although not yet conclusively demonstrated, the NB neurons seem to generate nitric oxide and also to use acetylcholine as a neurotransmitter (Fujie et al. 2005; Matsuo and Ito 2009; Matsuo et al. 2014a). At the morphological level, the NB neurons have small cell bodies and extend their protrusions into the TM and IM layers, whereas the B neurons have relatively larger cell bodies and project laterally within the CM layer (Fig. 5.1) (Watanabe et al. 1998). However, such a dichotomy may be an oversimplification. At the molecular level, the PC is composed of a more heterogeneous neuronal population, as indicated by the expression of nicotinic acetylcholine receptor subtypes (Matsuo et al. 2014a) or Phe-Met-Arg-Phe(FMRF)amide (Kobayashi et al. 2010). Moreover, the PC neurons exhibit various projection patterns to the other areas of the brain or to the other PC neurons (Chase and Tolloczko 1989; Ratté and Chase 1997). The complexity of the PC may be comparable to that of the cerebral cortex of the mammalian brain and seems to be sufficient to serve a mnemonic function, as described in the previous section.

The ventroposterior part of the circumesophageal ganglia (= brain) is called the subesophageal ganglia. In *Limax*, the subesophageal ganglia is composed of three pairs of ganglia (pleural, parietal, and pedal ganglia) and one visceral ganglion (Fig. 5.1). In other gastropod animal species (e.g., *Aplysia*, *Lymnaea*), the ganglia are separated spatially from each other in the body, and they are connected with bundles of nerves. Contrastingly, in *Limax* and *Helix* all the ganglia, including the cerebral ganglia (except for the buccal ganglia located on the surface of the buccal mass), assemble into a spatially more compact structure and fuse into one wherein the boundary between the constituent ganglia is blurred. Such a "cephalized" brain structure is one of the characteristics of the terrestrial pulmonates. In subesophageal ganglia, there are many large neurons. Most of them seem to be sensory, motor, or endocrine neurons, in clear contrast to the cerebral ganglia where a substantial part of the constituent neurons (i.e., PC neurons) are interneurons devoted to olfactory computations. The neurons with large cell bodies in the subesophageal ganglia contain polyploid genomic DNA within their nuclei, as described later in Section 5.4.

5.3.2 Structure and Function of the Tentacles

The slugs and snails (Stylommatophora) have two pairs of tentacles, that is, superior and inferior tentacles (Fig. 5.2a); these are also called the posterior and anterior tentacles, respectively. Both tentacles serve as olfactory and mechanosensory organs. The superior tentacle is different from the inferior tentacle in that it is equipped with an eye. Therefore, the superior tentacle also functions as a photosensory organ.

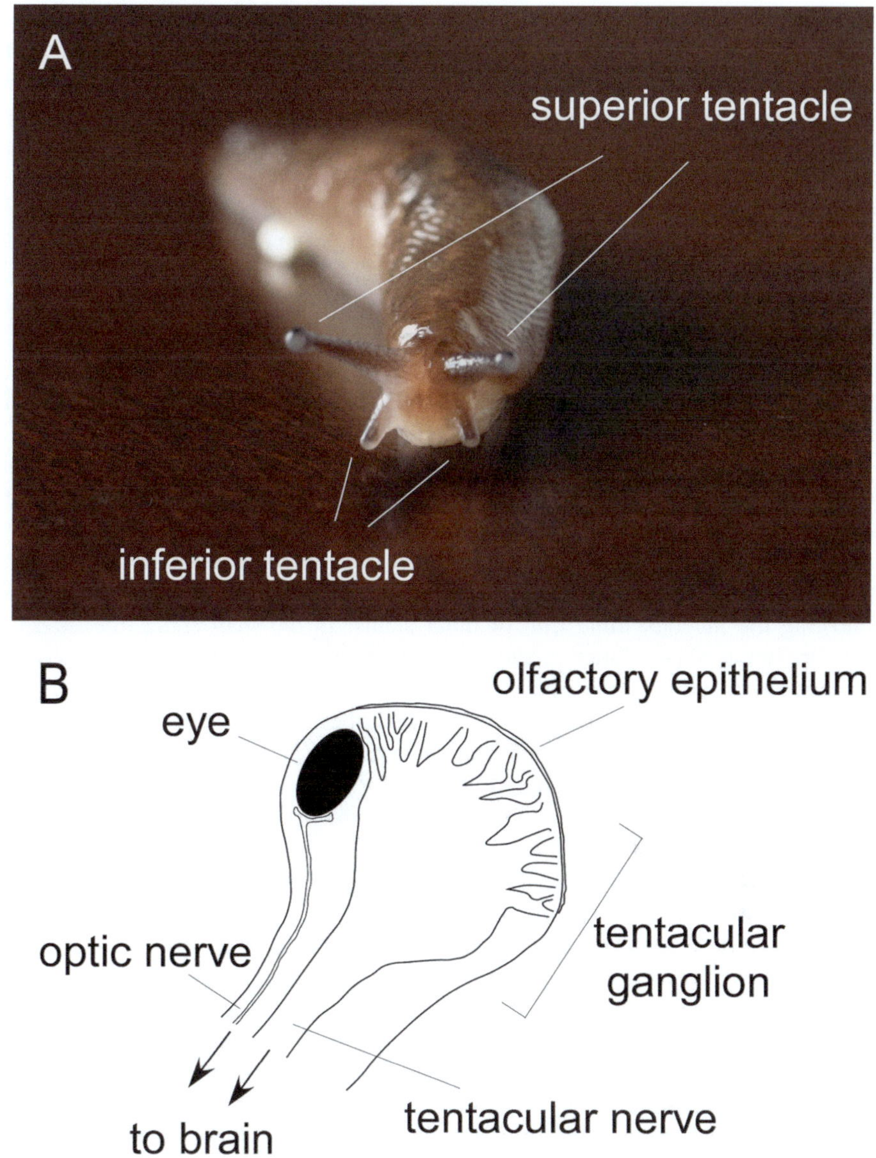

Fig. 5.2 Tentacles of stylommatophoran pulmonates. (**a**) A terrestrial slug, *Limax valentianus*, showing the two pairs of tentacles. (**b**) The structure of a superior tentacle

Odor molecules bind to as yet unidentified olfactory receptors expressed in the sensory neurons of the olfactory epithelium, which is located at the tip of the tentacles. Some of the olfactory sensory neurons in the epithelium send afferent projections directly to the PC, whereas others make synapses once or twice at the tentacular ganglion (the primary olfactory center located beneath the olfactory epithelium) before reaching the PC (Chase and Tolloczko 1993). The tentacular ganglion shares several common features with the PC, in that both contain multiple small globular neurons and exhibit field potential oscillation that is modified by olfactory inputs (Ito et al. 2001; Inokuma et al. 2002, and see Sect. 5.3.3). The structural features of the tentacular ganglia are very similar between those in the superior and inferior tentacles, but several differences have been reported (see Ito et al. 2000; Matsuo et al. 2011a for details). Not all the afferents in the tentacular nerve bundle (in the superior tentacle) project to the TM layers of the PC; some bypass the PC and go into the other parts of the cerebral ganglion, although what kinds of modalities (olfactory or mechanosensory) are conveyed by these nerves is not clear (Makinae et al. 2008; Matsuo et al. 2014a). The visual information from an eye, on the other hand, is conveyed through an optic nerve, a nerve bundle with a small diameter running in parallel with the thicker tentacular nerve bundle (Fig. 5.2b).

Besides the detection of light by an eye, are there any functional differences between the superior and inferior tentacles? Chase and Croll (1981) reported that in the snail *Achatina* the superior tentacles are necessary for anemotaxis (upwind movement in the air containing odor molecules) and also for orienting to a distant odor source in still air. On the other hand, the inferior tentacles are involved in following the mucous slime trail of other snails. In the snail *Helix*, the inferior and superior tentacles function during acquisition and retrieval, respectively, of appetitive food attraction memory (Friedrich and Teyke 1998). However, in the odor-aversion learning in *Limax* previously described, no functional difference was observed between the superior and inferior tentacles (Yamagishi et al. 2008). The memory acquired in the presence of both pairs of tentacles can be successfully retrieved with only one pair of tentacles. Furthermore, the slugs can acquire and retrieve the memory using either pair of tentacles. The two pairs of tentacles are therefore redundant and functionally equivalent with respect to odor-aversion learning (Yamagishi et al. 2008). It has not, however, been investigated yet whether the slugs can retrieve the odor-aversion memory with a pair of tentacles that were not used during the memory acquisition session. The answer awaits future studies.

Another interesting feature of the tentacular function is that the ascending projection to the cerebral ganglion is essentially ipsilateral: this has been demonstrated not only at the histological level (Kawahara et al. 1997; Kimura et al. 1998b) but also at the functional level. An appetitive food attraction memory acquired using unilateral tentacles cannot be retrieved with the tentacles on the other side of the head in *Helix* (Friedrich and Teyke 1998). The PC and the tentacle on the same side must be intact for odor-aversion learning to be formed in *Limax* (Matsuo et al. 2010a). There are supposed to be some physiological cross-interactions between the left and the right PCs or the cerebral ganglia through the cerebral commissure (Teyke et al. 2000; Matsuo et al. 2010a). The ipsilateral ascending projection and some bilateral cross interactions are also proposed in the optical information flow into the cerebral ganglion through an optic nerve, which is running in parallel along an olfactory nerve bundle within the superior tentacle (Matsuo et al. 2014b).

5.3.3 *Local Field Potential Oscillation and the Role of the Procerebrum in Olfactory Learning*

In 1990, Gelperin and Tank reported the rhythmic (0.7 Hz) activity of the local field potential (LFP) recorded on the surface of the PC (Gelperin and Tank 1990). This cortical electroencephalogram-like activity is produced by the synchronous activity of the constituent PC neurons. The minority B neurons exhibit periodic bursts, and thereby suppress the majority NB neurons around them (Kleinfeld et al. 1994; Watanabe et al. 1998). The LFP oscillation, therefore, reflects the synchronous outward currents generated in the local population of NB neurons. When the production of nitric oxide (probably generated by NB neurons) is pharmacologically suppressed, the LFP oscillation diminishes by a not yet fully understood mechanism (Gelperin 1994). At the same time, this manipulation abolishes the ability of the slug to discriminate similar odors (Teyke and Gelperin 1999; Sakura et al. 2004).

Another interesting feature is that the LFP wave propagates from the apical to the basal part of the PC. The phase difference between the apical and the basal regions of the PC is modulated by an application of odors to the olfactory epithelium on the tentacle (Delaney et al. 1994). The oscillatory frequency of LFP is also subject to change during odor detection (Gervais et al. 1996). Especially, an aversively or appetitively learned odor evokes the change in the oscillatory frequency of the LFP (Kimura et al. 1998c; Inoue et al. 2006; Samarova and Balaban 2006). Thus far several neuromodulatory transmitters have been identified that directly change the frequency of LFP oscillation by exogenous application: Dopamine, serotonin, glutamate, acetylcholine, and histamine increase the frequency (Gelperin et al. 1993; Watanabe et al. 2001; Inoue et al. 2001; Matsuo et al. 2009, 2014a, 2016a), whereas FMRFamide and octopamine decrease it (Kobayashi et al. 2010; Matsuo et al. 2016b). In fact, these transmitters are present endogenously within the PC (glutamate, acetylcholine, FMRFamide) or are used as neurotransmitters by projection nerves coming from outside the PC (dopamine, serotonin, histamine, octopamine) or from the tentacular nerves (acetylcholine, FMRFamide).

It is still not fully understood how the LFP oscillation and wave propagation contribute to olfactory information processing, but it has been proposed that these activities might function in assigning the space on the seemingly uniform cluster of multiple small PC neurons for the representation of the specific odors in a timing-dependent manner (Gelperin 1999). This suggestion means that the efficacy of synaptic transmission of olfactory input to NB neurons is dependent on the phase of ongoing LFP oscillation (Inoue et al. 2000) because the LFP wave reflects the coherent inhibitory currents generated in the local population of NB neurons. This condition is in clear contrast to the olfactory information processing system of insects' antennal lobe and the mammalian olfactory bulb, where the specific odor information is represented in built-in structural compartments (i.e., glomeruli).

The idea of the involvement of the PC in olfactory leaning has also been supported by the fact that the PC is labeled with a fluorescent molecule, Lucifer Yellow, injected into the body cavity just following the odor-aversion conditioning in *Limax* (Kimura et al. 1998b; Ermentrout et al. 2001). The labeling occurs specifically only when Lucifer Yellow was injected following CS–US pairing, and does not occur when the CS and US are delivered separately in time (i.e., no memory is formed). Although the physiological meaning of such labeling is not clear, this observation implies an enhanced neuronal activity in a spatially restricted area of the PC during the memory formation because Lucifer Yellow has been used as a neuronal activity marker in other experimental systems (Wilcox and Franceschini 1984). Interestingly, either one of the bilateral PCs is labeled following conditioning, but the two PCs are never labeled bilaterally.

Sekiguchi et al. (2010) recently found that the position of Lucifer Yellow labeling along the apical–basal axis depends on the valence of the odor (i.e., whether it is appetitively or aversively conditioned). They introduced a "van der Pol oscillator model" to explain this observation. In their model, multiple mutually interacting oscillators were assumed for respective CM and TM layers along the apical-to-basal axis of the PC, and only the TM layer was supposed to receive olfactory input. Lucifer Yellow were proposed to be incorporated only in the position where the phase difference between the CM and TM layers disappeared (i.e., synchronized). Their model could also replicate various other behaviors of the LFP oscillation in the PC, such as an odor input-dependent change in the oscillatory frequency (Gervais et al. 1996; Kimura et al. 1998c; Inoue et al. 2006; Samarova and Balaban 2006). We should, however, interpret their model cautiously because it was based on assumptions, some of which have not yet been experimentally demonstrated.

The importance of the PC in olfactory learning has been demonstrated more directly in a lesion experiment (Kasai et al. 2006). If the PC is surgically destroyed, the slugs cannot acquire an odor-aversion memory. Post-learning lesioning of the PC also abolished the retention or retrieval of the memory. The PC lesioning, however, did not result in the complete loss of the odor-sensing ability. These results strongly suggest the involvement of the PC in mnemonic function. Further studies presented data supporting the idea that either the right or left PC is used at any one time during odor learning. Unilateral PC destruction resulted in a mild learning deficit, and the learning score was intermediate between the slugs with bilateral PC destruction and those with a sham operation (Matsuo et al. 2010a). This result is consistent with the prediction based on the idea of unilateral use of the PC and agrees well with the observation of unilateral PC labeling with Lucifer Yellow injected immediately after conditioning (Kimura et al. 1998b; Ermentrout et al. 2001, see earlier).

It has not yet been understood what kind of neuronal plasticity underlies the mnemonic role of the PC. Recently, however, Sakura and Watanabe (2015) successfully induced an LTP-like change in the synaptic strength between NB and B neurons with high-frequency stimulation to the tentacular nerve (which conveys olfactory information to the PC). The importance of such synaptic plasticity is still elusive, and further investigation is awaited.

It is highly probable that the PC is the convergent site of the CS (odor information) and the US (aversive stimulus such as a bitter taste) during olfactory aversive conditioning. However, it is not unclear how the US information is transmitted to the PC, or which neurotransmitter carries the US information. Serotonin is one candidate, as in the case of the classical conditioning of gill withdrawal in *Aplysia* (Bailey et al. 1996), because serotonergic innervation into the PC has been reported from the cells located in the other part of the cerebral ganglion (Inoue et al. 2004; Elekes et al. 2012), and the depletion of serotonin impairs odor-aversion learning in *Limax* (Shirahata et al. 2006). Of course, there is no denying the possibility that other neurotransmitters, such as dopamine, histamine, or FMRFamide, are responsible for carrying the US information to the PC, taking into account that the nerves containing these neurotransmitters have projections into the PC (Makino and Yano 2010; Kobayashi et al. 2010; Matsuo et al. 2016a). Moreover, it is not known what is the primary neurotransmitter that carries olfactory information into the TM layer of the PC, although at least FMRFamidergic fibers run within the tentacular nerve into the TM layer of the PC to some extent (Matsuo et al. 2010b; Kobayashi et al. 2010).

From the aspect of brain evolution, it is very intriguing to discuss the origin of the PC. When compared to the mammalian brain, the PC resembles the olfactory bulb in that neurogenesis ensues in the adult in both structures. Furthermore, both structures exhibit changes in the oscillatory activity of the field potential in response to odor detection (Kimura et al. 1998a; Chabaud et al. 2000; Chase 2000; Ravel et al. 2003; Inoue et al. 2006). At the level of the connectivity, however, the PC seems to correspond to the olfactory cortices (e.g., piriform cortex) because it is the secondary olfactory center (but also the PC receives afferent projections as the primary and tertiary center to some extent; see Chase and Tolloczko 1993). Recently, it was shown in the mouse that the mammalian piriform cortex is involved in the learned response to odor (Choi et al. 2011). The olfactory bulb–piriform cortex–amygdala pathway is, therefore, responsible for the aversion to learned odors, whereas the more direct olfactory bulb–cortical amygdala pathway is involved in detecting innately aversive odors (Root et al. 2014); this is reminiscent of the fact that slugs can avoid innately aversive odorants (such as garlic) without a PC (Kasai et al. 2006).

5.4 Neuronal Polyploidization in the Brain of *Limax*

5.4.1 Body Growth-Dependent DNA Endoreplication in Brain Neurons

The brain of gastropods contains numerous giant neurons. In some cases, the diameter of the cell body is more than 1 mm (Moroz and Kohn 2010). The amount of genomic DNA in such neurons far exceeds that of somatic diploid cells. It has been estimated that the giant motor neurons of *Aplysia*, such as the R2 motor neuron, contain 200,000 times the amount of the haploid genome (Lasek and Dower 1971). Such amplified genomic DNA is generated through repeated DNA replication without cell division, that is, DNA endoreplication. DNA endoreplication is believed to occur to enhance the ability to synthesize macromolecules (proteins and the products of enzymatic reactions) in neurons, to meet the demand during the body growth of an animal. This belief would especially hold in the case of motor neurons because they directly innervate the tissues and organs that grow very large in gastropod animals (Gillette 1991).

To quantitatively examine the relationship between body growth and the neuronal DNA endoreplication, Yamagishi et al. (2011) studied the adult *Limax*. They divided the slugs into three groups. In the first group, the slugs were fed ad libitum for 44 days and body growth was enhanced (growth-promoted group). In the second group, the amount of the food was regulated so that the slugs grew at a normal rate (control group). In the third group, the slugs were completely starved for the same period (growth-suppressed group). As a result, the body weight of the growth-promoted slugs became nearly ten times that of the growth-suppressed slugs, and the body weight of the control group was intermediate. Brain volumes also differed; and there was a positive correlation between body weight and brain volume. The enlargement was especially prominent in the subesophageal ganglia where there are many motor and endocrine neurons. At the level of the neuron, cell body size was also enlarged, as shown by the increase in the volume of the cell body of the visceral giant cell (VGC), which is the largest peptidergic motor neuron located in the visceral ganglion of *Limax* (Matsuo et al. 2011b). The frequency of DNA endoreplication, which was quantified by counting the number of nuclei incorporating 5-bromo-2′-deoxyuridine (BrdU, a marker molecule for DNA synthesis), also increased in the brains of the growth-promoted slugs. The amount of mRNAs was also elevated concomitantly per a single neuron. These results support the idea proposed by Gillette (1991), in which body growth-dependent DNA endoreplication seems to be an adaptive response of the brain neurons to adjust to the increasing demand for macromolecule synthesis during body growth. The amplification of genomic DNA is an interesting strategy, considering that a change in the rate of transcription and other posttranscriptional regulation is the primary means of regulation in vertebrates for adjustment of the synthesis of biomolecules.

5.4.2 DNA Endoreplication Is Whole-Genome Polyploidization

The incorporation of BrdU in itself does not tell us anything about the mode of the DNA amplification. It is possible that the whole genome is replicated repeatedly (= polyploidy), but another possibility is that one or several necessary loci of the genome are specifically amplified (= polyteny). To distinguish between these two possibilities, the relative copy numbers of various genomic loci were quantified with real-time polymerase chain reaction (PCR) (Yamagishi et al. 2012).

There was no difference in the copy numbers among nine different genomic loci, irrespective of their transcriptional activities. Moreover, BrdU molecules were incorporated uniformly into the whole genomic regions of the endoreplicated DNA when analyzed in the magnified image of the large nuclei (Matsuo et al. 2012; Yamagishi et al. 2012). Therefore, it is highly probable that whole-genomic regions are equally replicated during endoreplication, as suggested by the experiment with Fleugen staining in the terrestrial snail *Succinea* (Anisimov 2005). In contrast, the PC neurons were demonstrated to be diploid, similar to most of the other somatic cells in the body (Yamagishi et al. 2012).

Why do neurons amplify whole genomic regions, most of which are useless for a single neuron? It would be actually difficult to evolve a mechanism of locus-specific genome amplification for each neuron because the sets of transcribed genes differ from neuron to neuron. Although the locus-specific DNA amplification might be an energetically more economical way of enhancing macromolecule synthesis, polyploidization must be an easier strategy because a neuron has only to skip the M-phase of the cell cycle to double the genomic DNA (Edgar and Orr-Weaver 2001). Locus-specific DNA amplification also requires the presence of specific replication origins in the gnome near the genes to be amplified. Furthermore, there would be little need to fine-tune the relative transcriptional activities among genes after the DNA amplification if the neuron proportionally enhances its morphological and functional parameters through polyploidization. Of course, DNA polyploidization and gene-specific transcriptional regulation may be mutually compatible, and can coexist in the same neuron.

5.4.3 Target Innervation and Neuronal Polyploidization

Body growth-dependent neuronal polyploidization implies that the brain neurons somehow sense the need to enhance the rate of macromolecule synthesis to meet the needs of the growing body. What kind of external signal dictates the initiation of DNA endoreplication? Neurons, especially motor and sensory neurons, might receive any retrograde signal from their growing target organs to trigger the initiation of DNA synthesis. Another possibility is that neurons might sense a favorable condition for body growth through the nutritional state of their surrounding body fluid (e.g., continued high glucose or insulin levels).

To distinguish these two possibilities, Matsuo et al. (2013) exploited the "brain transplantation" technique. A brain isolated from one animal is transplanted into the body cavity of another animal in this technique, sometimes used in gastropods (Cheng and Galloway 1970; Roubos 1976; Murphy and Kater 1980; Gomot et al. 1990). In this condition, a recipient slug has two "brains" in its body. One of them is an endogenous host brain innervating normally its whole part of the body, and the other transplant brain is devoid of target innervation. What happens to these two brains if the growth of the host slug is promoted by food supply? Assuming that some retrograde signal is necessary to initiate DNA endoreplication, it is predicted that only the host brain can undergo body growth-dependent DNA endoreplication. In contrast, if the enriched nutrition in the body fluid is sufficient, DNA endoreplication would be enhanced in both brains because they are both under the same body fluid condition of the host. The result was that the frequency of DNA endoreplication (measured as the number of $BrdU^+$ neurons) was elevated only in the endogenous host brain but was unchanged in the transplant brain (Matsuo et al. 2013). Therefore, neuronal DNA endoreplication seem to be triggered by some retrograde signal from the innervating targets, although the identity of the signal is not yet known.

The requirement of target innervation was further supported by a nerve dissection experiment in the endogenous brain. When one of the posterior pedal nerves was surgically cut, the frequency of DNA endoreplication was substantially reduced in the pedal ganglion ipsilateral to the dissected nerve (Matsuo et al. 2013). Taken together, these experiments suggest that neurons in the brain undergo DNA endoreplication by responding to some unidentified signal from the target tissues that are undergoing rapid growth.

5.5 Regenerative Ability of the Central Nervous System of *Limax*

5.5.1 Regeneration of the Tentacles

The regenerative ability of tentacles was reported in slugs and snails a long time ago. If the superior tentacle is surgically removed, all the elements of the tentacle, including the eye, are spontaneously regenerated during the life of the slug or snail (Eakin and Ferlatte 1973; Chase and Kamil 1983; Flores et al. 1983). Tentacle removal in *Limax* reduces the oscillatory frequency of the LFP in the PC, probably because of the loss of the afferent innervation of the tentacular nerves to the PC. As the tentacle regenerates, the innervation of olfactory afferents is also restored, and oscillatory frequency concomitantly returns to the normal frequency within 9 weeks (Matsuo et al. 2010b). The regenerated tentacles are functionally sound and are sufficient to function in an olfactory learning task, such as olfactory aversive learning (Matsuo et al. 2010b; Koga et al. 2016).

Regeneration of the tentacles would be advantageous for the survival of slugs and snails because the tentacle has very important roles in the animal's life as a multimodal sensory organ, including olfaction, mechanosensation, and vision (the last is applicable only to the superior tentacle). Tentacle regeneration is one example of the spontaneous regenerative ability of neuronal structures. However, a more striking manifestation of regenerative ability has been discovered in the brain of *Limax*, as described next.

5.5.2 Neurogenesis and Regeneration of the Procerebrum

In slugs and snails, new neurons are continuously added to the PC even in the adult (Zakharov et al. 1998; Watanabe et al. 2008). Neurogenesis occurs in the apical region of the PC, and the older neurons are slowly pushed forward toward the basal region (Fig. 5.3). Therefore, the age of a neuron is reflected in its position along the apical-to-basal axis of the PC. Such a neurogenic property is reminiscent of the olfactory bulb or hippocampus in mammals. In this context, it is intriguing that the PC is involved in both olfactory information processing and memory formation.

Although it has not been known how adult neurogenesis contributes to the normal function of the PC, it seems to be involved at least in the regeneration of the PC following injury. If the PC is surgically lesioned, the ability to acquire and retrieve olfactory aversion memory is abolished, as described earlier (Kasai et al. 2006). However, the slugs can normally acquire an olfactory aversion memory if a 31-day recovery period is given after the surgery (Matsuo and Ito 2008). Further analysis has revealed that the PC recovers its size spontaneously, and the oscillation of the LFP is also restored. Neurogenic activity seems to be enhanced because the number of BrdU-positive neurons increased in the injured PC when BrdU was injected into the body of the slug 7 days after the surgery (Matsuo et al. 2010c).

Recoverability of the PC also provides insight into the locus of memory storage. A memory acquired 7 days or 14 days before the surgery could not be retrieved even if a 32-day recovery period was given, whereas the recovered PC can serve a mnemonic role at this time because the slugs can acquire and retrieve a new odor-aversion memory 31 days after the surgery, as mentioned. This observation indicates that the surgical lesioning of the PC irreversibly abolished the memory that had probably been stored as a change in synaptic strength between the tentacular nerves and NB neurons, or between PC neurons. The PC is, therefore, the storage site of odor-aversion memory. Of course, it has been suggested that the tentacles are

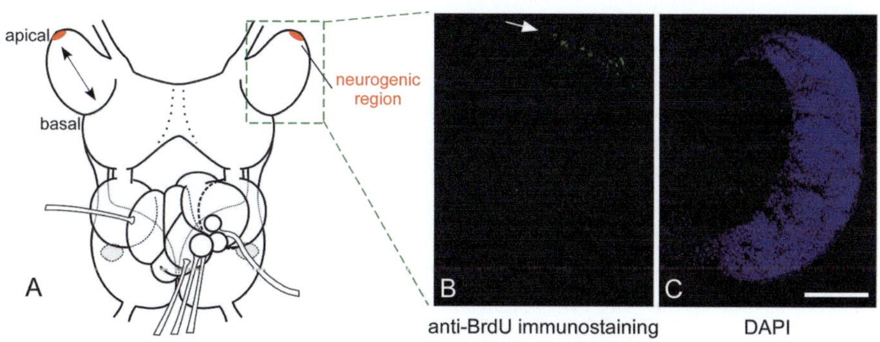

Fig. 5.3 Neurogenesis in the PC of an adult slug. (**a**) Neurogenesis occurs in the apical region of the PC (highlighted in *red*). Newborn neurons push the older neurons to the basal direction. (**b**) Newly generated PC neurons (*arrow*) with nuclei labeled with BrdU that was injected into the body of the slug 24 days earlier. (**c**) A fluorescence image of DAPI of (**b**). *Bar* 100 μm

also involved in memory storage and that the memory engram is distributed as the neuronal circuits encompassing both the tentacular ganglion and the PC (Inoue et al. 2004; Koga et al. 2016).

5.5.3 Evolution of the Regenerative Ability of the Brain

As illustrated here, gastropods are capable of regenerating an injured or lost part of the CNS, even in the adult (Moffett 1995), but the regenerative ability of the adult CNS is not a privilege of gastropods. Some urodele amphibians (newts and salamanders) can regenerate injured spinal cord and retinal ganglion cells in the adult (Butler and Ward 1967; Keefe 1973). The adult teleost fish *Apteronotus* also can regenerate retinal ganglion cells and cerebellum spontaneously (Hitchcock and Raymond 1992; Zupanc 1999; Zupanc and Sîrbulescu 2011). Adult neurogenesis is involved in such regeneration in most of these cases, but the presence of adult neurogenesis itself is not sufficient for the regeneration to occur in the CNS. It is well known that neurogenesis ensues in adult mammals in some parts of the brain, such as the hippocampus and olfactory bulb. The regenerative ability of the mammalian brain, however, is very limited. It would be necessary to enhance the neurogenic activity in response to injury for adult neurogenesis to suffice for functional and structural recovery (discussed in Ferretti 2011; Matsuo and Ito 2011), as in the PC of the slug (see earlier).

In addition to the presence of adult neurogenesis and its enhancement following injury, other physiological, ecological, and anatomical conditions must be relevant for the evolution of the regenerative capacity of the CNS (see discussion in Hulsebosch and Bittner 1980). Particularly, the following three conditions may be important: (1) a potential risk to suffer injury to a given part of the CNS during the animal's life, (2) the importance of this part of the CNS for the survival and/or reproductive success of the animal, and (3) the injury to this part is not fatal. These conditions may hold in the case of the tentacle of *Limax*. Otherwise, other parts of the CNS may also be at a potential risk of the damage caused by invasions of, for example, parasitic microbes.

Of course, it is essentially impossible to demonstrate experimentally the regenerative ability of the brain from an injury that is fatal to the animal. However, the "brain transplantation" technique (see Sect. 5.4.3) may be a good tool to tackle this problem and to prove the validity of that condition (3) because researchers can examine the potential regenerative ability of the isolated brain during incubation in the other animal's body cavity using this technique. Thus, even if a certain injury to the CNS is fatal to the slug, the potential regenerative ability can be examined for a long time if the injured brain is transplanted into the body cavity of another animal. Therefore, it is expected that the relationship between the fatality of an injury to any part of the brain and its regenerative capacity can be studied systematically using this technique, and this is one of the advantages of using the pulmonates for the study of neuronal function and its evolution.

5.6 Outlook and Conclusion

As seen in this chapter, terrestrial slugs and snails can demonstrate olfactory learning and perform complex logical operations such as second-order conditioning and blocking. The brain of these terrestrial pulmonates displays exquisite recoverability from an injury and increases the genomic DNA in neurons in the face of an increasing demand for macromolecules. These excellent abilities and the robustness of their brains are comparable, or even superior, to those of humans. It is likely we have not yet unveiled all the exquisite features of their brains, acquired during the long history of evolution. There remain vast frontiers to be explored in the field of the neurobiology of the terrestrial slugs and snails.

References

Anisimov AP (2005) Endopolyploidy as a morphogenetic factor of development. Cell Biol Int 29:993–1004

Bailey CH, Bartsch D, Kandel ER (1996) Toward a molecular definition of long-term memory storage. Proc Natl Acad Sci USA 93:1344–1352

Bull R, Ferrera E, Orrego F (1976) Effects of anisomycin on brain protein synthesis and passive avoidance learning in newborn chicks. J Neurobiol 7:37–49

Butler EG, Ward MB (1967) Reconstitution of the spinal cord after ablation in adult *Triturus*. Dev Biol 15:464–486

Chabaud P, Ravel N, Wilson DA, Mouly AM, Vigouroux M, Farget V, Gervais R (2000) Exposure to bahaviourally relevant odour reveals different characteristics in rat central olfactory pathways as studied through oscillatory activities. Chem Senses 25:561–573

Chase R (2000) Structure and function in the cerebral ganglion. Microsc Res Tech 49:511–520

Chase R, Croll RP (1981) Tentacular function in snail olfactory orientation. J Comp Physiol 143:357–362

Chase R, Kamil R (1983) Morphology and odor sensitivity of regenerated snail tentacles. J Neurobiol 14:43–50

Chase R, Tolloczko B (1989) Interganglionic dendrites constitute an output pathway from the procerebrum of the snail *Achatina fulica*. J Comp Neurol 283:143–152

Chase R, Tolloczko B (1993) Tracing neural pathways in snail olfaction: from the tip of the tentacles to the brain and beyond. Microsc Res Tech 24:214–230

Cheng TC, Galloway PC (1970) Transplantation immunity in mollusks: the histocompatibility of *Helisoma duryi normale* with allografts and xenografts. J Invertebr Pathol 15:177–192

Choi GB, Settler DD, Kallman BR, Bhaskar ST, Fleischmann A, Axel R (2011) Driving opposing behaviors with ensembles of piriform neurons. Cell 146:1004–1015

Debiec J, LeDoux JE, Nader K (2002) Cellular and systems reconsolidation in the hippocampus. Neuron 36:527–538

Delaney KR, Gelperin A, Fee MS, Flores JA, Gervais R, Tank DW, Kleinfeld D (1994) Waves and stimulus-modulated dynamics in an oscillating olfactory network. Proc Natl Acad Sci USA 91:669–673

Dudai Y, Eisenberg M (2004) Rites of passage of the engram: reconsolidation and the lingering consolidation hypothesis. Neuron 44:93–100

Eakin RM, Ferlatte MM (1973) Studies on eye regeneration in a snail, *Helix aspersa*. J Exp Zool 184:81–96

Edgar BA, Orr-Weaver TL (2001) Endoreplication cell cycles: more for less. Cell 105:297–306

Elekes K, Battonyai I, Kobayashi S, Ito E (2012) Organization of the procerebrum in terrestrial pulmonates (*Helix*, *Limax*) reconsidered: cell mass layer synaptology and its serotonergic input system. Brain Struct Funct 218:477–490
Ermentrout B, Wang JW, Flores J, Gelperin A (2001) Model for olfactory discrimination and learning in *Limax* procerebrum incorporating oscillatory dynamics and wave propagation. J Neurophysiol 85:1444–1452
Ferretti P (2011) Is there a relationship between adult neurogenesis and neuron generation following injury across evolution? Eur J Neurosci 34:951–962
Flood J, Bennett E, Rosenzweig M, Orme A (1972) Influence of training strength on amnesia induced by pretraining injections of cycloheximide. Physiol Behav 9:589–600
Flores V, Salas PJI, Saavedra JP (1983) Electroretinographic and ultrastructural study of regenerated eye of the snail *Cryptomphalus aspersa*. J Neurobiol 14:167–176
Friedrich A, Teyke T (1998) Identification of stimuli and input pathways mediating food-attraction conditioning in the snail, Helix. J Comp Physiol A 183:247–254
Fujie S, Yamamoto T, Murakami J, Hatakeyama D, Shiga H, Suzuki N, Ito E (2005) Nitric oxide synthase and soluble guanylyl cyclase underlying the modulation of electrical oscillations in a central olfactory organ. J Neurobiol 62:14–30
Fulton D, Kemenes I, Andrew RJ, Benjamin PR (2008) Time-window for sensitivity to cooling distinguishes the effects of hypothermia and protein synthesis inhibition on the consolidation of long-term memory. Neurobiol Learn Mem 90:651–654
Gelperin A (1975) Rapid food-aversion learning by a terrestrial mollusk. Science 189:567–570
Gelperin A (1994) Nitric oxide mediates network oscillations of olfactory interneurons in a terrestrial mollusc. Nature 369:61–63
Gelperin A (1999) Oscillatory dynamics and information processing in olfactory systems. J Exp Biol 202:1855–1864
Gelperin A, Tank DW (1990) Odour-modulated collective network oscillations of olfactory interneurons in a terrestrial mollusc. Nature 345:437–440
Gelperin A, Rhines LD, Flores J, Tank DW (1993) Coherent network oscillations by olfactory interneurons: modulation by endogenous amines. J Neurophysiol 69:1930–1939
Gervais R, Kleinfeld D, Delaney KR, Gelperin A (1996) Central and reflex neuronal responses elicited by odor in a terrestrial mollusk. J Neurophysiol 76:1327–1339
Gillette R (1991) On the significance of neuronal giantism in gastropods. Biol Bull 180:234–240
Gomot P, Gomot L, Marchand CR, Colard C (1990) Extirpation and transplantation of the brain of the snail *Helix aspersa*: a study of the survival of the animal and implant. Can J Zool 68:1505–1512
Hitchcock PF, Raymond PA (1992) Retinal regeneration. Trends Neurosci 15:103–108
Hulsebosch CE, Bittner GD (1980) Evolution of abilities to regenerate neurons in central nervous systems. Am Nat 115:276–284
Inokuma Y, Inoue T, Watanabe S, Kirino Y (2002) Two types of network oscillations and their odor responses in the primary olfactory center of a terrestrial mollusk. J Neurophysiol 87:3160–3164
Inoue T, Watanabe S, Kawahara S, Kirino Y (2000) Phase-dependent filtering of sensory information in the oscillatory olfactory center of a terrestrial mollusk. J Neurophysiol 84:1112–1115
Inoue T, Watanabe S, Kirino Y (2001) Serotonin and NO complimentarily regulate generation of oscillatory activity in the olfactory CNS of a terrestrial mollusk. J Neurophysiol 85:2634–2638
Inoue T, Inokuma Y, Watanabe S, Kirino Y (2004) In vitro study of odor-evoked behavior in a terrestrial mollusk. J Neurophysiol 91:372–381
Inoue T, Murakami M, Watanabe S, Inokuma Y, Kirino Y (2006) In vitro odor-aversion conditioning in a terrestrial mollusk. J Neurophysiol 95:3898–3903
Ito I, Nakamura H, Kimura T, Suzuki H, Sekiguchi T, Kawabata K, Ito E (2000) Neuronal components of the superior and inferior tentacles in the terrestrial slugs, *Limax marginatus*. Neurosci Res 37:191–200
Ito I, Kimura T, Ito E (2001) Odor responses and spontaneous oscillatory activity in tentacular nerves of the terrestrial slug, *Limax marginatus*. Neurosci Lett 304:145–148

Judge ME, Quartermain D (1982) Characteristics of retrograde amnesia following reactivation of memory in mice. Physiol Behav 28:585–590

Kamin LJ (1969) Predictability, surprise, attention, and conditioning. In: Church R, Campbell BA (eds) Punishment and aversive behavior. Appleton-Century-Crofts, New York, pp 279–296

Kasai Y, Watanabe S, Kirino Y, Matsuo R (2006) The procerebrum is necessary for odor-aversion learning in the terrestrial slug *Limax valentianus*. Learn Mem 13:482–488

Kawahara S, Toda S, Suzuki Y, Watanabe S, Kirino Y (1997) Comparative study of neural oscillation in the procerebrum of the terrestrial slugs *Incilaria bilineata* and *Limax marginatus*. J Exp Biol 200:1851–1861

Keefe JR (1973) An analysis of urodelian retinal regeneration. I. Studies of the cellular source of retinal regeneration in *Notophthalmus viridescens* utilizing ^{3}H-thymidine and colchicine. J Exp Zool 184:185–206

Kimura T, Toda S, Sekiguchi T, Kirino Y (1998a) Behavioral modulation induced by food odor aversive conditioning and its influence on the olfactory responses of an oscillatory brain networks in the slug *Limax marginatus*. Learn Mem 4:365–375

Kimura T, Suzuki H, Kono E, Sekiguchi T (1998b) Mapping of interneurons that contribute to food aversive conditioning in the slug brain. Learn Mem 4:376–388

Kimura T, Toda S, Sekiguchi T, Kawahara S, Kirino Y (1998c) Optical recording analysis of olfactory response of the procerebal lobe in the slug brain. Learn Mem 4:389–400

Kleinfeld D, Delaney KR, Fee MS, Flores JA, Tank DW, Gelperin A (1994) Dynamics of propagating waves in the olfactory network of a terrestrial mollusk: an electrical and optical study. J Neurophysiol 72:1402–1419

Kobayashi S, Hattori M, Elekes K, Ito E, Matsuo R (2010) FMRFamide regulates oscillatory activity of the olfactory center in the slug. Eur J Neurosci 32:1180–1192

Koga Y, Matsuo Y, Matsuo R (2016) Olfactory memory storage and/or retrieval requires the presence of the exact tentacle used during memory acquisition in the terrestrial slug *Limax*. Zool Sci 33:78–82

Lasek RJ, Dower WJ (1971) *Aplysia californica*: analysis of nuclear DNA in individual nuclei of giant neurons. Science 172:278–280

Lee JLC (2009) Reconsolidation: maintaining memory relevance. Trends Neurosci 32:413–420

Mactutus CF, Riccio DC, Ferek JM (1979) Retrograde amnesia for old (reactivated) memory: some anomalous characteristics. Science 204:1319–1320

Makinae H, Makino Y, Obara T, Yano M (2008) Specific spatio-temporal activities in the cerebral ganglion of *Incilaria fruhstorferi* in response to superior and inferior tentacle nerve stimulation. Brain Res 1231:47–62

Makino Y, Yano M (2010) Investigating the underlying intelligence mechanisms of the biological olfactory system. Adv Artif Intell 2010, 478107

Matsuo R (2015) Consolidation and reconsolidation of olfactory memory in the terrestrial pulmonates. In: Memory consolidation. Nova, New York, pp 53–64

Matsuo R, Ito E (2008) Recovery of learning ability after the ablation of the procerebrum in the terrestrial slug *Limax valentianus*. Acta Biol Hung 59(suppl):73–76

Matsuo R, Ito E (2009) A novel nitric oxide synthase expressed specifically in the olfactory center. Biochem Biophys Res Commun 386:724–728

Matsuo R, Ito E (2011) Spontaneous regeneration of the central nervous system in gastropods. Biol Bull 221:35–42

Matsuo R, Hitomi T, Watanabe S, Kirino Y (2002) Delayed-onset amnesia caused by protein synthesis inhibition in odor-taste associative memory of the terrestrial slug *Limax valentianus*. Neurosci Lett 334:201–205

Matsuo R, Kobayashi S, Watanabe S, Namiki S, Iinuma S, Sakamoto H, Hirose K, Ito E (2009) Glutamatergic neurotransmission in the procerebrum (olfactory center) of a terrestrial mollusk. J Neurosci Res 87:3011–3023

Matsuo R, Kawaguchi E, Yamagishi M, Amano T, Ito E (2010a) Unilateral memory storage in the procerebrum of the terrestrial slug *Limax*. Neurobiol Learn Mem 93:337–342

Matsuo R, Kobayashi S, Tanaka Y, Ito E (2010b) Effects of tentacle amputation and regeneration on the morphology and activity of the olfactory center of the terrestrial slug *Limax valentianus*. J Exp Biol 213:3144–3149
Matsuo R, Kobayashi S, Murakami J, Ito E (2010c) Spontaneous recovery of the injured olfactory center in the terrestrial slug *Limax*. PLoS One 5, e9054
Matsuo R, Kobayashi S, Yamagishi M, Ito E (2011a) Two pairs of tentacles and a pair of procerebra: optimized functions and redundant structures in the sensory and central organs involved in olfactory learning of terrestrial pulmonates. J Exp Biol 214:879–886
Matsuo R, Kobayashi S, Morishita F, Ito E (2011b) Expression of Asn-d-Trp-Phe-NH_2 in the brain of the terrestrial slug *Limax valentianus*. Comp Biochem Physiol B 160:89–93
Matsuo R, Yamagishi M, Ito E (2012) Analysis of DNA endoreplication in the brain neurons in the terrestrial slug *Limax valentianus*. Acta Biol Hung 63(suppl 2):171–178
Matsuo R, Yamagishi M, Wakiya K, Tanaka Y, Ito E (2013) Target innervation is necessary for neuronal polyploidization in the terrestrial slug *Limax*. Dev Neurobiol 73:609–620
Matsuo R, Kobayashi S, Wakiya K, Yamagishi M, Fukuoka M, Ito E (2014a) The cholinergic system in the olfactory center of the terrestrial slug *Limax*. J Comp Neurol 522:2951–2966
Matsuo Y, Uozumi N, Matsuo R (2014b) Photo-tropotaxis based on projection through the cerebral commissure in the terrestrial slug *Limax*. J Comp Physiol A 200:1023–1032
Matsuo R, Fukata R, Kumagai M, Kobayashi A, Kobayashi S, Matsuo Y (2016a) Distribution and modulatory effect of histaminergic neurons on the oscillatory activity in the olfactory center of the terrestrial slug *Limax*. J Comp Neurol 524:119–135
Matsuo R, Tanaka M, Fukata R, Kobayashi S, Aonuma H, Matsuo Y (2016b) Octopaminergic system in the central nervous system of the terrestrial slug *Limax*. J Comp Neurol 524:3849–3864
Maximova OA, Balaban PM (1984) Neuronal correlates of aversive learning in command neurons for avoidance behavior of *Helix lucorum* L. Brain Res 292:139–149
Milekic MH, Alberini CM (2002) Temporally graded requirement for protein synthesis following memory reactivation. Neuron 36:521–525
Misanin JR, Miller RR, Lewis DJ (1968) Retrograde amnesia produced by electroconvulsive shock after reactivation of consolidated memory trace. Science 160:554–555
Moffett S (1995) Neural regeneration in gastropod molluscs. Prog Neurobiol 46:289–330
Montarolo PG, Goelet P, Castellucci VF, Morgan J, Kandel ER, Schacher S (1986) A critical period for macromolecular synthesis in long-term heterosynaptic facilitation in *Aplysia*. Science 234:1249–1254
Moroz LL, Kohn AB (2010) Do different neurons age differently? Direct genome-wide analysis of aging in singe identified cholinergic neurons. Front Aging Neurosci 2, article 6
Murphy AD, Kater SB (1980) Sprouting and functional regeneration of an identified neuron in *Helisoma*. Brain Res 186:251–272
Nader K, Einarsson EO (2010) Memory reconsolidation: an update. Ann NY Acad Sci 1191:27–41
Nader K, Schafe GE, Le Doux JE (2000) Fear memories requires protein synthesis in the amygdala for reconsolidation after retrieval. Nature 406:722–726
Nakaya T, Kawahara S, Watanabe S, Lee D, Suzuki T, Kirino Y (2001) Identification and expression of a novel gene in odour-taste associative learning in the terrestrial slug. Genes Cells 6:43–56
Pedreira ME, Dimant B, Maldonado H (1996) Inhibitors of protein and RNA synthesis block context memory and long-term habituation in the crab *Chasmagnathus*. Pharmacol Biochem Behav 54:611–617
Ratté S, Chase R (1997) Morphology of interneurons in the procerebrum of the snail *Helix aspersa*. J Comp Neurol 384:359–372
Ratté S, Chase R (2000) Synapse distribution of olfactory interneurons in the procerebrum of the snail *Helix aspersa*. J Comp Neurol 417:366–384
Ravel N, Chabaud P, Martin C, Gaveau V, Hugues E, Tallon-Baudry C, Betrand O, Gervais R (2003) Olfactory learning modifies the expression of odour-induced oscillatory responses in

the gamma (60–90 Hz) and beta (15–40 Hz) bands in the rat olfactory bulb. Eur J Neurosci 17:350–358
Root CM, Denny C, Hen R, Axel R (2014) The participation of cortical amygdala in innate, odour-driven behavior. Nature 515:269–273
Roubos EW (1976) Neuronal and non-neuronal control of the neurosecretory caudo-dorsal cells of the freshwater snail *Lymnaea stagnalis* (L.). Cell Tissue Res 168:11–31
Sahley C, Gelperin A, Rudy JW (1981a) One-trial associative learning modifies food odor preferences of a terrestrial mollusc. Proc Natl Acad Sci USA 78:640–642
Sahley C, Rudy JW, Gelperin A (1981b) An analysis of associative learning in a terrestrial mollusc. J Comp Physiol 144:1–8
Sahley C, Martin KA, Gelperin A (1990) An analysis of associative learning in a terrestrial mollusc *Limax maximus*. II. J Comp Physiol 167:339–345
Sakura M, Watanabe S (2015) Olfactory oscillation and role for learning and discrimination of odors in the terrestrial slug. In: Memory consolidation. Nova, New York, pp 65–74
Sakura M, Kabetani M, Watanabe S, Kirino Y (2004) Impairment of olfactory discrimination by blockade of nitric oxide activity in the terrestrial slug *Limax valentianus*. Neurosci Lett 370:257–261
Samarova E, Balaban P (2006) Changes in frequency of spontaneous oscillations in procerebrum correlate to behavioral choice in terrestrial snails. Front Cell Neurosci 3, article 8
Sekiguchi T, Yamada A, Suzuki H (1997) Reactivation-dependent changes in memory states in the terrestrial slug *Limax flavus*. Learn Mem 4:356–364
Sekiguchi T, Furudate H, Kimura T (2010) Internal representation and memory formation of odor preference based on oscillatory activities in a terrestrial slug. Learn Mem 17:372–380
Shirahata T, Tsunoda M, Santa T, Kirino Y, Watanabe S (2006) Depletion of serotonin selectively impairs short-term memory without affecting long-term memory in odor learning in the terrestrial slug *Limax valentianus*. Learn Mem 13:267–270
Suenaga Y, Matsuo R (2016) Length of the memory retention period depends on the extent of protein synthesis in the terrestrial slug *Limax*. Neurosci Lett 630:222–227
Teyke T, Gelperin A (1999) Olfactory oscillations augment odor discrimination not odor identification by *Limax* CNS. Neuroreport 10:1061–1068
Teyke T, Wang JW, Gelperin A (2000) Lateralized memory storage and crossed inhibition during odor processing by *Limax*. J Comp Physiol A 186:269–278
Tully T, Preat T, Boynton SC, Del Vecchio M (1994) Genetic dissection of consolidated memory in *Drosophila*. Cell 79:35–47
Watanabe S, Kawahara S, Kirino Y (1998) Morphological characterization of the bursting and nonbursting neurones in the olfactory centre of the terrestrial slug *Limax marginatus*. J Exp Biol 201:925–930
Watanabe S, Kawahara S, Kirino Y (1999) Glutamate induces Cl^- and K^+ currents in the olfactory interneurons of a terrestrial slug. J Comp Physiol A 184:553–562
Watanabe S, Inoue T, Murakami M, Inokuma Y, Kawahara S, Kirino Y (2001) Modulation of oscillatory neural activities by cholinergic activation of interneurons in the olfactory center of a terrestrial slug. Brain Res 896:30–35
Watanabe S, Kirino Y, Gelperin A (2008) Neural and molecular mechanisms of microcognition in *Limax*. Learn Mem 15:633–642
Wilcox M, Franceschini N (1984) Illumination induces dye incorporation in photoreceptor cells. Science 225:851–854
Yabumoto T, Takanashi F, Kirino Y, Watanabe S (2008) Nitric oxide is involved in appetitive but not aversive olfactory learning in the land mollusk *Limax valentianus*. Learn Mem 15:229–232
Yamada A, Sekiguchi T, Suzuki H, Mizukami A (1992) Behavioral analysis of internal memory states using cooling-induced retrograde amnesia in *Limax flavus*. J Neurosci 12:729–735
Yamagishi M, Ito E, Matsuo R (2008) Redundancy of olfactory sensory pathways for odor-aversion memory in the terrestrial slug *Limax valentianus*. J Exp Biol 211:1841–1849
Yamagishi M, Ito E, Matsuo R (2011) DNA endoreplication in the brain neurons during the body growth of an adult slug. J Neurosci 31:5596–5604

Yamagishi M, Ito E, Matsuo R (2012) Whole genome amplification in large neurons of the terrestrial slug *Limax*. J Neurochem 122:727–737

Yasui K, Matsuo R, Kirino Y (2004) Onset of amnesia is delayed with a decreased inhibition of protein synthesis during odor-taste associative learning in the terrestrial slug *Limax valentianus*. Zool Sci 21:1163–1166

Zakharov IS, Hayes NL, Ierusalimsky VN, Nowakowski RS, Balaban PM (1998) Postembryonic neurogenesis in the procerebrum of the terrestrial snail, *Helix lucorum* L. J Neurobiol 35:271–276

Zupanc GK (1999) Neurogenesis, cell death and regeneration in the adult gymnotiform brain. J Exp Biol 202:1435–1446

Zupanc GK, Sîrbulescu RF (2011) Adult neurogenesis and neuronal regeneration in the central nervous system of teleost fish. Eur J Neurosci 34:917–929

Chapter 6
Insect Brains: Minute Structures Controlling Complex Behaviors

Michiyo Kinoshita and Uwe Homberg

Abstract Insects are the largest taxon of arthropods, characterized by a segmented body plan. They comprise the most abundant and diverse group of animals. Many insects show highly complex adaptive behaviors, including learning abilities, social interactions, and spatial orientation skills that, in simplified version, are reminiscent of the abilities of vertebrates and even humans. In contrast to their sophisticated behavior, their brain, however, is minute and simple compared to that of humans. Because of these features, many insects have become models for studies of the neuronal basis underlying specific behaviors.

The insect body is divided into three parts: the head, the thorax with wings and legs, and the abdomen. In most species, each part contains relatively autonomously operating neural circuits, which have functions in local sensing and motor control. The head contains the antennae, the compound eyes, the ocelli, various sense organs on the mouth parts, and, as part of the nervous system, the brain. The brain processes this multitude of sensory input and provides multisensory integration. In addition, it controls movements of the antennae and mouth parts and induces suitable behaviors by modifying the activity of the thoracic and abdominal nervous systems, which, likewise, provide sensory input and feedback to the brain. This chapter introduces the organization of the insect brain and then focuses on neural circuits underlying five aspects of insect behavior that are relatively well understood.

Keywords Sensory systems • Insect brain • Motion vision • Circadian clock • Learning and memory • Orientation • Courtship

M. Kinoshita (✉)
Laboratory of Neuroethology, SOKENDAI, The Graduate University for Advanced Studies, Shonan Village, Hayama 240-0193, Japan
e-mail: kinoshita_michiyo@soken.ac.jp

U. Homberg
Department of Biology, Animal Physiology, University of Marburg, D-35032 Marburg, Germany

S. Shigeno et al. (eds.), *Brain Evolution by Design*, Diversity and Commonality in Animals, DOI 10.1007/978-4-431-56469-0_6

6.1 Overview of the Insect Brain

The brain is the major part of the central nervous system in vertebrates and most invertebrates and is usually placed in the head. The obvious differences between vertebrate and invertebrate brains are their size and shape. In general, vertebrate brains are much bigger than those of invertebrates: the number of neurons is 10^5 to 10^6 for insect brains, 7×10^7 for a small mammal such as the mouse, and 10^{11} for humans (Menzel 2001; Herculano-Houzel et al. 2006; Azevedo et al. 2009).

Even though the shape and organization differ considerably between the vertebrate and insect brain, molecular genetic analysis suggests similarity in genetic patterning mechanisms in the formation of the brain in fruit fly and mouse (Arendt and Nübler-Jung 1996; Lichtneckert and Reichert 2005). Similarly, the role of the brain in processing sensory information, integrating multisensory modalities, and inducing behavior by controlling the motor system is similar in vertebrates, insects, and other invertebrates.

Insect species are highly abundant and diverse, impressively illustrating their phylogenetic success. Each species is adapted for a specific habitat, which results in a rich diversity of morphology, body size, physiology, and behavior (Chittka and Niven 2009; Strausfeld 2012). Considerable variations also exist in the organization of the brain. Neuropil morphology, such as shape, volume, and relative position in the brain, is highly species dependent, even though comparable brain neuropils can be found across taxa (Strausfeld 2012). This diversity in insect brain organization has resulted in different nomenclatures for different species, such as flies, bees, locusts, moths, beetles, cockroaches, and others (Strausfeld 1976; Mobbs 1985; Kurylas et al. 2008; el Jundi et al. 2009b; Dreyer et al. 2010; Strausfeld 2012). Recently, Ito et al. (2014) introduced a unifying systematic nomenclature for insect brain structures using the brain of the fruit fly *Drosophila melanogaster* as a model.

The insect brain consists of six neuromeres, which are arranged in two different ways. In hemimetabolous insects such as locusts and cockroaches, but also in beetles (holometabolous), the brain is separated into the cerebral ganglia (CRG) and the gnathal ganglia (GNG) (Fig. 6.1a, left). The CRG, containing the protocerebrum (PR), deutocerebrum (DE), and tritocerebrum (TR), lie above the esophagus (ES), whereas the mandibular (MN), maxillary (MX), and labial (L,B) ganglia, comprising the GNG, are below the esophagus and are separated from the CRG through connectives. In many holometabolous insects, such as flies, bees, and moths, however, the CRG and GNG are fused into one ganglionic mass (Fig. 6.1a right, b). Each neuromere contains neuropils defined as synapse-rich areas and a cell body rind surrounding the neuropils.

The CRG consist of the cerebrum between two visual processing centers, the optic lobes (OL). The cerebrum contains several prominent neuropils (Fig. 6.1c): the antennal lobe (AL), the mushroom body (MB), the lateral horn (LH), the anterior optic tubercle (AOTU), and the antennal mechanosensory and motor center (AMMC) in each brain hemisphere, and the central complex (CX) connected to two lateral complexes in the center. Other areas, separated by fiber bundles and

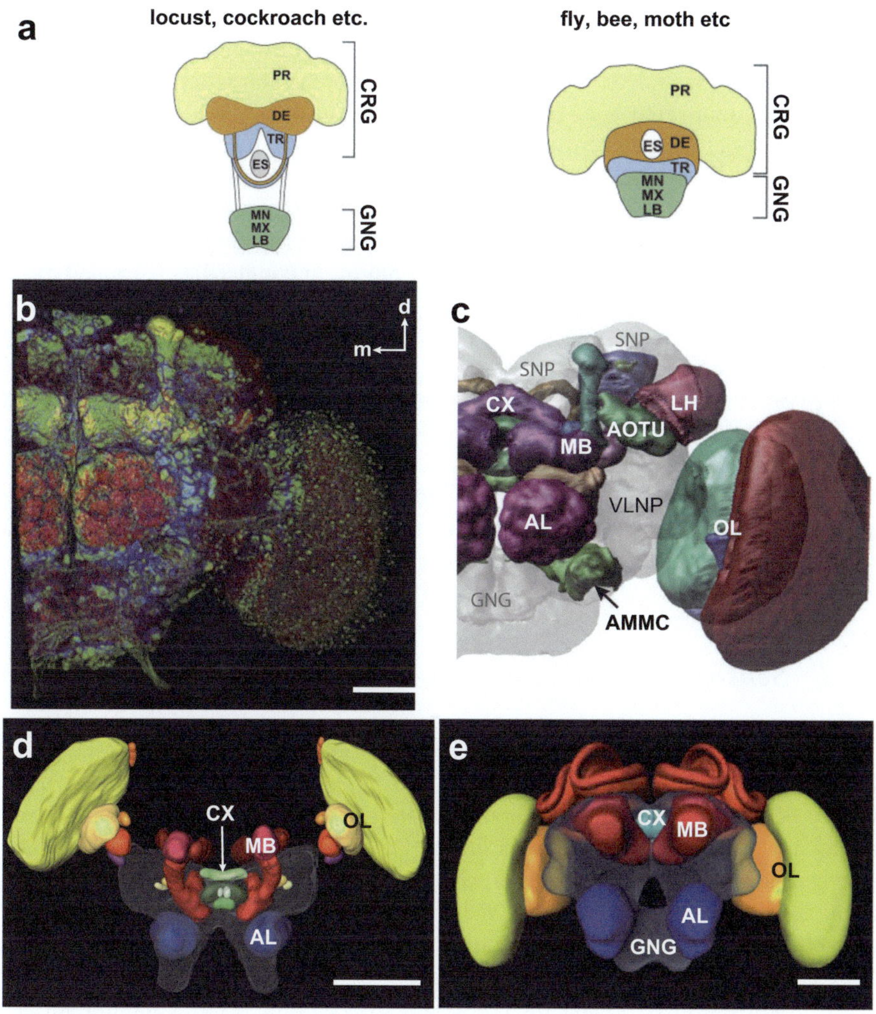

Fig. 6.1 Overview of insect brains. (**a**) Arrangement of cerebral ganglia (CRG) and gnathal ganglia (GNG) in many hemimetabolous (*left*) and holometabolous (*right*) insects. (**b**) 3D confocal scanning image of the *Drosophila* brain. *d* dorsal, *m* medial. *Scale* 50 μm ((**a**, **b**) Modified from Ito et al. (2014), with kind permission). (**c**) 3D reconstruction of the fly brain, frontal view. (**d**) The cerebral ganglia of the locust brain, posterior view. *Scale* 600 μm ((**c**, **d**) adapted from Kurylas et al. 2008, with kind permission). (**e**) 3D standard of the honeybee brain. *Scale* 200 μm. (Adapted from Brandt et al. (2005), with kind permission). See text for explanation of abbreviations

the prominent neuropils, are named based on relative position in the cerebrum, such as the superior (SNP), ventrolateral (VLNP), inferior, ventromedial, and periesophageal neuropils (Fig. 6.1c).

The relative position, shape, and volume of neuropils differ among insect brains and are related to the size and shape of the head and species-specific differences in the relative importance of different sensory systems (Farris 2015). In large insects such as the desert locust, the optic lobes are connected through a slender optic stalk with the cerebrum, and the whole brain appears stretched horizontally compared to the compact brain of the small fruit fly (Fig. 6.1d) (Kurylas et al. 2008). The neuraxis inside the insect head is bent upward, but in some taxa, such as flies, butterflies, and moths, is even tilted backward, partly reflecting a change in the position of the antennae, which in moths are not used as sensors exploring the ground but are held upward for wind-carried odor detection. As a consequence, the orientation of the cerebrum including MB and CX in flies, moths, and butterflies is rotated by as much as 90° in comparison to the brain of locusts and honeybees (Fig. 6.1). Finally, the size of different brain areas varies species specifically depending on the relative importance of different sensory modalities. This dictum applies most obviously to sensory brain areas such as the OLs, which are large in day-active dragonflies and butterflies, but relatively small in nocturnal cockroaches, and the ALs, which are large in cockroaches and moths but virtually nonexistent in dragonflies. Finally, central brain areas can vary in size, most dramatically seen in MBs: these are huge in social bees and ants, probably reflecting the importance of odors in foraging and social interactions requiring particularly high memory storage capacity.

6.1.1 Sensory Processing Regions in the Brain

The head contains a multitude of sensory organs and sensilla, including the ocelli and compound eyes, chemoreceptors on the antennae and mouth parts, a large variety of mechanoreceptors on the head capsule, mouth parts and antennae, and hygro- and thermoreceptors on the antennae. Sensory cells of the eyes and ocelli, the antennae, and parts of the head capsule project to the cerebral ganglia, and gustatory information is largely conveyed to the gnathal ganglia. Among these sensations, this section focuses on the most prominent neuropils involved in processing visual, olfactory, and mechanosensory signals.

Two light-receptive organs, the compound eyes and the ocelli, contain photoreceptors and convey visual signals to the optic lobe (compound eye) or directly to the central brain (ocelli) for further processing (Strausfeld 1970; Strausfeld 1976). The fly optic lobe (Fig. 6.2a) contains, from the retina, the lamina (LA; not shown in Fig. 6.2a), the medulla (ME), and the lobula complex. The lobula complex is subdivided further into the lobula (LO) and the thin lobula plate (LOP), which is attached to the posterior side of the LO (Fig. 6.2a, left top). In the lamina, photoreceptor axons form, together with lamina interneurons, units called cartridges. Fly LA cartridges contain photoreceptor axons from seven neighboring ommatidia, which is called neural superposition (Strausfeld 1989), whereas photoreceptor axons from single ommatidia innervate a single lamina cartridge in many other insect laminae (Ribi 1987; Takemura et al. 2005). The lamina cartridges appear as thin columnar structures along the axis of axons. This cartridge structure also exists in the ME, which, in addition, has a pronounced layered organization across the

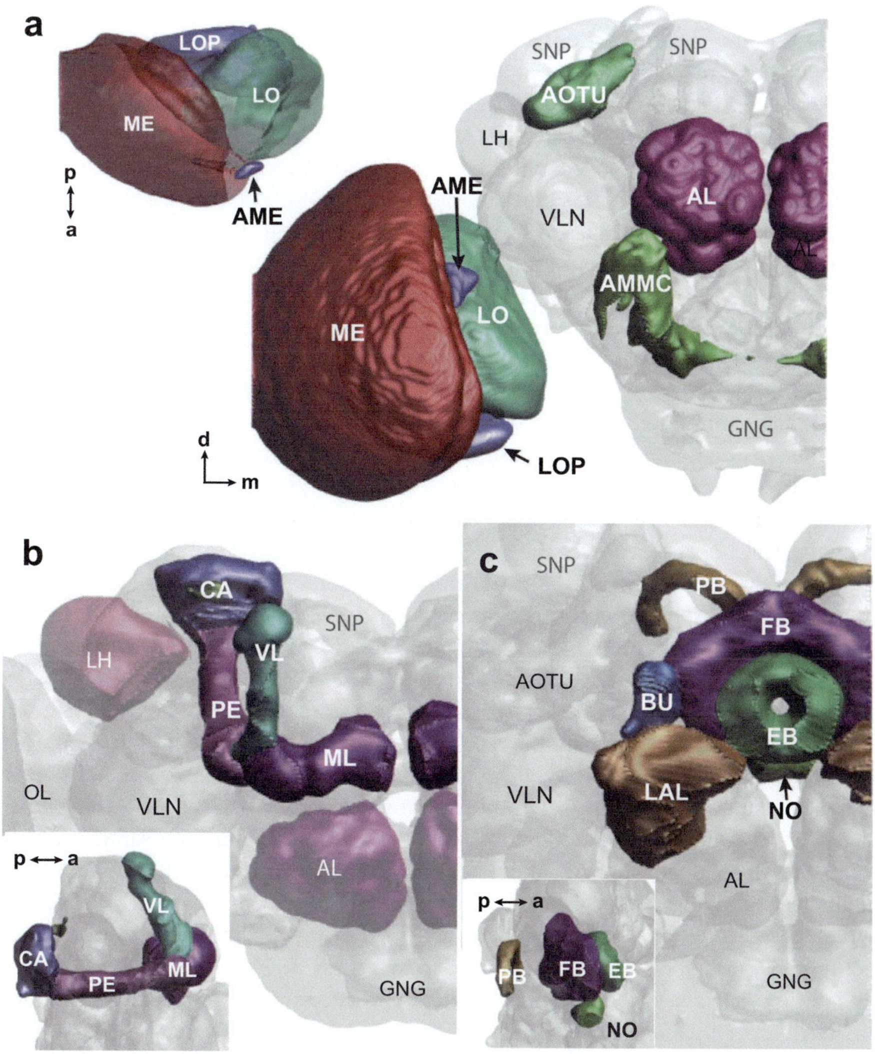

Fig. 6.2 Arrangement of prominent neuropils in the fly brain. (**a**) Neuropils for primary sensory processing. *Inset* shows dorsal view of the optic lobe. *a* anterior, *d* dorsal, *m* medial, *p* posterior. (**b**) The mushroom body. *Inset* shows lateral view of the mushroom body. (**c**) The central complex with the lateral accessory lobe (LAL) and bulb (BU). *Inset* shows lateral view of the central complex (Images taken from Ito et al. (2014), with kind permission). See text for explanation of abbreviations

cartridges. The ME of *Drosophila* contains ten layers (Fischbach and Dittrich 1989). In many insects, a thick layer in the middle of the ME, the so-called serpentine layer, separates the distal and proximal ME. In addition, a small neuropil, the accessory medulla (AME), lies at the anterior edge of the ME (Fig. 6.2a). The organization of the lobula complex is highly species specific. In the locust and praying mantis, the LO is composed of four or five substructures (Gouranton 1964; Leitinger et al. 1999), whereas in the bee it consists of a single ganglionic mass (Mobbs 1985; Paulk et al. 2008). Organization into cartridges and layers is common in the distal LO but is no longer present in proximal aspects or at the output side of the lobula complex. The number of layers in the ME and LO is species specific. The cartridge structure in the LA, ME, and LO indicates that retinotopic processing occurs in these neuropils.

In general, light information is processed in the optic lobe and is transferred to the central brain for integration with other sensory modalities or to the motor control system. Processing in the LOP strongly contributes to motion vision related to optomotor responses (Hausen and Egelhaaf 1989; Borst and Euler 2011), whereas the LO processes visual cues such as color, small object movement, and shape (Yang et al. 2004; Paulk et al. 2008; Dunbier et al. 2012; Okamura and Strausfeld 2007). Many neural tracts project from the ME and lobula complex to the cerebrum, and some of them continue as commissures to the contralateral optic lobe. Among neuropils in the cerebrum, the anterior optic tubercle (AOTU) receives visual information. In locusts, bees, and butterflies the AOTU consists of two major subunits, the upper and lower unit (Homberg et al. 2003a; Pfeiffer and Kinoshita 2012; Heinze and Reppert 2012). Additional smaller subunits appear to be present in certain insects, such as the monarch butterfly (Heinze and Reppert 2012). In *Drosophila* three subunits of the AOTU were identified based on the arborizations of visual projection neurons (Otsuna and Ito 2006).

Odorant information, characterized by a mixture of different volatile compounds, is first processed in the AL, located anteroventrally in the cerebrum (Fig. 6.2a). Olfactory receptor cells detect volatile compounds and send their axons to spheroidal condensations of neuropils, termed glomeruli, in the AL (Schachtner et al. 2005; Galizia 2014). Olfactory receptor cells expressing the same receptor molecule(s) send converging axons to the same glomerulus, thus providing a characteristic odor-response profile to each glomerulus (Galizia 2014). Processed olfactory information is conveyed from the AL by projection neurons to other brain areas, notably the MB and LH (Galizia and Rössler 2010). The number of glomeruli is species specific: 160 in the honeybee, about 60 in moths and butterflies (Schachtner et al. 2005), and 205 in the American cockroach (Watanabe et al. 2010). Male moths tracking sexual pheromones produced by females have particular, enlarged male-specific glomeruli, called macroglomerular complex (Rospars and Hildebrand 1992; Kazawa et al. 2009). The ALs are large in nocturnal insects and flower foragers relying on floral odors, whereas certain aquatic insects such as whirligig beetles lack ALs (Lin and Strausfeld 2013).

The antenna is usually covered with a large variety of mechanoreceptors, serving a variety of functions (Staudacher et al. 2005). Particularly prominent mechanoreceptors in the second antennal segment, the pedicel, form a mechanosensory organ called Johnston's organ that senses deflections and vibrations of the antennal flagellum. As such it has a variety of functions and has a role in gravity perception (honeybees, flies), detection of water surface vibrations (whirligig beetle; Kolmes 1983), wind and air vibrations for flight control (moths; Sane et al. 2007), and hearing (mosquitoes, flies, bees; Matsuo and Kamikouchi 2013). Its mechanoreceptors project into the antennal mechanosensory and motor center (AMMC) located lateral from the esophagus and above the GNG (Fig. 6.2a, c).

6.1.2 High-Order Brain Regions in the CRG

The mushroom bodies (MBs) and the central complex (CX) with the lateral complexes (LXs) are prominent neuropils in the cerebrum. The MBs are paired structures in the dorsal cerebrum, whereas the CX spans the brain midline, framed by the MB lobes.

The MB is constructed by intrinsic neurons, termed Kenyon cells, and is divided into three regions: the calyx (CA), pedunculus (PE), and lobes (Fig. 6.2b). In flies, the cup-shaped CA faces the posterior edge of the brain, the cylinder-like PE extends anteriorly, and the bifurcated lobes, the vertical lobe (VL), and the medial lobe (ML) are located in the anterior brain (Tanaka et al. 2008). These three regions in general correspond to the dendritic region, axon, and terminals of Kenyon cells. In the CA Kenyon cells receive sensory input, most prominently from antennal-lobe projection neurons, but in many species also from visual and gustatory fibers, whereas their main output sites are in the lobes.

The structure of the MB varies considerably among different insect species (Strausfeld et al. 1998; Strausfeld 2012). In butterflies, moths, cockroaches, bees, and ants, the CA is a double cup-shaped structure, whereas in secondarily anosmic species, such as backswimmers and damselflies, it is highly reduced or may even be completely absent (Strausfeld et al. 1998; Farris 2013). Inputs of different sensory modalities normally segregate in the CA (Gronenberg 1999; Ehmer and Gronenberg 2002; Kinoshita et al. 2015; Nishino et al. 2012). Two lobes, a VL and an ML, are present in most insects (Brandt et al. 2005; Sjöholm et al. 2006; Mobbs 1985; Fukushima and Kanzaki 2009; Kurylas et al. 2008), but only a single spherical lobe has been identified in butterflies (Heinze and Reppert 2012; Kinoshita et al. 2015). The MB in Lepidoptera has a secondary peduncle connected to the dorsal lobe (Sjöholm et al. 2006; Homberg et al. 1988; Fukushima and Kanzaki 2009; Heinze and Reppert 2012; Kinoshita et al. 2015). In the PE and lobes, Kenyon cells form concentric or laminar layers and make contacts with efferent and a smaller number of afferent neurons (Farris 2011).

Even though the CA receives inputs of multisensory modalities (Farris and Roberts 2005), the olfactory input is dominant in many insects. Olfactory information is conveyed by projection neurons from the antennal lobe to the CA and the lateral horn (LH). Therefore, the MB has been regarded primarily as a center for olfactory signal processing, specifically olfactory learning and memory (Strausfeld et al. 1998; Menzel 2014). However, the CA of honeybees, ants, and butterflies receive substantial visual input in addition to olfactory input. In bees, grasshoppers, and crickets, gustatory information also enters the CA (Farris 2008). Whirligig beetles have a distinct MB, which is dominated by visual input

(Lin and Strausfeld 2012). The shape and volume of the MB is highly diverse among species, which may relate to differences in foraging ecology, mating behavior (Farris and Roberts 2005), or other aspects of behavioral ecology requiring different amounts of memory storage capacity (Menzel 2014).

The CX lies in the center of the cerebrum and consists of four neuropils (Fig. 6.1c). The protocerebral bridge (PB), fan-shaped body (FB, also termed upper division of the central body, CBU), and ellipsoid body (EB, also termed lower division of the central body, CBL) are arranged from posterior to anterior whereas the paired noduli (NO) lie ventrally from the FB and EB (Fig. 6.2c, inset). The paired lateral accessory lobes (LAL) and bulbs (BU), comprising the LXs, are connected to both sides of the EB (Fig. 6.2c). These six neuropils are intimately interconnected by interneurons forming distinct neural circuits (Heinze and Homberg 2008; Wolff et al. 2015). The PB, FB, and EB are further subdivided into 16 slices (the PB of *Drosophila* into 18 slices), numbered 1–8 (1–9) from the midline to the lateral edges of both sides (Heinze and Homberg 2008; Wolff et al. 2015). The FB and EB, in addition, have a layered organization. The NO are also divided into subunits, usually four that are stacked on top of each other (Ito et al. 2014). The LAL consists of subregions (Ito et al. 2014), whereas the BU is composed of micro-glomeruli (Träger et al. 2008; Seelig and Jayaraman 2013).

The regular arrangement of slices and layers in the CX is highly conserved among insect species (el Jundi et al. 2009a; Heinze et al. 2012; el Jundi et al. 2014). The bulb is, in monarch butterflies, bumblebees, and locusts, further divided into two areas (Heinze et al. 2012; Pfeiffer and Kinoshita 2012; el Jundi et al. 2014). Evidence from locusts, flies, cockroaches, bees, and monarch butterflies suggests that the CX is dominated by visual input and has a key function in spatial orientation and memory during flight and walking (Pfeiffer and Homberg 2014).

6.2 Neural Networks Underlying Behaviors

This section outlines the neural networks and computations underlying five aspects of insect behavior: optomotor responses, olfactory learning and memory, sky compass navigation, circadian clock, and mating behavior. All of these are essential components of animal behavior across most taxa, suggesting that algorithms underlying those behaviors may be shared among many species.

One of the most important roles of the nervous system is to detect sensory signals from the environment, process this information, and, finally, induce appropriate behavior for a particular situation. Neural networks across several neuropils usually contribute to these tasks. To understand the neural operations involved, researchers traditionally studied a particular behavior of a favorable species such as the housefly, honeybee, locust, or cockroach, using behavioral assays, neuroanatomy, and electrophysiology. As the development of genetic tools, imaging, and in vivo patch recordings progressed, *Drosophila* became the widely used model system in insect neuroscience. New gene-editing techniques (TALEN, CRISPR) and RNAi may help to further studies on non-model insects, revealing increasing comparative and evolutionary insights into the organization and functioning of the insect nervous system in the future.

6.2.1 Optomotor Responses

Motion vision is an important fundamental function of the visual system. It serves a role in course control during self-motion, but also allows an animal to track other moving animals or objects. Directionally selective motion sensitivity is computed in the retina of mammals and in the optic lobes of insects (Borst and Helmstaedter 2015), and several models have been proposed underlying directional selectivity in motion vision (Barlow and Hill 1963; Reichardt 1987; Borst and Euler 2011; Borst and Helmstaedter 2015).

Optomotor responses are relatively simple behaviors based on motion vision and achromatic signal processing and have been studied in detail in flies. For studying optomotor responses, flies are tethered in the center of a striped drum (Fig. 6.3a). When the drum rotates clockwise or counterclockwise, the fly tries to rotate its body to follow the movement of the stripes (Fig. 6.3b). This behavior is important for body stabilization and course control during flight (Borst et al. 2010; Srinivasan 2011) and is elicited by optic flow stimulation, that is, by movement of the whole environment around the fly.

The lobula plate (LOP) in the fly brain has a key role in optic flow detection and optomotor responses. In *Musca* and *Calliphora* the LOP is innervated by 50 or more large field tangential neurons that are sensitive to optic flow stimulation and contribute critically to optomotor responses (Hausen and Egelhaaf 1989). These LOP tangential neurons are directionally selective: that is, motion in the preferred direction excites these neurons and motion in the opposite direction (null direction) inhibits them (Fig. 6.3c). Neurons sensitive to the four cardinal directions of motion (front to back, back to front, upward, downward) are organized in four layers of the LOP. The most thoroughly studied cell types are neurons of the horizontal system (HS), sensitive to horizontal motion from front to back, and neurons of the vertical system (VS) that are largely sensitive to downward motion (Fig. 6.3d). In *Musca* and *Calliphora*, the HS system consists of three neurons in the most anterior layer of the LOP with dendrites extending across dorsal, medial, and ventral regions (shown in blue in Fig. 6.3d), whereas the VS system consists of ten neurons, whose arborizations innervate narrow vertically extending regions in the most posterior layer of the LOP (shown in red; Fig. 6.3d). Axonal projections of these neurons make contact with neck motor neurons or descending neurons in the ventral lateral protocerebrum of the brain (blue and red arrowheads, respectively; Fig. 6.3d) (Strausfeld 1989). The descending neurons convey information to flight motor and locomotor circuits in the thoracic ganglia, largely through indirect connections (Strausfeld et al. 1984).

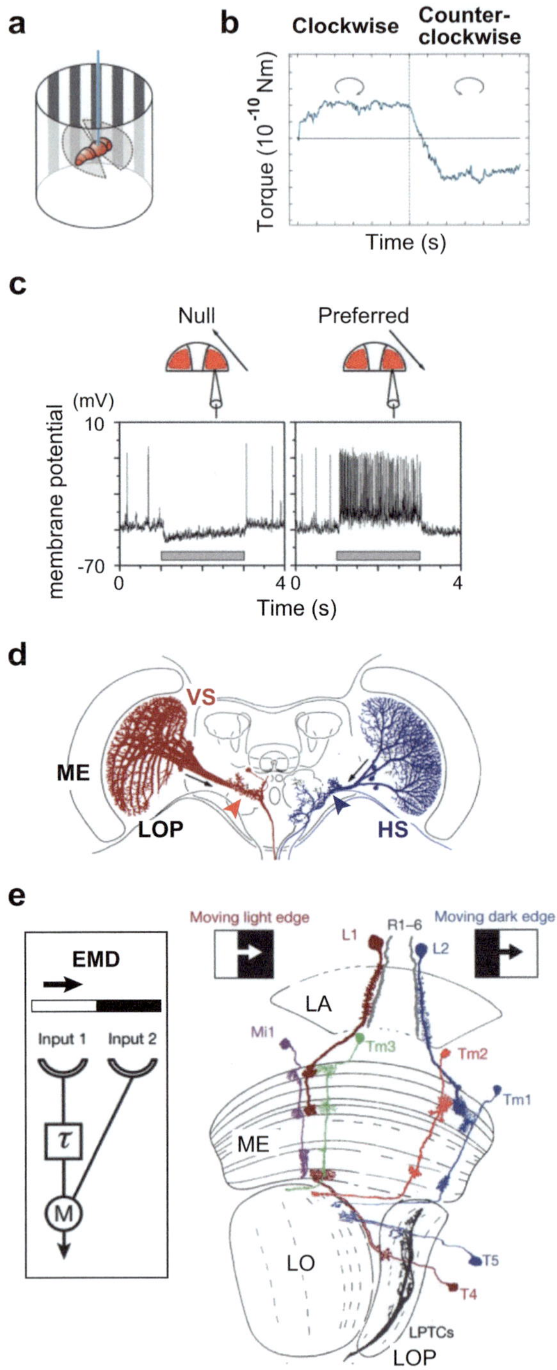

Fig. 6.3 Neural pathways underlying optomotor responses in flies. (**a**) A fly tethered to a torque meter inside a rotating striped drum. (**b**) Optomotor yaw response to clockwise and counterclockwise rotation of the drum (Modified from Borst et al. 2010, with kind permission). (**c**) Response of a directionally selective motion sensitive neuron. *Grey bar* motion stimulus (Modified from Borst and Haag 2002, with kind permission). (**d**) Morphology of LOP tangential neurons (Modified from Hausen and Egelhaaf 1989, with kind permission). *Me* medulla, *LOP* lobula plate, *HS* horizontal system, *VS* vertical system. (**e**) Elementary motion detector (EMD) (Modified from Reichardt 1987, with kind permission). Neuronal cell types in the optic lobe involved in optomotor responses. *M* multiplication, *L* lamina monopolar cell, *LA* lamina, *Mi* medulla intrinsic neuron, *Tm* transmedullary cell, *T* bushy T cell (Modified from Borst and Helmstaedter 2015, with kind permission)

The physiological properties of direction selectivity in the LOP tangential neurons can be explained by the Hassenstein–Reichardt model of elementary motion detectors (EMD) (Fig 6.3e, inset) (Reichardt 1987). When a striped pattern moves from left to right, the light part stimulates successively input 1 and 2, which are spatially separated. The signal from input 1, delayed (τ) through a temporal filter, and the non-delayed signal from input 2 simultaneously reach the next stage of nonlinear multiplication (M), which results in selectivity for rightward motion, but no response for leftward motion. In the correlation model, two mirror symmetrical units serve opposite directional signals [excite (+) and inhibit (−)] to a summation stage and, thereby, produce opposite responses to the oppositely directed movements.

Extensive recent research using fruit fly mutants identified the neuronal circuits in the optic lobe underlying the key elements of the EMD (Borst and Helmstaedter 2015). Motion is processed in two parallel processing pathways in the optic lobe, one for light edge movement (ON pathway) and the other for dark edge movement (OFF pathway) (Fig. 6.3e). The moving striped pattern first stimulates ommatidial arrays with photoreceptors in each ommatidium receiving light. The photoreceptors have synapses with different types of lamina monopolar cells (LMC). The first type of LMC (L1) is sensitive to a moving light edge (ON signal), whereas dark edge movement excites the other types of LMC (L2–L4, OFF signal; only L2 shown in Fig. 6.3e) (Joesch et al. 2010), indicating that light and dark edges are already processed separately in the lamina. These LMCs project to distal medulla layers and synapse upon intrinsic medulla cells (Mi1) and transmedullary cells (Tm1-3) projecting directly to the lobula (Takemura et al. 2013). The Mi1 and Tm3 cells are sensitive to light edge movement and Tm1 and Tm2 neurons are sensitive to dark edge movement. For producing direction selectivity with light edge movement, Mi1 produces a signal delay and Tm3 produces a non-delayed signal. In the pair of Tm1 and Tm2 detecting the dark edge, Tm1 produces the delay (Behnia et al. 2014). These delayed and non-delayed signals in those medulla neurons are the key input elements of the EMD. The Mi1 and Tm3 cells, then, make synaptic contacts with bushy T cells (T4) in the most proximal medulla layer, whereas Tm1 and Tm2 cells terminate on T5 cells in a distal lobula layer. T4 and T5 respond in a directionally selective manner and can be regarded as the multiplication stage of the EMD. T4 and T5 each consist of four neurons that respond maximally to one of the four cardinal directions of motion and project to one of the four layers of the LOP (Maisak et al. 2013). T4 and T5 cells responding to front-to-back motion provide excitatory input to horizontal motion-sensitive neurons in the anterior two layers of the LOP, and T4 and T5 cells responding to downward motion contact vertical motion-sensitive neurons in the posterior two layers. Neurons responsible for the inhibitory null directions have not been identified but are assumed to be local inhibitory interneurons signaling between adjacent LOP layers with oppositely tuned direction selectivity (Borst and Helmstaedter 2015).

Comparison of data from mammals and flies impressively revealed that in both systems motion detection follows highly similar rules and processing stages. In both systems it is based on parallel ON and OFF pathways, correlation of signals with different temporal dynamics, and convergence of these signals at the next stage in four parallel pathways corresponding to the four principal directions of motion (Borst and Euler 2011; Borst and Helmstaedter 2015).

6.2.2 Olfactory Learning and Memory

Animals change their behavior based on experience resulting in memory formation. Studies on the neural mechanisms of learning and memory formation ranging from analysis of synaptic plasticity in the sea hare *Aplysia* to studies of mammalian brain areas involved in memory formation, such as the hippocampus, amygdala, and cerebellum, increasingly show that mechanisms of synaptic plasticity underlying different forms of memory such as short- and long-term memory are widely shared in the animal kingdom (Kandel 2001; Roberts and Glanzman 2003; Glanzman 2010). In insects, a focus of research has been on the mechanisms of olfactory learning and memory especially in honeybees, crickets, locusts, cockroaches, and fruit flies (Menzel and Giurfa 2001; Mizunami et al. 2009; Watanabe et al. 2011; Stopfer 2014; Owald and Waddell 2015; Mizunami et al. 2015).

Flower-foraging honeybees have excellent capabilities for learning visual and olfactory cues of flowers by association with nectar (von Frisch 1950; Giurfa 2003). In appetitive classical conditioning in the laboratory, a harnessed honeybee receives a drop of sucrose solution as reward (unconditioned stimulus, US) shortly after presentation of a particular odor A (conditioned stimulus, CS+; Fig. 6.4a). After a single learning trial, the bee extends its proboscis without reward when the same odor is presented again (Fig. 6.4a, right). Local cooling of the mushroom body calyces (CAs) inhibits olfactory memory formation, whereas injection of octopamine into the mushroom body (MB) substitutes for the sugar water reward (Menzel and Erber 1978; Hammer and Menzel 1998). These experiments together with others indicate that the MB have a key role in olfactory learning and that octopaminergic neurons act in association between the US and the CS (Hammer and Menzel 1998).

Olfactory information is provided to the CAs of the MB by projection neurons from the antennal lobe (AL) via two different tracts (Fig. 6.4b, blue line). Mushroom body Kenyon cells integrate olfactory information with other sensory modalities. For associative odor learning in bees, the ventral unpaired median cell of the maxillary neuromere 1 (VUMmx1 neuron) provides information of the US (sucrose) to the MB (Fig. 6.4b, red line and inset) (Hammer 1993). VUMmx1 has its cell body at the midline of the gnathal ganglia (Fig. 6.4b) and dendrites in the dorsal GNG (not shown in Fig. 6.4b inset). It responds to sucrose stimulation and has axonal terminals in the AL, the CA, and the lateral horn, which covers all major olfactory brain areas. Hammer discovered that electrical stimulation of VUMmx1 substitutes for the US (sucrose) in association of the CS (odor) and the US (Hammer 1993). Taken together with the fact that VUMmx1 is an octopaminergic neuron (Schröter et al. 2007),

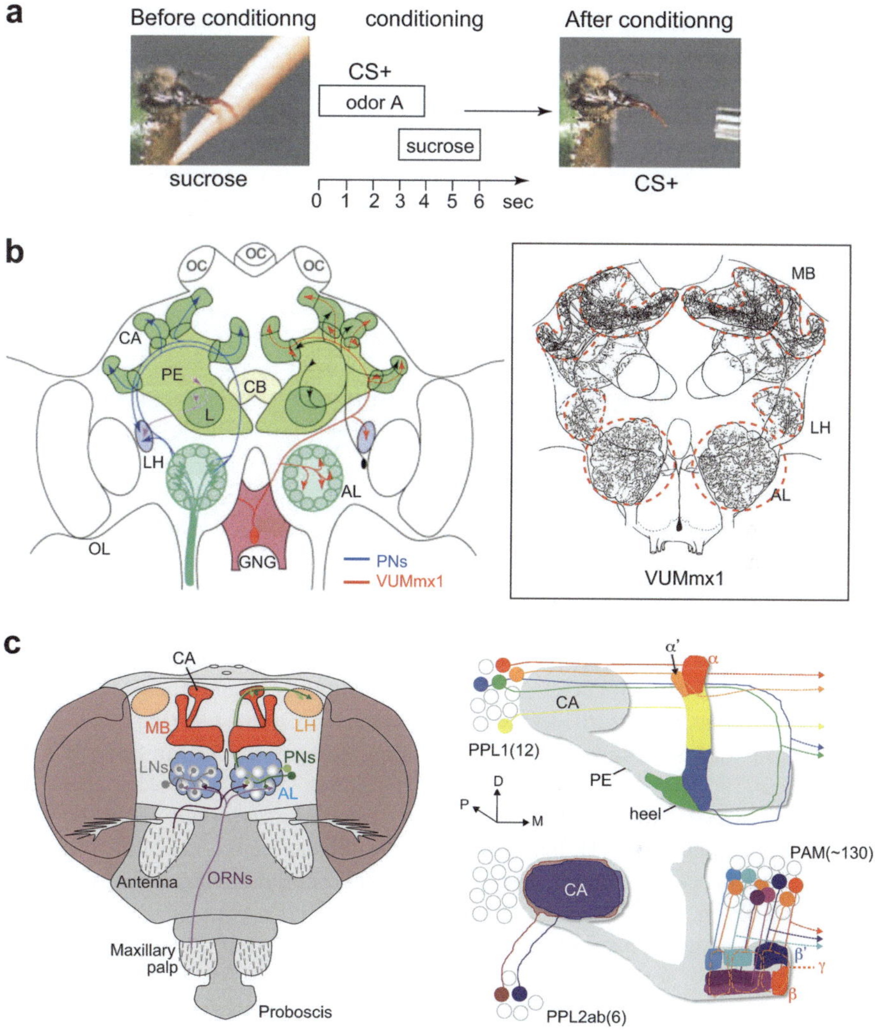

Fig. 6.4 Odor learning and memory in the honeybee and fruit fly. (**a**) Classical olfactory conditioning of the proboscis extension reflex in the honeybee. *CS+* conditioned stimulus. (**b**) Neural pathways for odor learning (*left*) and morphology of the ventral unpaired median cell of the maxillary neuromere 1, VUMmx1 (*right*) (Modified from Menzel and Giurfa 2001 and Hammer and Menzel 1995, with kind permission). *AL* antennal lobe, *CA* calyx of the mushroom body, *CB* central body, *MB* mushroom body, *OL* optic lobe, *PE* peduncle, *PNs* projection neurons, *LH* lateral horn. (**c**) Olfactory pathway (*left*) and dopaminergic neuron clusters for learning and memory (*right*) in the *Drosophila* mushroom body. Each protocerebral posterior lateral (PPL) neuron innervates a different compartment in the α, α′-lobe, and the heel or the CA; PAM neurons innervate different compartments in the β, β′-, and γ-lobe. Most PPL1 and some PAM neurons have projections to a similar zone in the contralateral MB (shown as *dotted lines* with *arrowheads*). *LNs* local interneurons of the AL, *ORNs* olfactory receptor neurons, *PAM* protocerebral anterior medial (Modified from Perisse et al. 2013a, b, and Waddell 2013, with kind permission)

VUMmx1 contributes to associative odor learning by transmitting octopamine to the projection neurons in the olfactory pathway and to the CA of the MB.

In addition to octopamine, dopamine also has a role in associative odor learning (Mizunami et al. 2009; Unoki et al. 2005; Liu et al. 2012; Aso et al. 2012; Burke et al. 2012). Especially in fruit flies, dopaminergic (DA) neurons contribute to both appetitive and aversive learning as in mammals (Aso et al. 2012; Liu et al. 2012; Burke et al. 2012; Perisse et al. 2013a, b; Guven-Ozkan and Davis 2014; Owald and Waddell 2015). In associative odor learning, flies learn in a T-maze assay to associate a particular odor (CS) with either a weak electric shock (punishment) or sucrose or water (reward) as unconditioned stimuli (US).

The CS pathway in the fly brain is basically the same as that in honeybees (Fig. 6.4c, left). Odor information (CS) is transferred to the CAs by projection neurons, which make synapses with the dendrites of intrinsic mushroom body neurons (Kenyon cells). The MB consists of three types of the Kenyon cells, α/β-, α′/β′-, and γ neurons, corresponding to different subregions in the lobes. These subregions are further divided into several compartments by innervation of different DA neurons and mushroom body output neurons (Owald and Waddell 2015; Yamagata et al. 2015).

The US information is conveyed to the lobes of the MB by extrinsic DA neurons. The DA neurons are defined by three different regions of cell bodies: the protocerebral anterior medial (PAM) and the protocerebral posterior lateral 1 and 2ab (PPL1, PPL2ab) clusters. These neurons mainly extend their processes to restricted compartments in the lobes of the MB, the heel, or the CA (Fig. 6.4c, right). Some PPL neurons mediate electric shock stimulation to the MB in odor learning (Waddell 2013; Aso et al. 2012). The PAM cluster provides mostly appetitive reinforcing signals such as sugar and water stimulation (Burke et al. 2012; Liu et al. 2012; Lin et al. 2014), but also aversive reinforcement (Aso et al. 2012). In addition, the US component of sugar is mediated by octopaminergic (OA) neurons (not shown in Fig. 6.4c) upstream of the DA neurons (Burke et al. 2012). VUMmx1-like OA neurons were also found in the fly brain (Busch et al. 2009), and mediate appetitive learning not only in the MB but also at the level of AL projection neurons (Keene and Waddell 2007). Other MB-associated neurons, such as the serotonergic dorsal paired medial neurons (DPMn) and the GABAergic anterior paired lateral neurons (APLn) innervate all lobes of the MB (DPMn) or the whole MB (APLn). The motivational state influences appetitive memory also through DA neurons innervating the MB (Krashes et al. 2009).

The *Drosophila* MB is also crucial for visual and gustatory learning (Vogt et al. 2014; Kirkhart and Scott 2015). Recent studies on MB outputs interestingly suggest that the MB also influences naive odor responses, independent of learning or memory (Lewis et al. 2015). Olfactory networks in insects and vertebrates share a number of similarities (Hildebrandt and Shepherd 1997; Leinwand and Chalasani 2011; Li and Liberless 2015), which may be best illustrated by the common chemotopic organization of processing units, the olfactory glomeruli, at the first stage of olfactory processing in the AL (insect) and olfactory bulb (vertebrates), sparse coding in secondary processing areas (piriform cortex, mushroom bodies), but may even extend to mechanisms of olfactory learning involving DA neurons in rodents as in *Drosophila* (Rosenkranz and Grace 2002).

6.2.3 Sky Compass Orientation

Many animals navigate over long distances in search for food, mating partners, or otherwise favorable habitats. Such traveling over long distances requires mechanisms of orientation based on environmental cues including sky compass cues, landmarks, olfactory cues, and even geomagnetic signals (Alerstam et al. 2001; Yamamoto et al. 2013; Lohmann et al. 2004; Biro et al. 2007).

Sky compass orientation refers to a spatial orientation behavior based on celestial cues. Except for the stars and the Milky Way, these cues are defined by the position of the sun or the moon (Gould 1998; Srinivasan 2011). Direct sunlight, which is unpolarized, is scattered in the atmosphere, resulting in a pattern of polarized light (POL), a spectral gradient, and an intensity gradient across the sky (Fig. 6.5a) (Coemans et al. 1994; Srinivasan 2011; el Jundi et al. 2014). The polarization pattern is characterized by electric field vectors arranged tangentially to concentric circles around the sun and a degree of polarization ranging from 0 % (direct sunlight) to a maximum of up to 75 % at an angular distance of 90° from the sun. Long-wavelength light is dominant in the sky, in particular around the sun, but its contribution relative to short-wavelength light decreases toward the anti-solar azimuth (Coemans et al. 1994). In addition, light intensity is highest around the sun and decreases with increasing distance from the sun.

After von Frisch demonstrated that homing honeybees relying on a sun compass orientation can read the sun‘s horizontal direction (azimuth) from the polarization pattern of the sky (von Frisch 1974), spatial orientation based on a sky compass using sky polarization and other celestial cues has been demonstrated in several insect species, including ants, monarch butterflies, and dung beetles (Rossel and Wehner 1984, 1986; Wehner 2003; Wehner and Müller 2006; Stalleicken et al. 2005; Dacke et al. 2003; el Jundi et al. 2015). These results suggest that celestial cues indicating spatial directions are represented somewhere in the insect brain. The identification of polarized-light sensitive neurons (POL neurons), arranged in a compass-like fashion in the locust brain, suggests that the central complex together with the lateral complex serves a role as an internal sky compass (Vitzthum et al. 2002; Heinze and Homberg 2007; Homberg et al. 2011). Figure 6.5b shows the morphologies of two tangential POL neurons in the lower division of the central body (CBL) of the locust central complex (CX). The neurons have dendrites in small areas in the lateral complex, the lateral and medial bulb (LBU, MBU), and send axons to a particular layer of the CBL. As illustrated for one of the two neurons (TL2), these neurons show polarization opponency, when stimulated with light through a rotating polarizer, resulting in *E*-vector-dependent sinusoidal modulation of spiking activity with excitatory and inhibitory responses at orthogonal *E*-vector orientations (Fig. 6.5b, bottom) (Vitzthum et al. 2002). These POL neurons also

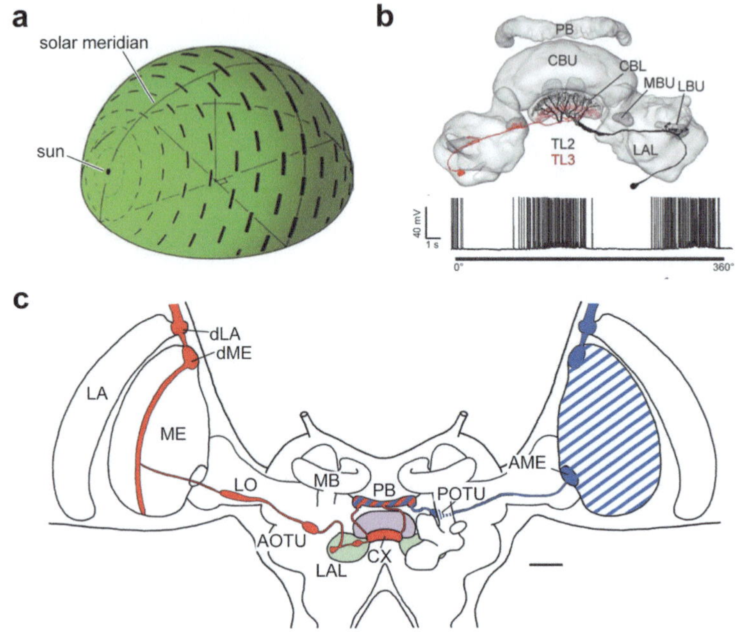

Fig. 6.5 (**a**) Visual cues for compass orientation in the sky (Modified from el Jundi et al. 2014, with kind permission). For explanation, see text. (**b**) Morphology of tangential neurons in the lower division of the central body (CBL) of the locust brain and response of a TL2 neuron to dorsal illumination of the locust though a rotating polarizer. Background activity of the neuron was at 10 imp/s (not shown). *CBU* upper division of the central body, *LAL* lateral accessory lobe, *LBU* lateral bulb, *MBU* medial bulb, *PB* protocerebral bridge, *TL* tangential neuron of the CBL (Modified from Vitzthum et al. 2002, with kind permission). (**c**) Neuronal pathways involved in sky compass orientation in the locust. *AME* accessory medulla, *AOTU* anterior optic tubercle, *CX* central complex, *dLA* dorsal rim area in the lamina, *dME* dorsal rim area in the medulla, *LA* lamina, *LAL* lateral accessory lobe, *LO* lobula, *MB* mushroom body, *ME* medulla, *POTU* posterior optic tubercle, *PB* protocerebral bridge. *Bar* 200 μm (Modified from (el Jundi and Homberg 2010, with kind permission)

respond to unpolarized light from a particular azimuth, likely representing direct sunlight.

In the locust brain (Fig. 6.5c), celestial polarization and spectral information is integrated in several steps in the optic lobe and is then fed to the CX (el Jundi et al. 2014). Photoreceptors in a specialized dorsal rim area of the eye are highly sensitive to the oscillation plane of polarized light and, furthermore, share the same spectral sensitivity (locust, blue sensitive) (Homberg et al. 2011; Schmeling et al. 2015). Their axons project to specific small areas in the lamina and medulla (the dorsal lamina, dLA, and the dorsal medulla, dME). A particular type of medulla tangential neurons has dendritic ramifications in the dME, probably receiving polarized light input, and a process extending along the dorsoventral axis of the medulla, where it might receive azimuth-dependent spectral information from photoreceptors in the remaining eye region (Homberg et al. 2003a). These tangential neurons send axonal fibers to the lower unit of the AOTU. AOTU neurons are sensitive to polarized blue light from zenithal directions and, in addition, show azimuth-dependent UV-green spectral opponency, when stimulated with unpolarized light spots from about 45° elevations (Kinoshita et al. 2007; Pfeiffer and Homberg 2007). One type of AOTU neurons, called TuLAL, feeds integrated celestial cues into the LBU and MBU (Pfeiffer et al. 2005; Pfeiffer and Homberg 2007). Their postsynaptic targets are TL neurons innervating the lower division of the central body (Fig. 6.5b). Columnar neurons transmit the celestial signals to the 16 slices of the protocerebral bridge (PB). Here, a topographic representation of celestial E-vectors covering roughly $2 \times 180°$ is established across the slices by multi-columnar neurons (Heinze and Homberg 2007). Neurons connecting the PB slices to the LAL constitute the outputs from the CX. Finally, descending neurons transfer the sky compass information to the wing and leg motor control centers in the thoracic ganglia (Träger and Homberg 2011).

A second pathway from the optic lobe (Fig. 6.5c, shown in blue) targets the CX via the PB (el Jundi and Homberg 2010). Medulla neurons with large dendritic fields including the dME make contact with the accessory medulla (AME) (el Jundi et al. 2011). AME projection neurons send axons to the POTU, which is connected by multi-columnar neurons to the PB. In flies and cockroaches, the AME is the site of the internal circadian clock (Helfrich-Förster et al. 1998). Therefore, this pathway might feed time information from the endogenous circadian clock to the sky compass network to compensate for daytime-dependent shifts in solar azimuth. The network for processing celestial cues in the locust is highly conserved among different insect species, as illustrated by data from monarch butterflies, bumblebees, and dung beetles, even though some physiological properties of the constituent neurons may be species specific (Heinze and Reppert 2011; el Jundi et al. 2015; Pfeiffer and Kinoshita 2012).

6.2.4 *Circadian System*

As an adaptation to the universal 24-h period of night and day, insects, similar to virtually all organisms, have an internal circadian clock roughly matching the 24-h periodicity on earth (Panda et al. 2002). The internal clock is entrained to the 24-h periodicity by external light or temperature cycles but shows a species-specific free running period under constant conditions. The internal circadian system allows animals to anticipate the day–night change, provides a temporal organization to all body functions, and is essential for a time-compensated sun compass orientation (see Sect. 6.2.3).

Circadian rhythms in fruit flies can be measured in behavioral laboratory assays. Fruit flies are placed in small tubes, their locomotor activity is measured by an infrared optical sensor, and the resulting signals are recorded over days in double-plotted actograms (Fig. 6.6a). Under a cycle of 12 h light and 12 h darkness

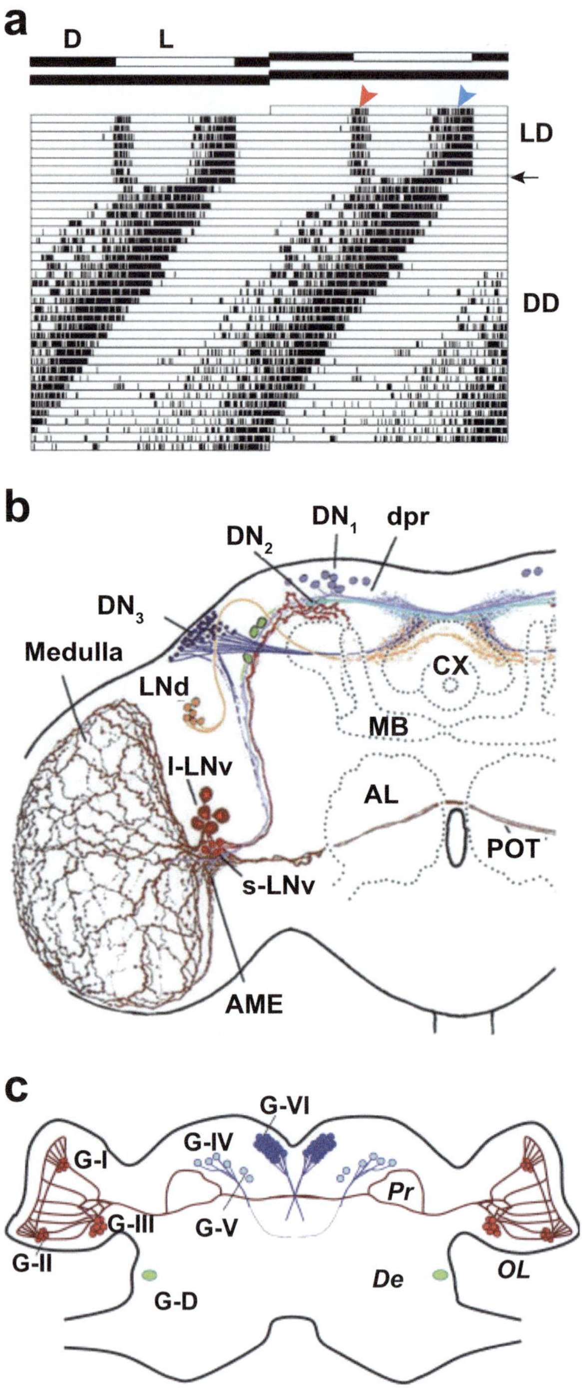

Fig. 6.6 Neural basis of the circadian clock. (**a**) Double-plotted actogram of a wild-type fruit fly. On day 10 (*arrow*), the light regime changes from a 12:12 h light–dark cycle (LD) cycle to DD (constant darkness, *black arrow*). The actogram shows morning (*red arrowhead*) and evening (*blue arrowhead*) activities (Modified from Helfrich-Förster 2004, with kind permission). (**b**) Clock neurons expressing the PERIOD protein in the fruit fly brain. *AL* antennal lobe, *aME* accessory medulla, *CX* central complex, *DN* dorsal neurons, *LN* lateral neurons, lLN_v large ventrolateral neurons, *MB* mushroom body, *POT* posterior optic tract, *dpr* dorsal protocerebrum, $s\text{-}LN_v$ small ventrolateral neurons (Modified from Helfrich-Förster 2004, with kind permission). (**c**) PER-expressing neurons in the cockroach brain. *G* group of cells, *Pr* protocerebrum, *De* deutocerebrum, *OL* optic lobe (From Tomioka and Matsumoto 2010, with kind permission)

(LD), the flies' activities increase around the time after 'lights on' (red arrowhead in Fig. 6.6a) and before 'lights off' (blue arrowhead in Fig. 6.6a), which refers to morning and evening activity, respectively (Helfrich-Förster 2000): this is the common activity pattern in crepuscular insects. In constant darkness (DD) the two activity components are fused and become unimodal. Behavioral assays with eyeless mutants indicate the importance of the optic lobe for maintenance of these activity patterns adjusted to the light regime. Studies on clock gene mutants revealed that two oscillators with different periods contribute to the morning and evening activities (Stoleru et al. 2004; Rieger et al. 2006).

Clock neurons in fruit flies that express *per* (*period*) and other clock genes are concentrated the optic lobe and in the dorsal and lateral protocerebrum. They were named the large and small ventrolateral neurons (l-LN_v, s-LN_v), the dorsal, and the lateral neurons ($DN_{1\text{–}3}$, LN_d) in accordance with the position of their cell bodies (Fig. 6.6b) (Helfrich-Förster 2004). The DNs make three clusters: two lie relatively anteriorly ($DN_{1,2}$), and a third lies more lateral posteriorly (DN_3). The l-LN_v and s-LN_v contain a neuropeptide, pigment-dispersing factor (PDF, shown as red in Fig. 6.6b). The l-LN_v have large dendritic fields in the distalmost layer of the entire medulla and the accessory medulla and project to the contralateral optic lobe via the posterior optic tract. The s-LN_v also arborize in the accessory medulla and project to the dorsal protocerebrum (Helfrich-Förster et al. 1998). The s-LN_v are essential for controlling circadian activity rhythms under DD conditions and release PDF rhythmically from their terminals. The l-LN_v, in contrast, mediate light input from the compound eye to the clock and probably feed light information to the s-LN_v. Mutants with *per* gene expression confined to a particular group of neurons showed that the s-LN_v control the morning activity and LN_d control the evening activity (Stoleru et al. 2004; Helfrich-Förster 2014). The PDF neurons and the LN_d are, therefore, pacemaker neurons for controlling circadian motor activity in fruit flies.

PDF receptors are expressed on most clock neurons, including the PDF-positive s-LN_v, a fifth PDF-negative s-LN_v, half the LN_d, and 7 of 17 DN_1. These neurons also express cryptochrome, a blue light-sensitive photopigment that directly contributes to light entrainment of the clock. The s-LN_v and half the $DN_{1,2}$ appear to control the morning activity by the fact that PDF speeds up the clock whereas the fifth s-LN_v, the remaining half of $DN_{1,2}$, and half of LN_d may contribute to the evening activity by the effect that PDF slows down the clock (Helfrich-Förster 2014).

Clock neurons expressing clock genes were identified in a number of other insects (Sandrelli et al. 2008; Tomioka and Matsumoto 2010). Neurons immunoreactive for PDF in other insects are homologous to the fly PDF neurons and have been characterized quite well in cockroaches. The brain of the cockroach *Blatella germanica* contains seven groups of PER-expressing neurons (Fig. 6.6c). Three of them, containing colocalized PDF, are in the optic lobe (shown in red; Fig. 6.6c). Three other groups are in the dorsal protocerebrum and one group lies in the deutocerebrum (shown in blue and green, respectively; Fig. 6.6c). The cell bodies of PDF neurons in the cockroach *Rhyparobia maderae* (formerly *Leucophaea maderae*) are at the anterior edge of the medulla, arborize in the accessory medulla as the master clock, the lamina, distal medulla, and send their axons toward wide areas in the cerebrum, where they probably induce the circadian rhythm in locomotion (Homberg et al. 2003b; Stengl and Arendt 2016). The branching pattern of PDF neurons in the cerebrum is diverse depending on species. Some PDF neurons in locusts project from the accessory medulla to the CX, possibly serving as time compensation of the internal sky compass (see Sect. 6.2.3).

6.2.5 Courtship Behavior and Sexual Dimorphism of the Brain

Reproductive success of an animal strongly depends on selecting the right partner. Species-specific mating behaviors, therefore, serve to recognize conspecifics, determine their sex, and communicate with the partner using several different sensory modalities. Usually, the behavior of males differs from that of females during mating, corresponding to sexual differences in brain structure and function (Cachero et al. 2010).

The courtship behavior of *Drosophila melanogaster* consists of six discrete steps (Fig. 6.7a) (Greenspan and Ferveur 2000). When a male fly finds a female, the male first orients his body axis toward the female and chases her (orientation). He then touches her abdomen with his forelegs (tapping). After following her, he opens and vibrates one of his wings, producing a courtship song (singing). In response, the female may stop walking, indicating her receptivity. The male then approaches the female from the back and licks her genitalia (licking), and eventually tries to mount her, bending his abdomen ventrally (attempt copulation). When the female accepts the male, she raises her wings and opens her vaginal plate for copulation.

As in other animals, a variety of sensory cues, including visual, olfactory, gustatory, and auditory, contribute to continue from one step to the next during courtship behavior (Yamamoto et al. 1997; Dickson 2008). For example, hydrocarbons on the body act as aphrodisiacs for the opposite sex and are sensed through the gustatory and olfactory system. Male flies transfer a pheromone (*cis*-vaccenyl acetate, cVA), produced in their accessory gland, to the female during copulation, and thereby reduce her sexual attractiveness to other males. The male courtship song contains two components, sine songs and pulse songs, which are species specific. Female flies, therefore, can use the song to discriminate conspecific males. Sensory signals during mating are, therefore, detected by a variety of sensory organs and, after processing in the optic lobe (vision), antennal lobe (olfaction), antennal mechanosensory and motor center (hearing, touch), and the gnathal ganglion (gustation) are transferred to the central brain for initiating a particular behavior (Fig. 6.6b).

Neural circuits underlying courtship behavior in the fruit fly were identified by tracing neurons expressing the *fruitless* (*fru*) gene. *Fruitless* mutant males show either abnormal courtship or complete lack of courtship behavior and often form male–male courtship chains (Ito et al. 1996; Ryner et al. 1996). The *fru* gene acts together with the doublesex (*dsx*) gene in the sex-determination cascade

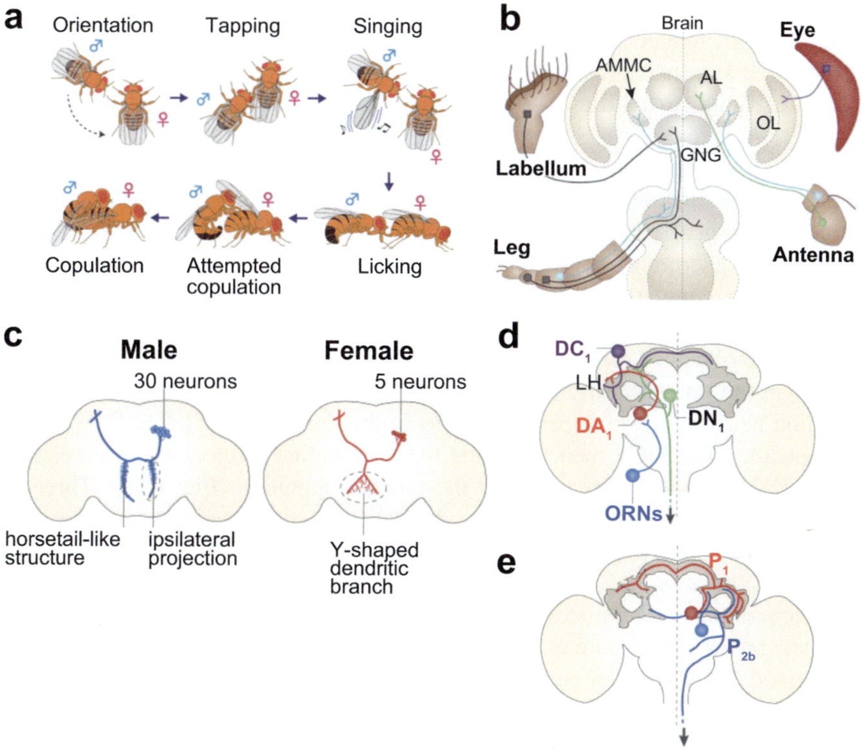

Fig. 6.7 Mating behavior and mating-related neurons in *Drosophila*. (**a**) Sequence of courtship behavior (Modified from Yamamoto et al. 2014, with kind permission). (**b**) Sensory inputs for courtship behavior. *AL* antennal lobe, *AMMC* antennal mechanosensory and motor center, *OL* optic lobe. (**c**) Sexual dimorphism of the medially located neurons just above the antennal lobe (mAL). (**d**) Pathway for male-specific pheromone processing. (**e**) Pathway for the initiation of courtship song ((**b**–**e**) Modified from Yamamoto and Koganezawa 2013, with kind permission)

(Ito et al. 1996; Ryner et al. 1996). Both Fru and Dsx proteins are putative transcription factors. Sex-specific splicing of the *fru* gene occurs in the brain under the control of the promoter P1, one of four promoters resulting in male-specific Fru (M) proteins but no counterpart proteins in females (Yamamoto and Koganezawa 2013); this leads to male-specific neural circuitry underlying courtship behavior and the development of a male-specific muscle, the muscle of Lawrence (Usui-Aoki et al. 2000). About 2000 *fru*-expressing neurons are widely distributed in the sensory, central, and motor system. About one third of these neurons show sex-specific morphological differences (Cachero et al. 2010; Yu et al. 2010). One of the best studied examples of sexual dimorphic *fru*-expressing interneurons is a cluster of cells located medially above the antennal lobe (mAL) (Kimura et al. 2005). The mAL neurons consist of 30 neurons in the male brain, but in females the mAL cluster contains only 5 cells. In males, mAL neurons send processes to the ipsilateral and contralateral GNG and a third fiber to the contralateral protocerebrum of the brain. In females, the mAL neurons lack the process to the ipsilateral side of the GNG, and their processes in the contralateral GNG have a Y-shaped form (Fig. 6.7c). This sexual dimorphism in the mAL neurons is induced by the Fru protein.

Knowledge about the circuits involved in courtship behavior from the sensory input to the motor output is still limited. An olfactory circuit contributing to a particular aspect of courtship consists of only four types of neuron detecting the male-specific pheromone cVA (Ruta et al. 2010). The cVA is transferred to the female during copulation and inhibits courtship behavior in a second male. A sexual dimorphic glomerulus of the antennal lobe, called DA1, receives input from cVA-specific olfactory receptor neurons (ORNs). The projection neurons from the DA1 target the lateral horn via the mushroom body similar to other antennal-lobe projection neurons and connect to neurons in the dorsal cluster1 (DC1), which is male specific. The DC1 then transmits the cVA-induced signal to a descending neuron, DN1, which projects toward the thoracic ganglion (Fig. 6.7d). Three of the four cell types involved, the ORNs, the DA1 projection neurons, and the DC1 interneurons, express the Fru protein.

A circuit including the male-specific P1 cluster likely contributes to induce the courtship song because females that had received a masculinized P1 clone exhibited male-type behavior (Kimura et al. 2008). In the P1 neurons, *fru* and *dsx* genes are coexpressed. The P1 cluster consists of 20 neurons. Their cell bodies lie in the dorsal posterior brain near the mushroom body and their branches extend bilaterally into both hemispheres of the protocerebrum (Fig. 6.7e). The P1 neurons can be activated by visual stimuli and by touching the female abdomen with the male's forelegs. Thus, the P1 neurons receive multimodal sensory information (Koganezawa et al. 2010; Kohatsu et al. 2011). The information in the P1 is apparently transmitted to descending neurons (not shown in Fig. 6.7e) in the lateral protocerebrum. These descending neurons project to the thoracic ganglion, containing the motor control center producing courtship song through wing vibrations.

6.2.6 Conclusions

Considerable advances during the past decades, especially in understanding the functional organization of the brain of *Drosophila*, have led to detailed knowledge on neuronal networks underlying a variety of behavioral elements, such as optomotor responses, as well as complete behavioral sequences, such as courtship behavior. For many of the network architectures and operations, striking similarities have been found between vertebrates and insects, pointing at commonly achieved optimal solutions or even suggesting common deep evolutionary origins. Similarities are highly evident when considering neural networks and mechanisms underlying particular elements of behavior such as optomotor responses (Borst and Helmstaedter 2015), circadian control (Helfrich-Förster 2004), or neural plasticity underlying memory formation (Glanzman 2010), and even point to strikingly similar network properties underlying more complex tasks such as spatial orientation (Seelig and Jayaraman 2013, 2015). Not surprisingly, however, differences across species are pronounced when considering behaviors that are highly species specific such as courtship behavior, which involves sequences of behavioral actions between the two mating partners that partly serve to recognize a conspecific and distinguish it from mates of related species. Nevertheless, even here, common principles might emerge when analyzing sexual differences in brain organization underlying mating behavior in different animal taxa (Forger and de Vries 2010).

References

Alerstam T, Gudmundsson GA, Green M, Hedenstrom A (2001) Migration along orthodromic sun compass routes by arctic birds. Science 291:300–303

Arendt D, Nübler-Jung K (1996) Common ground plans in early brain development in mice and flies. Bioessays 18:255–259

Aso Y, Herb A, Ogueta M, Siwanowicz I, Templier T, Friedrich AB, Ito K, Scholz H, Tanimoto H (2012) Three dopamine pathways induce aversive odor memories with different stability. PLoS Genet 8, e1002768

Azevedo FA, Carvalho LR, Grinberg LT, Farfel JM, Ferretti RE, Leite RE, Jacob Filho W, Lent R, Herculano-Houzel S (2009) Equal numbers of neuronal and nonneuronal cells make the human brain an isometrically scaled-up primate brain. J Comp Neurol 513:532–541

Barlow HB, Hill RM (1963) Selective sensitivity to direction of movement in ganglion cells of the rabbit retina. Science 139:412–414

Behnia R, Clark DA, Carter AG, Clandinin TR, Desplan C (2014) Processing properties of ON and OFF pathways for *Drosophila* motion detection. Nature 512:427–430

Biro D, Freeman R, Meade J, Roberts S, Guilford T (2007) Pigeons combine compass and landmark guidance in familiar route navigation. Proc Natl Acad Sci USA 104:7471–7476

Borst A, Euler T (2011) Seeing things in motion: models, circuits, and mechanisms. Neuron 71:974–994

Borst A, Haag J (2002) Neural networks in the cockpit of the fly. J Comp Physiol A 188:419–437

Borst A, Helmstaedter M (2015) Common circuit design in fly and mammalian motion vision. Nature Neurosci 18:1067–1076

Borst A, Haag J, Reiff DF (2010) Fly motion vision. Annu Rev Neurosci 33:49–70

Brandt R, Rohlfing T, Rybak J, Krofczik S, Maye A, Westerhoff M, Hege HC, Menzel R (2005) Three-dimensional average-shape atlas of the honeybee brain and its applications. J Comp Neurol 492:1–19

Burke CJ, Huetteroth W, Owald D, Perisse E, Krashes MJ, Das G, Gohl D, Silies M, Certel S, Waddell S (2012) Layered reward signalling through octopamine and dopamine in *Drosophila*. Nature 492:433–437

Busch S, Selcho M, Ito K, Tanimoto H (2009) A map of octopaminergic neurons in the *Drosophila* brain. J Comp Neurol 513:643 667

Cachero S, Ostrovsky AD, Yu JY, Dickson BJ, Jefferis GS (2010) Sexual dimorphism in the fly brain. Curr Biol 20:1589–1601

Chittka L, Niven J (2009) Are bigger brains better? Curr Biol 19:R995–R1008

Coemans MA, Vos Hzn JJ, Nuboer JF (1994) The relation between celestial colour gradients and the position of the sun, with regard to the sun compass. Vision Res 34:1461–1470

Dacke M, Nilsson DE, Scholtz CH, Byrne M, Warrant EJ (2003) Animal behaviour: insect orientation to polarized moonlight. Nature 424:33

Dickson BJ (2008) Wired for sex: the neurobiology of *Drosophila* mating decisions. Science 322:904–909

Dreyer D, Vitt H, Dippel S, Goetz B, el Jundi B, Kollmann M, Huetteroth W, Schachtner J (2010) 3D standard brain of the red flour beetle *Tribolium castaneum*: a tool to study metamorphic development and adult plasticity. Front Syst Neurosci 4:3

Dunbier JR, Wiederman SD, Shoemaker PA, O'Carroll DC (2012) Facilitation of dragonfly target-detecting neurons by slow moving features on continuous paths. Front Neural Circuits 6:79

Ehmer B, Gronenberg W (2002) Segregation of visual input to the mushroom bodies in the honeybee (*Apis mellifera*). J Comp Neurol 451:362–373

el Jundi B, Homberg U (2010) Evidence for the possible existence of a second polarization-vision pathway in the locust brain. J Insect Physiol 56:971–979

el Jundi B, Heinze S, Lenschow C, Kurylas A, Rohlfing T, Homberg U (2009a) The locust standard brain: a 3D standard of the central complex as a platform for neural network analysis. Front Syst Neurosci 3:21

el Jundi B, Huetteroth W, Kurylas AE, Schachtner J (2009b) Anisometric brain dimorphism revisited: implementation of a volumetric 3D standard brain in *Manduca sexta*. J Comp Neurol 517:210–225

el Jundi B, Pfeiffer K, Homberg U (2011) A distinct layer of the medulla integrates sky compass signals in the brain of an insect. PLoS One 6, e27855

el Jundi B, Pfeiffer K, Heinze S, Homberg U (2014) Integration of polarization and chromatic cues in the insect sky compass. J Comp Physiol A 200:575–589

el Jundi B, Warrant EJ, Byrne MJ, Khaldy L, Baird E, Smolka J, Dacke M (2015) Neural coding underlying the cue preference for celestial orientation. Proc Natl Acad Sci USA 112:11395–11400

Farris SM (2008) Tritocerebral tract input to the insect mushroom bodies. Arthropod Struct Dev 37:492–503

Farris SM (2011) Are mushroom bodies cerebellum-like structures? Arthropod Struct Dev 40:368–379

Farris SM (2013) Evolution of complex higher brain centers and behaviors: behavioral correlates of mushroom body elaboration in insects. Brain Behav Evol 82:9–18

Farris SM (2015) Evolution of brain elaboration. Philos Trans R Soc Lond B Biol Sci 370:20150054

Farris SM, Roberts NS (2005) Coevolution of generalist feeding ecologies and gyrencephalic mushroom bodies in insects. Proc Natl Acad Sci USA 102:17394–17399

Fischbach K-F, Dittrich APM (1989) The optic lobe of *Drosophila melanogaster*. I. A Golgi analysis of wild-type structure. Cell Tissue Res 258:441–475

Forger NG, de Vries GJ (2010) Cell death and sexual differentiation of behavior: worms, flies, and mammals. Curr Opin Neurobiol 20:776–783

Fukushima R, Kanzaki R (2009) Modular subdivision of mushroom bodies by Kenyon cells in the silkmoth. J Comp Neurol 513:315–330

Galizia CG (2014) Olfactory coding in the insect brain: data and conjectures. Eur J Neurosci 39:1784–1795

Galizia CG, Rössler W (2010) Parallel olfactory systems in insects: anatomy and function. Annu Rev Entomol 55:399–420

Giurfa M (2003) Cognitive neuroethology: dissecting non-elemental learning in a honeybee brain. Curr Opin Neurobiol 13:726–735

Glanzman DL (2010) Common mechanisms of synaptic plasticity in vertebrates and invertebrates. Curr Biol 20:R31–R36

Gould JL (1998) Sensory bases of navigation. Curr Biol 8:R731–R738

Gouranton J (1964) Contribution a l'étude de la structure des ganglions céréböides de *Locusta migratoria migratorioides*. Bull Soc Zool Fr 89:785–797

Greenspan RJ, Ferveur JF (2000) Courtship in *Drosophila*. Annu Rev Genet 34:205–232

Gronenberg W (1999) Modality-specific segregation of input to ant mushroom bodies. Brain Behav Evol 54:85–95

Guven-Ozkan T, Davis RL (2014) Functional neuroanatomy of *Drosophila* olfactory memory formation. Learn Mem 21:519–526
Hammer M (1993) An identified neuron mediates the unconditioned stimulus in associative olfactory learning in honeybees. Nature 366:59–63
Hammer M, Menzel R (1995) Learning and memory in the honeybee. J Neurosci 15:1617–1630
Hammer M, Menzel R (1998) Multiple sites of associative odor learning as revealed by local brain microinjections of octopamine in honeybees. Learn Mem 5:146–156
Hausen K, Egelhaaf M (1989) Neural mechanisms of visual course control in insects. In: Stavenga DG, Hardie RC (eds) Facets of vision. Springer, Berlin/Heidelberg, pp 391–424
Heinze S, Homberg U (2007) Maplike representation of celestial *E*-vector orientations in the brain of an insect. Science 315:995–997
Heinze S, Homberg U (2008) Neuroarchitecture of the central complex of the desert locust: intrinsic and columnar neurons. J Comp Neurol 511:454–478
Heinze S, Reppert SM (2011) Sun compass integration of skylight cues in migratory monarch butterflies. Neuron 69:345–358
Heinze S, Reppert SM (2012) Anatomical basis of sun compass navigation. I: The general layout of the monarch butterfly brain. J Comp Neurol 520:1599–1628
Heinze S, Florman J, Asokaraj S, el Jundi B, Reppert SM (2012) Anatomical basis of sun compass navigation. II: The neuronal composition of the central complex of the monarch butterfly. J Comp Neurol 521:267–298
Helfrich-Förster C (2000) Differential control of morning and evening components in the activity rhythm of *Drosophila melanogaster*: sex-specific differences suggest a different quality of activity. J Biol Rhythms 15:135–154
Helfrich-Förster C (2004) The circadian clock in the brain: a structural and functional comparison between mammals and insects. J Comp Physiol A 190:601–613
Helfrich-Förster C (2014) From neurogenetic studies in the fly brain to a concept in circadian biology. J Neurogenet 28:329–347
Helfrich-Förster C, Stengl M, Homberg U (1998) Organization of the circadian system in insects. Chronobiol Int 15:567–594
Herculano-Houzel S, Mota B, Lent R (2006) Cellular scaling rules for rodent brains. Proc Natl Acad Sci USA 103:12138–12143
Hildebrandt JG, Shepherd GM (1997) Mechanisms of olfactory discrimination: converging evidence for common principles across phyla. Annu Rev Neurosci 20:595–631
Homberg U (2008) Evolution of the central complex in the arthropod brain with respect to the visual system. Arthropod Struct Dev 37:347–362
Homberg U, Montague RA, Hildebrand JG (1988) Anatomy of antenno-cerebral pathways in the brain of the sphinx moth *Manduca sexta*. Cell Tissue Res 254:255–281
Homberg U, Hofer S, Pfeiffer K, Gebhardt S (2003a) Organization and neural connections of the anterior optic tubercle in the brain of the locust, *Schistocerca gregaria*. J Comp Neurol 462:415–430
Homberg U, Reischig T, Stengl M (2003b) Neural organization of the circadian system of the cockroach *Leucophaea maderae*. Chronobiol Int 20:577–591
Homberg U, Heinze S, Pfeiffer K, Kinoshita M, el Jundi B (2011) Central neural coding of sky polarization in insects. Philos Trans R Soc Lond B Biol Sci 366:680–687
Ito H, Fujitani K, Usui K, Shimizu-Nishikawa K, Tanaka S, Yamamoto D (1996) Sexual orientation in *Drosophila* is altered by the satori mutation in the sex-determination gene fruitless that encodes a zinc finger protein with a BTB domain. Proc Natl Acad Sci USA 93:9687–9692
Ito K, Shinomiya K, Ito M, Armstrong JD, Boyan G, Hartenstein V, Harzsch S, Heisenberg M, Homberg U, Jenett A, Keshishian H, Restifo LL, Rössler W, Simpson JH, Strausfeld NJ, Strauss R, Vosshall LB (2014) A systematic nomenclature for the insect brain. Neuron 81:755–765
Joesch M, Schnell B, Raghu SV, Reiff DF, Borst A (2010) ON and OFF pathways in *Drosophila* motion vision. Nature 468:300–304
Kandel ER (2001) The molecular biology of memory storage: a dialogue between genes and synapses. Science 294:1030–1038

Kazawa T, Namiki S, Fukushima R, Terada M, Soo K, Kanzaki R (2009) Constancy and variability of glomerular organization in the antennal lobe of the silkmoth. Cell Tissue Res 336:119–136
Keene AC, Waddell S (2007) *Drosophila* olfactory memory: signal genes to complex neural circuits. Nat Rev 8:341–354
Kimura K, Ote M, Tazawa T, Yamamoto D (2005) Fruitless specifies sexually dimorphic neural circuitry in the *Drosophila* brain. Nature 438:229–233
Kimura K, Hachiya T, Koganezawa M, Tazawa T, Yamamoto D (2008) Fruitless and doublesex coordinate to generate male-specific neurons that can initiate courtship. Neuron 59:759–769
Kinoshita M, Pfeiffer K, Homberg U (2007) Spectral properties of identified polarized-light sensitive interneurons in the brain of the desert locust *Schistocerca gregaria*. J Exp Biol 210:1350–1361
Kinoshita M, Shimohigasshi M, Tominaga Y, Arikawa K, Homberg U (2015) Topographically distinct visual and olfactory inputs to the mushroom body in the swallowtail butterfly, *Papilio xuthus*. J Comp Neurol 523:162–182
Kirkhart C, Scott K (2015) Gustatory learning and processing in the *Drosophila* mushroom bodies. J Neurosci 35:5950–5958
Koganezawa M, Haba D, Matsuo T, Yamamoto D (2010) The shaping of male courtship posture by lateralized gustatory inputs to male-specific interneurons. Curr Biol 20:1–8
Kohatsu S, Koganezawa M, Yamamoto D (2011) Female contact activates male-specific interneurons that trigger stereotypic courtship behavior in *Drosophila*. Neuron 69:498–508
Kolmes SA (1983) Ecological and sensory aspects of prey capture by the whirligig beetle *Dineutes discolor* (Coleoptera: Gyrinidae). J New York Entomol Soc 91:405–412
Krashes MJ, DasGupta S, Vreede A, White B, Armstrong JD, Waddell S (2009) A neural circuit mechanism integrating motivational state with memory expression in *Drosophila*. Cell 139:416–427
Kurylas AE, Rohlfing T, Krofczik S, Jenett A, Homberg U (2008) Standardized atlas of the brain of the desert locust, *Schistocerca gregaria*. Cell Tissue Res 333:125–145
Leinwand SG, Chalasani SH (2011) Olfactory networks: from sensation to perception. Curr Opin Gen Dev 21:806–811
Leitinger G, Pabst MA, Kral K (1999) Serotonin-immunoreactive neurones in the visual system of the praying mantis: an immunohistochemical, confocal laser scanning and electron microscopic study. Brain Res 823:11–23
Lewis LP, Siju KP, Aso Y, Friedrich AB, Bulteel AJ, Rubin GM, Grunwald Kadow IC (2015) A higher brain circuit for immediate integration of conflicting sensory information in *Drosophila*. Curr Biol 25:2203–2214
Li Q, Liberless SD (2015) Aversion and attraction through olfaction. Curr Biol 25:R120–R129
Lichtneckert R, Reichert H (2005) Insights into the urbilaterian brain: conserved genetic patterning mechanisms in insect and vertebrate brain development. Heredity (Edinb) 94:465–477
Lin C, Strausfeld NJ (2012) Visual inputs to the mushroom body calyces of the whirligig beetle *Dineutus sublineatus*: modality switching in an insect. J Comp Neurol 520:2562–2574
Lin C, Strausfeld NJ (2013) A precocious adult visual center in the larva defines the unique optic lobe of the split-eyed whirligig beetle *Dineutus sublineatus*. Front Zool 10:7
Lin S, Owald D, Chandra V, Talbot C, Huetteroth W, Waddell S (2014) Neural correlates of water reward in thirsty *Drosophila*. Nat Neurosci 17:1536–1542
Liu C, Placais PY, Yamagata N, Pfeiffer BD, Aso Y, Friedrich AB, Siwanowicz I, Rubin GM, Preat T, Tanimoto H (2012) A subset of dopamine neurons signals reward for odour memory in *Drosophila*. Nature 488:512–516
Lohmann KJ, Lohmann CM, Ehrhart LM, Bagley DA, Swing T (2004) Animal behaviour: geomagnetic map used in sea-turtle navigation. Nature 428:909–910
Maisak MS, Haag J, Ammer G, Serbe E, Meier M, Leonhardt A, Schilling T, Bahl A, Rubin GM, Nern A, Dickson BJ, Reiff DR, Hopp E, Borst A (2013) A directional tuning map of *Drosophila* elementary motion detectors. Nature 500:212–216
Matsuo E, Kamikouchi A (2013) Neural encoding of sound, gravity, and wind in the fruit fly. J Comp Physiol A 199:253–262

Menzel R (2001) Searching for the memory trace in a mini-brain, the honeybee. Learn Mem 8:53–62

Menzel R (2014) The insect mushroom body, an experience-dependent recoding device. J Physiol Paris 108:84–95

Menzel R, Erber J (1978) Learning and memory in bees. Sci Am 239(1):80–87

Menzel R, Giurfa M (2001) Cognitive architecture of a mini-brain: the honeybee. Trends Cogn Sci 5:62–71

Mizunami M, Unoki S, Mori Y, Hirashima D, Hatano A, Matsumoto Y (2009) Roles of octopaminergic and dopaminergic neurons in appetitive and aversive memory recall in an insect. BMC Biol 7:46

Mizunami M, Hamanaka Y, Nishino H (2015) Toward elucidating diversity of neural mechanisms underlying insect learning. Zool Lett 1:8

Mobbs PG (1985) Brain structure. In: Kerkut GA, Gilbert LI (eds) Comparative insect physiology, biochemistry and pharmacology, vol 5. Pergamon Press, New York, pp 299–370

Nishino H, Iwasaki M, Yasuyama K, Hongo H, Watanabe H, Mizunami M (2012) Visual and olfactory input segregation in the mushroom body calyces in a basal neopteran, the American cockroach. Arthropod Struct Dev 41:3–16

Okamura JY, Strausfeld NJ (2007) Visual system of calliphorid flies: motion- and orientation-sensitive visual interneurons supplying dorsal optic glomeruli. J Comp Neurol 500:189–208

Otsuna H, Ito K (2006) Systematic analysis of the visual projection neurons of *Drosophila melanogaster.* I. Lobula-specific pathways. J Comp Neurol 497:928–958

Owald D, Waddell S (2015) Olfactory learning skews mushroom body output pathways to steer behavioral choice in *Drosophila*. Curr Opin Neurobiol 35:178–184

Panda S, Hogenesch JB, Kay SA (2002) Circadian rhythms from flies to human. Nature 417:329–335

Paulk AC, Phillips-Portillo J, Dacks AM, Fellous JM, Gronenberg W (2008) The processing of color, motion, and stimulus timing are anatomically segregated in the bumblebee brain. J Neurosci 28:6319–6332

Perisse E, Burke C, Huetteroth W, Waddell S (2013a) Shocking revelations and saccharin sweetness in the study of *Drosophila* olfactory memory. Curr Biol 23:R752–R763

Perisse E, Yin Y, Lin AC, Lin S, Huetteroth W, Waddell S (2013b) Different Kenyon cell populations drive learned approach and avoidance in *Drosophila*. Neuron 79:945–956

Pfeiffer K, Homberg U (2007) Coding of azimuthal directions via time-compensated combination of celestial compass cues. Curr Biol 17:960–965

Pfeiffer K, Homberg U (2014) Organization and functional roles of the central complex in the insect brain. Annu Rev Entomol 59:165–184

Pfeiffer K, Kinoshita M (2012) Segregation of visual inputs from different regions of the compound eye in two parallel pathways through the anterior optic tubercle of the bumblebee (*Bombus ignitus*). J Comp Neurol 520:212–229

Pfeiffer K, Kinoshita M, Homberg U (2005) Polarization-sensitive and light-sensitive neurons in two parallel pathways passing through the anterior optic tubercle in the locust brain. J Neurophysiol 94:3903–3915

Reichardt W (1987) Evaluation of optical motion information by movement detectors. J Comp Physiol A 161:533–547

Ribi W (1987) Anatomical identification of spectral receptor types in the retina and lamina of the Australian orchard butterfly, *Papilio aegeus aegeus* D. Cell Tissue Res 247:393–407

Rieger D, Shafer OT, Tomioka K, Helfrich-Förster C (2006) Functional analysis of circadian pacemaker neurons in *Drosophila melanogaster*. J Neurosci 26:2531–2543

Roberts AC, Glanzman DL (2003) Learning in *Aplysia*: looking at synaptic plasticity from both sides. Trends Neurosci 26:662–670

Rosenkranz JA, Grace AA (2002) Dopamine-mediated modulation of odour-evoked amygdala potentials during pavlovian conditioning. Nature 417:282–287

Rospars JP, Hildebrand JG (1992) Anatomical identification of glomeruli in the antennal lobes of the male sphinx moth *Manduca sexta*. Cell Tissue Res 270:205–227

Rossel S, Wehner R (1984) Celestial orientation in bees: the use of spectral cues. J Comp Physiol A 155:605–613
Rossel S, Wehner R (1986) Polarization vision in bees. Nature 323:128–131
Ruta V, Datta SR, Vasconcelos ML, Freeland J, Looger LL, Axel R (2010) A dimorphic pheromone circuit in *Drosophila* from sensory input to descending output. Nature 468:686–690
Ryner LC, Goodwin SF, Castrillon DH, Anand A, Villella A, Baker BS, Hall JC, Taylor BJ, Wasserman SA (1996) Control of male sexual behavior and sexual orientation in *Drosophila* by the fruitless gene. Cell 87:1079–1089
Sandrelli F, Costa R, Kyriacou CP, Rosato E (2008) Comparative analysis of circadian clock genes in insects. Insect Mol Biol 17:447–463
Sane SP, Dieudonné A, Willis M, Daniel TL (2007) Antennal mechanosensors mediate flight control in moths. Science 315:863–866
Schachtner J, Schmidt M, Homberg U (2005) Organization and evolutionary trends of primary olfactory brain centers in Tetraconata (Crustacea + Hexapoda). Arthropod Struct Dev 34:257–299
Schmeling F, Tegtmeier J, Kinoshita M, Homberg U (2015) Photoreceptor projections and receptive fields in the dorsal rim area and main retina of the locust eye. J Comp Physiol A 201:427–440
Schröter U, Malun D, Menzel R (2007) Innervation pattern of suboesophageal ventral unpaired median neurones in the honeybee brain. Cell Tissue Res 327:647–667
Seelig JD, Jayaraman V (2013) Feature detection and orientation tuning in the *Drosophila* central complex. Nature 503:262–266
Seelig JD, Jayaraman V (2015) Neural dynamics for landmark orientation and angular path integration. Nature 521:186–191
Sjöholm M, Sinakevitch I, Strausfeld NJ, Ignell R, Hansson BS (2006) Functional division of intrinsic neurons in the mushroom bodies of male *Spodoptera littoralis* revealed by antibodies against aspartate, taurine, FMRF-amide, Mas-allatotropin and DC0. Arthropod Struct Dev 35:153–168
Srinivasan MV (2011) Honeybees as a model for the study of visually guided flight, navigation, and biologically inspired robotics. Physiol Rev 91:413–460
Stalleicken J, Mukhida M, Labhart T, Wehner R, Frost B, Mouritsen H (2005) Do monarch butterflies use polarized skylight for migratory orientation? J Exp Biol 208:2399–2408
Staudacher E, Gebhardt MJ, Dürr V (2005) Antennal movements and mechanoreception: neurobiology of active tactile sensors. Adv Insect Physiol 32:49–205
Stengl M, Arendt A (2016) Peptidergic circadian clock circuits in the Madeira cockroach. Curr Opin Neurobiol 41:44–52
Stoleru D, Peng Y, Agosto J, Rosbash M (2004) Coupled oscillators control morning and evening locomotor behaviour of *Drosophila*. Nature 431:862–868
Stopfer M (2014) Central processing in the mushroom bodies. Curr Opin Insect Sci 6:99–103
Strausfeld NJ (1970) The optic lobes of Diptera. Philos Trans R Soc Lond B 258:135–223
Strausfeld NJ (1976) Atlas of an insect brain. Springer, Berlin
Strausfeld NJ (1989) Beneath the compound eye: neuroanatomical analysis and physiological correlates in the study of insect vision. In: Stavenga DG, Hadie RC (eds) Facets of vision. Springer, Heidelberg/Berlin, pp 317–359
Strausfeld NJ (2012) Arthropod brains. Belknap Press/Harvard University Press, Cambridge
Strausfeld NJ, Bassemir U, Singh RN, Bacon JP (1984) Organizational principles of outputs from dipteran brains. J Insect Physiol 30:73–93
Strausfeld NJ, Hansen L, Li Y, Gomez RS, Ito K (1998) Evolution, discovery, and interpretations of arthropod mushroom bodies. Learn Mem 5:11–37
Takemura SY, Kinoshita M, Arikawa K (2005) Photoreceptor projection reveals heterogeneity of lamina cartridges in the visual system of the Japanese yellow swallowtail butterfly, *Papilio xuthus*. J Comp Neurol 483:341–350
Takemura SY, Bharioke A, Lu Z, Nern A, Vitaladevuni S, Rivlin PK, Katz WT, Olbris DJ, Plaza SM, Winston P, Zhao T, Horne JA, Fetter RD, Takemura S, Blazek K, Chang LA, Ogundeyi O,

Saunders MA, Shapiro V, Sigmund C, Rubin GM, Scheffer LK, Meinertzhagen IA, Chklovskii DB (2013) A visual motion detection circuit suggested by *Drosophila* connectomics. Nature 500:175–181

Tanaka NK, Tanimoto H, Ito K (2008) Neuronal assemblies of the *Drosophila* mushroom body. J Comp Neurol 508:711–755

Tomioka K, Matsumoto A (2010) A comparative view of insect circadian clock systems. Cell Mol Life Sci 67:1397–1406

Träger U, Homberg U (2011) Polarization-sensitive descending neurons in the locust: connecting the brain to thoracic ganglia. J Neurosci 31:2238–2247

Träger U, Wagner R, Bausenwein B, Homberg U (2008) A novel type of microglomerular synaptic complex in the polarization vision pathway of the locust brain. J Comp Neurol 506:288–300

Unoki S, Matsumoto Y, Mizunami M (2005) Participation of octopaminergic reward system and dopaminergic punishment system in insect olfactory learning revealed by pharmacological study. Eur J Neurosci 22:1409–1416

Usui-Aoki K, Ito H, Ui-Tei K, Takahashi K, Lukacsovich T, Awano W, Nakata H, Piao ZF, Nilsson EE, Tomida J, Yamamoto D (2000) Formation of the male-specific muscle in female *Drosophila* by ectopic fruitless expression. Nat Cell Biol 2:500–506

Vitzthum H, Müller M, Homberg U (2002) Neurons of the central complex of the locust *Schistocerca gregaria* are sensitive to polarized light. J Neurosci 22:1114–1125

Vogt K, Schnaitmann C, Dylla KV, Knapek S, Aso Y, Rubin GM, Tanimoto H (2014) Shared mushroom body circuits underlie visual and olfactory memories in *Drosophila*. Elife 3, e02395

von Frisch K (1950) Bees: their vision, chemical senses and language. Cornell University Press, Ithaca/London

von Frisch K (1974) Decoding the language of the bee. Science 185:663–668

Waddell S (2013) Reinforcement signalling in *Drosophila*: dopamine does it all after all. Curr Opin Neurobiol 23:324–329

Watanabe H, Nishino H, Nishikawa M, Mizunami M, Yokohari F (2010) Complete mapping of glomeruli based on sensory nerve branching pattern in the primary olfactory center of the cockroach *Periplaneta americana*. J Comp Neurol 518:3907–3930

Watanabe H, Matsumoto CS, Nishino H, Mizunami M (2011) Critical roles of mecamylamine-sensitive mushroom body neurons in insect olfactory learning. Neurobiol Learn Mem 95:1–13

Wehner R (2003) Desert ant navigation: how miniature brains solve complex tasks. J Comp Physiol A 189:579–588

Wehner R, Müller M (2006) The significance of direct sunlight and polarized skylight in the ant's celestial system of navigation. Proc Natl Acad Sci USA 103:12575–12579

Wolff T, Iyer NA, Rubin GM (2015) Neuroarchitecture and neuroanatomy of the *Drosophila* central complex: A GAL4-based dissection of protocerebral bridge neurons and circuits. J Comp Neurol 523:997–1037

Yamagata N, Ichinose T, Aso Y, Placais PY, Friedrich AB, Sima RJ, Preat T, Rubin GM, Tanimoto H (2015) Distinct dopamine neurons mediate reward signals for short- and long-term memories. Proc Natl Acad Sci USA 112:578–583

Yamamoto D, Koganezawa M (2013) Genes and circuits of courtship behaviour in *Drosophila* males. Nat Rev Neurosci 14:681–692

Yamamoto D, Jallon JM, Komatsu A (1997) Genetic dissection of sexual behavior in *Drosophila melanogaster*. Annu Rev Entomol 42:551–585

Yamamoto Y, Shibata H, Ueda H (2013) Olfactory homing of chum salmon to stable compositions of amino acids in natal stream water. Zool Sci 30:607–612

Yamamoto D, Sato K, Koganezawa M (2014) Neuroethology of male courtship in *Drosophila*: from the gene to behavior. J Comp Physiol A 200:251–264

Yang EC, Lin HC, Hung YS (2004) Patterns of chromatic information processing in the lobula of the honeybee, *Apis mellifera* L. J Insect Physiol 50:913–925

Yu JY, Kanai MI, Demir E, Jefferis GS, Dickson BJ (2010) Cellular organization of the neural circuit that drives *Drosophila* courtship behavior. Curr Biol 20:1602–1614

Chapter 7
Identifying Vertebrate Brain Prototypes in Deuterostomes

Takehiro G. Kusakabe

Abstract Vertebrates have a dorsal tubular central nervous system (CNS), the anterior part of which is a complex and highly organized brain. The invertebrate chordates (tunicates and cephalochordates) also have a CNS derived from the dorsal neural tube, but it is far simpler than the vertebrate CNS. The nervous system of ambulacrarians (hemichordates and echinoderms), the sister group of chordates, consists of a nerve net and multiple nerve cords with no discrete brain. Despite the poorly centralized organization of the ambulacrarian nervous system, genomics and molecular developmental biological studies have suggested that the major developmental programs that pattern the vertebrate brain already existed in the common ancestor of chordates and ambulacrarians. The CNS of cephalochordate amphioxus is a nerve cord with little anterior concentration, but has neuronal circuits with similarities to those of the vertebrate diencephalon-midbrain-hindbrain. The tadpole larva of the tunicate ascidian has a brain with sensory and motor control systems that shares many features with the vertebrate brain, including the retinal/hypothalamic territory, a locomotor central pattern generator, neural crests, and cranial placodes. Given the many shared characteristics among chordates, the CNS of invertebrate chordates should provide a unique platform to study the developmental and evolutionary bases underlying the emergence of the complex CNS of vertebrates.

Keywords Hemichordate • Echinoderm • Tunicate • Ascidian • Cephalochordate • Neurotransmitter • Locomotor neural circuit • Neural crest • Cranial placode

---

T.G. Kusakabe (✉)
Institute for Integrative Neurobiology & Department of Biology, Faculty of Science and Engineering, Konan University, 8-9-1 Okamoto, Higashinada-ku, Kobe 658-8501, Japan
e-mail: tgk@center.konan-u.ac.jp

S. Shigeno et al. (eds.), *Brain Evolution by Design*, Diversity and Commonality in Animals, DOI 10.1007/978-4-431-56469-0_7

7.1 Phylogeny and Body Plans of Deuterostomes

Currently, five groups of deuterostomes are recognized as phyla or subphyla: Vertebrata, Tunicata (Urochordata), Cephalochordata, Hemichordata, and Echinodermata (Nielsen 2012; Satoh et al. 2014) (Fig. 7.1). The Vertebrata, Tunicata, and Cephalochordata have long been recognized as the subphyla of the phylum Chordata, yet Satoh and colleagues (2014) recently proposed the superphylum status of Chordata, consisting of three phyla: Vertebrata, Tunicata, and Cephalochordata (Satoh et al. 2014). Molecular phylogenetic studies using large data sets have suggested that tunicates and vertebrates are the closest subphyla among chordates (Bourlat et al. 2006; Delsuc et al. 2006); the clade consisting of tunicates and vertebrates is called Olfactores (Jefferies 1991). Molecular phylogeny and genomics have unambiguously demonstrated that echinoderms and hemichordates form a clade called Ambulacraria, with similarities in coelomic systems and larvae (Metchnikoff 1881; Wada and Satoh 1994; Halanych 1995; Cameron et al. 2000; Perseke et al. 2013).

Ambulacraria has also been proposed to be a superphylum (Satoh et al. 2014). The adult body plans of echinoderms and hemichordates are fairly different. The

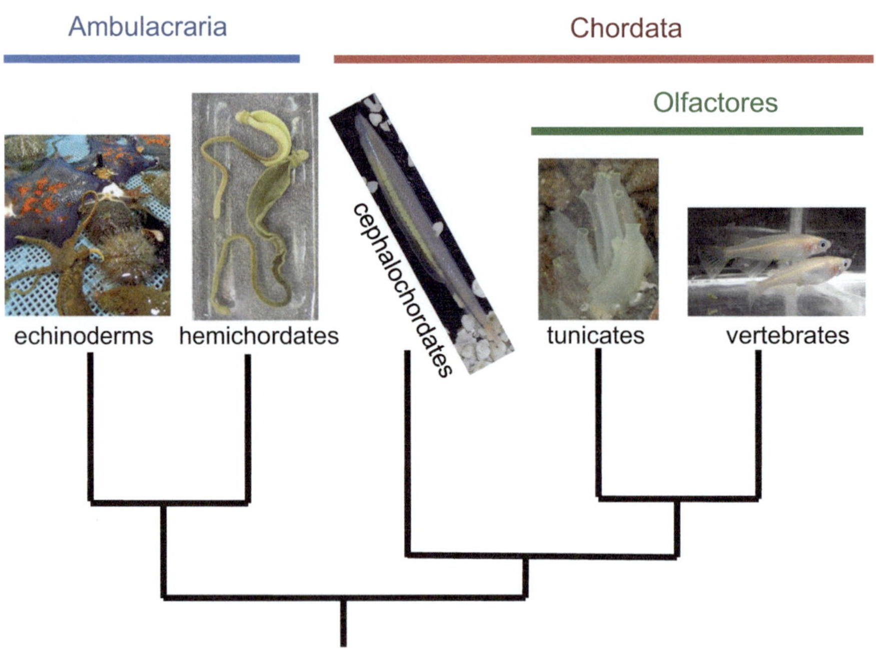

Fig. 7.1 Phylogenetic relationships among deuterostomes inferred from recent molecular phylogeny and genomics (See text for details) (The hemichordate photo and cephalochordate specimens are courtesy of Drs. Kuni Tagawa (Hiroshima University) and Takayuki Onai (RIKEN CDB), respectively)

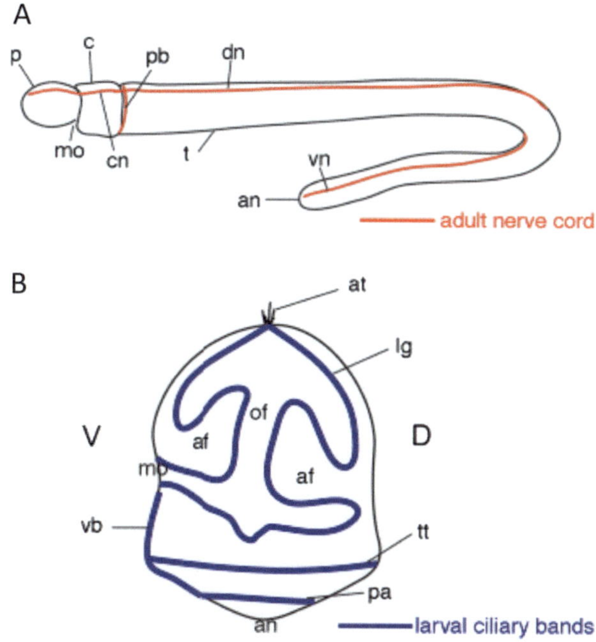

Fig. 7.2 Schematic drawing of the nervous system of an enteropneust hemichordate, *Balanoglossus simodensis*. (**a**) Schematic view of an adult acorn worm. Note the tripartite body structure of the proboscis, collar, and trunk. Dorsal and ventral nerve cords as well as the prebranchial nerve ring are shown in *red*. (**b**) Lateral view of a Metschnikoff-stage larva with fully grown ciliary bands. Larval ciliary bands are shown in *blue*. *af* aboral field, *an* anus, *at* apical tuft, *c* collar, *cn* collar nerve cord, *dn* dorsal nerve cord, *lg*, longitudinal ciliary band, *mo* mouth, *of* oral field, *p* proboscis, *pa* perianal ciliary ring, *pb* prebranchial nerve ring, *pc* protocoel, *t* trunk, *tt* telotroch, *vb* ventral ciliary band, *vn* ventral nerve cord (Reproduced with permission from Miyamoto et al. (2010). © 2010 Wiley Periodicals, Inc.)

body of echinoderms is radially symmetrical and pentameric. One of their unique features is a well-developed water circulatory system, which is connected to numerous tube feet that project through the body wall. Another unique feature is an internal calcium carbonate skeleton that protects the adult body. Adult hemichordates show a tripartite organization with bilateral symmetry. The body of the enteropneust hemichordates (acorn worms) is composed of an anterior proboscis, a middle collar region, and a long posterior trunk (Fig. 7.2). Although the adult body plans are thus distinct from each other, echinoderms and hemichordates have similar bilaterally symmetrical pelagic (dipleurula-type) larvae (Figs. 7.2 and 7.3).

In addition to Chordata and Ambulacraria, the phylum Xenacoelomorpha has recently been proposed to be a group of deuterostomes (Philippe et al. 2007, 2011). More recently, however, Xenacoelomorpha was proposed to be the sister group to Nephrozoa, namely, all remaining Bilateria (Cannon et al. 2016; Rouse et al. 2016).

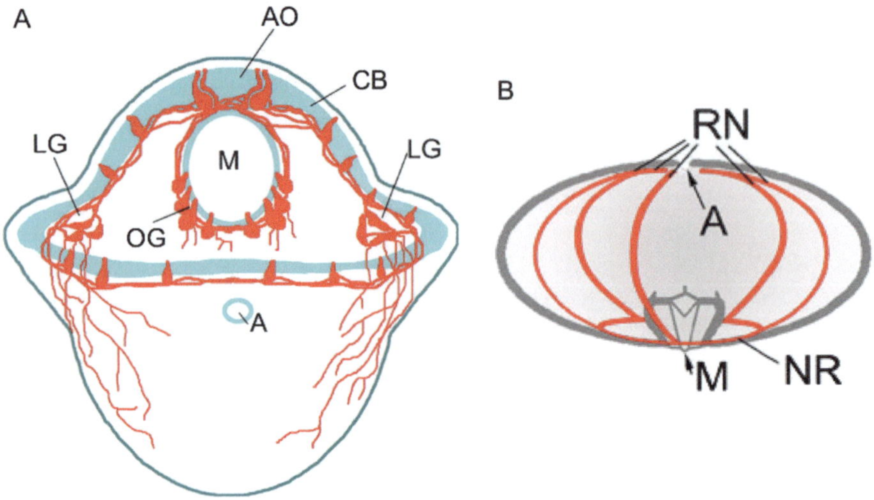

Fig. 7.3 Larval and adult nervous systems of sea urchins. Neural tissues are in *red*. (**a**) Larval nervous system of an early pluteus. (**b**) Lateral view of an adult sea urchin. (See text for details.) *A* anus, *AO* apical organ, *CB* ciliary band, *LG* lateral ganglia, *M* mouth, *NR* nerve ring, *OG* oral ganglia, *RN* radial nerves (Reproduced with permission from Burke et al. (2006). © 2006 Elsevier Inc. All rights reserved)

Thus, the position of the phylum Xenacoelomorpha has been controversial, and therefore their nervous system is not discussed here.

7.2 Larval Nervous Systems of Ambulacrarians

The ambulacrarian dipleurula-type larvae have a bilaterally symmetrical nervous system consisting of an apical organ and ciliary bands associated with neurons and nerve tracts (Nieuwenhuys 2002; Burke et al. 2006; Nakano et al. 2009). In both echinoderms and hemichordates, the nervous system of the pelagic larvae degenerates and an adult nervous system is newly formed during metamorphosis (Chia and Burke 1978; Lester 1988; Heinzeller and Welsch 2001; Nakano et al. 2006, 2009; Byrne et al. 2007; Hirokawa et al. 2008; Raff 2008; Nomaksteinsky et al. 2009; Miyamoto et al. 2010). In the late nineteenth century, Garstang proposed that the chordate CNS evolved by the dorsal fusion of ciliary band nerve tracts of the dipleurula-type ancestral deuterostome larva, which was similar to the auricularia larva of holothurians (Garstang 1894). This hypothesis, however, is not in favor with the recent molecular developmental biology data, which rather support another hypothesis, namely, that the hemichordate adult CNS is homologous to the vertebrate CNS (Morgan 1894; Bateson 1886; Kappers 1929; Brown et al. 2008;

Lowe 2008; Nomaksteinsky et al. 2009; Kaul and Stach 2010; Luttrell et al. 2012; Pani et al. 2012; Miyamoto and Wada 2013; Holland 2015).

How the ambulacrarian larval nervous system functions has been deduced from neuroanatomy and behavior (Nieuwenhuys 2002). The tornaria larva of acorn worms has a well-developed apical organ, called the apical complex, situated at the anterior pole. The apical complex contains a thickened epithelium with long cilia and a ganglion that contains clusters of serotonergic neurons (Hay-Schmidt 2000). The serotonergic axons extend from the ganglion cells into and along the ciliary band throughout its length (Hay-Schmidt 2000). Some species have a pair of eyes at the sides of the apical complex, consisting of cups of ectoderm cells surrounded by pigment (Morgan 1891). Thus, the apical complex is thought to serve as a sensory organ.

The tornaria larva has two ciliated band systems: a circumoral and circumanal band system (Nieuwenhuys 2002) (Fig. 7.2). The circumoral band is composed of uniciliated cells and has basal ciliary nerves. Nerve cells, provided with multiple apical and basal processes, occur regularly along the band (Lacalli and West 1993). The circumoral band system combines feeding and locomotor functions (Lacalli and West 1993). The circumanal band consists of multiciliated cells and serves as a principal locomotor organ (Lacalli and West 1993).

Similar to the hemichordate tornaria, the echinoderm bipinnaria (sea stars), pluteus (sea urchins and brittle stars), and auricularia (sea cucumber) larvae have a circumoral ciliary band, which is their principal feeding and locomotor organ, but a circumanal band is lacking (Nieuwenhuys 2002). Their apical organ is less complex than that of tornaria; there is no tuft of cilia nor eyes (Nieuwenhuys 2002). The apical ganglion contains serotonergic cells, which send their axons into the anterodorsal parts of the ciliary band (Burke et al. 1986; Hay-Schmidt 2000). The ciliated cells of the ciliary band and musculature of the esophagus and mouth appear to be the principal effectors (Strathmann 1975). It is assumed that coordinated reversals of cilia, reversals of peristalsis in the esophagus, and contraction of the muscles that open the mouth and flex the arms are controlled by the nervous system (Bisgrove and Burke 1986). The electrical activity from ciliary bands was shown to be coincident with ciliary reversals (Mackie et al. 1969).

Three types of neurons, sensory, bipolar, and multipolar, are embedded in the ciliary bands of echinoderm larvae (Lacalli et al. 1990; Lacalli and West 1993; Nieuwenhuys 2002). The sensory and bipolar cells are uniciliated elements. The multipolar cells are flask shaped with a slender tapering process that extends to the surface of the band. In addition to the serotonergic fibers, the ciliary nerves contain axons of catecholaminergic neurons, which are also localized along the bands (Burke et al. 1986; Bisgrove and Burke 1987; Nakajima 1987). Another neurotransmitter that seems to be used in the echinoderm larval nervous system is γ-amino butyric acid (GABA). GABA induces the settlement of the sea urchin *Strongylocentrotus droebachiensis* larvae (Pearce and Scheibling 1990). It has been suggested that the swimming activity of sea urchin larvae is regulated by a combination of serotonergic, dopaminergic, and GABAergic systems (Katow et al. 2007, 2010, 2013).

7.3 Adult Nervous Systems of Echinoderms

Extant echinoderms are divided into two subphyla, the Eleutherozoa and the Pelmatozoa. The Eleutherozoa include sea cucumbers, sea urchins, starfish, and brittle stars, and the extant Pelmatozoa are the crinoids, namely, feather stars and sea lilies. The adults of both the Eleutherozoa and the Pelmatozoa have a pentaradial nervous system containing a circumoral nerve ring and radial nerve cords (Hyman 1955; Smith 1965; Nieuwenhuys 2002; Burke et al. 2006; Nakano et al. 2009) (Fig. 7.3). Studies on the expression of developmental patterning genes have failed to find evidence that the nerve ring or radial nerves are homologous to any part of the brain or nerve cord in bilaterians (Sly et al. 2002; Nielsen 2006; Cisternas and Byrne 2009; Holland 2015). In addition to the nerve ring and the radial nerve cords, the Pelmatozoa crinoids have an aboral nerve center, a ganglion situated at the aboral region (Hyman 1955; Bohn and Heinzeller 1999; Nakano et al. 2009). Fossil records suggest an aboral ganglion was the dominant nervous system in the ancestor of echinoderms (Paul and Smith 1984). Therefore, the evolutionary relationship of the crinoid aboral ganglion to the nervous system of other deuterostomes is of particular interest, although no information is available regarding their homology at present.

Although many studies have been conducted on the morphological aspects of the adult echinoderm nervous system, the functional and physiological aspects remain poorly understood, except for the wealth of information on neurotransmitters. For example, acetylcholine is the excitatory neurotransmitter mediating motor responses in the echinoderms (Devlin 2001). Both GABA and acetylcholine have been shown to cause contractions of the muscles of the isolated tube feet of sea urchins (Florey et al. 1975) and starfish (Protas and Muske 1980). GABA also shows an inhibitory action on the longitudinal muscle of the body wall of sea cucumbers (Devlin 2001). The neuropeptides known as SALMFamides have been identified and shown to act as muscle relaxants in starfish and sea cucumbers (Elphick et al. 1991; Elphick and Melarange 2001). Recent studies using sea urchin genome information have identified components for pathways of major neurotransmitters, such as serotonin (5-hydroxytryptamine, or 5-HT), dopamine, acetylcholine, GABA, glutamate, and glycine, suggesting that sea urchins use a broad range of neurotransmitters known in vertebrates (Burke et al. 2006). Another interesting piece of information also comes from the genome sequences. Orthologues of mammalian cannabinoid, lysophospholipid, and melanocortin receptors are not present in the *Drosophila*, *Caenorhabditis elegans*, and sea urchin genomes (Elphick and Egertova 2001, 2005; Burke et al. 2006), whereas the *Ciona* genome contains an orthologue of mammalian cannabinoid and lysophospholipid receptors (Elphick et al. 2003), suggesting that this group of receptors is unique to chordates (Burke et al. 2006).

7.4 Adult Nervous Systems of Hemichordates

There are two classes of hemichordates, the Enteropneusta (acorn worms) and the Pterobranchia. Both enteropneusts and pterobranchs possess a similar tripartite body organization but are characterized by distinct feeding mechanisms (Lowe 2008). The pterobranchs use a lophophore, a ciliated extension from the mesosome, to filter feed (Halanych 1995), whereas enteropneusts use their highly muscular and ciliated proboscis for particle ingestion and filter feeding (Cameron 2002). Most of the data in hemichordate molecular developmental biology have been obtained in enteropneust species. Enteropneust worms have nerve cords in the dorsal midline and ventral midline (Fig. 7.2). The dorsal nerve cord is divided into the proboscis stalk region, the collar cord, and the dorsal nerve cord in the trunk region (Nomaksteinsky et al. 2009). The ventral nerve cord exists only in the trunk region and is connected to the dorsal nerve cord via the prebranchial nerve ring. Among them, only the collar cord has tubular organization and thus is proposed to be a homologous organ to the neural tube (Bateson 1886; Morgan 1894; Brown et al. 2008; Luttrell et al. 2012; Kaul and Stach 2010; Miyamoto and Wada 2013), although this homology is still controversial (Nomaksteinsky et al. 2009; Kaul and Stach 2010).

Lowe and colleagues examined the expression of orthologues of 22 transcription factors that have conserved roles in the patterning of the brain and spinal cord of vertebrates along the anteroposterior axis. Expression patterns of many of these genes in the hemichordate juvenile ectoderm are similar to the patterns found in the developing vertebrate CNS (Lowe et al. 2003). Furthermore, genetic programs homologous to the three vertebrate signaling centers [the anterior neural ridge (ANR), zona limitans intrathalamica (ZLI), and midbrain–hindbrain boundary (MHB)/isthmic organizer (IsO)] that direct anteroposterior patterning in the vertebrate anterior neural plate are shown to be present in hemichordates (Pani et al. 2012). Thus, hemichordates and vertebrates share the genetic mechanisms for anteroposterior patterning of the neuroectoderm. Recent histological and gene expression analyses suggest that the hemichordate collar cord is subdivided into dorsoventral domains (Miyamoto and Wada 2013). In vertebrate embryos, Hedgehog signaling from the notochord to the neural plate is essential for floor plate induction and for patterning of the neural tube along the dorsoventral axis (Echelard et al. 1993). Similarly to the dorsoventral patterning of the vertebrate neural tube, the dorsoventral patterning of collar cord neurulation may be regulated by Hedgehog signaling from the dorsal endoderm (Miyamoto and Wada 2013). Thus, molecular mechanisms for both anteroposterior and dorsoventral patterning may be conserved between vertebrates and hemichordates.

Neuronal types were examined in enteropneust worms. Numerous giant 5-HT neurons, in the absence of 5-HT fibers, were observed in the collar cord of *Glossobalanus berkeleyi* (Brown et al. 2008), whereas the arrangement was the opposite in *Ptychodera flava*, whose collar cord was rich in 5-HT fibers but devoid of 5-HT cells (Nomaksteinsky et al. 2009). However, 5-HT cells were abundant in the epidermal peripheral nervous system (PNS) of *P. flava*. The distributions of GABAergic and cholinergic neurons were also examined in *P. flava* (Nomaksteinsky et al. 2009). The highest density of GABAergic neurons was found in the neural plate at the base of the proboscis, but some were also detected in the ventral and dorsal cords and in the PNS of the proboscis. A low density of cholinergic neurons was observed in the nerve cords and a dense dorsal cluster was observed in the anterior neural plate, at the level where it surrounds the proboscis stem. Interestingly, no cholinergic cells were detected in the PNS (Nomaksteinsky et al. 2009).

Nomaksteinsky et al. (2009) also investigated neuronal types in *P. flava* by examining the expression of two neuron type-specific transcription factors, Drg11 and Hb9. In vertebrates, Drg11 is largely specific for somatic sensory neurons of the CNS and PNS (Saito et al. 1995; Rebelo et al. 2007). Inside the vertebrate nervous system, Hb9 is specific for somatic motor neurons (Tanabe et al. 1998) and a subclass of interneurons (Wilson et al. 2005) in the spinal cord and hindbrain. It is also expressed in somatic motor neurons in ascidians (Dufour et al. 2006) and possibly cephalochordates (Ferrier et al. 2001), as well as in protostomes (Thor and Thomas 2002; Denes et al. 2007). In *P. flava*, Dgr11+ neurons and Hb9+ neurons were found in the collar cord, proboscis stem, and ventral and dorsal cords. An occasional Drg11+ neuron but no Hb9+ neurons were found in the PNS of the proboscis or collar folds. Thus, Drg11+, Hb9+, and cholinergic neurons are preferentially or exclusively located in the CNS, whereas 5-HT neurons are restricted to the PNS. Based on these observations, Nomaksteinsky et al. (2009) pointed out that the entire cord system or CNS of *P. flava* would correspond to the spinal and hindbrain levels of the vertebrate CNS, where Hb9+ and Dgr11+ cells are confined.

Homology of hemichordate nerve cords to the chordate spinal cord and hindbrain has also been suggested by morphological studies. Giant neurons were found in the collar cord of enteropneust worms of the family Ptychoderidae, especially balanoglossids and glossobalanids (Brown et al. 2008) (Fig. 7.4). These unipolar cells are probably serotonergic and send their axons across the midline and posteriorly into the dorsal cord of the trunk (Bullock 1944; Brown et al. 2008). Bullock (1944, 1965) has suggested that the giant neurons may be homologous to Mauthner cells of the lamprey and Rhode cells of amphioxus (Brown et al. 2008; Holland et al. 2013). The giant neurons have been proposed to be involved in hemichordate escape responses, the rapid contraction of the posterior segments of the worm (Bullock 1944, 1965).

7.5 Nervous Systems of Cephalochordates

Cephalochordate amphioxus (also called lancelets) are morphologically homogeneous animals that are classified into only three genera: *Branchiostoma*, *Epigonichthys*, and *Asymmetron* (Kon et al. 2007). Despite their striking morphological

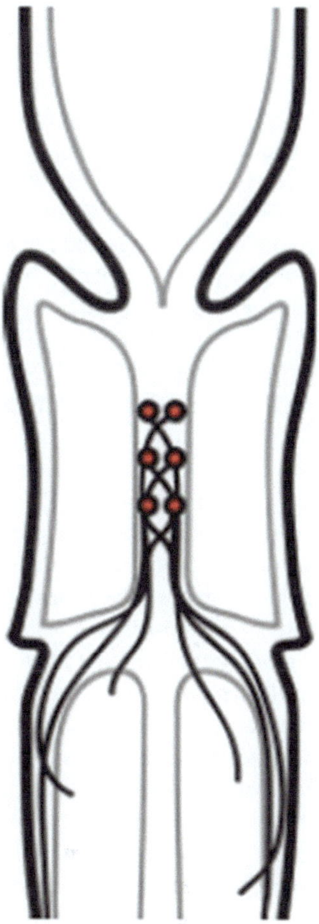

Fig. 7.4 Giant cells in the collar nerve cord of a balanoglossid enteropneust worm, redrawn from Bullock (1944). The giant neurons (*red*) located in the collar nerve cord send their axons across the midline and posteriorly into the dorsal cord of the trunk (Reproduced with permission from Brown et al. (2008). Copyright © 2008 Wiley-Liss, Inc.)

similarity, the divergence time of the last common ancestor of the three extant genera is estimated to be 162 million years ago (Nohara et al. 2005). Furthermore, extant cephalochordates exhibit body plans similar to early chordates, such as *Cathaymyrus* (Shu et al. 1996) and *Pikaia* (Conway Morris 1982).

The body structure of both larval and adult amphioxus is vertebrate like, but much simpler (Nieuwenhuys 2002; Bertrand and Escriva 2011). They have typical chordate characters, such as a dorsal hollow neural tube and notochord, a ventral gut and a pharynx with gill slits, segmented paraxial muscles, and the thyroid gland homologue (endostyle). The free-living planktonic larvae show an asymmetrical body plan: the mouth forms on the left side, the gill slits form on the right ventral side, and the right series of somites form half a segment posterior to the left ones. During metamorphosis, much of this asymmetry disappears, although the axial muscles retain their asymmetry (Bertrand and Escriva 2011).

The notochord extends from the anterior tip of the body to the end of the tail. The CNS of the amphioxus consists of the tubular nerve cord, located immediately dorsal to the notochord, and extends throughout almost the entire length of the body. In contrast to the CNSs of the other groups of chordates (tunicates and vertebrates), an anterior accumulation of the nervous tissue is not conspicuous in the nerve cord of amphioxus, especially at the adult stage (Nieuwenhuys 2002; Wicht and Lacalli 2005); a transient anterior swelling (the cerebral vesicle) is recognizable only in young larvae (Wicht and Lacalli 2005). The anteroposterior organization of the CNS is not evident externally; however, regional differences along the anteroposterior axis can be recognized by developmental gene expression (Castro et al. 2006; Irimia et al. 2010; Holland et al. 2013) and cytoarchitectural differences in both adults and larvae (Wicht and Lacalli 2005).

The expression patterns of developmental regulatory genes in the CNS show conserved features between amphioxus and vertebrates (Holland 2009; Bertrand and Escriva 2011; Holland et al. 2013). For example, the expression of *Otx* marks the anterior part of the CNS, and posterior to the *Otx*-expressing domain *Gbx* is expressed, suggesting the presence of a region homologous to the vertebrate MHB at the boundary between the two expression domains (Castro et al. 2006; Holland et al. 2013). However, the genes that confer organizer properties on the vertebrate MHB, namely, the *Engrailed*, *Wnt1*, *Pax2/5/8*, and *FGF* genes, are not expressed in this region, suggesting that the amphioxus MHB lacks organizer ability (Holland 2009; Holland et al. 2013). Thus, the organizer properties of this region may have evolved later during the evolution of ancestral olfactores or early vertebrates. The presence of ANR and ZLI, the signaling centers for anteroposterior brain organization, in addition to MHB/IsO, has been controversial. Holland and colleagues argue that amphioxus retain these signaling centers (Holland et al. 2013), whereas Pani and colleagues claim that amphioxus and tunicates have lost all or part of these three signaling centers during evolution (Pani et al. 2012).

Another example of anteroposterior conservation of gene expression is that of *Hox* genes. The amphioxus genome contains a single archetypal *Hox* gene cluster (Garcia-Fernandez and Holland 1994; Amemiya et al. 2008). The colinear expression of *Hox* genes is also conserved in amphioxus (Wada et al. 1999). The gene expression data collectively suggest that the anterior portion of the amphioxus nerve cord is homologous to the vertebrate brain. Regional homologies within the CNS have been proposed based on these gene expression data (Castro et al. 2006; Lacalli 2008; Holland et al. 2013). In addition to the conservation along the anteroposterior axis, gene networks for the dorsoventral patterning also seem to be conserved between vertebrates and amphioxus to a certain extent (Yu et al. 2007; Holland 2009).

Neuroanatomical organization of the amphioxus nerve cord has been investigated in detail in the young larvae (Wicht and Lacalli 2005; Lacalli 2008) (Fig. 7.5). The anterior portion of the nerve cord shows distinct features compared to the typical organization of the rest of the nerve cord. The anteriormost region is called the cerebral vesicle (CV). Immediately caudal to the CV is the primary motor center

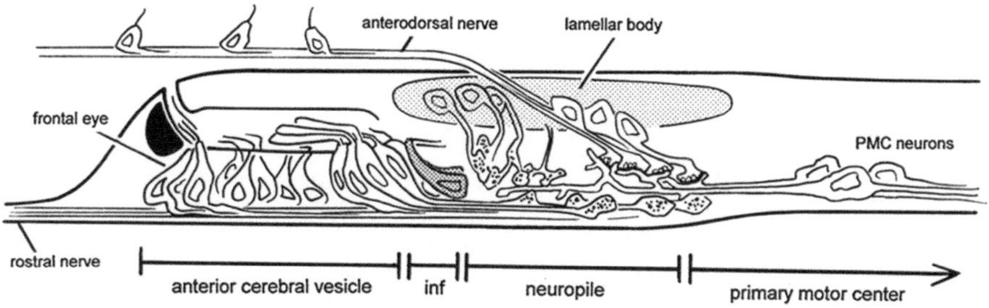

Fig. 7.5 Organization of the anterior nerve cord in a 12-day amphioxus larva, lateral view, anterior to the *left*. The central nervous system (CNS) region shown in this figure contains about 150 neurons at this stage. Inf (*dark shading*), infundibular cells. (See text for details.) (Reproduced with permission from Lacalli (2008).)

(PMC), which contains the anteriormost motor neurons and a set of interneurons with caudal projections that are thought to be involved in locomotor control.

The CV can be further divided into the anterior and posterior portions, each exhibiting a distinct organization (Wicht and Lacalli 2005; Lacalli 2008). The anterior CV contains the frontal eye, which consists of pigment cells, putative photoreceptor cells, and some neurons. The posterior CV contains a ventral neuropil and the dorsal lamellar body, which is another putative photoreceptor organ. The lamellar body has been implicated in circadian control (Wicht and Lacalli 2005). There is no direct experimental evidence for such rhythms in either adults or larvae, but the larvae have diurnal patterns of vertical migration in the plankton (Wickstead and Bone 1959). A ventral cluster of secretory infundibular cells serves as a useful landmark at the transition between the anterior and posterior CV. Sensory receptor cells of the anterior CV and the infundibular region have neurites projecting to the neuropil of the posterior CV, and neurites from the lamellar body and peripheral sensory neurons also terminate at the post-infundibular neuropil. These sensory terminals form varicosities but not synapses, suggesting that these anterior neurons act largely through paracrine release of transmitters (presumably neuropeptides), which further suggests that the post-infundibular neuropil is mainly a modulatory center (Lacalli 2008).

In contrast to the paucity of synaptic connections in the post-infundibular neuropil, conventional synaptic transmissions predominate in the PMC, suggesting that the locomotor circuits depend on fast transmission with acetylcholine and amino acid transmitters (Wicht and Lacalli 2005; Lacalli 2008). The PMC contains three pairs of large paired interneurons (LPN1-3 s), which are innervated by various sensory inputs (Fig. 7.6). The third pair of neurons, the LPN3s, are contralaterally innervated by each other, implying that they are mutually inhibitory and could serve as key components of a central pattern generator (Lacalli and Kelly 2003). Dendrites of these neurons extend to the post-infundibular neuropil, where they receive mechanosensory inputs via well-developed synapses (Lacalli 2008). The nerve targets of LPN3s are ventral "fast" motor neurons via synapses and dorsal

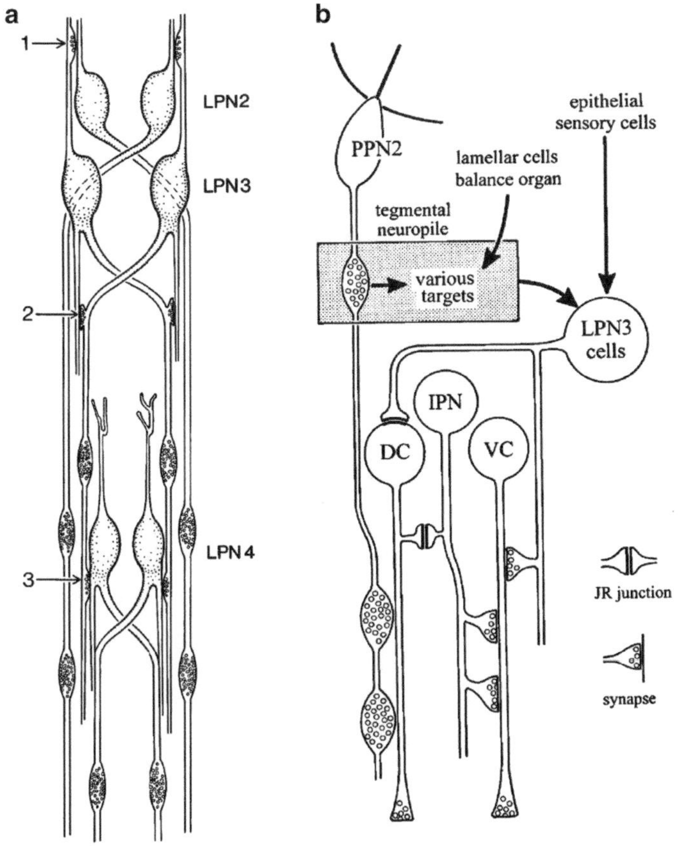

Fig. 7.6 Pacemaker circuits in young amphioxus larvae. (**a**) Schematic diagram shows the inferred pattern of synaptic contacts between large paired neurons (LPN2s, LPN3s, and LPN4s) in the anterior nerve cord of amphioxus larvae. For the output of LPN2, LPN3, and LPN4 to be in phase, according to Lacalli (2003), synapse 1 must be excitatory and synapses 2 and 3 must both be inhibitory (Reproduced with permission from Lacalli (2003). © 2003 Wiley-Liss, Inc.) (**b**) Schematic diagram shows inputs to motor neurons in the anterior nerve cord of amphioxus larvae. The third pair of large paired neurons (LPN3) are suggested to having a pacemaker function. Junctions are either chemical synapses (with vesicles) or juxtareticular (JR) junctions (parallel lines, no vesicles). *DC* dorsal compartment motor neurons, *VC* ventral compartment motor neurons, *IPN* ipsilateral projection neurons, *PPN2* type 2 preinfundibular projection neurons (PPN2) (Reproduced with permission from Lacalli (2002). © 2002 Wiley-Liss, Inc.)

"slow" motor neurons via an unusual class of intercellular junctions (Fig. 7.6). Young larvae respond very strongly to touch by initiating a fast escape response, although they are also capable of more prolonged periods of slow swimming, which probably drives the diurnal vertical migrations; the escape response may be suppressed during slow migration. The structural features of PMC neuronal circuits suggest that they act as a central pattern generator for swimming locomotion as well as a switch between fast and slow modes of swimming behavior (Wicht and Lacalli 2005; Lacalli 2008).

The *Otx*-expressing domain roughly corresponds to the CV and an anterior portion of the PMC, suggesting that this region is homologous to the forebrain and midbrain of vertebrates (Lacalli 2008). The boundary between the forebrain/midbrain and hindbrain counterparts is located somewhere between somites 1 and 2, although the exact position is uncertain. Correspondingly, the expression patterns of *Hox* genes suggest that a portion of the nerve cord, beginning at about the level of somite 3, is homologous to the vertebrate hindbrain (Wada et al. 1999; Nieuwenhuys 2002). As to further regional homologies within the forebrain/midbrain counterpart, it has been proposed that the frontal eye and the lamellar body are homologous to the paired eyes and the pineal organ, respectively (Nieuwenhuys 2002; Wicht and Lacalli 2005; Lacalli 2008; Vopalensky et al. 2012). In addition, the middle CV region immediately anterior to the cluster of infundibular cells has been proposed to be homologous to the hypothalamus (Lacalli and Kelly 2003). The PMC may be a rudimentary homologue of the ventral part of the vertebrate midbrain because the anteriormost motor neurons and the beginning of the reticulospinal system are located in these regions and also because they are the posteriormost expression zone of *Otx* (Lacalli and Kelly 2003). The homologies described here have been proposed based on similarities not only in development and gene expression but also in neuroanatomical organization and expected physiological functions (Nieuwenhuys 2002; Wicht and Lacalli 2005; Lacalli 2008). For example, a variety of modulatory inputs, including signals from sensory neurons located in the hypothalamus or its amphioxus equivalent, converge on a ventral locomotor control center (the midbrain tegmentum and reticulospinal system in vertebrates and the PMC in amphioxus), and this initiates a locomotor response (Wicht and Lacalli 2005).

In conclusion, although the anterior nerve cord of amphioxus larvae is not prominently enlarged as in the brains of vertebrates and tunicates, it seems to have sophisticated neuronal circuits with overall similarities to those of the vertebrate diencephalon-midbrain-hindbrain. Current understanding of the functional aspects, however, is largely dependent on anatomical observations. In addition, the neurons of amphioxus have been only partially identified. Therefore, neurophysiological studies as well as precise identification of neurons are needed to elucidate the precise structure and function of the neuronal circuitry in amphioxus larvae.

7.6 Nervous Systems of Tunicates

Tunicates or urochordates comprise three classes: the Ascidiacea (ascidians), the Appendicularia (larvaceans), and the Thaliacea (salps, pyrosomes, and doliolids). The class Ascidiacea consists of two orders: the Enterogona (representative genera include *Ciona* and *Phallusia*) and the Pleurogona (representative genera are *Styela*, *Botryllus*, and *Halocynthia*). Recent molecular phylogeny suggests that thaliaceans are included in the Enterogona clade of ascidians (Swalla et al. 2000). The phylogenic position of larvaceans is still controversial; some authors place larvaceans at a basal position among tunicates while others place them within the Pleurogona clade of ascidians (Nishino and Satoh 2001; Stach et al. 2008; Satoh 2009). Among tunicates, the development, organization, and function of the nervous system have been most well studied in ascidians. Therefore, this section focuses mainly on the nervous system of ascidians.

The life history of ascidians consists of distinct larval and adult stages. The adult ascidians are sessile animals, bearing an extremely modified version of the chordate body plan with a simple nervous system. In contrast, the body plan of the free-swimming, tadpole-like larvae shares basic features with the body plan of vertebrates, including a CNS derived from a dorsal neural tube (Nicol and Meinertzhagen 1988). The CNS of the ascidian larva is, however, far simpler than that of vertebrates; it consists of fewer than 200 neurons (Cole and Meinertzhagen 2004). The expression patterns of transcription factors that regulate the patterning of the neural tube along the anterior–posterior axis are conserved between ascidians and vertebrates, suggesting that the developmental mechanisms of the CNS are conserved among chordates (Wada et al. 1998; Imai et al. 2002; Takahashi and Holland 2004; Ikuta and Saiga 2007). The adult CNS, a simple cerebral ganglion (Fig. 7.7), is formed from the larval CNS during metamorphosis (Dufour et al. 2006; Horie et al. 2011); this provides a good contrast with the situation in ambulacrarians, in which no direct link has been found between the larval and adult nervous systems (Chia and Burke 1978; Lester 1988; Heinzeller and Welsch 2001; Nakano et al. 2006, 2009; Byrne et al. 2007; Hirokawa et al. 2008; Raff 2008; Nomaksteinsky et al. 2009; Miyamoto et al. 2010). In this regard, the tadpole-like larva stage of

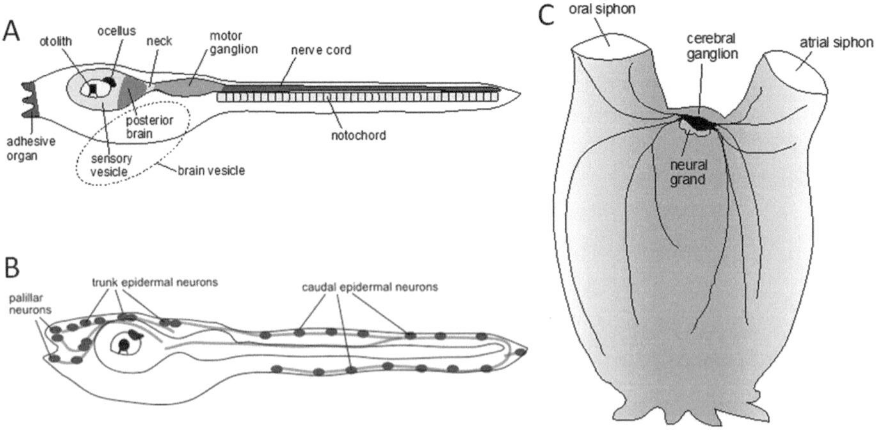

Fig. 7.7 Schematic illustration of the nervous systems of ascidians. (**a**) The central nervous system of the larva. (**b**) The peripheral nervous system of the larva. (**c**) The simple nervous system of the adult (See text for details)

ascidians is not correspondent to the dipleurula-type larva stage of ambulacrarians, further contradicting Garstang's hypothesis (Garstang 1894).

The CNS of the ascidian larva is divided into three parts, a brain vesicle (also called a sensory vesicle), a motor ganglion (also called a visceral ganglion or trunk ganglion), and a caudal nerve cord, from anterior to posterior (Katz 1983; Meinertzhagen and Okamura 2001; Meinertzhagen et al. 2004; Horie et al. 2009) (Fig. 7.7). In addition to these three regions, a slender "neck" region containing six cells has been morphologically recognized between the brain vesicle and the motor ganglion in *Ciona* (Nicol and Meinertzhagen 1991), although it is not apparent in *Halocynthia*. Recent studies on the structure and function of sensory organs and neurons in the *Ciona* brain vesicle have suggested that the brain vesicle can be functionally divided into an anterior region consisting of a ventricular cavity and surrounding cells, including sensory organs, and a posterior region containing a number of diverse interneurons, which presumably form neural circuits for sensory information processing and motor regulation. Taking this difference into account, Horie and colleagues proposed to further divide the brain vesicle into the anterior "sensory vesicle" and the posterior brain (Horie et al. 2009).

The sensory vesicle contains two conspicuous pigmented sensory organs. The pigmented sensory organ located anteriorly is an otolith, and used for the perception of gravity, whereas the posterior one is an ocellus, used for light reception (Tsuda et al. 2003b; Sakurai et al. 2004). Electron microscopic observation showed that the otolith is a spherical mass of pigment granules connected to the midline of the floor of the sensory vesicle by a narrow stalk (Dilly 1961, 1964; Eakin and Kuda 1971; Ohtsuki 1990; Sakurai et al. 2004). The otolith-associated neurons extend their axon to the posterior brain (Horie et al. 2008a). The *Ciona* ocellus is composed of three lens cells, one pigment cup cell, and a group of photoreceptor cells (Kusakabe and Tsuda 2007; Horie et al. 2008b). In addition to the pigmented ocellus, which is located at the right side of the brain vesicle, the *Ciona* larva has a nonpigmented ocellus at the ventromedial side of the brain vesicle (Horie et al. 2008b). The photoreceptor cells of the conventional pigmented ocellus are called the group I and II photoreceptor cells, and the unique photoreceptor cells constituting the nonpigmented ocellus are called the group III photoreceptor cells (Horie et al. 2008b).

Regional homology within the CNS between ascidians and vertebrates has been discussed based on regulatory gene expression. The ascidian motor ganglion/nerve cord, in which *Hox* genes are expressed, has been suggested to be homologous to the vertebrate hindbrain/spinal cord (Wada et al. 1998; Imai et al. 2002; Takahashi and Holland 2004; Ikuta and Saiga 2007). The neck region has been proposed to be homologous to MHB/IsO because this region expresses *Pax-2/5/8* (Wada et al. 1998; Imai et al. 2002). The existence of the region corresponding to MHB/IsO in ascidians, however, has been controversial, because some of the key regulatory genes are not expressed in the expected patterns or are missing in the genome (Lacalli 2006; Pani et al. 2012). The *Pax-2/5/8*-expressing neck cells also express a paired-like homeobox gene *Phox2*, which is a marker for cranial motor neurons of the vertebrate hindbrain (Dufour et al. 2006). The neck cells expressing *Phox2* develop into branchial motor neurons in the adult CNS. Based on this observation, Dufour and colleagues proposed that the neck is homologous to the hindbrain and the motor ganglion is homologous to the spinal cord (Dufour et al. 2006). The *Ciona Otx* gene is expressed in the CNS region anterior to the *Pax-2/5/8*-expressing cells, suggesting that this most anterior region corresponds to the forebrain/midbrain of vertebrates (Wada et al. 1998; Imai et al. 2002; Takahashi and Holland 2004). However, there seems to be no midbrain counterpart in the ascidian CNS, because the midbrain marker *Dmbx* is not expressed in the region anterior to the *Pax-2/5/8*-expressing cells (Takahashi and Holland 2004). Within the *Otx*-expressing "forebrain" region of *Ciona*, a region homologous to the hypothalamus and retina has been suggested based on the commonality in both regulatory gene expression and differentiated cell types (Horie et al. 2008b; Razy-krajka et al. 2012).

7.7 Neuronal Types in the Ascidian Larva

Recent studies have revealed neuronal types and networks in the CNS of the ascidian larva (Takamura 1998; Yoshida et al. 2004; Imai and Meinertzhagen 2007; Horie et al. 2008a, 2009, 2010, 2011; Nishino et al. 2010; Takamura et al. 2002, 2010; Stolfi and Levine 2011; Razy-Krajka et al. 2012; Kusakabe et al. 2012) (Fig. 7.8). The functions of neuronal networks have been investigated by pharmacological and genetic manipulations of the larval CNS (Inada et al. 2003; Brown et al. 2005; Nishino et al. 2010; Razy-Krajka et al. 2012). These studies have revealed commonalities of neurotransmitters and neuronal cell types between tunicates and vertebrates. In the following paragraphs, neuronal types known in the ascidian larva are summarized with reference to their vertebrate counterparts.

Glutamatergic Neurons In the *Ciona* larva, glutamatergic neurons are present in the CNS and PNS (Horie et al. 2008a). Most sensory neurons, including papillar neurons, epidermal neurons, the otolith cell, and ocellus photoreceptor cells are glutamatergic. Most of these glutamatergic axons terminate in the posterior brain. In the posterior brain, a group of glutamatergic interneurons are present and send their axons to the motor ganglion, suggesting that these neurons control the activity of the motor system. Thus, glutamatergic neurotransmission probably has a major role in sensory systems and in the integration of the sensory inputs of the ascidian larva (Horie et al. 2008a).

Cholinergic Neurons Cholinergic neurons are present in the posterior brain, motor ganglion, and nerve cord of the ascidian larva (Takamura et al. 2002, 2010; Yoshida et al. 2004; Horie et al. 2010). A few cholinergic interneurons located in the posterior brain extend their axons to the motor ganglion. As in vertebrates, acetylcholine is the transmitter at the neuromuscular junction in the ascidian larva (Whittaker 1973; Ohmori and Sasaki 1977). Cholinergic motor neurons are located in the motor ganglion, and extend their axons posteriorly to form the nerve terminals on the tail muscle (Horie et al. 2010).

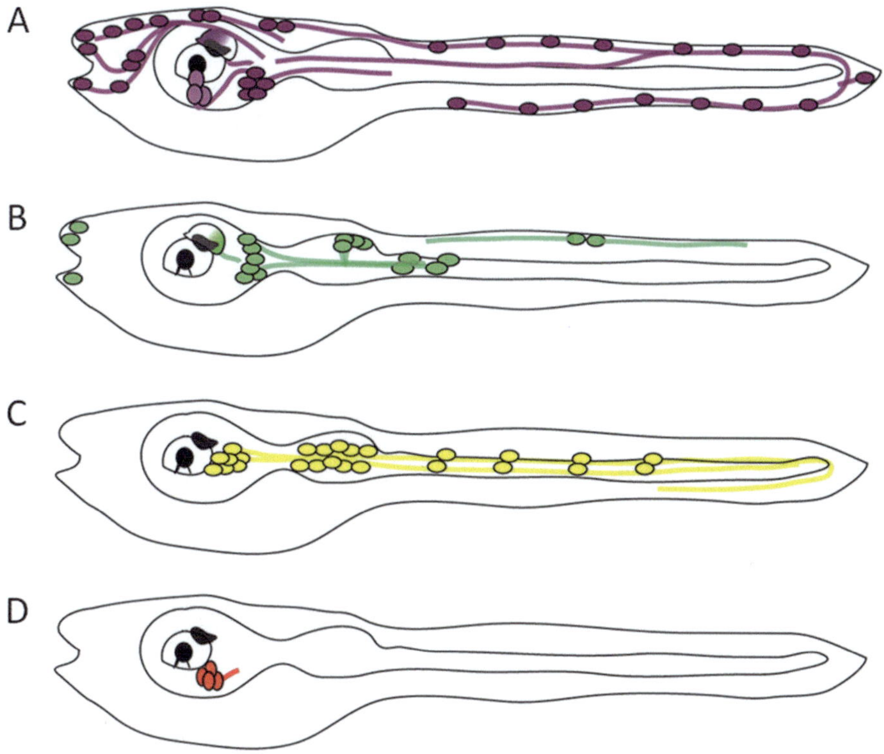

Fig. 7.8 Summary diagrams showing distribution patterns of neurotransmitter phenotypes of neurons in the *Ciona intestinalis* larva. (**a**) Glutamatergic neurons (Horie et al. 2008a). (**b**) GABAergic/glycinergic neurons (Yoshida et al. 2004; Horie et al. 2010; Nishino et al. 2010). (**c**) Cholinergic neurons (Takamura et al. 2002, 2010; Yoshida et al. 2004; Horie et al. 2010). (**d**) Dopaminergic neurons (Moret et al. 2005a; Razy-Krajka et al. 2012) (See text for details)

GABA/glycinergic Neurons GABA is a major inhibitory neurotransmitter in the CNS of both vertebrates and invertebrates (Varju et al. 2001; Alford et al. 2003; Schuske et al. 2004), whereas glycine is a vertebrate-specific inhibitory neurotransmitter (Legendre 2001). Vesicular GABA/glycine transporter (VGAT), which uptakes both GABA and glycine into the synaptic vesicle, is used as a GABA/glycinergic neuron marker in the *Ciona* larva (Yoshida et al. 2004; Horie et al. 2010; Nishitsuji et al. 2012). GABA/glycinergic neurons are present in the adhesive organ, sensory vesicle, posterior brain, motor ganglion, and dorsal regions of the tail (Yoshida et al. 2004). Among these neurons, two pairs of neurons located in the anterior nerve cord (anterior caudal inhibitory neurons; ACINs) are probably glycinergic, because ACINs do not express glutamic acid decarboxylase (GAD), a GABAergic neuron-specific marker, and are labeled by immunostaining using an anti-glycine antibody (Nishino et al. 2010). Conversely, VGAT-positive neurons in the brain vesicle and motor ganglion are probably GABAergic because they are

GAD positive (Zega et al. 2008) and are also labeled by immunostaining using an anti-GABA antibody (Brown et al. 2005). The inhibitory neurons in the posterior brain and motor ganglion may be involved in sensory information processing or motor activity control or both. In vertebrates, GABA is mainly used in the brain and glycine is mainly used in the spinal cord and brainstem. It should be noted that the situation is similar in ascidians: the brain uses GABA and the nerve cord contains glycinergic ACINs.

Pharmacological and genetic manipulation of glycine receptors has revealed that alternating swimming movements are controlled by glycinergic inhibitory neurotransmission (Nishino et al. 2010). In addition, glycine receptors are present on both cholinergic motor neurons in the motor ganglion and, unexpectedly, an anterior set of tail muscle cells, suggesting that muscle contraction activity may be controlled both directly and indirectly by ACINs. Thus, the glycinergic control system in the ascidian larva exhibits both conserved and unique features when compared with vertebrates (Nishino et al. 2010).

Serotoninergic Neurons Tryptophan hydroxylase (TPH) is a rate-limiting enzyme in serotonin (5-HT) synthesis. In *Ciona* larvae, the *TPH* gene is expressed in the motor ganglion and muscle cells (Pennati et al. 2007). In *Phallusia mammillata*, 5-HT was detected in the neurons located near the ocellus, in papilla, and in caudal epidermal neurons (Pennati et al. 2001). In this species, 5-HT and dopamine have been implicated in the onset of metamorphosis (Zega et al. 2005). In the *Ciona* larva, however, the localization of 5-HT has not been clearly demonstrated (Stach 2005; Pennati et al. 2007). Interestingly, a homologue of the 5-HT transporter (SERT) is expressed in the *TH*-positive dopaminergic neurons in the brain vesicle of the *Ciona* larva, and 5-HT uptake by these cells was observed when the larvae were treated with exogenous 5-HT (Razy-Krajka et al. 2012).

Catecholaminergic Neurons Dopamine, noradrenaline, and adrenaline are called catecholamines, and are synthesized from tyrosine by the catecholamine synthetic pathway. TH is a rate-limiting enzyme in catecholamine synthesis and used as a marker for catecholaminergic neurons. In the *Ciona* larva, the *TH* gene is expressed in a small population of cells in the ventral region of the sensory vesicle (Moret et al. 2005a). Immunoreactivity of dopamine was detected in the *TH*-positive cells, but no staining was observed by immunostaining with an anti-noradrenaline antibody (Moret et al. 2005a). Because *Ciona* lacks the machinery to synthesize adrenaline, adrenergic neurons are probably absent in *Ciona* (Dehal et al. 2002). Therefore, the *TH*-positive cells in the ascidian larva are thought to be dopaminergic neurons. In vertebrates, dopaminergic neurons are involved in the regulation of motor behavior. In *Ciona*, the growth of the axon of dopaminergic neurons occurs a few hours after hatching. Moret et al. (2005a) speculated that dopaminergic neurons could contribute to the age-dependent change in the swimming behavior of the larva. Razy-Krajka and colleagues further examined the role of dopamine in the swimming behavior of *Ciona* larvae (Razy-Krajka et al. 2012). The dopaminergic cells are located in the vicinity of photoreceptor cells and correspond to the coronet cells, which resemble the dopaminergic coronet cells of the saccus vasculosus of cartilaginous and teleost fishes. Pharmacological and behavioral analyses suggest that dopaminergic cells modulate the light-off-induced swimming behavior of ascidian larvae by acting on alpha2-like receptors, supporting a role of dopamine in the modulation of the photic response.

GnRH Neurons Gonadotropin-releasing hormone (GnRH) is a key neuropeptide responsible for the control of reproductive function in vertebrates (Oka 1997). In the *C. intestinalis* genome, two GnRH genes are present, each of which encodes three GnRH peptides and four GnRH receptor (GnRHR) genes (Adams et al. 2003; Kusakabe et al. 2003; Tello et al. 2005). In the adult ascidian, GnRH neurons are distributed along the dorsal strand and cerebral ganglion (Powell et al. 1996; Tsutsui et al. 1998; Ohkuma et al. 2000). Injection of GnRH into mature adults of *C. intestinalis* was able to induce gamete release, suggesting that GnRH plays a role in reproduction as in vertebrates (Terakado 2001).

The two GnRH genes are conspicuously expressed in the *C. intestinalis* larval CNS through the entire anteroposterior body axis (Kusakabe et al. 2012). Correspondingly, the GnRH receptor genes are specifically expressed in the phylotypic axial structures along the CNS (Kusakabe et al. 2012). The larvae of *C. intestinalis* occupy a nonreproductive dispersal stage; the gonad develops after metamorphosis and reproduction occurs much later. Conspicuous expression of the genes for GnRHs and GnRHRs in the nervous system and tail tissues in the larva strongly suggests that GnRHs regulate biological processes that are not directly related to reproduction. Interestingly, similar expression patterns of a GnRH gene have been suggested in vertebrates (Kusakabe et al. 2012). Thus, a nonreproductive GnRH system may be evolutionarily conserved between tunicates and vertebrates. One possible nonreproductive role of *Ciona* GnRHs is the control of locomotion and swimming behavior (Kusakabe et al. 2012). Another possible role of GnRH in the ascidian larva may be the control of metamorphosis (Kusakabe et al. 2012; Kamiya et al. 2014). The results of treatment of *Ciona* larvae with GnRHs suggested that GnRHs are responsible for the control of two major metamorphic events, namely, tail absorption and adult organ growth (Kamiya et al. 2014).

7.8 Behavior and Neuronal Function of Ascidian Larvae

Ciona larvae show different behavior patterns during larval life (Kajiwara and Yoshida 1985; Svane and Young 1989; Nakagawa et al. 1999; Tsuda et al. 2003a; Zega et al. 2006). In the early stage until 3 h after hatching, the larvae swim upwards (first-phase behavior). In the following stage, the larvae start swimming when light intensity decreases and stop swimming when light intensity increases (second-phase behavior) (Nakagawa et al. 1999; Tsuda et al. 2003a). Although the photoreceptor cells have already differentiated, including expression and localization of the visual pigment opsin (Kusakabe et al. 2001; Horie et al. 2008b), the photoresponse does not emerge until 3 h after hatching (Nakagawa et al. 1999). Horie et al. (2005)

demonstrated that the development of the nerve terminal of the photoreceptors was correlated with the photoresponse. By 3 h after hatching, the nerve terminals had expanded to a remarkable degree, concomitant with the emergence of a photoresponse.

The ascidian larva swims by rhythmically oscillating its tail. This movement is generated by alternate contraction and relaxation of the lateral muscles. This type of stereotyped repetitive motion is produced by a central pattern generator (CPG) in various animals (Wilson 1961; Satterlie 1985; Grillner et al. 1995; Roberts 2000). A simple CPG is organized into a reciprocal inhibitory neural circuit. In the vertebrate locomotor CPGs located in the spinal cord, commissural interneurons provide reciprocal coordination between the left and right sides of the spinal cord (Soffe et al. 1984; Dale 1985; Buchanan 1999; Quinlan and Kiehn 2007). A motor control system similar to that of the vertebrate CPGs has been proposed in the swimming of the tadpole larva of ascidians (Horie et al. 2009, 2010; Nishino et al. 2010). The CPG of the ascidian larva has been proposed to contain two major components: several bilateral pairs of cholinergic motor neurons in the motor ganglion and two bilateral pairs of inhibitory interneurons in the anterior part of the tail nerve cord (Fig. 7.9). The inhibitory neurons (ACINs) send axons to contralateral cholinergic neurons (Horie et al. 2009, 2010; Nishino et al. 2010). The organization of the CPG in the ascidian larva seems simpler but resembles that of the spinal CPGs for vertebrate locomotion. Takamura and colleagues further proposed an extended version of the neural circuit of the ascidian larva by incorporating other interneurons and sensory

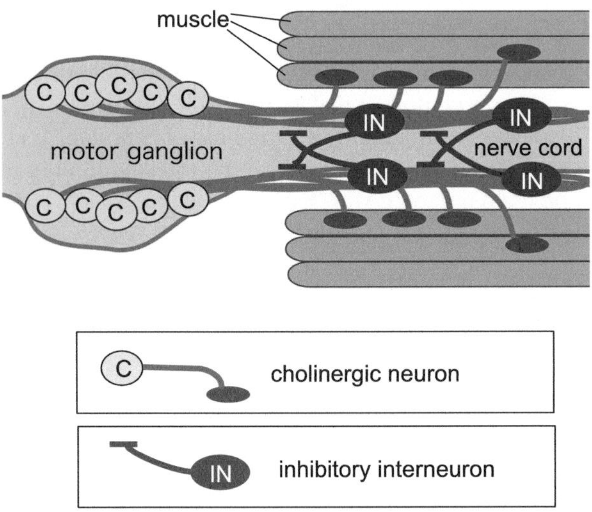

Fig. 7.9 Schematic diagram shows a neuronal circuit for larval swimming locomotion of *C. intestinalis*. The neural circuit comprises cholinergic motor neurons in the motor ganglion and inhibitory interneurons, called *ACINs*, in the anterior nerve cord (See text for details)

inputs from the brain sensory organs and peripheral sensory neurons (Takamura et al. 2010).

7.9 Photoreceptors and Cerebrospinal Fluid (CSF)-Contacting Neuron Homologues in Ascidians

The morphology and electrophysiological response of retinal photoreceptors differ between vertebrates and invertebrates. Vertebrate retinal photoreceptors are ciliary photoreceptors, and they hyperpolarize in response to light. On the other hand, most photoreceptors of invertebrate eyes are rhabdomeric and depolarize in response to light. There are a number of important differences in phototransduction between ciliary photoreceptors and rhabdomeric photoreceptors (Tsuda 1987; Hardie and Raghu 2001). Ciliary photoreceptors and rhabdomeric photoreceptors use different types of opsins, which are clearly distinguished by their primary structures.

The photoreceptor cells of the ascidian larva show morphological and electrophysiological properties that are similar to those of the visual cells of the vertebrate retina (Kusakabe and Tsuda 2007). The photoreceptor cells of the pigmented ocellus are ciliary and hyperpolarized photoreceptors (Eakin and Kuda 1971; Gorman et al. 1971). Ci-opsin1, the photopigment of ascidian ocelli, is a member of the vertebrate ciliary opsin subfamily (Kusakabe et al. 2001). The Ci-opsin1-knockdown larvae have lost the photic behavior, demonstrating that Ci-opsin1 is responsible for photoreception in swimming behavior (Inada et al. 2003). Targeted knockdown of the *Ciona* orthologue of *rx*, a homeobox gene that regulates photoreceptor development in vertebrates, resulted in the loss of ocellus pigment and photoreceptor-specific gene expression, suggesting that the developmental program is also similar between vertebrate and ascidian photoreceptors (D'Aniello et al. 2006). The outer segments of photoreceptor cells of *Ciona* larval ocelli are exposed to the lumen of the brain vesicle, a space homologous to the vertebrate ventricular cavity (Horie et al. 2008b). This positional arrangement of photoreceptors is common to the retinal, pineal, and deep brain photoreceptors of vertebrates. These similarities between the ascidian ocelli and vertebrate photoreceptors suggest their common ancestry.

The ocelli of the ascidian larva may have some sort of evolutionary link to CSF-contacting neurons in vertebrates. CSF-contacting neurons are sensory neurons located in the wall of the brain ventricle that send a ciliated process into the CSF (Vigh et al. 2002). Various opsins and phototransduction cascade proteins have been demonstrated in telencephalic and hypothalamic groups of CSF-contacting neurons. These specialized neurons are called "deep brain photoreceptors" (Vigh et al. 2002). The nonpigmented ocellus is located in the region that has been proposed to be homologous to the vertebrate hypothalamus (Moret et al. 2005a, 2005b). Thus, the photoreceptor cells of the nonpigmented ocellus might be counterparts of the deep brain photoreceptors of vertebrates.

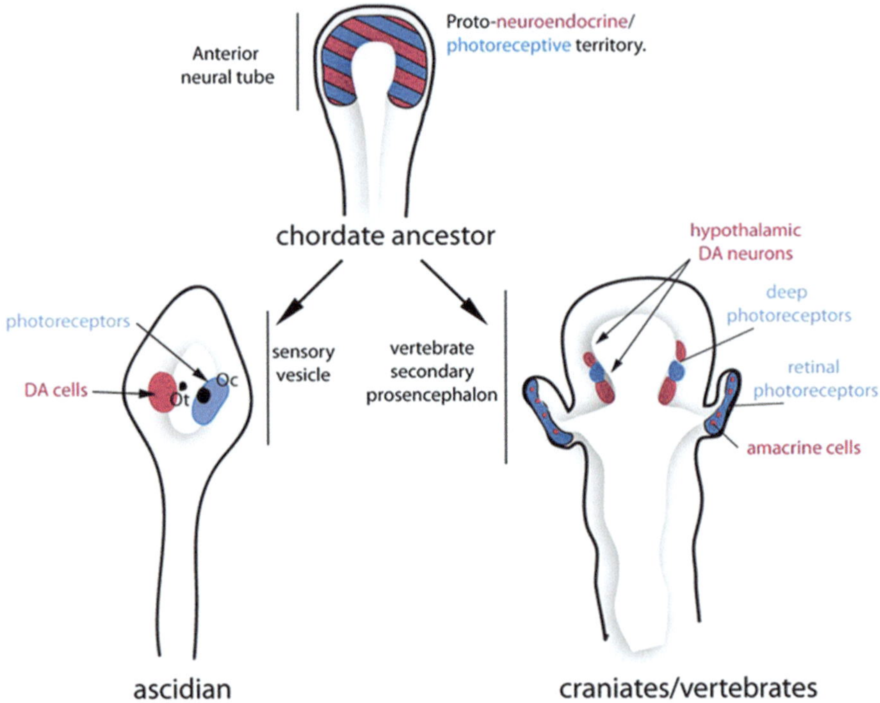

Fig. 7.10 Schematic representation of the proposed evolutionary scenario for the emergence of the retina and hypothalamus in craniates/vertebrates. The anterior neural tube of a chordate ancestor contained periventricular photoreceptor cells intermingled with neuroendocrine cells synthesizing dopamine and neuropeptides. These cells could also be connected to the CNS. These cell types line the anterior neural ventricle and contact the cerebrospinal fluid (CSF). A reminiscent but derived situation is found in ascidians, where photoreceptor cells line the ventricle and are adjacent to the dopaminergic cells of the sensory vesicle, which form coronets inside the ventricle. Dopaminergic cells are able to modulate the motor response to light. In craniates/vertebrates, the optic vesicle becomes separated from the anterior hypothalamus at the end of the neurulation process and bulged out of the neural tube to reach the lateral neuroectodermal epithelium and the lens placode, leading to new morphogenetic movements and to the inversion of the retina. The retina comprises several cell types inherited from the protochordate ancestor, including photoreceptor cells, pigmented epithelium, and dopaminergic amacrine cells (Reproduced from Razy-Krajka et al. (2012). © Razy-Krajka et al.; licensee BioMed Central, Ltd. 2012)

Other putative homologues of the CSF-contacting neurons are the dopaminergic coronet cells of the ascidian larva, which share traits with the CSF-contacting dopamine neurons of the vertebrate hypothalamus (Moret et al. 2005a; Razy-Krajka et al. 2012) (Fig. 7.10). As a result of the ventral location of the dopaminergic cells and their protrusions (coronets) in the lumen of the sensory vesicle, these cells resemble the dopaminergic coronet cells of the saccus vasculosus of cartilaginous and teleost fishes, a paraventricular organ of the hypothalamus that is secondarily lost in tetrapods, in addition to resembling the dopamine-synthesizing cells of the caudal hypothalamus in teleosts (Joly et al. 2007). On the other hand, the dopaminergic coronet cells are located in the vicinity of photoreceptor cells and share traits with dopaminergic amacrine cells of the vertebrate retina (Razy-Krajka et al. 2012). Based on these observations, Razy-Krajka and colleagues proposed that that the dopaminergic coronet cells of the ascidian larva were derived from an ancestral multifunctional cell population located in the periventricular, photoreceptive field of the anterior neural tube of chordates, which also gave rise to both the anterior hypothalamus and the retina in vertebrates (Fig. 7.10).

7.10 Rudimentary Neural Crest and Cranial Placodes in Invertebrate Chordates

It has been proposed that the emergence of the vertebrate head during evolution was accompanied by the acquisition of two types of embryonic ectodermal tissues, the neural crest and the cranial placodes (Northcutt and Gans 1983; Northcutt 2005). The neural crest, which undergoes an epithelial–mesenchymal transition and subsequent extensive migration, produces a diverse array of cell types, including ectomesenchyme derivatives that elaborate the vertebrate head (Bronner and Le Douarin 2012; Green et al. 2015). The cranial placodes give rise to the anterior pituitary gland, sensory organs [nose, ear, eye (lens), and lateral line], neuroendocrine cells in the brain, and sensory ganglia of the vertebrate head (Schlosser 2006; Begbie 2008; Schlosser 2010). Both the neural crest and cranial placodes originate in the neural plate border region (Schlosser 2008).

Indisputable homologues of the neural crest and placodes are lacking in amphioxus (Holland 2009), although they have populations of cells that have been proposed to be homologues of the neural crest and cranial placodes (Holland et al. 1996; Schlosser 2005; Kaltenbach et al. 2009; Ivashkin and Adameyko 2013). In ascidians, neural crest-like cells of three distinct lineages have been reported (Jeffery et al. 2004, 2008; Abitua et al. 2012; Stolfi et al. 2015). One type of the proposed homologues of the neural crest in ascidians are the migratory cells that give rise to pigmented cells (Jeffery et al. 2004; Jeffery 2006). In *Ciona intestinalis*, the neural crest-like cells of this type are shown to be the trunk lateral cells, which are migratory mesenchymal cells of mesodermal lineage (Jeffery et al. 2008). The pigment cells of the brain sensory organs, the otolith and the ocellus, are neural crest-like cells of another type reported in *Ciona intestinalis* (Abitua et al. 2012). These pigment cells arise at the neural plate border and express many neural crest-related genes. Moreover, their development is regulated by Wnt signaling in the manner of the vertebrate neural crest. Interestingly, these cells can be reprogrammed into migrating ectomesenchyme by the targeted misexpression of *Twist*. Based on these results, it has been proposed that that the co-option of mesenchyme determinants, such as Twist, into the neural plate ectoderm was crucial to the emergence of the vertebrate head (Abitua et al. 2012).

Cranial placode homologues have been proposed in ascidians (Wada et al. 1998; Manni et al. 2004a, 2004b, 2005; Mazet and Shimeld 2005; Mazet et al. 2005; Abitua et al. 2015). These putative placodal regions include cells giving rise to oral and atrial siphon primordia, the neurohypophysial duct, the larval adhesive palps, and preoral sensory neurons. Among these, the rostral palp-forming ectoderm has been proposed to be an olfactory placode homologue (Mazet et al. 2005), but it does not express *Pitx*, *Six3/6*, *Six1/2*, *Six4/5*, or *Eya* at the neural plate stage, and its position relative to the oral opening is different from that of the vertebrate olfactory placode. The oral siphon primordium and the neurohypophyseal duct derived from the anterior neural plate have been proposed to be olfactory and adenohypophyseal homologues (Manni et al. 2004a, 2005; Mazet and Shimeld 2005; Mazet et al. 2005) and express a set of key transcription factors for placode development. These regions, however, have not been demonstrated to be "neurogenic," which is an important characteristic of the vertebrate olfactory placode. The bilaterally paired atrial primordia also express a set of key placodal transcription factors and have been proposed to be homologues of otic or lateral line placodes (Wada et al. 1998; Manni et al. 2004b; Mazet and Shimeld 2005; Mazet et al. 2005). The atrial primordia, however, do not develop from the anterior neural plate border, and they contribute adult tissues associated with the anal opening. Thus, neither their embryonic origin nor future position in the adult body correspond to the vertebrate counterparts (Abitua et al. 2015).

Recently, the preoral ectoderm of *C. intestinalis* was proposed to be a proto-neurogenic placode that is homologous to the vertebrate neurogenic placodes (Abitua et al. 2015). This embryonic region was shown to give rise to dual-functional neurons that harbor characteristics of both chemosensory and GnRH-generating neurons, cell types developed from the olfactory placode of vertebrates (Fig. 7.11). In addition, these cells express the key regulatory determinants of cranial placodes, *Six1/2* and *Eya*, and their development is regulated by bone morphogenetic protein (BMP) signaling in a manner similar to the development of cranial placodes of vertebrates. These observations provide evidence that the putative sensory neurons arise from a placodal-like territory in a non-vertebrate. In vertebrates, GnRH neuroblasts migrate along the axon tracts of chemosensory neurons to a final destination in the brain, such as the hypothalamus. Olfactory and GnRH neurons form a coherent neuronal circuit, and pheromones detected by chemosensory neurons cause a release of gonadotropins via hypothalamic GnRH neurons, a function crucial to sexual reproduction. Thus, the intimate connection between olfaction and reproductive control likely arose long ago. The proposed chemosensory and neurosecretory functions of these neurons may have become segregated into dedicated cell types that work together within a coherent circuit during early vertebrate evolution. This type of cellular subfunctionalization might be an important mechanism of neuronal circuit evolution in vertebrates (Abitua et al. 2015). A similar subfunctionalization process has also been proposed for the evolution of the proto-hypothalamo-retinal territory of chordates (Razy-Krajka et al. 2012).

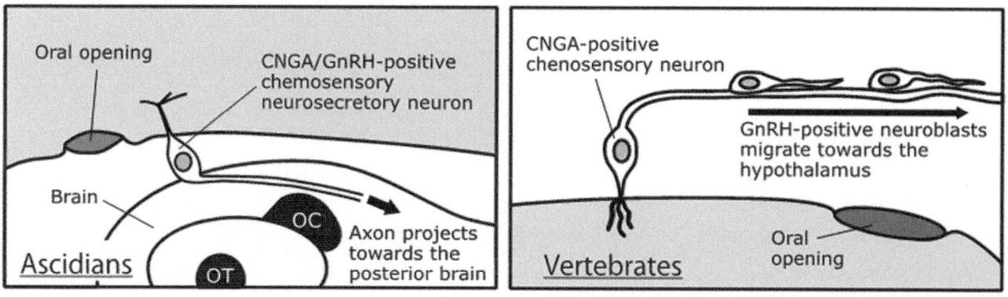

Fig. 7.11 Putative sensory neurons arising from a placodal-like territory in a tunicate embryo (Abitua et al. 2015). The ascidian larva (*left*) possesses a group of GnRH neurons, located behind the oral opening, derived from the proto-placodal ectoderm. These neurons possess cilia that have a $9 + 2$ microtubule arrangement, a shared characteristic attributed to nonmotile cilia of vertebrate olfactory neurons, and express a functional cyclic nucleotide-gated channel (CNGA), which is also a characteristic of vertebrate olfactory neurons. In vertebrates (*right*), neurosecretory and chemosensory cells both arise from the olfactory placode and are intimately linked. GnRH neuroblasts use the axon tracts of chemosensory neurons to guide them to their final destination in the hypothalamus (Modified from Abitua et al. 2015)

7.11 Conclusions

Comparison of the nervous systems among the extant deuterostome taxa illustrates both their remarkable diversity and shared basic plans of neural architecture. For example, irrespective of whether they are homologous or analogous, a motor control system containing a pair(s) of contralaterally inhibitory interneurons seems to be common among hemichordates, cephalochordates, tunicates, and vertebrates. Considering the highly conserved expression patterns of anteroposterior patterning genes in the CNS between protostomes (arthropods and annelids) and deuterostomes (chordates) (Arendt et al. 2008; Holland et al. 2013), the simple organization of the nervous system in some deuterostome taxa, such as hemichordates, echinoderms, and tunicates, may be a feature evolved from an ancestor with a more complex CNS. If so, the simple CNS in such taxa could serve as an excellent model system for elucidating the basic mechanisms of the vertebrate CNS descended from the common ancestor.

It seems highly plausible that the simple tunicate CNS was evolved from the more complex CNS of a common ancestor of the Olfactores. Tunicates have photoreceptive and gonadotropin-releasing hormone systems, which are clearly homologous to those of vertebrates, whereas equivalent systems seem to be absent or highly diverged in cephalochordate amphioxus (Kusakabe et al. 2001, 2009, 2012; Tello and Sherwood 2009; Roch et al. 2014). The putative homologues of neural crest and neurogenic cranial placodes have been identified in ascidians (Abitua et al. 2012, 2015; Stolfi et al. 2015). The less cephalized CNS of cephalochordates also has neuronal circuits with similarities to those of the vertebrate diencephalon-midbrain-hindbrain. With the shared characters among chordates, the CNS of invertebrate chordates should provide a unique platform for studying the developmental and evolutionary bases underlying the emergence of the complex CNS of vertebrates.

References

Abitua PB, Wagner E, Navarrete IA, Levine M (2012) Identification of a rudimentary neural crest in a non-vertebrate chordate. Nature 492:104–107

Abitua PB, Gainous TB, Kaczmarczyk AN, Winchell CJ, Hudson C, Kamata K, Nakagawa M, Tsuda M, Kusakabe TG, Levine M (2015) The pre-vertebrate origins of neurogenic placodes. Nature 524:462–465

Adams BA, Tello JA, Erchegyi J, Warby C, Hong DJ et al (2003) Six novel gonadotropin-releasing hormones are encoded as triplets on each of two genes in the protochordate, *Ciona intestinalis*. Endocrinology 144:1907–1919

Alford S, Schwartz E, Viana di Prisco G (2003) The pharmacology of vertebrate spinal central pattern generators. Neuroscientist 9:217–228

Amemiya CT, Prohaska SJ, Hill-Force A, Cook A, Wasserscheid J, Ferrier DE, Pascual-Anaya J, Garcia-Fernandez J, Dewar K, Stadler PF (2008) The amphioxus Hox cluster: characterization, comparative genomics, and evolution. J Exp Zool B Mol Dev Evol 310:465–477

Arendt D, Denes AS, Jékely G, Tessmar-Raible K (2008) The evolution of nervous system centralization. Philos Trans R Soc Lond B Biol Sci 363:1523–1528

Bateson W (1886) The ancestry of the chordata. Q J Microsc Sci 26:535–571

Begbie J (2008) Migration of neuroblasts from neurogenic placodes. Dev Neurosci 30:33–35

Bertrand S, Escriva H (2011) Evolutionary crossroads in developmental biology: amphioxus. Development 138:4819–4830

Bisgrove BW, Burke RD (1986) Development of serotonergic neurons in embryos of the sea urchin *Strongylocentrotus purpuratus*. Dev Growth Differ 28:569–574

Bisgrove BW, Burke RD (1987) Development of the nervous system of the pluteus larva of *Strongylocentrotus droebachiensis*. Cell Tissue Res 248:335–343

Bohn JM, Heinzeller T (1999) Morphology of the bourgueticrinid and isocrinid aboral nervous system and its possible phylogenetic implications (Echinodermata, Crinoidea). Acta Zool 80:241–249

Bourlat SJ, Juliusdottir T, Lowe CJ, Freeman R, Aronowicz J, Kirschner M, Lander ES, Thorndyke M, Nakano H, Khon AB, Heyland A, Moroz LL, Copley RR, Telford MJ (2006) Deuterostome phylogeny reveals monophyletic chordates and the new phylum Xenoturbellida. Nature 444:85–88

Bronner ME, Le Douarin NM (2012) Evolution and development of the neural crest: an overview. Dev Biol 366:2–9

Brown ER, Nishino A, Bone Q, Meinertzhagen IA, Okamura Y (2005) GABAergic synaptic transmission modulates swimming in the ascidian larva. Eur J Neurosci 22:2541–2548

Brown FD, Prendergast A, Swalla BJ (2008) Man is but a worm: chordate origins. Genesis 46:605–613

Buchanan JT (1999) Commissural interneurons in rhythm generation and intersegmental coupling in the lamprey spinal cord. J Neurophysiol 81:2037–2045

Bullock TH (1944) The giant nerve fiber system in balanoglossids. J Comp Neurol 80:355–367

Bullock TH (1965) The nervous system of hemichordates. In: Bullock TH, Horridge GA (eds) Structure and function in the nervous systems of invertebrates. Freeman, San Francisco/London, pp 1567–1577

Burke RD, Brand DG, Bisgrove BW (1986) Structure of the nervous system of the auricularia larva of *Parasticopus californicus*. Biol Bull 170:450–460

Burke RD, Angerer LM, Elphick MR, Humphrey GW, Yaguchi S, Kiyama T, Liang S, Mu X, Agca C, Klein WH, Brandhorst BP, Rowe M, Wilson K, Churcher AM, Taylor JS, Chen N, Murray G, Wang D, Mellott D, Olinski R, Hallböök F, Thorndyke MC (2006) A genomic view of the sea urchin nervous system. Dev Biol 300:434–460

Byrne M, Nakajima Y, Chee FC, Burke RD (2007) Apical organs in echinoderm larvae: insights into larval evolution in the Ambulacraria. Evol Dev 9:432–445

Cameron CB (2002) Particle retention and flow in the pharynx of the enteropneust worm *Harrimania planktophilus*: the filter-feeding pharynx may have evolved before the chordates. Biol Bull 202:192–200

Cameron CB, Garey JR, Swalla BJ (2000) Evolution of the chordate body plan: new insights from phylogenetic analyses of deuterostome phyla. Proc Natl Acad Sci USA 97:4469–4474

Cannon JT, Vellutini BC, Smith J III, Ronquist F, Jondelius U, Hejnol A (2016) Xenacoelomorpha is the sister group to Nephrozoa. Nature 530:89–93

Castro LFC, Rasmussen SLK, Holland PWH, Holland ND, Holland LZ (2006) A Gbx homeobox gene in amphioxus: insights into ancestry of the ANTP class and evolution of the midbrain/hindbrain boundary. Dev Biol 295:40–51

Chia FS, Burke RD (1978) Echinoderm metamorphosis: fate of larval structures. In: Chia FS, Rice ME (eds) Settlement and metamorphosis of marine invertebrate larvae. Elsevier-North Holland, New York, pp 219–234

Cisternas P, Byrne M (2009) Expression of Hox4 during development of the pentamerous juvenile sea star, *Parvulastra exigua*. Dev Genes Evol 219:613–618

Cole AG, Meinertzhagen IA (2004) The central nervous system of the ascidian larva: mitotic history of cells forming the neural tube in the late embryonic *Ciona intestinalis*. Dev Biol 271:239–262

Conway Morris S (1982) Atlas of the Burgess shale. Palaeontological Association, London

Dale N (1985) Reciprocal inhibitory interneurones in the *Xenopus* embryo spinal cord. J Physiol 363:61–70

D'Aniello S, D'Aniello E, Locascio A, Memoli A, Corrado M, Russo MT, Aniello F, Fucci L, Brown ER, Branno M (2006) The ascidian homolog of the vertebrate homeobox gene Rx is essential for ocellus development and function. Differentiation 74:222–234

Dehal P, Satou Y, Campbell RK, Chapman J, Degnan B et al (2002) The draft genome of *Ciona intestinalis*: insights into chordate and vertebrate origins. Science 298:2157–2167

Delsuc F, Brinkmann H, Chourrout D, Philippe H (2006) Tunicates and not cephalochordates are the closest living relative of vertebrates. Nature 439:965–968

Denes AS, Jekely G, Steinmetz PR, Raible F, Snyman H, Prud'-homme B, Ferrier DE, Balavoine G, Arendt D (2007) Molecular architecture of annelid nerve cord supports common origin of nervous system centralization in Bilateria. Cell 129:277–288

Devlin CL (2001) The pharmacology of γ-aminobutyric acid and acetylcholine receptors at the echinoderm neuromuscular junction. J Exp Biol 204:887–896

Dilly N (1961) Electron microscope observations of the receptors in the sensory vesicle of the ascidian tadpole. Nature 191:786–787

Dilly N (1964) Studies on the receptors in the cerebral vesicle of the ascidian tadpole. 2. The ocellus. Q J Microsc Sci 105:13–20

Dufour HD, Chettouh Z, Deyts C, de Rosa R, Goridis C, Joly JS, Brunet JF (2006) Precraniate origin of cranial motoneurons. Proc Natl Acad Sci USA 103:8727–8732

Eakin RM, Kuda A (1971) Ultrastructure of sensory receptors in ascidian tadpoles. Z Zellforsch 112:287–312

Echelard Y, Epstein DJ, St.-Jacques B, Shen L, Mohler J, McMahon JA, McMahon AP (1993) Sonic hedgehog, a member of a family of putative signaling molecules, is implicated in the regulation of CNS polarity. Cell 75:1417–1430

Elphick MR, Egertova M (2001) The neurobiology and evolution of cannabinoid signalling. Philos Trans R Soc Lond B Biol Sci 356:381–408

Elphick MR, Egertova M (2005) The phylogenetic distribution and evolutionary origins of endocannabinoid signalling. Handb Exp Pharmacol 168:283–297

Elphick MR, Melarange R (2001) Neural control of muscle relaxation in echinoderms. J Exp Biol 204:875–885

Elphick MR, Price DA, Lee TD, Thorndyke MC (1991) The SALMFamides: a new family of neuropeptides isolated from an echinoderm. Philos Trans R Soc Lond B Biol Sci 243:121–127

Elphick MR, Satou Y, Satoh N (2003) The invertebrate ancestry of endocannabinoid signalling: an orthologue of vertebrate cannabinoid receptors in the urochordate *Ciona intestinalis*. Gene 302:95–101

Ferrier DEK, Brooke NM, Panopoulou G, Holland PWH (2001) The Mnx homeobox gene class defined by *HB9*, *MNR2* and amphioxus *AmphiMnx*. Dev Genes Evol 211:103–107

Florey E, Cahill MA, Rathmayer M (1975) Excitatory actions of GABA and of acetylcholine in sea urchin tube feet. Comp Biochem Physiol 51C:5–12

Garcia-Fernandez J, Holland PWH (1994) Archetypal organization of the amphioxus Hox gene cluster. Nature 370:563–566

Garstang W (1894) Preliminary note on a new theory of the ancestry of the Chordata. Zool Anz 17:122–125

Gorman AL, McReynolds JS, Barnes SN (1971) Photoreceptors in primitive chordates: fine structure, hyperpolarizing receptor potentials, and evolution. Science 172:1052–1054

Green SA, Simoes-Costa M, Bronner ME (2015) Evolution of vertebrates as viewed from the crest. Nature 520:474–482

Grillner S, Deliagina T, Ekeberg O, El Manira A, Hill RH, Lansner A, Orlovsky GN, Wallen P (1995) Neural networks that co-ordinate locomotion and body orientation in lamprey. Trends Neurosci 18:270–279

Halanych KM (1995) The phylogenetic position of the pterobranch hemichordates based on 18S rDNA sequence data. Mol Phylogenet Evol 4:72–76

Hardie RC, Raghu P (2001) Visual transduction in *Drosophila*. Nature 413:186–193

Hay-Schmidt A (2000) The evolution of the serotonergic nervous system. Proc R Soc Lond B 267:1071–1079

Heinzeller T, Welsch U (2001) The echinoderm nervous system and its phylogenetic interpretation. In: Roth G, Wulliman WF (eds) Brain evolution and cognition. Wiley-Spektrum, New York/Heidelberg, pp 41–75

Hirokawa T, Komatsu M, Nakajima Y (2008) Development of the nervous system in the brittle star *Amphipholis kochii*. Dev Genes Evol 218:15–21

Holland LZ (2009) Chordate roots of the vertebrate nervous system: expanding the molecular toolkit. Nat Rev Genet 10:736–746

Holland LZ (2015) Evolution of basal deuterostome nervous systems. J Exp Biol 218(pt 4):637–645

Holland ND, Panganiban G, Henyey EL, Holland LZ (1996) Sequence and developmental expression of *AmphiDll*, an amphioxus distal-less gene transcribed in the ectoderm, epidermis and nervous system: insights into evolution of craniate forebrain and neural crest. Development 122:2911–2920

Holland LZ, Carvalho JE, Escriva H, Laudet V, Schubert M, Shimeld SM, Yu JK (2013) Evolution of bilaterian central nervous systems: a single origin? Evodevo 4:27

Horie T, Orii H, Nakagawa M (2005) Structure of photoreceptors in the ascidian *Ciona intestinalis* larva revealed by an anti-arrestin antibody. J Neurobiol 65:241–250

Horie T, Kusakabe T, Tsuda M (2008a) Glutamatergic networks in the *Ciona intestinalis* larva. J Comp Neurol 508:249–263

Horie T, Sakurai D, Ohtsuki H, Terakita A, Shichida Y, Usukura J, Kusakabe T, Tsuda M (2008b) Pigmented and nonpigmented ocelli in the brain vesicle of the ascidian larva. J Comp Neurol 509:88–102

Horie T, Nakagwa M, Sasakura Y, Kusakabe TG (2009) Cell type and function of neurons in the ascidian nervous system. Dev Growth Differ 51:207–220

Horie T, Nakagawa M, Sasakura Y, Kusakabe TG, Tsuda M (2010) Simple motor system of the ascidian larva: neuronal complex comprising putative cholinergic neurons and GABAergic/glycinergic neurons. Zool Sci 27:181–190

Horie T, Shinki R, Ogura Y, Kusakabe TG, Satoh N, Sasakura Y (2011) Ependymal cells of chordate larvae are stem-like cells that form the adult nervous system. Nature 469:525–528

Hyman LH (1955) The invertebrates. Echinodermata: the coelomate Bilateria, vol 4. McGraw-Hill, New York

Ikuta T, Saiga H (2007) Dynamic change in the expression of developmental genes in the ascidian central nervous system: revisit to the tripartite model and the origin of the midbrain-hindbrain boundary region. Dev Biol 312:631–643

Imai JH, Meinertzhagen IA (2007) Neurons of the ascidian larval nervous system in *Ciona intestinalis*: I. Central nervous system. J Comp Neurol 501:316–334

Imai KS, Satoh N, Satou Y (2002) Region specific gene expressions in the central nervous system of the ascidian embryo. Gene Expr Patterns 2:319–321

Inada K, Horie T, Kusakabe T, Tsuda M (2003) Targeted knockdown of an opsin gene inhibits the swimming behaviour photoresponse of ascidian larvae. Neurosci Lett 347:167–170

Irimia M, Piñeiro C, Maeso I, Gómez-Skarmeta JL, Casares F, Garcia-Fernàndez J (2010) Conserved developmental expression of fezf in chordates and *Drosophila* and the origin of the zona limitans intrathalamica (ZLI) brain organizer. Evodevo 1:7

Ivashkin E, Adameyko I (2013) Progenitors of the protochordate ocellus as an evolutionary origin of the neural crest. Evodevo 4:12

Jefferies RPS (1991) In: Bock GR, Marsh J (eds) Biological asymmetry and handedness. Wiley, Chichester, pp 94–127

Jeffery WR (2006) Ascidian neural crest-like cells: phylogenetic distribution, relationship to larval complexity, and pigment cell fate. J Exp Zool B Mol Dev Evol 306:470–480

Jeffery WR, Strickler AG, Yamamoto Y (2004) Migratory neural crest-like cells form body pigmentation in a urochordate embryo. Nature 431:696–699

Jeffery WR, Chiba T, Razy-Krajka F, Deyts C, Satoh N, Joly JS (2008) Trunk lateral cells are neural crest-like cells in the ascidian *Ciona intestinalis*: insights into the ancestry and evolution of the neural crest. Dev Biol 324:152–160

Joly JS, Osorio J, Alunni A, Auger H, Kano S, Retaux S (2007) Windows of the brain: towards a developmental biology of circumventricular and other neurohemal organs. Semin Cell Dev Biol 18:512–524

Kajiwara S, Yoshida M (1985) Change in behavior and ocellar structure during the larval life of solitary ascidians. Biol Bull 169:565–577

Kaltenbach SL, Yu JK, Holland ND (2009) The origin and migration of the earliest-developing sensory neurons in the peripheral nervous system of amphioxus. Evol Dev 11:142–151

Kamiya C, Ohta N, Ogura Y, Yoshida K, Horie T, Kusakabe TG, Satake H, Sasakura Y (2014) Nonreproductive role of gonadotropin-releasing hormone in the control of ascidian metamorphosis. Dev Dyn 243:1524–1535

Kappers CUA (1929) The evolution of the nervous system in invertebrates, vertebrates and man. Bohn, Haarlem

Katow H, Yaguchi S, Kyozuka K (2007) Serotonin stimulates $[Ca^{2+}]_i$ elevation in ciliary ectodermal cells of echinoplutei through a serotonin receptor cell network in the blastocoel. J Exp Biol 210:403–412

Katow H, Suyemitsu T, Ooka S, Yaguchi J, Jin-Nai T, Kuwahara I, Katow T, Yaguchi S, Abe H (2010) Development of a dopaminergic system in sea urchin embryos and larvae. J Exp Biol 213:2808–2819

Katow H, Abe K, Katow T, Zamani A, Abe H (2013) Development of the GABA-ergic signaling system and its role in larval swimming in sea urchin. J Exp Biol 216:1704–1716

Katz MJ (1983) Comparative anatomy of the tunicate tadpole *Ciona intestinalis*. Biol Bull 164:1–27

Kaul S, Stach T (2010) Ontogeny of the collar cord: neurulation in the hemichordate *Saccoglossus kowalevskii*. J Morphol 271:1240–1259

Kon T, Nohara M, Yamanoue Y, Fujiwara Y, Nishida M, Nishikawa T (2007) Phylogenetic position of a whale-fall lancelet (Cephalochordata) inferred from whole mitochondrial genome sequences. BMC Evol Biol 7:127

Kusakabe T, Tsuda M (2007) Photoreceptive systems in ascidians. Photochem Photobiol 83:248–252

Kusakabe T, Kusakabe R, Kawakami I, Satou Y, Satoh N, Tsuda M (2001) *Ci-opsin1*, a vertebrate-type opsin gene, expressed in the larval ocellus of the ascidian *Ciona intestinalis*. FEBS Lett 506:69–72

Kusakabe T, Mishima S, Shimada I, Kitajima Y, Tsuda M (2003) Structure, expression, and cluster organization of genes encoding gonadotropin-releasing hormone receptors found in the neural complex of the ascidian *Ciona intestinalis*. Gene 322:77–84

Kusakabe TG, Takimoto N, Jin M, Tsuda M (2009) Evolution and the origin of the visual retinoid cycle in vertebrates. Philos Trans R Soc Lond B Biol Sci 364:2897–2910

Kusakabe TG, Sakai T, Aoyama M, Kitajima Y, Miyamoto Y, Takigawa T, Daido Y, Fujiwara K, Terashima Y, Sugiuchi Y, Matassi G, Yagisawa H, Park MK, Satake H, Tsuda M (2012) A conserved non-reproductive GnRH system in chordates. PLoS One 7:e41955

Lacalli TC (2002) The dorsal compartment locomotory control system in amphioxus larvae. J Morphol 252:227–237

Lacalli TC (2003) Ventral neurons in the anterior nerve cord of amphioxus larvae. II. Further data on the pacemaker circuit. J Morphol 257:212–218

Lacalli TC (2006) Prospective protochordate homologs of vertebrate midbrain and MHB, with some thoughts on MHB origins. Int J Biol Sci 2:104–109

Lacalli TC (2008) Basic features of the ancestral chordate brain: a protochordate perspective. Brain Res Bull 75:319–323

Lacalli TC, Kelly SJ (2003) Ventral neurons in the anterior nerve cord of amphioxus larvae. I. An inventory of cell types and synaptic patterns. J Morphol 257:190–211

Lacalli TC, West JE (1993) A distinctive nerve cell type common to diverse deuterostome larvae: comparative data from echinoderms, hemichordates and amphioxus. Acta Zool 74:1–8

Lacalli TC, Gilmour THJ, West JE (1990) Ciliary band innervation in the bipinnaria larva of *Pisaster ochraceus*. Philos Trans R Soc B Biol Sci 330:371–390

Legendre P (2001) The glycinergic inhibitory synapse. Cell Mol Life Sci 58:760–793

Lester S (1988) Settlement and metamorphosis of *Rhabdopleura normani* (Hemichordata: Pterobranchia). Acta Zool 69:111–120

Lowe CJ (2008) Molecular genetic insights into deuterostome evolution from the direct-developing hemichordate *Saccoglossus kowalevskii*. Philos Trans R Soc Lond B Biol Sci 363:1569–1578

Lowe CJ, Wu M, Salic A, Evans L, Lander E, Stange-Thomann N, Gruber CE, Gerhart J, Kirschner M (2003) Anteroposterior patterning in hemichordates and the origins of the chordate nervous system. Cell 113:853–865

Luttrell S, Konikoff C, Byrne A, Bengtsson B, Swalla BJ (2012) Ptychoderid hemichordate neurulation without a notochord. Integr Comp Biol 52:829–834

Mackie GO, Spencer AN, Strathmann R (1969) Electrical activity associated with ciliary reversal in an echinoderm larva. Nature 223:1384–1385

Manni L, Lane NJ, Gasparini F, Tiozzo S, Caicci F, Zaniolo G, Burighel P (2004a) Neurogenic and nonneurogenic placodes in ascidians. J Exp Zool B Mol Dev Evol 302:483–504

Manni L, Caicci F, Gasparini F, Zaniolo G, Burighel P (2004b) Hair cells in ascidians and the evolution of lateral line placodes. Evol Dev 6:379–381

Manni L, Agnoletto A, Zaniolo G, Burighel P (2005) Stomodeal and neurohypophysial placodes in *Ciona intestinalis*: insights into the origin of the pituitary gland. J Exp Zool B Mol Dev Evol 304:324–339

Mazet F, Shimeld S (2005) Molecular evidence from ascidians for the evolutionary origin of vertebrate cranial sensory placodes. J Exp Zool B Mol Dev Evol 304:340–346

Mazet F, Hutt JA, Milloz J, Millard J, Graham A, Shimeld SM (2005) Molecular evidence from *Ciona intestinalis* for the evolutionary origin of vertebrate cranial placodes. Dev Biol 282:494–508

Metchnikoff E (1881) Über die systematische Stellung von Balanoglossus. Zool Anz 4:153–157

Meinertzhagen IA, Okamura Y (2001) The larval ascidian nervous system: the chordate brain from its small beginnings. Trends Neurosci 24:401–410

Meinertzhagen IA, Lemaire P, Okamura Y (2004) The neurobiology of the ascidian tadpole larva: recent developments in an ancient chordate. Annu Rev Neurosci 27:453–485

Moret F, Christiaen L, Deyts C, Blin M, Joly JS, Vernier P (2005a) The dopamine-synthesizing cells in the swimming larva of the tunicate *Ciona intestinalis* are located only in the hypothalamus-related domain of the sensory vesicle. Eur J Neurosci 21:3043–3055

Moret F, Christiaen L, Deyts C, Blin M, Vernier P, Joly JS (2005b) Regulatory gene expressions in the ascidian ventral sensory vesicle: evolutionary relationships with the vertebrate hypothalamus. Dev Biol 277:567–579

Morgan TH (1891) The growth and development of tornaria. J Morphol 5:407–458

Morgan TH (1894) Development of *Balanoglossus*. J Morphol 9:1–86

Miyamoto N, Wada H (2013) Hemichordate neurulation and the origin of the neural tube. Nat Commun 4:2713

Miyamoto N, Nakajima Y, Wada H, Saito Y (2010) Development of the nervous system in the acorn worm *Balanoglossus simodensis*: insights into nervous system evolution. Evol Dev 12:416–424

Nakagawa M, Miyamoto T, Ohkuma M, Tsuda M (1999) Action spectrum for the photophobic response of *Ciona intestinalis* (Ascidiacea, Urochordata) larvae implicates retinal protein. Photochem Photobiol 70:359–362

Nakajima Y (1987) Localization of catecholaminergic nerves in larval echinoderms. Zool Sci 4:293–299

Nakano H, Murabe N, Amemiya S, Nakajima Y (2006) Nervous system development of the sea cucumber *Stichopus japonicus*. Dev Biol 292:205–212

Nakano H, Nakajima Y, Amemiya S (2009) Nervous system development of two crinoid species, the sea lily *Metacrinus rotundus* and the feather star *Oxycomanthus japonicus*. Dev Genes Evol 219:565–576

Nicol D, Meinertzhagen IA (1988) Development of the central nervous system of the larva of the ascidian, *Ciona intestinalis* L. II. Neural plate morphogenesis and cell lineages during neurulation. Dev Biol 130:737–766

Nicol D, Meinertzhagen IA (1991) Cell counts and maps in the larval central nervous system of the ascidian *Ciona intestinalis* (L.). J Comp Neurol 309:415–429

Nielsen C (2006) Homology of echinoderm radial nerve cords and the chordate neural tube? Evol Dev 8:1–2

Nielsen C (2012) Animal evolution: interrelationships of the living phyla, 3rd edn. Oxford University Press, New York

Nieuwenhuys R (2002) Deuterostome brains: synopsis and commentary. Brain Res Bull 57:257–270

Nishino A, Satoh N (2001) The simple tail of chordates: phylogenetic significance of appendicularians. Genesis 29:36–45

Nishino A, Okamura Y, Piscopo S, Brown ER (2010) A glycine receptor is involved in the organization of swimming movements in an invertebrate chordate. BMC Neurosci 11:6

Nishitsuji K, Horie T, Ichinose A, Sasakura Y, Yasuo H, Kusakabe TG (2012) Cell lineage and *cis*-regulation for a unique GABAergic/glycinergic neuron type in the larval nerve cord of the ascidian *Ciona intestinalis*. Dev Growth Differ 54:177–186

Nohara M, Nishida M, Miya M, Nishikawa T (2005) Evolution of the mitochondrial genome in Cephalochordata as inferred from complete nucleotide sequences from two *Epigonichthys* species. J Mol Evol 60:526–537

Nomaksteinsky M, Röttinger E, Dufour HD, Chettouh Z, Lowe CJ, Martindale MQ, Brunet JF (2009) Centralization of the deuterostome nervous system predates chordates. Curr Biol 19:1264–1269

Northcutt RG (2005) The new head hypothesis revisited. J Exp Zool B Mol Dev Evol 304B:274–297

Northcutt RG, Gans C (1983) The genesis of neural crest and epidermal placodes: a reinterpretation of vertebrate origins. Q Rev Biol 58:1–28

Ohkuma M, Katagiri Y, Nakagawa M, Tsuda M (2000) Possible involvement of light regulated gonadotropin-releasing hormone neurons in biological clock for reproduction in the cerebral ganglion of the ascidian, *Halocynthia roretzi*. Neurosci Lett 293:5–8

Ohmori H, Sasaki S (1977) Development of neuromuscular transmission in a larval tunicate. J Physiol 269:221–254

Ohtsuki H (1990) Statocyte and ocellar pigment cell in embryos and larvae of the ascidian, *Styela plicata* (Lesueur). Dev Growth Differ 32:85–90

Oka Y (1997) The gonadotropin-releasing hormone (GnRH) neuronal system of fish brain as a model system for the study of peptidergic neuromodulation. In: Parhar IS, Sakuma Y (eds) GnRH neurons: genes to behavior. Brain Shuppan, Tokyo, pp 245–276

Pani AM, Mullarkey EE, Aronowicz J, Assimacopoulos S, Grove EA, Lowe CJ (2012) Ancient deuterostome origins of vertebrate brain signalling centres. Nature 483:289–294

Paul CRC, Smith AB (1984) The early radiation and phylogeny of echinoderms. Biol Rev 59:443–481

Pearce CM, Scheibling RE (1990) Induction of metamorphosis of larvae of the green sea urchin, *Strongylocentrotus droebachiensis*, by coralline red algae. Biol Bull 179:304–311

Pennati R, Groppelli S, Sotgia C, Candiani S, Pestarino M, De Bernardi F (2001) Serotonin localization in *Phallusia mammillata* larvae and effects of 5-HT antagonists during larval development. Dev Growth Differ 43:647–656

Pennati R, Candiani S, Biggiogero M, Zega G, Groppelli S, Oliveri D, Parodi M, De Bernardi F, Pestarino M (2007) Developmental expression of tryptophan hydroxylase gene in *Ciona intestinalis*. Dev Genes Evol 217:307–313

Perseke M, Golombek A, Schlegel M, Struck TH (2013) The impact of mitochondrial genome analyses on the understanding of deuterostome phylogeny. Mol Phylogenet Evol 66:898–905

Philippe H, Brinkmann H, Martinez P, Riutort M, Baguñà J (2007) Acoel flatworms are not platyhelminthes: evidence from phylogenomics. PLoS ONE 2:e717

Philippe H, Brinkmann H, Copley RR, Moroz LL, Nakano H, Poustka AJ, Wallberg A, Peterson KJ, Telford MJ (2011) Acoelomorph flatworms are deuterostomes related to *Xenoturbella*. Nature 470:255–258

Powell JFF, Reska-Skinner SM, Prakash MO, Fischer WH, Park M et al (1996) Two new forms of gonadotropin-releasing hormone in a protochordate and the evolutionary implications. Proc Natl Acad Sci USA 93:10461–10464

Protas LL, Muske GA (1980) The effects of some transmitter substances on the tube foot muscles of the starfish, *Asterias amurensis* (Lutken). Gen Pharmacol 11:113–118

Quinlan KA, Kiehn O (2007) Segmental, synaptic actions of commissural interneurons in the mouse spinal cord. J Neurosci 27:6521–6530

Raff RA (2008) Origin of the other metazoan body plans: the evolution of larval forms. Philos Trans R Soc Lond B Biol Sci 363:1473–1479

Razy-Krajka F, Brown ER, Horie T, Callebert J, Sasakura Y, Joly J-S, Kusakabe TG, Vernier P (2012) Monoaminergic modulation of photoreception in ascidian: evidence for a proto-hypothalamo-retinal territory. BMC Biol 10:45

Rebelo S, Reguenga C, Osorio L, Pereira C, Lopes C, Lima D (2007) DRG11 immunohistochemical expression during embryonic development in the mouse. Dev Dyn 236:2653–2660

Roberts A (2000) Early functional organization of spinal neurons in developing lower vertebrates. Brain Res Bull 53:585–593

Roch GJ, Tello JA, Sherwood NM (2014) At the transition from invertebrates to vertebrates, a novel GnRH-like peptide emerges in amphioxus. Mol Biol Evol 31:765–778

Rouse GW, Wilson NG, Carvajal JI, Vrijenhoek RC (2016) New deep-sea species of *Xenoturbella* and the position of *Xenacoelomorpha*. Nature 530:94–97

Saito T, Greenwood A, Sun Q, Anderson DJ (1995) Identification by differential RT-PCR of a novel paired homeodomain protein specifically expressed in sensory neurons and a subset of their CNS targets. Mol Cell Neurosci 6:280–292

Sakurai D, Goda M, Kohmura Y, Horie T, Iwamoto H, Ohtsuki H, Tsuda M (2004) The role of pigment cells in the brain of ascidian larva. J Comp Neurol 475:70–82

Satoh N (2009) An advanced filter-feeder hypothesis for urochordate evolution. Zool Sci 26:97–111
Satoh N, Rokhsar D, Nishikawa T (2014) Chordate evolution and the three-phylum system. Proc R Soc B 281:20141729
Satterlie RA (1985) Reciprocal inhibition and postinhibitory rebound produce reverberation in a locomotor pattern generation. Science 229:402–404
Schlosser G (2005) Evolutionary origins of vertebrate placodes: insights from developmental studies and from comparisons with other deuterostomes. J Exp Zool B Mol Dev Evol 304B:347–399
Schlosser G (2006) Induction and specification of cranial placodes. Dev Biol 294(2):303–351
Schlosser G (2008) Do vertebrate neural crest and cranial placodes have a common evolutionary origin? BioEssays 30:659–672
Schlosser G (2010) Making senses development of vertebrate cranial placodes. Int Rev Cell Mol Biol 283:129–234
Schuske K, Beg AA, Jorgensen EM (2004) The GABA nervous system in *C. elegans*. Trends Neurosci 27:407–414
Shu DG, Conway Morris S, Zhang XL (1996) A *Pikaia*-like chordate from the Lower Cambrian of China. Nature 348:157–158
Sly BJ, Hazel JC, Popodi EM, Raff RA (2002) Patterns of gene expression in the developing adult sea urchin central nervous system reveal multiple domains and deep-seated neural pentamery. Evol Dev 4:189–204
Smith JE (1965) Echinodermata. In: Bullock TH, Horridge GA (eds) Structure and function in the nervous systems of invertebrates. Freeman, London, pp 1519–1558
Soffe SR, Clarke JDW, Roberts A (1984) Activity of commissural interneurons in the spinal cord of *Xenopus* embryos. J Neurophysiol 51:1257–1267
Stach T (2005) Comparison of the serotonergic nervous system among Tunicata: implications for its evolution within Chordata. Org Divers Evol 5:15–24
Stach T, Winter J, Bouquet JM, Chourrout D, Schnabel R (2008) Embryology of a planktonic tunicate reveals traces of sessility. Proc Natl Acad Sci USA 105:7229–7234
Stolfi A, Levine M (2011) Neuronal subtype specification in the spinal cord of a protovertebrate. Development 138:995–1004
Stolfi A, Ryan K, Meinertzhagen IA, Christiaen L (2015) Migratory neuronal progenitors arise from the neural plate borders in tunicates. Nature 527:371–374
Strathmann RR (1975) Larval feeding in echinoderms. Am Zool 15:717–730
Svane I, Young CM (1989) The ecology and behavior of ascidian larvae. Oceanogr Mar Biol Rev 27:45–90
Swalla BJ, Cameron CB, Corley LS, Garey JR (2000) Urochordates are monophyletic within the deuterostomes. Syst Biol 49:122–134
Takahashi T, Holland PWH (2004) Amphioxus and ascidian Dmbx homeobox genes give clues to the vertebrate origin of midbrain development. Development 131:3285–3294
Takamura K (1998) Nervous network in larvae of the ascidian *Ciona intestinalis*. Dev Genes Evol 208:1–8
Takamura K, Egawa T, Ohnishi S, Okada T, Fukuoka T (2002) Developmental expression of ascidian neurotransmitter synthesis genes. I. Choline acetyltransferase and acetylcholine transporter genes. Dev Genes Evol 212:50–53
Takamura K, Minamida N, Okabe S (2010) Neural map of the larval central nervous system in the ascidian *Ciona intestinalis*. Zool Sci 27:191–203
Tanabe Y, William C, Jessell TM (1998) Specification of motor neuron identity by the MNR2 homeodomain protein. Cell 95:67–80
Tello JA, Sherwood NM (2009) Amphioxus: beginning of vertebrate and end of invertebrate type GnRH receptor lineage. Endocrinology 150:2847–2856
Tello JA, Rivier JE, Sherwood NM (2005) Tunicate gonadotropin-releasing hormone (GnRH) peptides selectively activate *Ciona intestinalis* GnRH receptors and the green monkey type II GnRH receptor. Endocrinology 146:4061–4067

Terakado K (2001) Induction of gamete release by gonadotropin-releasing hormone in a protochordate, *Ciona intestinalis*. Gen Comp Endocrinol 124:277–284

Thor S, Thomas JB (2002) Motor neuron specification in worms, flies and mice: conserved and 'lost' mechanisms. Curr Opin Genet Dev 12:558–564

Tsuda M (1987) Photoreception and phototransduction in invertebrate photoreceptors. Photochem Photobiol 45:915–931

Tsuda M, Kawakami I, Shiraishi S (2003a) Sensitization and habituation of the swimming behavior in ascidian larvae to light. Zool Sci 20:13–22

Tsuda M, Sakurai D, Goda M (2003b) Direct evidence for the role of pigment cells in the brain of ascidian larvae by laser ablation. J Exp Biol 206:1409–1417

Tsutsui H, Yamamoto N, Ito H, Oka Y (1998) GnRH-immunoreactive neuronal system in the presumptive ancestral chordate, *Ciona intestinalis* (ascidian). Gen Comp Endocrinol 112:426–432

Varju P, Katarova Z, Madarasz E, Szabo G (2001) GABA signaling during development: new data and old questions. Cell Tissue Res 305:239–246

Vigh B, Manzano MJ, Zadori A, Frank CL, Lukats A, Rohlich P, Szel A, David C (2002) Non visual photoreceptors of the deep brain, pineal organs and retina. Histol Histopathol 17:555–590

Vopalensky P, Pergner J, Liegertova M, Benito-Gutierrez E, Arendt D, Kozmik Z (2012) Molecular analysis of the amphioxus frontal eye unravels the evolutionary origin of the retina and pigment cells of the vertebrate eye. Proc Natl Acad Sci USA 109:15383–15388

Wada H, Satoh N (1994) Details of the evolutionary history from invertebrates to vertebrates, as deduced from the sequences of 18S rDNA. Proc Natl Acad Sci USA 91:1801–1804

Wada H, Saiga H, Satoh N, Holland PWH (1998) Tripartite organization of the ancestral chordate brain and the antiquity of placodes: insights from ascidian *Pax-2/5/8*, *Hox* and *Otx* genes. Development 125:1113–1122

Wada H, Garcia-Fernàndez J, Holland PWH (1999) Colinear and segmental expression of amphioxus Hox genes. Dev Biol 213:131–141

Whittaker JR (1973) Segregation during ascidian embryogenesis of egg cytoplasmic information for tissue-specific enzyme development. Proc Natl Acad Sci USA 70:2096–2100

Wicht H, Lacalli TC (2005) The nervous system of amphioxus: structure, development, and evolutionary significance. Can J Zool 83:122–150

Wickstead JH, Bone Q (1959) Ecology of acraniate larvae. Nature 184:1849–1851

Wilson DM (1961) The central nervous control of flight in a locust. J Exp Biol 38:471–479

Wilson JM, Hartley R, Maxwell DJ, Todd AJ, Lieberam I, Kaltschmidt JA, Yoshida Y, Jessell TM, Brownstone RM (2005) Conditional rhythmicity of ventral spinal interneurons defined by expression of the Hb9 homeodomain protein. J Neurosci 25:5710–5719

Yoshida R, Sakurai D, Horie T, Kawakami I, Tsuda M, Kusakabe T (2004) Identification of neuron-specific promoters in *Ciona intestinalis*. Genesis 39:130–140

Yu JK, Satou Y, Holland ND, Shin-i T, Kohara Y, Satoh N, Bronner-Fraser M, Holland LZ (2007) Axial patterning in cephalochordates and the evolution of the organizer. Nature 445:613–617

Zega G, Penatti R, Groppelli S, Sotgia C, De Bernardi F (2005) Dopamine and serotonin modulate the onset of metamorphosis in the ascidian *Phallusia mammillata*. Dev Biol 282:246–256

Zega G, Thorndyke MC, Brown ER (2006) Development of swimming behavior in the larva of the ascidian *Ciona intestinalis*. J Exp Biol 209:3405–3412

Zega G, Biggiogero M, Groppelli S, Candiani S, Oliveri D, Parodi M, Prestarino M, Bernardi F, Pennati R (2008) Delopmental expression of glutamic acid decarboxylase and of γ-aminnobutyric acid type B receptors in the ascidian *Ciona intestinalis*. J Comp Neurol 506:489–505

Chapter 8
Genome and Transcriptome-Wide Research of Brain Evolution

Atsushi Ogura

Abstract Genomics and transcriptomics research of the brain, accelerated by the development of sequencing technologies and genomic analysis methods, can reveal the genetic mechanisms underlying brain function, evolution, and development in various animals. This chapter first introduces the background of the recent technology of sequencing machines and analytical methods by bioinformatics, which makes possible the large-scale study of brains, and then presents recent results and achievements for brain function and evolution. Recent studies utilizing these emerging technologies are also introduced to demonstrate the power of large-scale analysis of genome and transcriptome for brain research. A perspective of international research projects for brain function is also introduced in the last section.

Keywords Genomics • Transcriptomics • Omics • Gene regulatory network • Next-generation sequencing • Genetic variants • Brain function • Brain evolution • Brain diversification

8.1 Introduction

8.1.1 Overview

Genomics is a research field for understanding the genetic background underlying species by analyzing the target species genome. Transcriptomics is a research method to understand functional context underlying organisms, tissues, and cells. Genomics can reveal the function and structure of genomes and genes. Genomic information is usually static and considered to be identical to individual organisms, although recent research has revealed that there are many epigenetic changes in cells and tissues that often cause disorders. Transcriptomic information, on the other hand, varies in tissues and conditions even though in the same individuals.

A. Ogura (✉)
Nagahama Institute of Bio-Science and Technology, Nagahama, Japan
e-mail: aogu@whelix.info

S. Shigeno et al. (eds.), *Brain Evolution by Design*, Diversity and Commonality in Animals, DOI 10.1007/978-4-431-56469-0_8

A combination of genomics and transcriptomics is necessary to understand the mechanism underlying functions in cells and organisms. The genomics approach is first applied to the historically important model organisms such as bacteria, yeast, and humans, and to small-sized-genome model species such as the worm and fly (Blattner et al. 1997; Goffeau et al. 1997; International Human Genome Sequencing Consortium 2001; The *C. elegans* Sequencing Consortium 1998; Adams et al. 2000). As the number of genome-sequenced species is few for conducting a comparative analysis of brain evolution, it was difficult to identify genetic changes required for the development of the brain. However, modern technologies represented by next-generation sequencers changed the situation drastically (Schuster 2008). With these emerging technologies, we could obtain genomic information required for tracing the evolutionary history of brain development easily. The transcriptomics approach, on the other hand, is first attempted by sequencing a small portion of expressed genes and tags and calculating gene expression frequencies in various samples (Adams et al. 1991). As this information depends on samples and conditions, it is required to compare several samples with biological and technical replicates to show statistically significant differences. Spatiotemporal differences in gene expression between regions and developmental processes is an essential approach to understand how the brains of animals developed and evolved (Kang et al. 2011). Transcriptome studies give us comprehensive data on the animal brain transcriptome and insights into the basis of neurodevelopment.

Application of genomics and transcriptomics for brain function is usually carried out by comparative analysis of genome data among individuals (Kang et al. 2011). A fundamental role of this large-scale analysis in brain research is to extract genes that are related to brain function. By comparing individuals with various phenotypes related to brain functions, it is possible to locate genes corresponding to the phenotype. This approach is also taken for deciphering genes that cause disorders in the brain (Hang et al. 2014). The other approach to understanding brain function is transcriptome analysis. One of the primary targets of transcriptome study for brain function is to clarify the gene regulatory network that controls brain function (Carro et al. 2010). The transcriptome reflects environmental changes so that it is possible to relate genes and functions.

It is essential to compare genomes and transcriptomes of brains among species for brain research (Cahoy et al. 2008). For instance, comparing human and chimpanzee could reveal differences to form human brains from apes (Khaitovich et al. 2005). Comparing remotely related but highly organized brain such as primates and sea primates, cephalopods, could unveil the evolutionary mechanism and diversification of brains (Albertin et al. 2012). The advantages of these large-scale genomic and transcriptomic studies of animal brains are that they can trace the evolutionary process of not only the key regulatory genes but also genes related to brain functions through the analyses of orthologous gene sets found in animals.

A question arises how to utilize next-generation sequencers (NGS) for brain research. NGS could characterize DNA and RNA sequence information deeply, and comparing NGS data including genome and transcriptome of various brain samples could reveal insight into brain function as noted. NGS also accelerated analyses of genes and the gene regulatory networks in brains that explain the evolution and diversification of central nervous systems among animals. Advances in library construction methods from a small amount of samples make it possible to examine tissues and cells in the brain for clarifying cell–cell interactions. This chapter introduces recent studies utilizing emerging novel technologies of NGS and the novel targets of brain evolution and diversification.

8.1.2 The Rise of Next-Generation Sequencers

Development of the next- or second-generation sequencers has been triggered by a pyrosequencing method reported by 454 life sciences articles in 2005 (Margulies et al. 2005). In comparison with the first-generation sequencers (Sanger method), the second-generation sequencers utilize slides or plates with millions of wells, in which nucleotide incorporation occurs, and imaging is used to identify the nucleotide sequence. As an important characteristic, NGS technology can produce shorter, but various, sequences whereas the Sanger method can produce relatively longer pieces. One of the latest sequencers, the Hiseq 2500 sequencer by Illumina, Inc., could generate more than 600 gigabases (Gb) per experiment that can cover a 200-fold sequence of the human genome whereas the Sanger method sequencer could generate fewer than 1 megabase (Mb) per experiment. The running costs of NGS machines are becoming lower and lower, and a draft genome project of an animal having 1-Gb-sized genome can be performed for a few hundred dollars (http://www.genome.gov/sequencingcosts/). The cost of the human genome project was hundreds of thousands of dollars, but with the latest techniques it might be reduced to a $10 genome project shortly (Fig. 8.1; Table 8.1).

Lately, single-molecule real-time sequencers have been developed and released by many companies (Levene 2003). DNA polymerase fixed at the bottom of an apparatus could start to replicate a sample DNA molecule as a template. One of the preceding products is Pacbio RS2 (Pacific Biosciences) makes it possible to handle more than a 7-kb sequence for a read. This long sequence read could easily overcome repetitive elements in the genome that often remain as gaps in the genome assembly. In the case of the transcriptome, long sequence reads are also useful as they can distinguish different splicing variants processed by alternative splicing mechanisms directly (Chin et al. 2014).

Bioinformatics has advanced together with the development of NGS technologies. Assembling genome and transcriptome data is one of the most difficult problems in bioinformatics (Miller et al. 2010). Sequence reads consist of an enormous number of short sequences with tiny errors, heterozygosity, single-nucleotide polymorphisms (SNPs), and mutations that interfere with a concise genome and

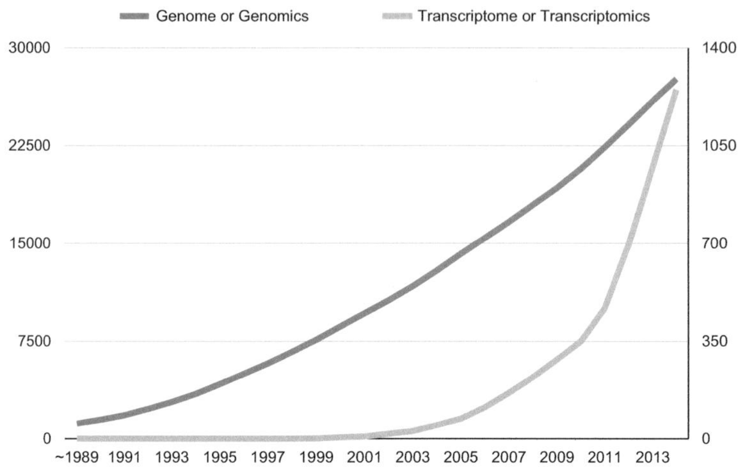

Fig. 8.1 Research activities show expansion of brain genomics and transcriptomics. The number of publications regarding brain genomics is represented in a darker line with *left* axis, and the number of publications regarding brain transcriptomics is represented in a lighter line with *right* axis. Although the publications for brain transcriptomics are fewer in number, growth is rapid since the mid-2000s

Table 8.1 Currently available next-generation sequencers for genomics and transcriptomics. Cost does not include library preparation and varies depending on time and region

Products	Read length	Data	Technology	Cost/Gb
Illumina Hiseq	2 × 100–150 bp	−60 Gbp/lane	Sequencing by synthesis	$50
Illumina Miseq	2 × 150–300 bp	−9 Gbp/lane	Sequencing by synthesis	$150
Ion Proton	200 bp	8 Gbp	Ion semiconductor sequencing	$1000
Ion PGM	200 bp	0.8 Gbp	Ion semiconductor sequencing	$1000
Pacbio RSII	5,000–10,000 bp	0.5 Gbp	Single molecule real-time sequencing	$1000

2 × ***bp represents sequence reads produced from pair-end library that allows us to sequence both ends of a fragment. Gb represents giga basepairs for sequence reads

gene assembly and produce chimeric products. Gene expression frequencies are calculated as reads per kilobase per million reads (RPKM). In contrast to the genome sequence assembly, transcriptome data include different types of transcripts including transcripts with alternative splicing. Therefore, an assembly algorithm designed for a transcriptome but not for a genome is required (Martin and Wang 2011). In comparative genome and transcriptome analyses, it is also important to distinguish orthologues and paralogues (Koonin 2005). Homologous genes would be orthologues if they were separated by a speciation event, whereas they are paralogues if a duplication event separated them. Orthologues are useful not only for phylogenetic analysis but also for functional annotation because orthologous genes tend to have the same function. On the other hand, paralogues tend to have different functions as duplicated genes are often redundant in species, and neo-functionalization or sub-functionalization occurs to gain novel or substitute existing functions (Ohno 1970, Stoltzfus 1999). One of the main motivations of meta-transcriptome studies is to clarify the dynamics of gene regulation and interaction, and in this respect discrimination between orthologues and paralogues and brief annotations of genes are essential.

These NGS technologies and bioinformatics shed light on brain research (Fig. 8.2). Regarding the comprehensive scale analysis of gene expressions, microarray technologies based on DNA–DNA hybridization on designated probes have been utilized so far. However, they can detect only gene expression intensities but cannot identify sequence directly. In contrast, NGS can determine the sequence of samples that lead to further detailed analysis of alternative splicing, methylation, RNA editing, and micro-exons that are known to be related to brain function.

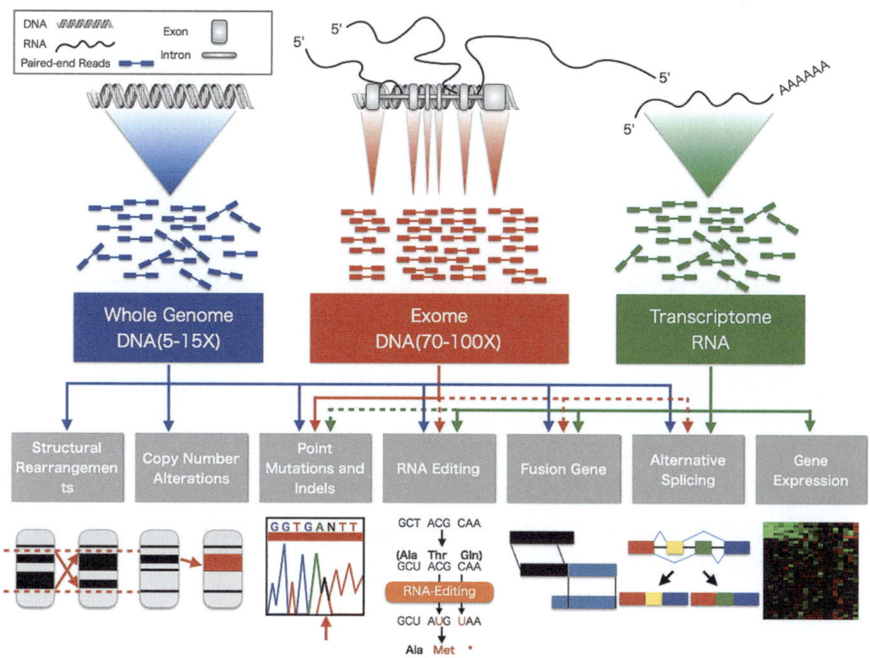

Fig. 8.2 Various applications of next-generation sequencer (NGS) technology in brain research. Whole-genome sequencing is useful for structural rearrangements of chromosomes and copy number variations. As whole-genome sequencing is expensive, exome sequencing that could sequence only exon regions is useful to detect point mutations. Together with the transcriptome, we could clarify the gene regulatory system working in the brain, such as alternative splicing, gene fusion, and RNA editing

For instance, alternative splicing (AS) contributes to the functional diversity of most genes in multicellular organisms by producing various proteins by generating multiple transcript isoforms from the same gene. This AS has a significant role in the brain as AS regulates the development of the brain; more than 20,000 AS forms are functionarily regulated in the brain (Yeo et al. 2004, Meshorer et al. 2005, Goymer 2007). Sequence analysis of genomic DNA is widely used to identify responsible genes and mutations for certain phenotypes, such as single-nucleotide polymorphisms (SNPs). There are various applications for NGS. Exome sequencing is a method to sequence only the exon region by capturing probes to exons so that we can quickly access gene information (Sarah et al. 2009). Exome sequencing has advantages regarding cost and coverage quality. Whole-genome resequencing can clarify noncoding regions of genomes that are known to be important for regulation of the gene expression network. Sequence analysis of transcripts is also accepted to understand the dynamics of gene expression profiles in cells and tissues that are essential to understanding the higher layers of the living system. Phenotypes often directly link to gene expression profiles. Micro-RNA (miRNA) and lincRNA (long internecine noncoding RNA) are also targeted for RNA-seq analysis (Kapranov et al. 2007). Research for gene regulatory systems has also been analyzed via NGS technologies. ChIP-Seq is a technique to combine chromatin immunoprecipitation and NGS to clarify protein–DNA interaction such as the target region of particular transcription factors in the genome. As data generated from NGS are extensive, bioinformatics approaches by utilizing high-end servers are required to understand the functionality of genes working in the brain. Understanding the gene regulatory network and interactions are critical targets for genomics and transcriptomics approaches. For examination of the functional representation of such networks, it is required to identify neighborhoods of functionally related genes, well-represented pathways that can be highlighted which are functional in the environment. Adopting a network analysis could identify node genes are mediating essential roles within the network. Network analysis of this type can be performed using tools and algorithms designed for RNA-seq studies, such as Cytoscape (Shannon et al. 2003) and Ingenuity Pathways Analysis (Ingenuity® Systems).

8.2 Genome- and Transcriptome-Wide Research on the Brain

Recent studies have made significant achievements in finding genes related to functions of the human brain. However, it is still difficult to understand the genetic basis of brain evolution and brain diversification. By utilizing comparative genomics and transcriptomics, which involves comparisons of gene sequences and their expressions across many species, we could trace the evolutionary history of genetic changes among animals. To understand brain evolution, gene expression comparisons and genome structural variations between species are also essential.

These evolutionary studies have become possible not only because of the availability of large amounts of sequence data but also by the development of modern methods and computational tools.

Genomics studies for the brain have increased lately. As of 2014, more than 25,000 research articles related to brain genomics were identified using a PubMed title/abstract search under the keywords "genome" and "brain," but the real number of studies related to the brain will be even higher. Most of these studies were published within the past 10 years, indicating that this field of research has grown rapidly (Fig. 8.1). This rapid increase was prompted by recent advancements in next-generation sequencing (NGS) technologies and high-throughput methods for genomic analysis (Table 8.1). Genomics offers a valuable approach to study the functional and genetic diversity of animal brains; however, this approach cannot provide information on active functional genes and gene regulatory networks in the living systems. In fact, a growth rate of transcriptomics studies for brain function is higher than that of genomics research, even though the number itself is much lower, which might reflect the difficulties in performing a transcriptomic study in the brain.

8.2.1 Genomics Approach

To understand the evolutionary process of the animal brain, it is essential to identify the biological basis of the genomic differences among various animal brains that would be responsible for brain size, brain organization, brain development, brain function, and behavior. There are several way to estimate regulatory sequences in the genome to trace the evolutionary processes of animal brains. One is to utilize sequence conservation across animal genomes, which is important for controlling one regulatory network and gene expressions. Another approach is to perform experiments to seek for binding sites of certain transcription factors and by comparing binding information and spatial and temporal regulation of the gene expression that is inscribed in the genome and is predictive of regulatory function.

Application of genomics to brain research varies, such as genome-wide association study (GWAS) for quantitative trait loci (QTLs) and variant analysis for indels, which are underlying a particular phenotype in brain functions. GWAS is a method of mapping QTLs that can link phenotypes to genotypes, usually conducted by microarray analysis. GWAS is a popular approach for brain research so far, and more than ten publications are frequently cited (Table 8.2). According to the GWAS catalog developed by the European Molecular Biology Laboratory (http://www.ebi.ac.uk/gwas/) (Fig. 8.3), more than 2000 QTLs related to brain function have been identified so far. For example, GWAS analysis for Alzheimer's disease was performed using quality-controlled genotype and scan data including more than 620 K SNPs and found several new QTLs related to the disease (Table 8.2, Shen et al. 2010). In another case, six genes or chromosomal regions were identified by GWAS analysis that relates to the pathways involved in neurodevelopment and response to stress (Table 8.2, Potkin et al. 2009). Thus, a genome-wide whole-brain

Table 8.2 Selected publications with high numbers of citations regarding genomics and transcriptomics for brain research

Title	Authors	Year	Journal	Citation
An anatomically comprehensive atlas of the adult human brain transcriptome	Hawrylycz MJ, Lein ES, Guillozet-Bongaarts AL et al.	2012	Nature	312
Spatiotemporal transcriptome of the human brain	Kang HJ, Kawasawa YI, Cheng F et al.	2011	Nature	448
Whole-transcriptome sequencing reveals gene expression and splicing differences in brain regions affected by Alzheimer's disease	Twine NA, Janitz K, Wilkins MR et al.	2011	PLOS ONE	121
Divergence of human and mouse brain transcriptome highlights Alzheimer disease pathways	Miller JA, Horvath S, Geschwind DH.	2010	PNAS	154
Whole-genome association study of brain-wide imaging phenotypes for identifying quantitative trait loci in MCI and AD: A study of the ADNI cohort	Shen L, Kim S, Risacher SL et al.	2010	Neuroimage	160
Genome-wide analysis reveals mechanisms modulating autophagy in normal brain aging and in Alzheimer's disease	Lipinski MM, Zheng B, Lu T et al.	2010	PNAS	157
Brain function in carriers of a genome-wide supported bipolar disorder variant	Erk S, Meyer-Lindenberg A, Schnell K et al.	2010	Arch Gen Psychiatry	122
Functional and evolutionary insights into human brain development through global transcriptome analysis	Johnson MB, Kawasawa YI, Mason CE et al.	2009	Neuron	286
A genome-wide association study of schizophrenia using brain activation as a quantitative phenotype	Potkin SG, Turner JA, Guffanti G et al.	2009	Schizophr Bulletin	123
Transgenerational epigenetic programming of the brain transcriptome and anxiety behavior	Skinner MK, Anway MD, Savenkova MI et al.	2008	PLOS ONE	158
Evolution of the aging brain transcriptome and synaptic regulation	Loerch PM, Lu T, Dakin KA et al.	2008	PLOS ONE	140
A transcriptome database for astrocytes, neurons, and oligodendrocytes: a new resource for understanding brain development and function	Cahoy JD, Emery B, Kaushal A et al.	2008	The Journal of Neuroscience	1161

(continued)

Table 8.2 (continued)

Title	Authors	Year	Journal	Citation
Functional organization of the transcriptome in human brain	Oldham MC, Konopka G, Iwamoto K et al.	2008	Nature Neuroscience	308
HITS-CLIP yields genome-wide insights into brain alternative RNA processing	Licatalosi DD, Mele A, Fak JJ et al.	2008	Nature	622
Transcriptome-wide identification of novel imprinted genes in neonatal mouse brain	Wang X, Sun Q, McGrath SD et al.	2008	PLOS ONE	145
Genome-wide atlas of gene expression in the adult mouse brain	Lein ES, Hawrylycz MJ, Ao N, Ayres M et al.	2007	Nature	1809
From the genome to the proteome: uncovering peptides in the *Apis* brain	Hummon AB, Richmond TA, Verleyen P et al.	2006	Science	198
High-resolution genome-wide mapping of genetic alterations in human glial brain tumors	Bredel M, Bredel C, Juric D et al.	2005	Cancer Research	140
A genome-wide in situ hybridization map of RNA-binding proteins reveals anatomically restricted expression in the developing mouse brain	McKee AE, Minet E, Stern C et al.	2005	BMC Dev Biology	177
Mouse brain organization revealed through direct genome-scale tf expression analysis	Gray PA, Fu H, Luo P et al.	2004	Science	325
Sister grouping of chimpanzees and humans as revealed by genome-wide phylogenetic analysis of brain gene expression profiles	Uddin M, Wildman DE, Liu G et al.	2004	PLOS ONE	207
Blood–brain barrier genomics	Li JY, Boado RJ, Pardridge WM.	2001	J Cerebral Blood Flow & Metabolism	119
Rapid accumulation of genome rearrangements in liver but not in brain of old mice	Dollé ME, Giese H, Hopkins CL et al.	1997	Nature Genetics	205
Genetic correlates of brain aging on MRI and cognitive test measures: a genome-wide association and linkage analysis in the Framingham Study	Seshadri S, DeStefano AL, Au R et al.	2007	BMC Medical Genetics	140

(continued)

Table 8.2 (continued)

Title	Authors	Year	Journal	Citation
The mouse blood–brain barrier transcriptome: a new resource for understanding the development and function of brain endothelial cells	Daneman R, Zhou L, Agalliu D et al.	2010	PLOS ONE	128
Genome-wide analysis reveals differences in brain gene expression patterns associated with caste and reproductive status in honey bees (*Apis mellifera*)	Grozinger CM1, Fan Y, Hoover SE, Winston ML.	2007	Wiley Online Library	108
3'tag digital gene expression profiling of human brain and universal reference RNA using Illumina Genome Analyzer	Asmann YW, Klee EW, Thompson EA et al.	2009	BMC Genetics	104
Transcriptome analysis of channel catfish (*Ictalurus punctatus*): genes and expression profile from the brain	Ju Z, Karsi A, Kocabas A et al.	2000	Gene	100

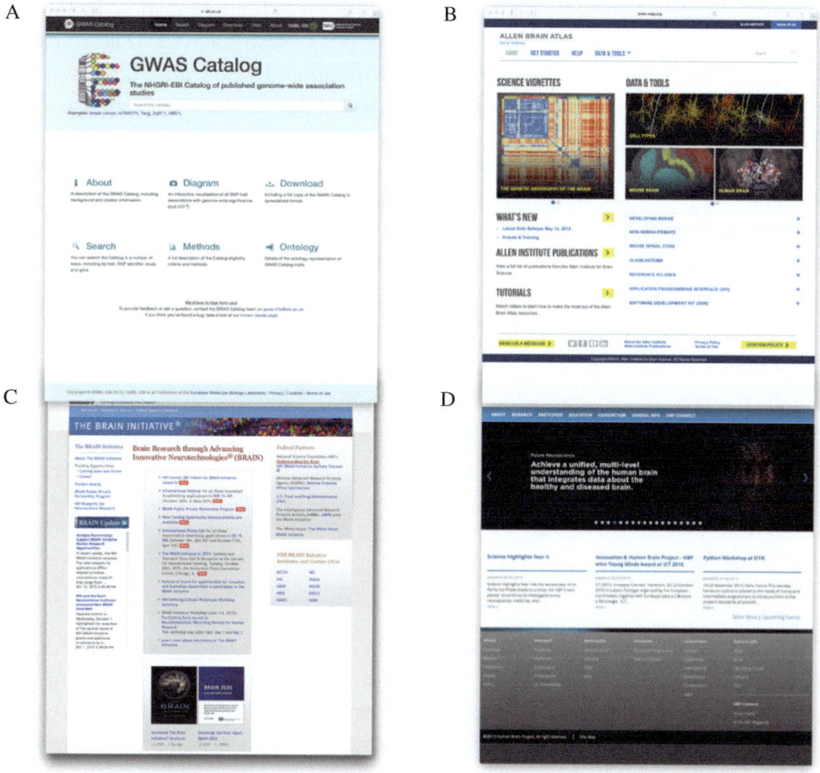

Useful web resources regarding brain genomics and transcriptomics.

A. GWAS catalog (https://www.ebi.ac.uk/gwas/)
B. Allen Brain Atlas (http://www.brain-map.org)
C. The Brain Initiative (http://braininitiative.nih.gov)
D. Human Brain Project (https://www.humanbrainproject.eu)

Fig. 8.3 Useful web resources and projects for brain research: useful web resources regarding brain genomics and transcriptomics. (**a**) GWAS catalog (https://www.ebi.ac.uk/gwas/). (**b**) Allen Brain Atlas (http://www.brain-map.org). (**c**) The Brain Initiative (http://braininitiative.nih.gov). (**d**) Human Brain Project (https://www.humanbrainproject.eu)

search strategy could reveal novel candidate genes and loci for a specific diseases or functions (Table 8.2, Lipinski et al. 2010; Twine et al. 2011; Erk et al. 2010; Bredel et al. 2005; Li et al. 2001).

The variant analysis is an approach to detect nucleotide variation in individuals by whole-exome sequencing (Fig. 8.2; Table 8.2). Mutations in somatic cells that are caused by radiation, chemicals, and infection by viruses often have an effect on epigenetic changes in a genome and alter physiological characteristics in cells and tissues. These changes are a leading cause of cancer, and sequencing of these responsible genes and regions can be a primary strategy for cancer research. The number of reports is not so large up to 2015 as this technology is new to brain research, but it should be increased shortly.

The genome of somatic cells has been often regarded as static and unchanged. However, many epigenetic changes have been accumulated. These epigenetic changes cause diversification and differentiation of cells and tissues, and neural cells can acquire characteristics to form the neuron and brain. The primary molecular mechanism of such epigenetic changes is methylation and histone modification. It has long been possible to survey these changes across genomes, but recent technologies of sequencing made it possible to analyze methylations and histone modifications across the whole genome by bisulfite sequencing, ChIP-seq. The International Human Epigenome Consortium (IHEC) (http://ihec-epigenomes.org) is an international consortium for studying the methylome, that stands for all methylation in the genome, for 1000 kinds of cells in the human, and 18 samples from adult or fetal brain have been studied already. There are only a few studies for large-scale epigenetic analysis across species so that it is a bit difficult to trace evolutionary changes in an epigenomic system so far.

8.2.2 *Transcriptomics Approach*

Transcriptomic approaches to understanding brain function are also often conducted by many researchers (Table 8.2). Application of transcriptomics to brain research is intensively carried out in model organisms including the human. Researchers at the Allen Institute for Brain Science have reported a comprehensive study of gene expression in the brain tissues obtained by laser microdissection (Table 8.4, Hawrylycz et al. 2012). Their results allow us to compare gene expression profiles between humans and other animals, and to conduct neurogenetic studies of normal and abnormal human brain function (http://www.brain-map.org) (Fig. 8.3). Several groups performed a transcriptome study for multiple brain regions and neocortical areas of developing and adult human brains and found that 70–90 % of genes were differentially regulated across brain regions or time (Table 8.4, Johnson et al. 2009; Kang et al. 2011).

In other animals, such as mice and primates, transcriptomic studies are also conducted to elucidate the developmental process and functions in the brain (Table 8.2, Lein et al. 2007; McKee et al. 2005; Gray et al. 2004; Uddin et al. 2004). These data are valuable as we could compare gene expression profiles of animals to identify genes and interactions responsible for the higher class of brain functions.

The same as the genomics approach, transcriptomics is also a powerful tool to identify genes responsible for disorders. To understand how brain aging has evolved, Loerch et al. compared age-related gene expression profiles in the cortex of humans, rhesus macaques, and mice by utilizing a microarray, and found that repression of gene expression in the neuron is a key to understanding the evolution of brain aging in humans and rhesus macaques (Table 8.2, Loerch et al. 2008). There are many projects for brain transcriptomes, and outcomes of more than 3000 projects

Table 8.3 Brain transcriptome projects

Species	Experiments	Run	Total size (MB)
Mus musculus	4479	5258	3, 857, 483
Homo sapiens	3560	4409	10, 106, 284
Rattus norvegicus	2770	2805	640, 853
Platynereis dumerilii	213	285	65, 297
Drosophila melanogaster	95	103	253, 679
Cerapachys biroi	90	90	155, 063
Rabies virus	84	84	21, 299
Macaca mulatta	83	83	92, 492
Pan troglodytes	59	59	70, 081
West Nile virus	44	58	16, 432
Bos taurus	39	43	129, 242
Meleagris gallopavo	39	39	23, 800
Mus musculus musculus	37	37	73, 383
Aedes aegypti	32	32	224, 134
Ovis aries	32	40	57, 042
Nothobranchius furzeri	31	31	42, 684
Macaca fascicularis	30	30	96, 457
Danio rerio	28	30	46, 656
Oryctolagus cuniculus	27	29	45, 204
Apis mellifera	26	198	64, 895
All other taxa	757	980	1, 943, 559

The number of brain transcriptome projects is collected from Sequence Read Archives (http://www.ncbi.nlm.nih.gov/sra) with the search keyword "brain" with RNA sources, as of April 2015. The column "species" is shown as represented in the database, although there are strange rows such as *Mus musculus* and *Mus musculus musculus*, and viruses that should be the pathogen to animals. The column "Run" represents the number of sequencing samples. Total size represents the total length of sequence reads

are already submitted to the Sequence Read Archives (SRA) that are maintained by National Center for Biotechnology Information, National Library of Medicine (Table 8.3). These transcriptomics approaches reveals many genes that involve in various functions in the brain (Table 8.4). These genes listed in Table 8.4 were found to be functional in the brain through large-scale transcriptomics studies. Most of those genes seem to be conserved only in humans and mice but not in other primitive animals such as the fruit fly and worm, because these genes are mainly studied in the mouse so that there are few genes only found in human. Transcriptomics using human cell lines can tell us about human brain specific genes shortly.

Table 8.4 The list of genes related to brain functions and their conservation among animals

Gene	Description	*H. sapiens* ID	*M. musculus* ID	*D. melanogaster* ID	*C. elegans* ID	*A. mellifera* ID	References
DRD1	Dopamine receptor D1	ENSG00000184845	ENSMUSG00000021478				Hawrylycz MJ et al
DRD2	Dopamine receptor D2	ENSG00000149295	ENSMUSG00000032259		CELE_T14E8.3		Hawrylycz MJ et al
DRD3	Dopamine receptor D3	ENSG00000151577	ENSMUSG00000022705		CELE_K09G1.4		Hawrylycz MJ et al
DRD4	Dopamine receptor D4	ENSG00000069696	ENSMUSG00000025496				Hawrylycz MJ et al
DRD5	Dopamine receptor D5	ENSG00000169676	ENSMUSG00000039358				Hawrylycz MJ et al
TH	Tyrosine hydroxylase	ENSG00000180176	ENSMUSG00000000214	FLYBASE:FBgn0005626	CELE_B0432.5		Hawrylycz MJ et al
DOC2A	Double C2-like domains, alpha	ENSG00000149927					Hawrylycz MJ et al
SLC6A2	Solute carrier family 6 (neurotransmitter transporter), member 2	ENSG00000103546	ENSMUSG00000055368	FLYBASE:FBgn0034136	CELE_T23G5.5		Hawrylycz MJ et al
SLC6A3	Solute carrier family 6 (neurotransmitter transporter), member 3	ENSG00000142319	ENSMUSG00000021609				Hawrylycz MJ et al
SLC18A1	Solute carrier family 18 (vesicular monoamine transporter), member 1	ENSG00000036565	ENSMUSG00000036330				Hawrylycz MJ et al
SLC18A2	Solute carrier family 18 (vesicular monoamine transporter), member 2	ENSG00000165646	ENSMUSG00000025094	FLYBASE:FBgn0260964	CELE_W01C8.6		Hawrylycz MJ et al
MAOA	Monoamine oxidase A	ENSG00000189221	ENSMUSG00000025037				Hawrylycz MJ et al
MAOB	Monoamine oxidase B	ENSG00000069535	ENSMUSG00000040147				Hawrylycz MJ et al
COMT	Catechol-*O*-methyltransferase	ENSG00000093010	ENSMUSG00000000326				Hawrylycz MJ et al
DLGAP2	Discs, large (*Drosophila*) homolog-associated protein 2	ENSG00000198010	ENSMUSG00000047495				Hawrylycz MJ et al

DLGAP3	Discs, large (*Drosophila*) homolog-associated protein 3	ENSG00000116544	ENSMUSG00000042388				Hawrylycz MJ et al
PDE2A	Phosphodiesterase 2A, cGMP-stimulated	ENSG00000186642	ENSMUSG00000030653				Hawrylycz MJ et al
SYT1	Synaptotagmin I	ENSG00000067715	ENSMUSG00000035864	FLYBASE:FBgn0004242	CELE_F31E8.2		Hawrylycz MJ et al; Oldham MC et al
SNAP25	Synaptosomal-associated protein, 25 kDa	ENSG00000132639	ENSMUSG00000027273	FLYBASE:FBgn0266720	CELE_Y22F5A.3		Hawrylycz MJ et al
STX1A	Syntaxin 1A (brain)	ENSG00000106089	ENSMUSG00000007207	FLYBASE:FBgn0013343	CELE_F56A8.7		Hawrylycz MJ et al
HOMER1	Homer homolog 1 (*Drosophila*)	HPRD:09211					Hawrylycz MJ et al
VSNL1	Visinin-like 1	ENSG00000163032					Hawrylycz MJ et al
AMPH	Amphiphysin	ENSG00000078053	ENSMUSG00000021314		CELE_F58G6.1		Hawrylycz MJ et al
GABBR2	γ-Aminobutyric acid (GABA) B receptor, 2	ENSG00000136928					Hawrylycz MJ et al
NEFL	Neurofilament, light polypeptide	ENSG00000277586					Hawrylycz MJ et al
NEFM	Neurofilament, medium polypeptide	ENSG00000104722	ENSMUSG00000022054				Hawrylycz MJ et al
NEFH	Neurofilament, heavy polypeptide	ENSG00000100285	ENSMUSG00000020396				Hawrylycz MJ et al
ANK1	Ankyrin 1, erythrocytic	ENSG00000029534	ENSMUSG00000031543				Hawrylycz MJ et al
PLP1	Proteolipid protein 1	ENSG00000123560	ENSMUSG00000031425				Hawrylycz MJ et al; Oldham MC et al
MOG	Myelin oligodendrocyte glycoprotein	ENSG00000204655	ENSMUSG00000076439				Hawrylycz MJ et al; Oldham MC et al
MBP	Myelin basic protein	ENSG00000197971	ENSMUSG00000041607				Hawrylycz MJ et al

(continued)

Table 8.4 (continued)

Gene	Description	*H. sapiens* ID	*M. musculus* ID	*D. melanogaster* ID	*C. elegans* ID	*A. mellifera* ID	References
SCN1A	Sodium channel, voltage-gated, type I, alpha subunit	ENSG00000144285	ENSMUSG00000064329				Hawrylycz MJ et al
SCN1B	Sodium channel, voltage-gated, type I, beta subunit	ENSG00000105711	ENSMUSG00000019194				Hawrylycz MJ et al
TTC18 (CFAP70)	Tetratricopeptide repeat domain 18 (cilia and flagella associated protein 70)	ENSG00000156042	ENSMUSG00000039543				Hawrylycz MJ et al
DLEC1	Deleted in lung and esophageal cancer 1	ENSG00000008226	ENSMUSG00000038060				Hawrylycz MJ et al
DNALI1	Dynein, axonemal, light intermediate chain 1	ENSG00000163879	ENSMUSG00000042707	FLYBASE:FBgn0037962			Hawrylycz MJ et al
TYROBP	TYRO protein tyrosine kinase binding protein	ENSG00000011600	ENSMUSG00000030579				Hawrylycz MJ et al
C1QA	complement component 1, q subcomponent, A chain	ENSG00000173372	ENSMUSG00000036887				Hawrylycz MJ et al
C1QB	Complement component 1, q subcomponent, B chain	ENSG00000173369	ENSMUSG00000036905				Hawrylycz MJ et al
AQP4	Aquaporin 4	ENSG00000171885	ENSMUSG00000024411	FLYBASE:FBgn0015872			Hawrylycz MJ et al; Oldham MC et al
HEPH	Hephaestin	ENSG00000089472	ENSMUSG00000031209				Hawrylycz MJ et al
VDAC1	Voltage-dependent anion channel 1	ENSG00000213585	ENSMUSG00000020402				Hawrylycz MJ et al
CLTC	Clathrin, heavy chain (Hc)	ENSG00000141367	ENSMUSG00000047126	FLYBASE:FBgn0000319	CELE_T20G5.1		Hawrylycz MJ et al
RGS9	Regulator of G-protein signaling 9	ENSG00000108370	ENSMUSG00000020599				Hawrylycz MJ et al

RPL19	Ribosomal protein L19	ENSG00000108298	ENSMUSG00000017404	FLYBASE:FBgn0002607			Hawrylycz MJ et al
RPS26	Ribosomal protein S26	ENSG00000197728	ENSMUSG00000025362	FLYBASE:FBgn0261597	CELE_F39B2.6		Hawrylycz MJ et al
EEF1B2	Eukaryotic translation elongation factor 1 beta 2	ENSG00000114942	ENSMUSG00000025967	FLYBASE:FBgn0028737	CELE_F54H12.6		Hawrylycz MJ et al
ENPP2	ectonucleotide pyrophos-phatase/phosphodiesterase 2	ENSG00000136960	ENSMUSG00000022425				Hawrylycz MJ et al; Oldham MC et al
PCDH11Y	Protocadherin 11 Y-linked	ENSG00000099715					Kang HJ et al
RPS4Y1	Ribosomal protein S4, Y-linked 1	ENSG00000279950					Kang HJ et al
USP9Y	Ubiquitin0specific peptidase 9, Y-linked	ENSG00000114374	ENSMUSG00000069044				Kang HJ et al
DDX3Y	DEAD (Asp-Glu-Ala-Asp) box helicase 3, Y-linked	ENSG00000067048	ENSMUSG00000069045		CELE_Y54E10A.9		Kang HJ et al
NLGN4Y	Neuroligin 4, Y-linked	ENSG00000165246		FLYBASE:FBgn0031866	CELE_C40C9.5		Kang HJ et al
UTY	Ubiquitously transcribed tetratricopeptide repeat containing, Y-linked	ENSG00000183878	ENSMUSG00000068457				Kang HJ et al
EIF1AY	Eukaryotic translation initiation factor 1A, Y-linked	ENSG00000198692		FLYBASE:FBgn0026250	CELE_H06H21.3		Kang HJ et al
ZFY	Zinc-finger protein, Y-linked	ENSG00000067646					Kang HJ et al
ZFX	Zinc-finger protein, X-linked	ENSG00000005889	ENSMUSG00000079509				Kang HJ et al
MAG	Myelin-associated glycoprotein	ENSG00000105695	ENSMUSG00000036634				Oldham MC et al
OLIG2	Oligodendrocyte lineage transcription factor 2	ENSG00000205927	ENSMUSG00000039830				Oldham MC et al

(continued)

Table 8.4 (continued)

Gene	Description	*H. sapiens* ID	*M. musculus* ID	*D. melanogaster* ID	*C. elegans* ID	*A. mellifera* ID	References
MOBP	Myelin-associated oligodendrocyte basic protein	ENSG00000168314	ENSMUSG00000032517				Oldham MC et al
CNP	2',3'-Cyclic nucleotide 3′-phosphodiesterase	ENSG00000173786	ENSMUSG00000006782				Oldham MC et al
GFAP	Glial fibrillary acidic protein	ENSG00000131095	ENSMUSG00000020932				Oldham MC et al
GJA1	Gap junction protein, alpha 1, 43 kDa	ENSG00000152661	ENSMUSG00000050953				Oldham MC et al
GLUL	Glutamate-ammonia ligase	ENSG00000135821	ENSMUSG00000026473				Oldham MC et al
GLUD1	Glutamate dehydrogenase 1	ENSG00000148672	ENSMUSG00000021794	FLYBASE:FBgn0001098	CELE_ZK829.4		Oldham MC et al
SLC1A2	Solute carrier family 1 (glial high affinity glutamate transporter), member 2	ENSG00000110436	ENSMUSG00000005089		CELE_C12D12.2		Oldham MC et al
MAP2	Microtubule-associated protein 2	ENSG00000078018	ENSMUSG00000015222				Oldham MC et al
MAP1B	Microtubule-associated protein 1B	ENSG00000131711	ENSMUSG00000052727				Oldham MC et al
NRXN1	Neurexin 1	ENSG00000179915	ENSMUSG00000024109		CELE_C29A12.4		Oldham MC et al
SLC1A1	Solute carrier family 1 (neuronal/epithelial high affinity glutamate transporter, system Xag), member 1	ENSG00000106688	ENSMUSG00000024935				Oldham MC et al
NRCAM	Neuronal cell adhesion molecule	ENSG00000091129	ENSMUSG00000020598	FLYBASE:FBgn0264975	CELE_Y54G2A.25		Oldham MC et al

PLLP	Plasmolipin	ENSG00000102934	ENSMUSG00000031775				Oldham MC et al
MAL	Mal, T-cell differentiation protein	ENSG00000172005	ENSMUSG00000027375				Oldham MC et al
HSPA2	Heat shock 70-kDa protein 2	ENSG00000126803	ENSMUSG00000059970				Oldham MC et al
TF	Transferrin	ENSG00000091513	ENSMUSG00000032554				Oldham MC et al
GPR37	G-protein-coupled receptor 37 (endothelin receptor type B-like)	HPRD:03992					Oldham MC et al
ENPP2	Ectonucleotide pyrophos-phatase/phosphodiesterase 2	ENSG00000136960	ENSMUSG00000022425				Oldham MC et al
FA2H	Fatty acid 2-hydroxylase	ENSG00000103089	ENSMUSG00000033579	FLYBASE:FBgn0050502	CELE_C25A1.5		Oldham MC et al
C11orf9 (MYRF)	Myelin regulatory facto	ENSG00000124920	ENSMUSG00000036098				Oldham MC et al
AHCYL1	Adenosyl homocysteinase-like 1	ENSG00000168710	ENSMUSG00000027893				Oldham MC et al
NTRK2	Neurotrophic tyrosine kinase, receptor, type 2	ENSG00000148053	ENSMUSG00000055254				Oldham MC et al
SOX9	SRY (sex-determining region Y)-box 9	ENSG00000125398	ENSMUSG00000000567				Oldham MC et al
PDLIM5	PDZ and LIM domain 5	ENSG00000163110	ENSMUSG00000028273				Oldham MC et al
PPAP2B	Phosphatidic acid phosphatase type 2B	ENSG00000162407	ENSMUSG00000028517				Oldham MC et al
PLSCR4	Phospholipid scramblase 4	ENSG00000114698	ENSMUSG00000032377				Oldham MC et al
TP53BP2	Tumor protein p53 binding protein 2	ENSG00000143514	ENSMUSG00000026510				Oldham MC et al
METTL7A	Methyltransferase-like 7A	ENSG00000185432	ENSMUSG00000054619				Oldham MC et al
PREPL	Prolyl encopeptidase-like	ENSG00000138078	ENSMUSG00000024127				Oldham MC et al

(continued)

Table 8.4 (continued)

Gene	Description	*H. sapiens* ID	*M. musculus* ID	*D. melanogaster* ID	*C. elegans* ID	*A. mellifera* ID	References
SYNJ1	Synaptojanin 1	ENSG00000159082	ENSMUSG00000022973	FLYBASE:FBgn0034691	CELE_JC8.10		Oldham MC et al
TPD52	Tumor protein D52	ENSG00000076554	ENSMUSG00000027506				Oldham MC et al
G3BP2	GTPase-activating protein (SH3 domain) binding protein 2	ENSG00000138757	ENSMUSG00000029405				Oldham MC et al
YWHAZ	Tyrosine 3-monooxygenase/tryptophan 5-monooxygenase activation protein, zeta	ENSG00000164924	ENSMUSG00000022285	FLYBASE:FBgn0004907	CELE_F52D10.3		Oldham MC et al
PAFAH1B1	Platelet-activating factor acetylhydrolase 1b, regulatory subunit 1 (45 kDa)	ENSG00000007168	ENSMUSG00000020745		CELE_T03F6.5		Oldham MC et al
GABRG2	γ-Aminobutyric acid (GABA) A receptor, gamma 2	ENSG00000113327	ENSMUSG00000020436				Oldham MC et al
SCAMP1	Secretory carrier membrane protein 1	HPRD:06072	ENSMUSG00000021687	FLYBASE:FBgn0040285	CELE_M01D7.2		Oldham MC et al
ANK2	Ankyrin 2, neuronal	ENSG00000145362					Oldham MC et al
GLRB	Glycine receptor, beta	ENSG00000109738	ENSMUSG00000028020	FLYBASE:FBgn0003011			Oldham MC et al
CD24	CD24 molecule	ENSG00000272398					Oldham MC et al
DPYSL3	Dihydropyrimidinase-like 3	ENSG00000113657	ENSMUSG00000024501				Oldham MC et al
ASCL1	Achaete-scute family bHLH transcription factor 1	ENSG00000139352	ENSMUSG00000020052				Oldham MC et al
IQCG	IQ motif containing G	ENSG00000114473	ENSMUSG00000035578				Oldham MC et al
STK38L	Serine/threonine kinase 38 like	ENSG00000211455	ENSMUSG00000001630		CELE_R11G1.4		Oldham MC et al
CETN2	Centrin, EF-hand protein, 2	HPRD:02051	ENSMUSG00000031347	FLYBASE:FBgn0040010			Oldham MC et al

FLJ22167 (TMEM231)	Transmembrane protein 231	ENSG00000205084	ENSMUSG00000031951				Oldham MC et al
NEK1	NIMA-related kinase 1	ENSG00000137601	ENSMUSG00000031644				Oldham MC et al
NPC1	Niemann–Pick disease, type C1	ENSG00000141458	ENSMUSG00000024413	FLYBASE:FBgn0024320			Oldham MC et al
PMP22	Peripheral myelin protein 22	ENSG00000109099	ENSMUSG00000018217				Oldham MC et al
CRYAB	Crystallin, alpha B	ENSG00000109846	ENSMUSG00000032060				Oldham MC et al
PRRG1	Proline-rich Gla (G-carboxyglutamic acid) 1	ENSG00000130962	ENSMUSG00000047996				Oldham MC et al
TUBB2B	Tubulin, beta 2B class IIb	ENSG00000137285	ENSMUSG00000045136	FLYBASE:FBgn0003889	CELE_ZK154.3		Oldham MC et al
SLC1A3	Solute carrier family 1 (glial high affinity glutamate transporter), member 3	ENSG00000079215	ENSMUSG00000005360		CELE_Y53C12A.2		Oldham MC et al
PON2	Paraoxonase 2	ENSG00000105854					Oldham MC et al
SDC4	Syndecan 4	ENSG00000124145	ENSMUSG00000017009				Oldham MC et al
EDG1 (S1PR1)	Sphingosine-1-phosphate receptor 1	ENSG00000170989	ENSMUSG00000045092				Oldham MC et al
MAPK1	Mitogen-activated protein kinase 1	ENSG00000100030	ENSMUSG00000063358	FLYBASE:FBgn0003256	CELE_F43C1.2		Oldham MC et al
FGF12	Fibroblast growth factor 12	ENSG00000114279					Oldham MC et al
NUDT21	Nudix (nucleoside diphosphate linked moiety X)-type motif 21	ENSG00000167005		FLYBASE:FBgn0035987	CELE_F43G9.5		Oldham MC et al
RAB5A	RAB5A, member RAS oncogene family	ENSG00000144566	ENSMUSG00000017831		CELE_F26H9.6		Oldham MC et al
DNM1L	Dynamin 1-like	ENSG00000087470	ENSMUSG00000022789	FLYBASE:FBgn0026479	CELE_T12E12.4		Oldham MC et al
SYN2	Synapsin II	ENSG00000157152	ENSMUSG00000009394				Oldham MC et al

(continued)

Table 8.4 (continued)

Gene	Description	*H. sapiens* ID	*M. musculus* ID	*D. melanogaster* ID	*C. elegans* ID	*A. mellifera* ID	References
PITPNA	Phosphatidylinositol transfer protein, alpha	ENSG00000174238	ENSMUSG00000017781				Oldham MC et al
PLTP	Phospholipid transfer protein	ENSG00000100979	ENSMUSG00000017754				Oldham MC et al
TuJ1	Not found						Oldham MC et al
ALDH1L1	Aldehyde dehydrogenase 1 family, member L1	ENSG00000144908	ENSMUSG00000030088		CELE_F36H1.6		Oldham MC et al
AKH	Adipokinetic hormone			FLYBASE:FBgn0004552		BEEBASE:GB53230	Hummon AB et al
AST	Allatostatin			FLYBASE:FBgn0015591		BEEBASE:GB47928	Hummon AB et al
Apid1	Apidaecin 1					BEEBASE:GB47546	Hummon AB et al
Burs	Bursicon			FLYBASE:FBgn0038901		BEEBASE:GB45446	Hummon AB et al
DH31	Diuretic hormone 31			FLYBASE:FBgn0032048		BEEBASE:GB47217	Hummon AB et al
CRZ	Corazonin			FLYBASE:FBgn0013767		BEEBASE:GB53951	Hummon AB et al
CCAP	Crustacean cardioactive peptide			FLYBASE:FBgn0039007		BEEBASE:GB50604	Hummon AB et al
ITP	Ion-transport peptide					BEEBASE:GB47095	Hummon AB et al
ETH	Ecdysis-triggering hormone			FLYBASE:FBgn0028738		EEBASE:GB40094	Hummon AB et al
EH	Eclosion hormone			FLYBASE:FBgn0000564		BEEBASE:GB49646	Hummon AB et al
FLRF	FLRFamide-like			FLYBASE:FBgn0000715			Hummon AB et al
LOC409241	IDLSRFYGHFNT-containing neuropeptide (prohormone-4)					BEEBASE:GB45263	Hummon AB et al

INS	Insulin	ENSG00000254647	ENSMUSG00000000215				Hummon AB et al
LOC726472	ITGQGNRIF-containing neuropeptide (prohormone-3)					BEEBASE:GB50651	Hummon AB et al
LOC725616	LRNQLDIGDLQ-containing neuropeptide (prohormone-1)					BEEBASE:GB43201	Hummon AB et al
LOC409634	MVPVPVHHMADELL-RNGPDTVI-containing neuropeptide					BEEBASE:GB45265	Hummon AB et al
MS	Myosuppressin			FLYBASE:FBgn0011581			Hummon AB et al
NPF	Neuropeptide F			FLYBASE:FBgn0027109		BEEBASE:GB50693	Hummon AB et al
NPFF	Neuropeptide FF-amide peptide precursor	ENSG00000139574	ENSMUSG00000023052				Hummon AB et al
Neuroparsin	Not found						Hummon AB et al
NPLP-1	Neuropeptide-like precursor 1			FLYBASE:FBgn0035092			Hummon AB et al
NPLP-2	Neuropeptide-like precursor 2			FLYBASE:FBgn0040813			Hummon AB et al
NPLP-3	Neuropeptide-like precursor 3			FLYBASE:FBgn0042201			Hummon AB et al
LOC409314	NVPIYQEPRF-containing neuropeptide (prohormone-2)					BEEBASE:GB44988	Hummon AB et al
Ork	Orcokinin	Not found					Hummon AB et al

(continued)

Table 8.4 (continued)

Gene	Description	*H. sapiens* ID	*M. musculus* ID	*D. melanogaster* ID	*C. elegans* ID	*A. mellifera* ID	References
Capa	AFGLLTYPRIa-containing (Periviscerokinin)			FLYBASE:FBgn0039722			Hummon AB et al
PBAN	Pheromone biosynthesis-activating neuropeptide					BEEBASE:GB46057	Hummon AB et al
PDH	Pigment-dispersing hormone	Not found					Hummon AB et al
RFamide-like1	Not found	Not found					Hummon AB et al
RFamide-like2	Not found	Not found					Hummon AB et al
sNPF	Short neuropeptide F precursor			FLYBASE:FBgn0032840			Hummon AB et al
SIFa	SIFamide			FLYBASE:FBgn0053527			Hummon AB et al
SK	Sulfakinin			FLYBASE:FBgn0000500			Hummon AB et al
TK	Tachykinin			FLYBASE:FBgn0037976		BEEBASE:GB49248	Hummon AB et al
LOC724564	TWKSPDIVIRFa-containing neuropeptide					BEEBASE:GB41295	Hummon AB et al
FOXP2	Forkhead box P2	ENSG00000128573	ENSMUSG00000029563				Preuss TM

8.3 Remarks

There are large projects for elucidating the brain and its function from various angles including genomics and transcriptomics. One is the BRAIN Initiative (Brain Research through Advancing Innovative Neurotechnologies) that was announced on April 2, 2013, as a national project by the Obama administration, the White House Office of Science and Technology Policy (OSTP), as part of a broader White House Neuroscience Initiative (http://braininitiative.nih.gov) (Fig. 8.3). This project aims at the development of novel technologies for understanding brain function and brain disorders such as Alzheimer's and Parkinson's disease. The other is the Human Brain Project led by EU that aims to help researchers clarify brain functions and disorders (https://www.humanbrainproject.eu) (Fig. 8.3). These projects pursue a "functional connectome" comprising genomics as well as other experimental approaches that would be developed in the progress of the projects. Japan and China also focus on brain research as national projects in various research fields such as medical science, neuroscience, computer science, and life sciences. With these projects, a huge amount of data including genomics and transcriptomics will be available, so that we could prepare a novel diagram to understand brain evolution and diversification.

References

Adams MD et al (1991) Complementary DNA sequencing: expressed sequence tags and human genome project. Science 252(5013):1651–1656

Adams MD et al (2000) The genome sequence of *Drosophila melanogaster*. Science 287(5461):2185–2195

Albertin CB et al (2012) Cephalopod genomics: a plan of strategies and organization stand. Genomic Sci 7:175–188

Blattner FR et al (1997) The complete genome sequence of *Escherichia coli* K-12. Science 277:1453–1462

Cahoy JD et al (2008) A transcriptome database for astrocytes, neurons, and oligodendrocytes: a new resource for understanding brain development and function. J Neurosci 28(1):264–278

Carro MS et al (2010) The transcriptional network for mesenchymal transformation of brain tumours. Nature 463:318–325

Chin J et al (2014) SMRT® sequencing solutions for large genomes and transcriptomes. J Biomol Tech 25(suppl):S15

Goffeau A et al (1997) The yeast genome directory. Nature 387:5–105

Goymer P (2007) Alternative splicing switches on the brain. Nat Rev Genet 8:572

Hang G et al (2014) Genomics in neurological disorders. Genomics Proteomics Bioinformatics 12(4):156–163

International Human Genome Sequencing Consortium (2001) Initial sequencing and analysis of the human genome. Nature 409(6822):860–921

Kang HJ et al (2011) Spatio-temporal transcriptome of the human brain. Nature 478:483–489

Kapranov P et al (2007) RNA maps reveal new RNA classes and a possible function for pervasive transcription. Science 316(5830):1484–1488

Khaitovich P et al (2005) Parallel patterns of evolution in the genomes and transcriptomes of humans and chimpanzees. Science 309:1850
Koonin EV (2005) Orthologs, paralogs, and evolutionary genomics. Annu Rev Genet 39:309–338
Levene MJ (2003) Zero-mode waveguides for single-molecule analysis at high concentrations. Science 299(5607):682–686
Margulies M et al (2005) Genome sequencing in microfabricated high-density picolitre reactors. Nature 437:376–380
Martin JA, Wang Z (2011) Next-generation transcriptome assembly. Nat Rev Genet 12:671–682
Meshorer E et al (2005) Chronic cholinergic imbalances promote brain diffusion and transport abnormalities. FASEB J 19:910–922
Miller JR et al (2010) Assembly algorithms for next-generation sequencing data. Genomics 95(6):315–327
Ohno S (1970) Evolution by gene duplication. Springer, Berlin/New York
Sarah BN et al (2009) Targeted capture and massively parallel sequencing of 12 human exomes. Nature 461:272–276
Schuster SC (2008) Next-generation sequencing transforms today's biology. Nat Methods 5(1):16–18
Shannon P et al (2003) Cytoscape: a software environment for integrated models of biomolecular interaction networks. Genome Res 13:2498–2504
Stoltzfus A (1999) On the possibility of constructive neutral evolution. J Mol Evol 49(2):169–181
The C. elegans Sequencing Consortium (1998) Genome sequence of the nematode *C. elegans*: a platform for investigating biology. Science 282:2012–2018
Yeo G et al (2004) Variation in alternative splicing across human tissues. Genome Biol 5:R74

Part III
Cognitive Systems

Chapter 9
The Origin of Vertebrate Brain Centers

Yasunori Murakami

Abstract Vertebrate brains have diversified into a variety of forms, probably because of lineage-specific modifications of the neural tube. By contrast, the anteroposterior arrangement of brain compartments from the telencephalon to the rhombencephalon is observed in all extant vertebrates, and the stereotyped framework of longitudinal and commissural tracts are also conserved among vertebrates. Thus, vertebrate brains are thought to have inherited their basic organization in the course of evolution, although the size and functions of brain subregions may have diverged in different vertebrate lineages. In this evolutionary process, spatially and temporally regulated gene expression is thought to have a crucial role. Recent evolutionary developmental biology (Evo-Devo) studies using bilateral animals suggested that combinatory expression of regulatory genes which are involved in the patterning of the neural tube may be inherited from an early stage of animal evolution. By contrast, integrative centers including the cerebellum, mesencephalon, and telencephalon are thought to have been established in vertebrates as an evolutionary novelty.

Keywords Brain centers • Central nervous system • Cyclostomes • Development • Gene expression • Lamprey • Vertebrates • Telencephalon

9.1 Introduction

Vertebrate brains have evolved a variety of forms in each group to adapt to many environments. In particular, the relative size of integrative centers in the brain appears to link to the physiology or body morphology of animals. For example, sharks or rays, which are sensitive to chemicals dissolved in water, possess a well-developed olfactory bulb to process olfactory information, and birds, which mainly use visual cues in searching for prey, have a large orbit and well-organized optic tectum in the midbrain, which have an important integrative role for the visual

Y. Murakami (✉)
Graduate School of Science and Engineering, Ehime University, 2-3 Bunkyo cho, Matsuyama, Ehime 790-8577, Japan
e-mail: muakami.yasunori.mu@ehime-u.ac.jp

S. Shigeno et al. (eds.), *Brain Evolution by Design*, Diversity and Commonality in Animals, DOI 10.1007/978-4-431-56469-0_9

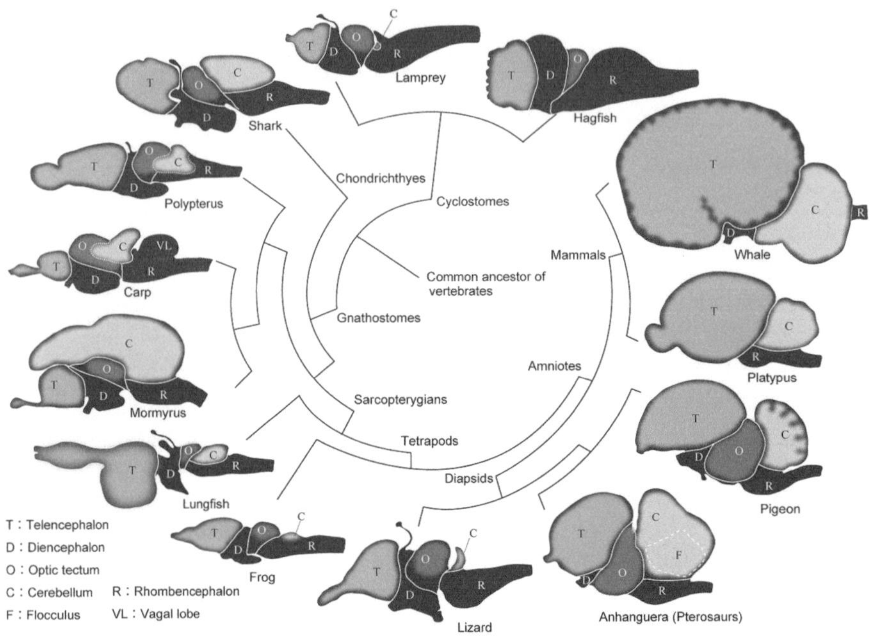

Fig. 9.1 Schematic drawings of vertebrate brains showing their diversified morphology. Brain anatomy is based on Nieuwenhuys 1997 and Witmer et al. (2003)

system (Fig. 9.1). Thus, it is important to note that the morphology and function of brain centers in vertebrates may have evolved symbiotically with other body elements such as the skeletal, muscular, and vascular systems.

Conversely, vertebrate brains also represent a conserved organization among groups. Namely, the anteroposterior arrangement of brain compartments including the telencephalon, diencephalon, midbrain (mesencephalon), and hindbrain (rhombencephalon) are conserved in all extant vertebrates (Fig. 9.2). In addition to brain compartments, many longitudinal and commissural neuronal circuits connecting brain regions in the central nervous system (CNS) are also highly conserved in vertebrates. Thus, vertebrate brains are thought to have inherited their basic organization during the course of evolution, whereas the size and functions of brain subregions may have diverged in different lineages depending on the adaptive radiation of vertebrates. Modification of the brain could arise from small changes in conserved developmental mechanisms during embryogenesis.

In this chapter, we search for the origin of the basic organization of the brain regions and for the modification process in the early stage of vertebrate evolution. To reveal these evolutionary processes, we focused mainly on the cyclostomes (lampreys and hagfishes), which diversified from the jawed vertebrate (gnathostomes) lineage in the early evolutionary period. By comparing the brains of cyclostomes and gnathostomes, we hope we can identify which part of the brain is ancestral (plesiomorphic) or derivative (apomorphic), and discuss possible evolutionary

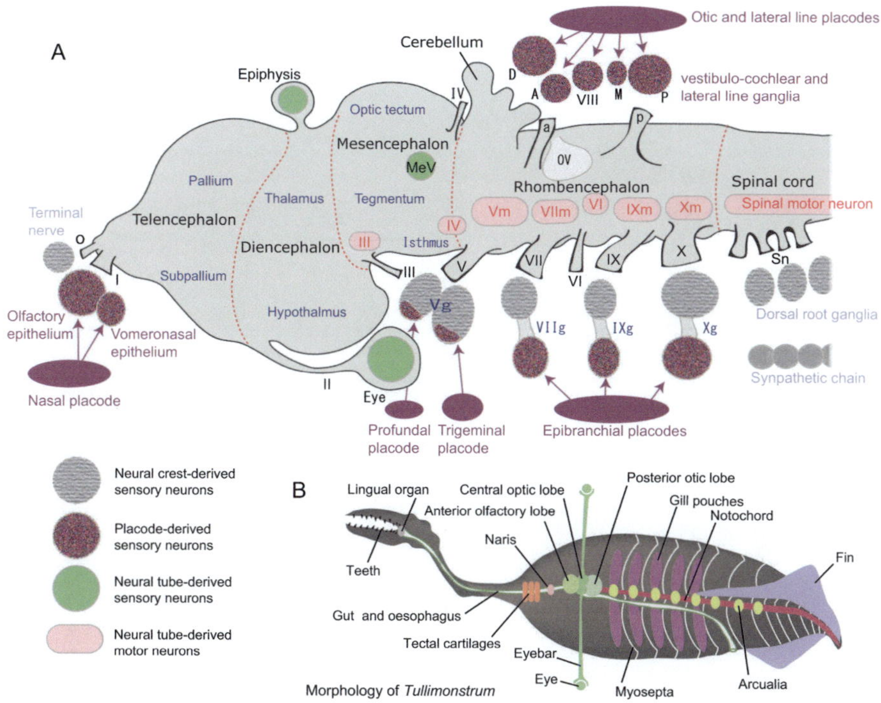

Fig. 9.2 (**a**) Basic organization of developing vertebrate brain. Subregions of the neural tube and developmental origin of neurons. (**b**) Morphology of *Tullimonstrum* (After McCoy et al. 2016). *OV* otic vesicle, *a* anterior lateral line nerve, *A* ganglion of the anterior lateral line nerve, *D* dorsal ganglion of the anterior lateral line nerve, *M* ganglion of the middle lateral line nerve, *MeV* mesencephalic trigeminal nucleus, *p* posterior lateral line nerve, *P* ganglion of the posterior lateral line nerve, *Sn* spinal nerves, *0* terminal nerve, *I* olfactory nerve, *II* optic nerve, *III* oculomotor nerve, *IV* trochlear nerve, *V* trigeminal nerve, *VI* abducens nerve, *VII* facial nerve, *VIII* cochlear nerve, *IX* glossopharyngeal nerve, *X* vagus nerve

processes in the brain developmental programs that led to the diversification of the vertebrate brain.

9.2 Origin of Vertebrate-Specific Traits

The brains of protochordates (tunicates and amphioxus) and vertebrates share many fundamental features (see Chap. 7). In addition to the neural tube, tissues giving rise to the neural crest, the mid-hindbrain boundary (MHB, also known as the isthmic organizer; Fig. 9.3), and neurogenic placodes have been identified in tunicates (Mazet et al. 2005). Because chordates and vertebrates share several orthologous genes specifying the vertebrate nervous system, new vertebrate-specific structures

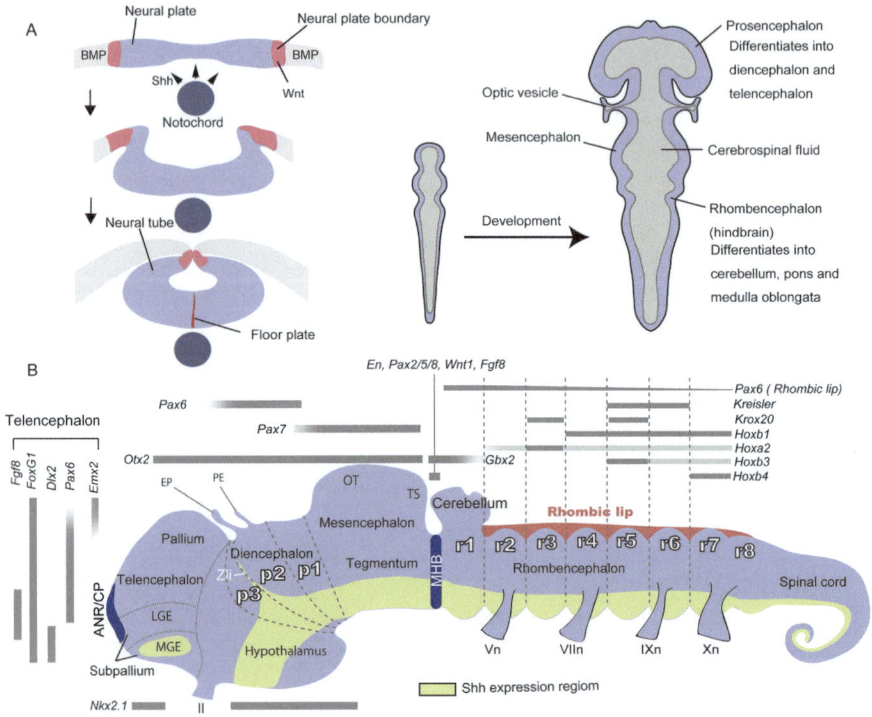

Fig. 9.3 (**a**) Development of the neural tube. Dermal ectoderm differentiates into the neural plate by the signal from the notochord. Then, neural plate invaginates to form the neural tube. The lateral wall of the neural tube then enlarged and differentiates into brain vesicles and the internal cavity, which subsequently will be filled with cerebrospinal fluid and gives rise to ventricles. (**b**) Basic organization of the developing CNS. Developing brain is subdivided into series of segmental units called rhombomres (r1–r8) and prosomeres (p; see below). Mid-hindbrain boundary (*MHB*) locates between mesencephalon and metencephalon (MHB is also known as the ithmic organizer). Zona limitans intrathalamica (*Zli*) lies in the diencephalon primordium between prosomere2 (*p2*) and p3. Anterior neural ridge (*ANR*) or commissural plate (*CP*) locates in a rostral end of the telencephalon primordium. *EP* epiphysis, *LGE* lateral medial ganglionic eminence, *MGE* medial ganglionic eminence, *OT* optic tectum, *PE* parietal eye, *TS* torus semicircularis, *II* optic nerve, *Vn* trigeminal nerve root, *VIIn* facial nerve root

may have evolved through co-option of additional copies of genes into existing gene networks (Ohno 1970; Holland 2013). In addition, whole-genome duplication (WGD) events appear to have occurred several times in vertebrate evolution: the common ancestor of vertebrates, cyclostomes, chondrosteans, and teleosts (Sidow 1996; Miyata and Suga 2001; Abi-Rached et al. 2002; McLysaght et al. 2002; Meyer and Van de Peer 2005; Crow et al. 2012). These events may have provided additional duplicated genes or enhancer elements for preexisting gene networks and accelerated vertebrate diversification.

9.3 Brains of Early Vertebrates

The vertebrate nervous system can be divided into the CNS and the peripheral nervous system (PNS), with the CNS divided further into the brain and the spinal cord (Fig. 9.2). A comparative approach using fossil records sometimes provides important insights into brain evolution because well-preserved fossils indicate signs of nervous system morphology. Recent studies identified the earliest vertebrates, *Myllokunmingia fengjiaoa*, *Haikouichthys ercaicunensis*, and *Metaspriggina walcotti*, from sediment of the Cambrian period (Shu et al. 2003; Butler and Hodos 2005; Morris and Caron 2014). Thus, the origin of vertebrates may date back 540 million years. Based on morphological characters such as paired eyes, segmented myotomes, a series of pharyngeal arches, and undifferentiated jaws, these animals are classified as agnathans (jawless animals). It is important to note that signs of eyes or nasal pits imply the presence of neurogenic placodes, which give rise to sensory cells, and the evidence of pharyngeal arches also indicates the presence of neural crest cells, which produce mesenchymal and skeletal elements in the craniofacial region. Some fossil osteostracans, which possess a skeletal head shield, have preserved casts of many internal organs, including the brain. A fossil specimen of *Norselaspis*, a kind of osteostracan that inhabited Earth during the Devonian period, presents remarkable signs of both a CNS and a PNS (Janvier 2002). Furthermore, recent advanced techniques using synchrotron radiation X-ray tomography or propagation phase-contrast synchrotron microtomography images enable us to see the brain cavity in the skull without disturbing fossils, and we can reconstruct the external morphology of the ancient brain (Witmer et al. 2003; Witmer and Ridgely 2009; Rowe et al. 2011; Balanoff et al. 2013; Tanaka et al. 2013; Cong et al. 2014). Using this new technique, the brain morphology of *Shuyu*, one of the earliest and most primitive galeaspid genera, has been reported (Gai et al. 2011). The brain of *Shuyu* shows signs of cerebral hemispheres with olfactory bulbs and the earliest evidence for a clear separation of the olfactory organs from the hypophysial duct. Recently, Dupret and colleagues argued that the morphology of brains appears to be changed in crown gnathostomes (Dupret et al. 2014), that is, telencephalons extend anteriorly in these lineages (Fig. 9.4). This extension may possibly result from a modification of the developmental position of the nasal and hypophysial placodes: a short telencephalon was associated with separate but closely spaced nasal and hypophysial placodes, whereas a long telencephalon could be supported and protected by a horizontal shelf that was created between the nasal sacs and hypophysis.

Although endocast analysis provides valuable insights into the size and external morphology of brains, as just noted, the brain cast sometimes can provide an overestimate of the size of the brain because of the presence of space between the brain and bone (and meninges) in the live condition. In addition, we cannot identify the internal structures (cellular organization and neuronal connections) of ancient brains. A possible approach is comparative analyses of extant animals. If particular traits are shared by both groups, they may be inherited from a common ancestor of the two groups of vertebrates and hence may be plesiomorphic

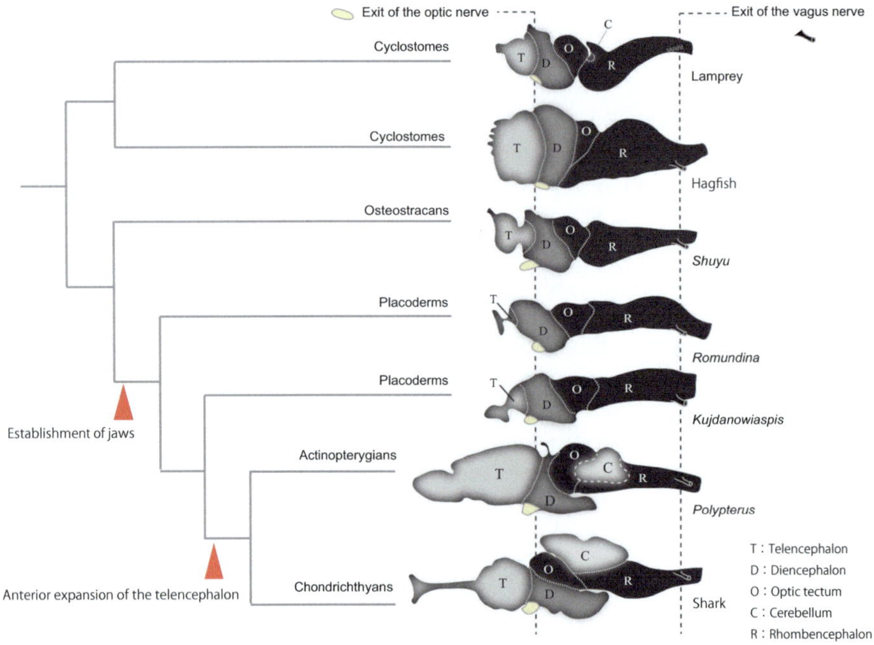

Fig. 9.4 Brain of early vertebrates: schematic drawings of brains of agnathans (cyclostomes and osteostracans) and gnathostomes (placoderms, actinopterygians, and chondrichthyans). Brains are aligned based on the position of optic and vagus nerve exits. Telencephalons of gnathostomes extend anteriorly before the diversification between actinopterygians and chondrichthyans (After Dupret et al. 2014)

characters. Using these morphological characters, we could reconstruct the ancient brain architectures. In addition, recent studies have shown that various transcription factors and morphogens are expressed in embryonic mouse brains in a region-specific manner (Shimamura et al. 1995; Wilson and Rubenstein 2000; Mallamaci and Stoykova 2006; Guillemot 2007; Pierani and Wassef 2009). Therefore, comparison of molecular cues underlying brain morphogenesis in various species may provide insight into the origins and diversification processes of vertebrate brains. Because cyclostomes (lamprey and hagfish), belonging to agnathans, bifurcated from gnathostomes at an early evolutionary stage (Kuratani et al. 2002; Kuraku and Kuratani 2006; Heimberg et al. 2010; Oisi et al. 2013), the brains of cyclostomes will shed light on the origin of vertebrate brain centers.

9.4 Cyclostomes

Brain structures shared between cyclostomes and gnathostomes are expected to provide important plesiomorphic signs about the brain elements. Extant cyclostomes (lampreys and hagfishes) thus possibly inherit some important traits that are characteristic not only in extinct agnathans but also in the common ancestor of vertebrates. Generally, the brains of cyclostomes are relatively small, and plotting brain weight against body size reveals a less-developed brain in cyclostomes than in gnathostomes (Jerison 1985; Striedter 2005).

9.4.1 Lampreys

There are 38 species of lamprey, 29 of which inhabit freshwater (Nelson 2006). The adult animal possesses a sucker-like oral funnel with rings of horny teeth, a single nostril, a single pair of eyes, and seven pairs of gill openings. They have dual semicircular canals in the vestibular organ, in contrast to gnathostomes, which have three canals. Although lampreys lack a lateral horizontal canal, they appear to have evolved in parallel an anatomically distinct horizontal duct system (Maklad et al. 2014). They lack paired appendages, although they possess unpaired dorsal and tail fins. The nervous system of the lamprey exhibits some cyclostome-specific features as well; it lacks sympathetic chain ganglia and the nerve axons lack myelin sheaths. Adult animals breed in shallow rivers, and their hatched larvae, which are called ammocoetes, have eye spots embedded in their skin and an endostyle that is thought to be homologous to the thyroid gland in other vertebrates. After metamorphosis, some species remain in freshwater, whereas marine species move to a saltwater environment. During spring and early summer, marine lampreys migrate to shallow freshwater streams to breed. After spawning, they usually die.

The most ancient lamprey (*Priscomyzon riniensis* gen. et sp. nov.), which possesses many of the key characteristics of modern forms, has been found in the fossil record from the Devonian period (360 million years ago). Therefore, it is evident that the lamprey lineage has a long evolutionary history (Gess et al. 2006).

A recent study showed that *Tullimonstrum gregarium*, known as the Tully monster, possesses the buccal apparatus containing teeth, the paired eyes extending laterally on a long rigid bar, the tri-lobed brain (the anterior olfactory lobe, the central optic lobe, and the posterior otic lobe), and the segmented body with caudal fin. This combination of characters indicates *Tullimonstrum* as a vertebrate and places it on the stem lineage to lampreys (Fig. 9.2; McCoy et al. 2016).

9.4.2 Hagfishes

Extant hagfishes live in a marine environment. They possess three pairs of short tentacles around the mouth. They have from one to six pairs of pharyngeal openings and a mucous gland on the body surface to prevent attacks by enemies. Their semicircular canal appears to be single, but a recent morphological study suggests that they originally had two canals similar to lampreys, after which one canal degenerated (Jorqensen et al. 1998). In contrast to lampreys, all living hagfishes lack larval stages in their life cycle (Jorqensen et al. 1997). Because they breed in the deep sea, their developmental process was not observed for a long time. However, recent studies have successfully obtained fertilized eggs and living embryos from the Japanese hagfish *Eptatretus burgeri* (Ota and Kuratani 2006; Ota et al. 2007) and have revealed that they share a common body plan with lampreys (Ota et al. 2007, 2011; Oisi et al. 2013).

9.4.3 Cyclostome-Specific Characters

Cyclostomes may possibly have inherited plesiomorphic traits (e.g., a neural tube, paired eyes, a nasal organ, inner ears, and a pineal organ), which are established in the common ancestor of vertebrates. By contrast, they have their own morphological characters such as the first arch-derived oral apparatus and a distinct immune system (Hirano et al. 2013). These cyclostome-specific traits may have been established after the divergence between agnathans and gnathostomes by the modification of their developmental plan. Notably, recent studies revealed that lampreys possess distinct Hox clusters (Mehta et al. 2013), suggesting that cyclostomes experienced their own WGD. This genomic modification may have contributed to the evolution of cyclostome-specific traits.

9.5 Origin of the Vertebrate CNS

9.5.1 Neural Tube

The vertebrate brain is divided into several components including the telencephalon, diencephalon, midbrain (mesencephalon), cerebellum, and hindbrain (rhombencephalon). These regions originate from the neural tube, which is initially specified from the dorsal ectoderm (Fig. 9.3). In the process of neural tube formation, some signaling molecules, such as Hedgehog, are secreted from the notochord and later in the floor plate, and induce ventral differentiation in the neural tube. Dorsalizing factors, such as bone morphogenetic protein (BMPs) are expressed in the ectoderm overlying the neural tube. The ventro-dorsal axis of the neural tube is determined by antagonistic interaction between these two signals (Fig. 9.3) (Yamada et al. 1991; Echelard et al. 1993; Echevarria et al. 2003; Sanes 2006; Suzuki et al. 2012). The lateral wall of the neural tube then enlarges and differentiates into three brain vesicles, known as the procecephalon, mesencephalon, and rhombencephalon (Fig. 9.3), along the anteroposterior axis of the tube. It has been believed that the second brain vesicle gives rise to the mesencephalon. However, a recent study suggests that the second brain vesicle differentiates into the mesencephalon and a rostral region of rhombencephalon, at least in some vertebrates (Ishikawa et al. 2012). The internal cavity of the neural tube, which will be filled with cerebrospinal fluid, gives rise to the ventricles. The ventricles are initially surrounded by neuroepithelial cells that produce neurons and are then replaced with ependymal cells (Jacobson 1991). The neural tube could be divided into dorsal (alar) and ventral (basal) plates. Within the brainstem (from the diencephalon to the hindbrain) and the spinal cord, sensory centers tend to be dorsal, autonomic centers are lateral and intermediate, and motor centers are ventral. This regionalization is patterned by signaling molecules such as SHH secreted from the floor plate and BMP from the roof plate (Yamada et al. 1991; Echelard et al. 1993; Roelink et al. 1995; Lee and Jessell 1999). This neural tube-based patterning mechanism appears to serve as a basic plan for the vertebrate brain. However, a tubular CNS is also found in chordates, including the cephalochordates and urochordates. Comparative studies have shown that the expression patterns of genes that specify the neural plate and its border in amphioxus and vertebrates are highly conserved (Beccari et al. 2013; Holland 2013). Moreover, a number of genes that pattern the region of the neural tube have been expressed similarly between vertebrates and chordates, although some genes (e.g., *Gbx*) are absent in the tunicate genome (Passamaneck and Di Gregorio 2005; Holland 2013). Thus, the developmental mechanism underlying neural tube formation has been established, at least in the common ancestor of chordates. Interestingly, a recent developmental study showed that the collar cord of the hemichordate, in which a tubular nerve cord could be observed, shared gene expression patterns common to that seen during chordate neurulation. In addition, it receives a Hedgehog signal from the dorsal endoderm of the buccal tube and the stomochord, which lie beneath the collar cord. This finding suggests that the endoderm functions as an organizer to pattern the overlying collar cord, similar to the relationship between the notochord and the neural tube in chordates (Miyamoto and Wada 2013). Taken together, the origin of the core genetic mechanisms underlying the development of the neural tube date back to the last common ancestor of deuterostomes. Because this regulatory gene expression has also been observed in insects (Lichtneckert and Reichert 2005), the gene regulatory framework involved in brain formation may originate from the urbilaterians, a presumptive ancestral form of bilaterians (De Robertis and Sasai 1996). However, despite the considerable similarities between the developmental plans of chordates and vertebrates, the brain morphology of chordates shows some remarkable differences from that of vertebrates. For example, the brains of amphioxus and tunicates apparently lack brain regions such as the telencephalon and cerebellum (Wicht and Lacalli 2005). Moreover, a recent study has shown that the amphioxus neural tube appears to lack a mesencephalic region (Suzuki et al. 2015b). Therefore, some brain compartments observed in extant vertebrates may possibly have been established after the split between chordates and vertebrates. The vertebrate-specific modification of the gene regulatory network involved in brain morphogenesis may have caused the innovation in the vertebrate brain. In vertebrates, despite the diversity of brain regions, the anteroposterior or dorsoventral organization of the neural tube is quite similar throught different lineages: this indicates that the developmental plan of the vertebrate CNS is highly conserved for more than 500 million years under a constraint (evolutionary constraint).

9.5.2 Organizing Centers

In the development of the vertebrate brain, some specific regions act as important signaling centers called organizers, one of which is the MHB that is located between the mesencephalon and metencephalon (Fig. 9.3). The other organizer is a *zona limitans intrathalamica* (*Zli*), which lies in the diencephalon primordium in which the boundary between prosomere 2 (p2) (see following) and p3 is established. Finally, the anterior neural ridge (ANR) or commissural plate (CP), located at the rostral end of the telencephalon primordium, is crucial in the patterning of the telencephalon.

These organizing centers are characterized by the combinatory expression of transcription factors and signaling molecules. For example, the MHB is patterned in the expression boundary between *Otx2* and *Gbx2*, and FGF8 secreted in the boundary increases the level of expression of *Pax2/5/8*, *En1/2*, and *Wnt1* (Broccoli et al. 1999; Millet et al. 1999; Katahira et al. 2000; Matsunaga et al. 2002). *Zli* is marked by the expression of *Shh*, which differentiates alar plate regions of p2 and p3 into the thalamus and the prethalamus, respectively (Rubenstein et al. 1994; Kobayashi et al. 2002; Echevarria et al. 2003). ANR/CP secretes FGF8 and regulates the size of the telencephalon (Cecchi 2002; O'Leary and Sahara 2008; Medina and Abellan 2009). Importantly, this gene expression has been partially identified in the cephalochordates and urochordates (see Chap. 7). Surprisingly, gene expression that characterizes these centers is also found in the ectoderm of hemichordates (Pani et al. 2012). This finding suggests that the developmental basis that gives rise to these centers might have emerged in an ancestor of hemichordates as an ectodermal patterning mechanism. At that time, in chordates, these organizers might have been diverted to a brain-patterning program. However, chordates lack some of the brain regions that are observed in vertebrates. Thus, it is likely that many, but not all, of the gene regulatory frameworks for constructing organizing centers were already present in a common ancestor of chordates, although their neural tube was only partially regionalized into brain vesicles that were comparable to those of vertebrates. This issue may be explained by a fascinating concept called "deep homology" (see following).

9.6 Origin of the Vertebrate PNS

9.6.1 Cranial Placodes and Neural Crests

The PNS essentially consists of cranial, spinal, sympathetic, and enteric nerves, which include axons of sensory and motor neurons (Figs. 9.2 and 9.5). The sensory peripheral nerves are induced during developmental periods by specific cells such as sensory placodes (neurogenic placodes) or neural crests. Sensory placodes are formed in specific parts of an epidermal thickening in the craniofacial

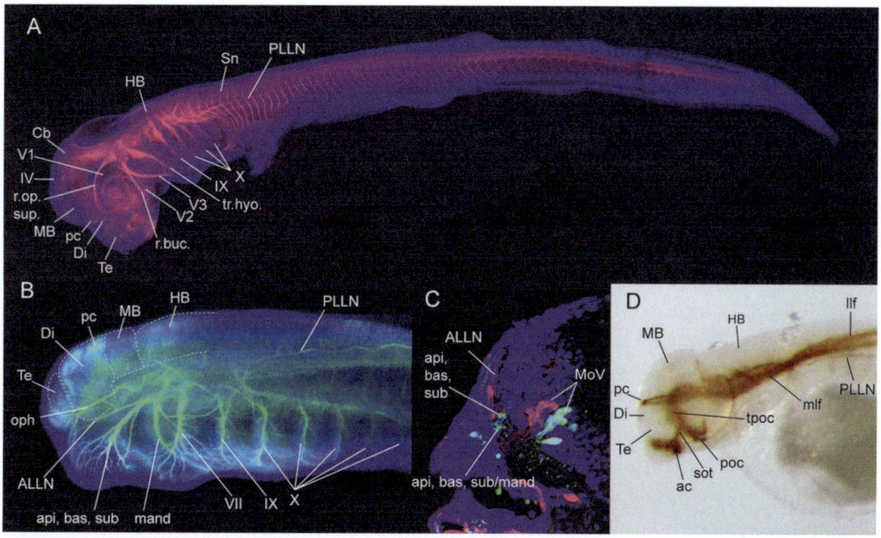

Fig. 9.5 Morphology of the peripheral nervous system (*PNS*) in shark (**a**) and lamprey (**b**). (**c**) Coronal section of the lamprey hindbrain shows projection of the trigeminal and lateral line nerves. Trigeminal nerve branches innervating the *upper lip* (*green*) and *lower lip* (*red*) are segregated and enter a specific part of the hindbrain. (**d**) Early axonal scaffold of the embryo of red seabream (*Pagrus major*). *ac* anterior commissure, *ALLN* anterior lateral line nerve, *api* apical nerve, *bas* basilar nerve, *Cb* cerebellum, *Di* diencephalon, *HB* hindbrain, *llf* lateral longitudinal fascicle, *mand* mandibular nerve, *MB* midbrain, *mlf* medial longitudinal fascicle, *MoV* motor nucleus of the trigeminal nerve, *oph* ophthalmic ramus of the trigeminal nerve, *pc* posterior commissure, *PLLN* posterior lateral line nerve, *poc* postoptic commissure, *r. buc* buccal ramus of the lateral line nerve, *r. op. sup* superficial ramus of the ophthalmic lateral line nerve, *sot* supraoptic tract, *sub* suborbital nerve, *Te* telencephalon, *tpoc* tract of the postoptic commissure, *tr. hyo.* hyomandibular nerve, *V1* ophthalmic ramus of the trigeminal nerve, *V2* maxillary ramus of the trigeminal nerve, *V3* mandibular ramus of the trigeminal nerve, *IV* trochlear nerve, *VII* facial nerve, *IX* glossopharyngeal nerve, *X* vagus nerve

region surrounding the brain (Schlosser 2006). These placodes differentiate into sensory receptor cells, supporting cells, and sensory neurons. In vertebrates, nasal, hypophysial, lens, profundal, trigeminal, otic, lateral line, and epibranchial placodes can be observed (Fig. 9.2).

Neural crest cells develop in a boundary between the epidermis and the neural plate and subsequently migrate to several parts of the body. The sympathetic and enteric nervous systems are derived from these cells. Because neural crests generate pigment cells and many parts of the head skeleton and pharyngeal arches (including jaws), they have a key role, not only in the body's morphogenesis, but also in the morphological evolution of vertebrates.

Molecular mechanisms underlying the formation of sensory placodes and neural crests have been extensively studied. The cranial placodes that are formed surrounding the neural tube are known as the preplacodal region (PPR), in which *Six1*, *Six4*, and *Eya1* are expressed (Schlosser 2006; Streit 2007; McCabe and Bronner-Fraser 2009). Later, other genes are additionally expressed in PPR to form specific placodes. Alternatively, *Sox9*, *Snail2*, *Foxd3*, and other genes are important in the differentiation of the neural crests (Betancur et al. 2010; Milet and Monsoro-Burq 2012). Accordingly, complicated gene regulatory networks are involved in the patterning of neural crests (Betancur et al. 2010; Theveneau and Mayor 2012). In the later stages, axon guidance molecules including semaphorin and ephrin are involved in the migration of these cells. Recent studies have revealed that interactions between semaphorin3A and neuropilin1, a receptor for semaphorin ligands, are involved in the formation of the sympathetic chain (Theveneau and Mayor 2012). In an evolutionary aspect, many PNS components are present in the lamprey, although a sympathetic chain cannot be observed in this animal. In the process of lamprey development, placodes and neural crests, which are thought to be involved in the differentiation of sensory ganglia, are identified (Langille and Hall 1988; McCauley and Bronner-Fraser 2003; Modrell et al. 2014), and the general pathways of cranial neural crest migration are conserved through the vertebrates, although a lamprey-specific migratory route could be identified (McCauley and Bronner-Fraser 2003). These lines of evidence suggest that the developmental mechanism underlying the production of the PNS may have been established before the diversification between cyclostomes and extant gnathostomes occurred.

9.6.2 Origins of Placodes and Neural Crest Cells

Protochordates apparently lack convincing homologues of neurogenic (sensory) placodes in vertebrates, except possibly for the olfactory epithelium in tunicates (Bassham and Postlethwait 2005). However, amphioxus and tunicates have some ectodermal sensory cells that are considered homologues of neurogenic placodes in vertebrates. These cells express some genes such as *Six* and *Eya*, similar to vertebrate neurogenic placodes (Bassham and Postlethwait 2005; Holland 2013). Thus, the evolutionary origin of some neurogenic placodes may date back to the ancestor of chordates. In the case of neural crests, homologues of genes specifying the vertebrate neural crest are not expressed at the edges of the amphioxus neural plate, although homologues of genes that specify the neural plate and its edges are similarly expressed in amphioxus and vertebrates (Yu et al. 2002; Meulemans and Bronner-Fraser 2004; Holland 2013, 2015). By contrast, tunicates appear to have a homologous neural crest cell (Mazet et al. 2005; Abitua et al. 2012; Stolfi et al. 2015). Therefore, the developmental plan for initializing neural crests may be incorporated into the gene network after the split between cephalochordates and other chordates. Consequently, the increased number or rearrangement of genes by genome duplication in the vertebrate lineage may have a key function in the establishment of multipotency of neural crest cells in vertebrates.

Some aquatic vertebrates such as fishes and amphibians possess neural crest-derived neurons called Rohon–Bared cells (RB cells) (Beard 1889; Bernhardt et al. 1990; Coghill 1914; Hughes 1957). These unique neurons have their cell body in the spinal cord and extend their axons within or beyond the spinal cord (Bernhardt et al. 1990; Hartenstein 1993). Generally, in the course of their development, RB cells decrease their numbers by apoptosis and are replaced by dorsal root ganglia (DRG), which are also derived from neural crests (Lamborghini 1987). Because RB cell-like neurons are observed in amphioxus, they must have emerged in the early stage of chordate evolution. In lampreys, both RB cells and extramedullary sensory neurons (DRG homologues) are present, although their spinal roots are primitive and asymmetrical (Fritzsch and Northcutt 1993). Thus, in the common ancestor of chordates, sensory stimuli were presumably received by neurons whose cell bodies are located in the spinal cord. Then, in the course of the evolution of vertebrates, this type of cell was replaced by neurons whose cell body is located outside of the spinal cord; that is, sensory stimuli are perceived by DRG neurons located outside the spinal cord. This evolutionary switching of neurons is a characteristic event in the evolution of the PNS. A recent study has shown that this developmental switching from RB cells to DRG neurons is regulated by Six1 in *Xenopus* (Yajima et al. 2014).

9.7 Regionalization of the Vertebrate CNS

9.7.1 Brain Segmentation

In the course of brain development, the neural tube differentiates into a series of vesicles representing unique morphology. Namely, the wall of the neural tube shows segmental bulges, resembling a lepidopteran caterpillar. These structures were initially identified by von Baer in the chick embryo (von Baer 1828). Since then, segmental bulges have been identified in all vertebrate embryos studied so far. Several studies have shown that these bulges contain specific neurons or nuclei, and for that reason these segments were thought to provide morphological units that are important in eliciting a specific function. Thus, Bergquist and Källén (1953) referred to these segments as "Grundgebiete (basic domains)." These brain segments are now called neuromeres (Rubenstein et al. 1998), and those in the procencephalon and the hindbrain are named prosomeres and rhombomeres, respectively (Fig. 9.3). Prosomeres have been observed in many gnathostomes, including teleosts (Wullimann and Puelles 1999; Diaz-Regueira and Anadon 2000), chickens (Redies et al. 2000), frogs (Javier-Milan and Puelles 2000), and mice (Puelles and Rubenstein 1993; Shimamura et al. 1995; Puelles and Rubenstein 2003). Rhombomeres have attracted the attention of developmental biologists because of their topographical relationships with peripheral nerves and segmental boundary-related gene expression patterns (Orr 1887; Vaage 1969; Lumsden and Keynes 1989; Clarke and Lumsden 1993; Kuratani and Eichele 1993; Kontges and Lumsden 1996). It is also important to note that rhombomeres represent an arrangement of serially homologous segments that can give rise to a series of similar neuronal elements. For example, the motor nuclei for the branchial neurons including trigeminal, facial, glossopharyngeal, and vagal motor nuclei develop in corresponding rhombomere segments (Neal 1896; Lumsden and Keynes 1989; Noden 1991; Gilland and Baker 1993). Furthermore, in some vertebrates, rhombomeres generate similar repeated sets of neurons, namely, "serial homologues" (Metcalfe et al. 1986; Mendelson 1986; Kimmel et al. 1988; Hanneman et al. 1988; Lee et al. 1993; Clarke and Lumsden 1993), one of which is an anteroposteriorly arranged set of reticulospinal neurons that forms part of the pattern-generating circuits for a variety of behavioral patterns including swimming. The Mauthner neuron, a type of reticulospinal neuron, is important for escape behavior and appears in rhombomere 4 in many anamniotes including the lamprey, and similar neurons are observed in rhombomere 5 as serial homologues in some aquatic species (Mendelson 1986; Hanneman et al. 1988; Murakami et al. 2004).

9.7.2 Molecular Basis for Constructing Rhombomeres

Rhombomeres are characterized by the expression of some homeodomain-containing transcription factors. In particular, rhombomere segments are identified by the combinatory expression of *Hox* or *Krox* genes (Fig. 9.6) (Wilkinson et al. 1989; Schneider-Maunoury et al. 1997; Hunt et al. 1991; Krumlauf et al. 1993; Rijli

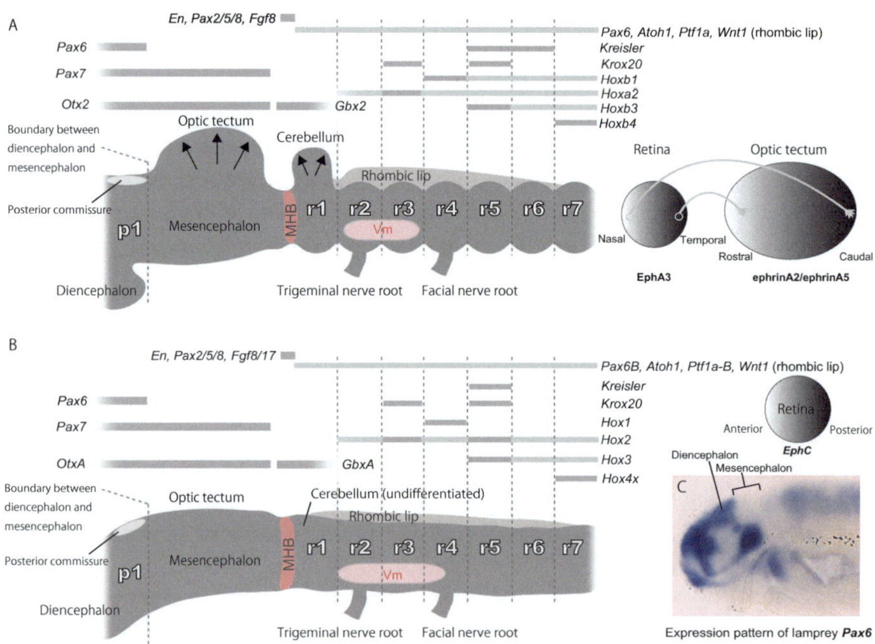

Fig. 9.6 Developmental plan of vertebrate mesencephalon and hindbrain. (**a**) Expression patterns of regulatory genes in developing gnathostomes. *Right* drawing shows scheme of the retino-tectal projection and expression domains of axon guidance molecules (EphA3 and ephrinsA5/epharinA2) in the retina and the tectum. (**b**) Expression patterns of regulatory genes in the developing lamprey (after Takio et al. 2007; Parker et al. 2014; Sugahara et al. 2016). Expression domain of lamprey EphC (*LjEphC*) is indicated in the *left*. (**c**) *Pax6* (*LjPax6*) in the lamprey embryo at stage 27. *LjPax6* mRNA is expressed in specific parts of the CNS, whereas it is absent in the mesencephalon. *Vm* motor nucleus of trigeminal nerve. See Fig. 9.3 for abbreviations

et al. 1998; Schilling and Knight 2001). *Hox* family genes are expressed colinearly in the hindbrain primordium and are involved in the formation of rhombomere compartments. The colinearity of *Hox* gene expression is found in the amphioxus (Schubert et al. 2006), although this animal has no morphological bulges in its neural tube. In cyclostomes, neuromeres similar to those of gnathostomes have been identified in the lamprey (Pombal and Puelles 1999; Pombal et al. 2001, 2009; Murakami et al. 2001, 2004; Osorio et al. 2005). In the hagfish *Eptatretus burgeri*, the embryonic hindbrain shows six rhombomeres (Oisi et al. 2013). Thus, the molecular framework for generating neuromeres is thought to have been acquired in a chordate-like ancestor. Then, morphological segments may be established in the vertebrate lineage. This observation indicates the establishment of a molecular mechanism that restricts cell proliferation or migration into specific compartments (Cooke et al. 2001), and establishment of neuromeres as module elements may be a key innovation of the vertebrate brain.

9.7.3 Early Neuronal Scaffolds

Some specific neuronal circuits are formed in the developmental period. These early tracts, called "early axonal scaffolds," act as a guidepost for later developing axons and provide a template for the subsequent development of complex neural circuitry (Fig. 9.5). Importantly, the framework of these tracts is highly conserved in vertebrates (Anderson and Key 1999; Chitnis and Kuwada 1990; Doldan et al. 2000; Easter et al. 1993; Ishikawa et al. 2004; Ross et al. 1992; Barreiro-Iglesias et al. 2008). Given that the basic axonal connection of adult animals is similar between vertebrates, this conservation may be the result of stereotyped scaffolds that provide basic frameworks for axons. These early tracts consist of longitudinal (extending along the anteroposterior axis) and commissural (connecting the left and right side of the brain) types. Longitudinal tracts include the lateral longitudinal fascicle, the tract of the postoptic commissure, and the supraoptic tract, and the commissural tracts include the anterior, habenular, and posterior commissures.

In the cyclostomes, although the basic framework of the axonal scaffold is similar to those of gnathostomes (Barreiro-Iglesias et al. 2008), the anterior commissure is poorly developed. Instead, they have a well-developed interbulbar commissure (coib), which is thought to be homologous to the hippocampal commissure in gnathostomes (Wicht and Northcutt 1992). In gnathostomes, axons of the olfactory system enter the anterior commissure, but in cyclostomes, the olfactory fibers enter into the coib. This observation indicates that the axon guidance mechanism underlying the olfactory tract might have been differentially modified between cyclostomes and gnathostomes in relationship to the evolution of the anterior telencephalon. Given that the morphology of the telencephalon changed dramatically in the early stage of vertebrate evolution (Dupret et al. 2014), the gross modification of the olfactory tract might have occurred during the agnathan–gnathostome transition.

These early formed tracts provide a landmark, not only in the adult brain, but also in the developing brain compartments. For example, the posterior commissure defines the boundary between prosomere 1 (pretectum) and mesencephalon. These highly conserved tracts imply a strict neurodevelopmental program that guides extending axons to their correct targets. Some transcription factors and axon guidance molecules such as the Slit2 ligand and Robo2 receptor are important in the formation of networks in the early tracts (Plump et al. 2002; Shu and Richards 2001; Shu et al. 2003; Lopez-Bendito et al. 2007; Devine and Key 2008; Hocking et al. 2010; Ricano-Cornejo et al. 2011, Tosa et al. 2015).

Interestingly, orthologues of these genes are involved in the formation of commissures in the insect brain, which are seen to be "not homologous" in the vertebrate brain (Rothberg et al. 1990). Why are orthologous genes involved in the formation of nonhomologous brains? A concept raised in recent evolutionary morphology called "deep homology" may be able to explain this incongruity.

9.8 Deep Homology

Classical morphology has established that body plans for vertebrates and insects are quite different. For example, the insect brain originated from a placode-like ectodermal enlargement, whereas the vertebrate brain is formed from a neural tube. Insects have a compound eye that is different from the camera eyes in vertebrates. However, recent studies have revealed that the genes involved in body morphogenesis are sometimes quite similar in these animals (Fig. 9.7). For example, Pax6 is involved in the morphogenesis of compound eyes in insects and camera eyes in vertebrates. Interestingly, if vertebrate Pax6 is expressed ectopically in fruit

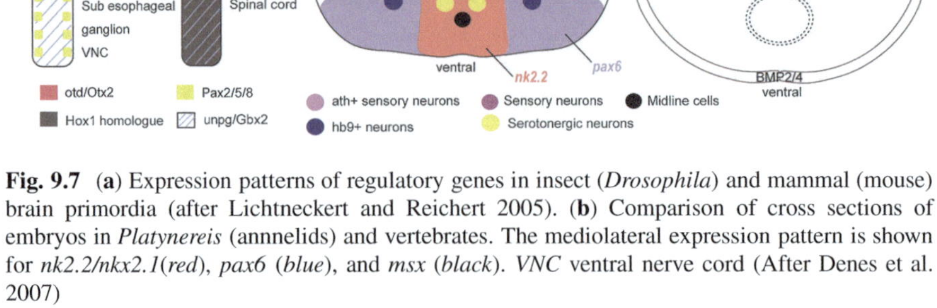

Fig. 9.7 (**a**) Expression patterns of regulatory genes in insect (*Drosophila*) and mammal (mouse) brain primordia (after Lichtneckert and Reichert 2005). (**b**) Comparison of cross sections of embryos in *Platynereis* (annelids) and vertebrates. The mediolateral expression pattern is shown for *nk2.2/nkx2.1*(*red*), *pax6* (*blue*), and *msx* (*black*). *VNC* ventral nerve cord (After Denes et al. 2007)

flies, they will still develop compound eyes instead of camera eyes (Gehring 1996). This finding indicates that Pax6 has a key role in the formation of nonhomologous optic organs among animals, suggesting that a fundamental molecular mechanism is shared between vertebrates and insects, after which the downstream target genes for Pax6 were modified in the course of evolution, causing different eye morphology. Thus, even though there seems to be no morphological similarity, a common gene regulatory network may be involved in the formation of apparently nonhomologous structures. A fascinating concept that may be useful in addressing this issue is "deep homology," in which highly conserved genetic regulatory networks across phyla might be derived from ancient regulatory systems established in a common ancestor (Carroll 2008; Shubin et al. 1997). Because this is often identified in several areas of brain formation, including in higher cognitive centers (Fig. 9.7) (Denes et al. 2007; Strausfeld and Hirth 2013), it is a very important concept with which to explore brain evolution at the level of gene regulation.

9.9 Evolution of the Vertebrate CNS

9.9.1 Hindbrain (Rhombencephalon)

The hindbrain is defined as the region that is located posterior (caudal) to the midbrain and rostral to the spinal cord. Although some researchers regard the cerebellum as a part of the hindbrain, it presents as a distinct region in this issue (see following). Although the vertebrate hindbrain shows diversified morphology in some groups such as teleosts (Fig. 9.1), it also represents highly conserved organization between vertebrates. Namely, the vertebrate hindbrain is characterized by the series of segmental neurons such as branchial motor nuclei and reticulospinal neurons (see brain segmentation). In addition to those neurons, nuclei of the general or special somatic sensory system are also arranged anteroposteriorly corresponding to rhombomeres in the vertebrate hindbrain (Marin and Puelles 1995; Gaufo et al. 2004; Oury et al. 2006). This highly conserved columnar organization in vertebrates implies the presence of a strict developmental program that imposes selective pressure or evolutionary constraint. A number of studies have revealed that the *Hox* gene family displays nested, segmentally restricted expression patterns with sharp anterior boundaries, and has a crucial role in the morphogenesis of the hindbrain (Fig. 9.6) (Hunt et al. 1991; Krumlauf et al. 1993; Rijli et al. 1998; Schilling and Knight 2001; Gaufo et al. 2004; Kiecker and Lumsden 2005; Geisen et al. 2008; Erzurumlu et al. 2010).

The vertebrate hindbrain is also characterized by the number of nerve roots through which cranial nerves connect to the CNS. Actually, these roots function as an important window for "plug in" of sensory nerves and "output" of motor nerves (Fig. 9.6). These cranial nerves have been identified in cyclostomes and fossil agnathans in which the anteroposterior arrangements are similar to those of gnathostomes (Janvier 2002; Gai et al. 2011; Dupret et al. 2014).

Among the cranial nerves, the development of the trigeminal nerve has been extensively studied because it is one of the largest peripheral nerves and has an important function in somatosensory perception in the craniofacial region (Erzurumlu et al. 2010). In vertebrates, cells in the trigeminal ganglion generally receive input through the peripheral processes innervating the skin or sensory organs and send axons to the hindbrain through the trigeminal nerve root in rhombomere 2 (Figs. 9.3, 9.5, and 9.6); exceptionally, the trigeminal nerve root in the shark shifts to r3 during its development (Kuratani and Horigome 2000). These afferents then connect with the trigeminal sensory nuclei located in the anterior hindbrain.

Lamprey hindbrains display some specific features. For example, the trigeminal motor nucleus is located at r2–r3 and the posterior limit expands to the middle part of r4, differeing from gnathostomes, in which the nucleus is generally restricted to r2 and r3 (Fig. 9.6) (Murakami et al. 2004). In the lamprey, a homologue of the principal nucleus of the trigeminal nerve (PrV) has not been identified (Koyama et al. 1987). In addition, the maxillary and mandibular branches of the maxillomandibular nerve (mmV) are not comparable between lampreys and gnathostomes, and spacial segregation of mmV neruron precursor territories may represent lamprey-specific feature (Kuratani et al. 2004; Higashiyama and Kuratani 2014; Modrell et al. 2014). Despite these morphological differences, the expression pattern of signaling molecules or transcription factors that is involved in the formation of the hindbrain is quite similar between lampreys and other gnathostomes. The regulatory elements that are involved in Hox gene expression are also similar between lampreys and gnathostomes (Fig. 9.6) (Parker et al. 2014). Thus, the origin of the gene regulatory network for hindbrain patterning may date back to a common ancestor of vertebrates.

The trigeminal system in hagfishes has five sensory ganglia and nerves and represents a unique feature because it is thought to facilitate the chemosensory and cutaneous sensation (Braun 1998). Similar to lampreys, hagfishes appear to lack a PrV, and their trigeminal afferents project to the hindbrain, where five columns of fibers surrounded by cell bodies can be observed (Nishizawa et al. 1988).

9.9.2 Cerebellum

In gnathostomes, the cerebellum is located between the isthmus in the midbrain and the medulla oblongata (posterior part of the hindbrain) (Fig. 9.2). Generally, the vertebrate cerebellum receives proprioceptive input from the spinal cord via a spinocerebellar tract and vestibular input from the inner ear. However, in mammals, the cerebellum receives input from a higher center, such as the neocortex, relayed through pontine nuclei, and has an important role in cognitive function. In the vertebrate lineage, the cerebellum seems to have evolved in relationshp to sensory perception and motor regulation because some vertebrates that inhabit aquatic or aierial environments have been found to have an enlarged cerebellum. In particular, the cerebellum in some vertebrates, including chondrichthyans, avians, and pterosaurs, is highly expanded and has elongated features (Fig. 9.1) (Voogd and Glickstein 1998; Witmer et al. 2003). In those animals, the cerebellum receives strong input from the vestibular and spinocerebellar tracts. Pterosaurs also have a quite enlarged flocculus, which might receive inputs from well-developed semicircular canals (Witmer et al. 2003). Because of this specialized cerebellum, pterosaurs with an extremely large body size may have been able to fly. Interestingly, mormyrids, which are a kind of actinopterygian, have a remarkably enlarged cerebellar region (valvula cerebelli), which is thought to have evolved with their electroreception (Nieuwenhuys 1967a; Voogd and Glickstein 1998). Mormyrids also process signals from an electrosensory organ in the electrosensory lobe of the medulla oblongata and in a region of the midbrain called the exterolateral nucleus, where the timing of responses of knollenorgans, a kind of electroreceptors, is compared to extract information about electric signals (Carlson et al. 2011). By contrast, in agnathans, the cerebellum is less prominent (Fig. 9.6) (Nieuwenhuys 1967a), although fossil osteostraci *Norselaspis* or the galeaspid *Shuyu* have a bulge on the dorsal metencephalon (Fig. 9.4) (Janvier 2002; Gai et al. 2011). Because these animals appear to be more closely related to gnathostomes than to cyclostomes (Fig. 9.4) (Davis et al. 2012; Oisi et al. 2013), a gnathostome-type cerebellum might have evolved in early vertebrates after a split from the cyclostome lineage. Extant lampreys possess a small bulge on the roof of the hindbrain called a corpus cerebelli in which Purkinje-like cells can be observed. However, these cells are intermingled with the granule-like cells and are not stained by antibodies to zebrin, which recognize gnathostome Purkinje cells (Lannoo and Hawkes 1997). Moreover, hagfishes have only the uncertain sign of a cerebellum in the hindbrain (Larsell 1947; Bone 1963; Nieuwenhuys 1967a; Kusunoki et al. 1982), and cyclostomes lack some cerebellum-related nuclei, such as the red nucleus, that are identified in gnathostomes (Butler and Hodos 2005). Because the red nucleus or precerebellar nuclei have been identified in sharks (Pose-Mendez et al. 2014), organization of a cerebellum that possesses molecular, Purkinje cell, and granular layers with a precerebellar system may possibly have been established after a split between cyclostomes and gnathostomes. The evolution of paired appendages, which enable animals to move accurately in three-dimensional space, may have accelerated the evolution of the cerebellum.

In developing gnathostomes, the cerebellum emerges from the dorsal side of the hindbrain, which corresponds to the first rhombomere segment (rhombomere 1; Figs. 9.3 and 9.6), and cells in the cerebellar primordium are marked by some transcription factors (Butts et al. 2014). In mammals, *Ptf1a*-expressing cells migrate dorsally and differentiate into Purkinje cells. Conversely, Atoh1-positive cells migrate tangentially in the rhombic lip to form an external granular layer. Subsequently, these cells expressing NeuroD migrate ventrally through the Purkinje cell layer and form an internal granular layer. The cellular organization and expression of *Atoh1* orthologues in the cerebellar primordium have been observed in many gnathostome species including amniotes, teleosts, and sharks (Chaplin et al. 2010), suggesting that the developmental plan to produce a cerebellar cytoarchitecture could date back at least to the common ancestor of gnathostomes.

However, presumptive cognates of *Atoh1* and *Ptf1a* could be identified in the lamprey, and their expression domains are located in the dorsal side of the hindbrain, which corresponds to the rhombic lip (Fig. 9.6) (Sugahara et al. 2016). Thus, a certain genetic background underlying the acquisition of the cerebellum might have been already established in the latest common ancestor of vertebrates, although the cerebellum proper is less prominent in extant cyclostomes.

9.9.3 Midbrain (Mesencephalon)

The midbrain is located anterior to the rhombencephalon and is composed mainly of three parts: the isthmus, the optic tectum, and the tegmentum (Fig. 9.2) (Butler and Hodos 2005). Among these regions, the optic tectum (a region homologous to the mammalian superior colliculus) is characterized by a laminated structure in which extensive retinal input enters the superficial layers, whereas other sensory inputs, such as the somatosensory system, terminate in the deep layers. The optic tectum is highly developed in some vertebrate groups such as teleosts. In amniotes, including birds and reptiles, it represents highly expanded and elongated features (Figs. 9.1 and 9.6). Thus, the optic tectum in those animals must have evolved to serve as a processing center for several sensory systems during the process of adaptation to terrestrial environments. The torus semicircularis, which is located on the ventral side of the optic tectum, is thought to be a homologue of the mammalian inferior colliculus. In aquatic anamniotes, it receives input mainly from the lateral line and auditory systems (in fish, it also receive inputs from somatosensory signals coming from the spinal cord and the trigeminal system) (Yamamoto et al. 2010). In some avian species, such as owls, the midbrain center of the auditory system possesses an elegant system to identify the position of prey in a dark environment (Konishi 2006). Tegmental regions contain some nuclei such as the red nucleus and substantia nigra, both of which are important in motor coordination, and are connected to the cerebellum and basal ganglia, respectively. In vertebrates studied so far, the midbrain is induced by an FGF8 signal that is secreted from the anterior part of the MHB, in which *Otx2* is expressed (Fig. 9.6). In the early developmental period, expression domains of *Otx2* and *Gbx2* overlap in the presumptive MHB. Then, in the course of development, these two genes restrict their expression and represent a clear expression boundary in MHB (Broccoli et al. 1999; Millet et al. 1999; Katahira et al. 2000; Matsunaga et al. 2002). Subsequently, the midbrain and forebrain are induced in the *Otx2* expression domain, whereas the hindbrain and cerebellum are differentiated in the *Gbx2* domain (Fig. 9.6). The midbrain is also characterized by the presence of *Pax7* and the absence of *Pax6* transcripts (Ferran et al. 2007). Thus, these *Pax* expressions represent an important landmark in the vertebrate midbrain primordium (Fig. 9.6).

In the lamprey, the cognate of *Fgf8/17* (*LjFgf8/17*) is restricted to the MHB, and the orthologues of *Otx2* (*LjOtxA*) and *Gbx2* (*LjGbxA*) represent a clear expression boundary in the isthmic region in tailbud-stage embryos (stages 26–27) (Fig. 9.6)

(Murakami et al. 2001; Takio et al. 2007). In addition, *LjPax6* is absent but *LjPax3/7* is present in the *LjOtxA* expression domain. These observations strongly indicate that the developing lamprey brain has a midbrain primordium similar to those of other vertebrates (Fig. 9.6). Although the lamprey optic tectum has less clear lamination in the ammocoete larva, the adult animal, which develops camera eyes during metamorphosis, has an expanded optic tectum in which a clearly laminated organization is visible, as in other vertebrates (Kennedy and Rubinson 1977; Nieuwenhuys 1997). Moreover, in the hagfish, three layers can be observed in the optic tectum (Iwahori et al. 1996). Thus, a laminated optic tectum is thought to have been established in the common ancestor of vertebrates. The torus semicircularis is observed in the lamprey and receives input from the acousticolateral line system, optic tectum, diencephalon, and reticular formation (Gonzalez et al. 1999). Conversely, it is not observed in the hagfish (Butler and Hodos 2005). These observations indicate that it might have been established in the common ancestor of vertebrates, and then secondarily degenerated in the hagfish lineage because of a regression of the lateral line system. In the tegmentum of cyclostomes, the substantia nigra has been identified in the lamprey and is thought to link to the telencephalon and be essential in motor coordination (Fig. 9.8) (Stephenson-Jones et al. 2011).

The mesencephalic nucleus of the trigeminal nerve that regulates motoneurons innervating jaw muscles has not been identified in either the lamprey or the hagfish (Nieuwenhuys 1997). Because this nucleus has been identified in many gnathostomes, including chondrichthyes (Fig. 9.2) (Witkovsky and Roberts 1975; MacDonnell 1980), it is considered to have evolved in accordance with the establishment of jaws.

9.9.4 Retinotectal Projection

In the chick, retinal axons project topographically in the specific region of the optic tectum. In this retinotectal projection process, repulsive interaction between the Eph receptor and the ephrin ligand is crucial (Fig. 9.6). Specifically, *EphAs* and *ephrinAs* are expressed in the retina and the optic tectum, respectively, with anteroposterior gradients in each region. The formation of an *ephrinA* gradient in the tectum is mediated by En1/2 expressed in the MHB (Logan et al. 1996; Shigetani et al. 1997). In the developmental process, EphA3-expressing retinal axons are repelled by ephrinA5-expressing tectal neurons and result in the topographic sorting of retinal axons. A recent study of lampreys showed that orthologues of *EphA* familiy (*LjEphC* and *LjEphB*) are expressed in the retina during metamorphosis, simultaneously with the development of their camera eyes (Fig. 9.6; Suzuki et al. 2015a). This finding indicates that the molecular mechanism underlying the retinotectal projection may have been established in the common ancestor of vertebrates. Interestingly, both the lateral eye in the larval lamprey (ammnocoete larva) and the frontal eye in amphioxus project to the caudal prosencephalic region

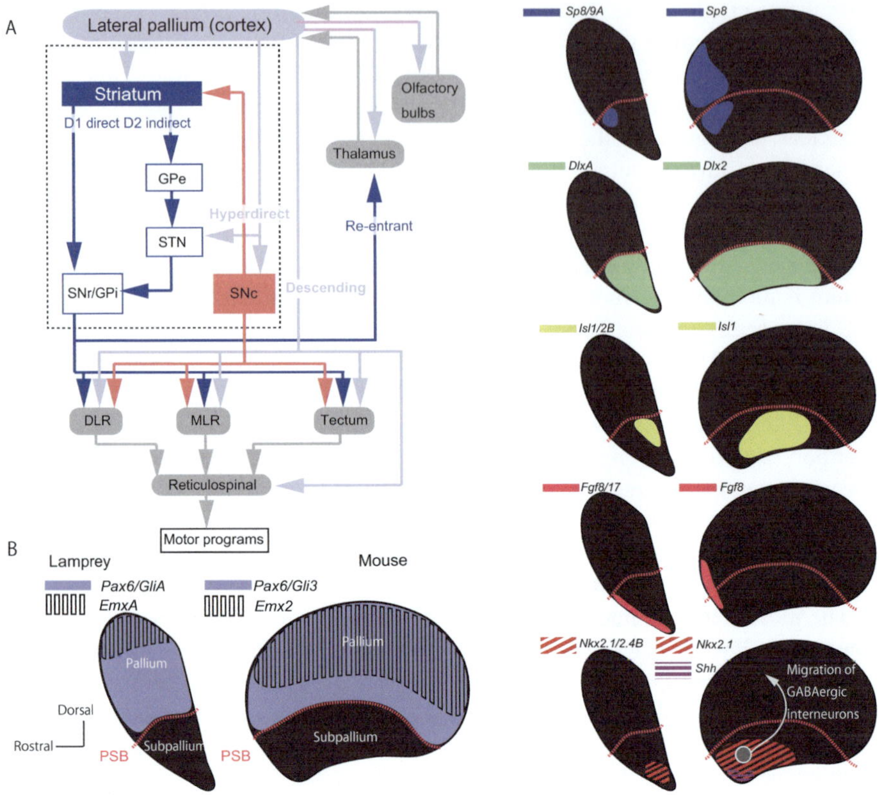

Fig. 9.8 (**a**) Presumptive neuronal connections of the lamprey subpallium (After Stephenson-Jones et al. 2011; Ocana et al. 2015). (**b**) Expression patterns of transcription factors and signaling morphogens in the developing mouse and lamprey (After Murakami et al. 2001; Sugahara et al. 2013, 2016). *DLR* diencephalic locomotor region, *GPe* globus pallidus externa, *GPi* globus pallidus interna, *MLR* mesencephalic locomotor region, *SNr* substantia nigra pars reticulata, *SNc* substantia nigra pars compacta, *STN* subthalamic nucleus

marked by *Pax6*, indicating that a light-detecting visual center in the ancestral vertebrates is not located in the optic tectum, where *Pax6* expression is negative (Suzuki et al. 2015b). Thus, the visual system of the larval lamprey provides an important model to study the evolutionary transition of visual centers.

9.9.5 Diencephalon

In early embryogenesis, the vertebrate forebrain (the anterior end of the neural tube) differentiates into a rostral telencephalon (including the pallium or cortex and subpallium or basal ganglia) and a diencephalon. Thus, the diencephalon is located posterior (caudal) to the telencephalon and rostral to the midbrain (Fig. 9.2). In recent morphological and embryological points of view, the vertebrate diencephalon has been subdivided into the pretectum, thalamus, prethalamus, and hypothalamus. However, it is important to note that a recent prosomeric model proposed that the hypothalamus belongs to the same prosomeres as hypothlamo-telencephalic prosomeres (Puelles and Rubenstein 2003, 2015). To maintain continuity with the terms in previous anatomical studies, this issue follow the traditional view, in which the hypothalamus is included in the diencephalon.

The pretectum is closely associated with the visual system and is also marked by the posterior commissure on the dorsal side (Figs. 9.5 and 9.6). The thalamus has an epiphysis that is identified in fossil agnathans (Janvier 2002) and has a parapineal organ or parietal eye in some lineages. The thalamus also represents important relay nuclei, which connect nuclei of several sensory system to the telencephalon. Thus, the vertebrate thalamus may have evolved to serve as a relay center for several sensory modalities, although the olfactory sense reaches the telencephalon directly without a relay through the thalamus. These diencephalic regions have originated from prosomeres (p), in which P1, P2, and P3 give rise to the pretectum, thalamus, and prethalamus, respectively (Fig. 9.3) (Puelles and Rubenstein 2003). As noted earlier (see *5.2*), the boundary between P2 and P3 corresponds to the *Zli*, in which SHH is crucial in the patterning of P2 and P3 compartments (Fig. 9.3) (Echevarria et al. 2003; Kiecker and Lumsden 2005). The hypothalamus regulates hormone production and involuntary autonomic responses. The primordium of the hypothalamus is defined by the expression of *Nkx2.1* (Kimura et al. 1996).

Prosomeric compartments have been identified in the adult lamprey by detailed morphological observation (Pombal et al. 2009). Moreover, in the development period, the lamprey has a posterior commissure on the posterior side of the diencephalon and epiphysis and habenular commissure on the thalamus, as do the gnathostomes (Barreiro-Iglesias et al. 2008). In addition to these morphological traits, the expression pattern of regulatory genes specifying the diencephalon primordium in vertebrates is similar in lampreys and gnathostomes. Specifically, *Pax6* is expressed on the dorsal side of the diencephalic primordium, overlapping with *Dlx* in the anterodorsal part (Murakami et al. 2001). The prethalamic eminence, a part of the presumptive P3, has been shown by the expression of *Lhx1/5* transcripts (Osorio et al. 2006). In particular, a lamprey cognate of *Shh* (*LjHhA*) is expressed in the presumptive *Zli*, as occurs in gnathostomes (Fig. 9.3). In the hypothalamus, a lamprey homologue of *Nkx2.1* (*LjNkx2.1*) is expressed in the primordium of the hypothalamic region in which *LjHhA* and *LjNkx2.1* expression domains overlap, as seen in the mouse and the zebrafish (Rohr et al. 2001; Wullimann and Mueller 2004). Moreover, molecular asymmetries between the left and right developing habenulae were identified in catshark and lamprey (Lagadec et al. 2015). These observations strongly suggest that the lamprey has a diencephalic compartment comparable to those of gnathostomes, although the diencephalon proper is less prominent in cyclostomes than in gnathostomes.

9.10 Telencephalon

9.10.1 *Pallium and Subpallium*

Evolution of the cognitive center is thought to be a crucial phase in the adaptive radiation of animals. In vertebrates, the morphology of brains and neuron circuits seems to have evolved in relationship to their sensory and behavioral diversification. In particular, the telencephalon, located at the most anterior part of the CNS, was found to show remarkable morphological diversity among vertebrate groups (Fig. 9.1). For example, the telencephalon in some vertebrates, including hagfishes, chondrichthyes, and amniotes, has highly expanded and elongated features (Nieuwenhuys 1997). Because the telencephalon of cyclostomes or chondrichthyes has large olfactory bulbs with a number of olfactory fibers, it may have evolved to serve as a processing center for olfaction in the initial stage of vertebrate evolution. By contrast, the highly expanded telencephalon in amniotes appears to be correlated with their well-organized integrative and cognitive functions, which lead to physiological and behavioral adaptations to the terrestrial environment. In mammals and birds, the telencephalon (also called the cerebrum) is crucial in memory, attention, sensory integration, and voluntary motor control (Medina and Reiner 2000; Butler and Hodos 2005; Jarvis et al. 2005, 2013). During development, the telencephalon of gnathostomes is subdivided into dorsal and ventral portions, called the pallium and the subpallium, respectively (Figs. 9.2 and 9.8). In mammals, the pallium generally gives rise to the hippocampus, neocortex, olfactory cortex, and a part of the amygdala (cortical amygdala). Alternatively, the subpallium induces the subpallial part of the amygdala, striatum, and pallidum (Puelles et al. 2000). Pallial and subpallial components are marked by differential expression patterns of some transcription factor-encoding genes. Among these, *Pax6* and *Dlx2* are expressed in primordial pallium and subpallium, respectively (Puelles et al. 2000; Bishop et al. 2000, 2002).

9.10.2 *Origin of the Telencephalon*

Although a telencephalon is identified in all extant vertebrate species, its evolutionary origin remains unknown. Some researchers suggest that its origin dates back to the common ancestor of deuterostomes and protostomes (Tomer et al. 2010) because of the similarities in cellular composition and molecular architecture of the pallium and mushroom body in annelids. These similarities may be explained by the concept of deep homology (see earlier). In chordates, the amphioxus has a tiny bulge called a cerebral vesicle, which is located in the most anterior part of the neural tube (Lacalli et al. 1994). Based on the expression patterns of some key marker genes, this bulge has been regarded as a homologue of the vertebrate brain (Holland and Short 2008; Venkatesh et al. 1999; Irimia et al. 2010). As a signature for the telencephalic primordium, the expression of *FoxG1*, a marker for the vertebrate telencephalon, has been identified in the amphioxus cerebral vesicle. However, its domain of expression is restricted to the surrounding of the frontal eye, which does not appears to be homologous to the telencephalon (Toresson et al. 1998). Moreover, the amphioxus cerebral vesicle seems to lack an olfactory bulb, a characteristic feature of the vertebrate telencephalon. In addition, although some orthologous genes encoding vertebrate olfactory receptor proteins are found in the amphioxus, the olfactory epithelium that connects to the olfactory bulb in vertebrates is less prominent in this animal (Lacalli 2004; Niimura 2009). Taken together, the vertebrate telencephalon probably represents an evolutionary novelty (or a synapomorphy) in the vertebrate lineage. Fossil records show that some fossil vertebrates represent signatures of the telencephalon: the osteostracan *Norselaspis* possesses an olfactory organ, as in lampreys (Janvier 2002), and the galeaspid *Shuyu* apparently possesses a paired olfactory bulb (Gai et al. 2011). Moreover, the earliest agnathan has nasal sacs between its eyes, implying the presence of an olfactory epithelium (Shu et al. 2003; Morris and Caron 2014). Thus, the origin of the telencephalon should date back to the last common ancestor of vertebrates.

9.10.3 Telencephalon of Cyclostomes

The telencephalons of lampreys and hagfishes are characterized by the huge olfactory bulb that is connected to the olfactory epithelium. In the lamprey olfactory bulb, glomeruli lie in a single layer, and mitral cell-like neurons are located in it, although these neurons have a different morphology and position from those in gnathostomes. In addition, granule cell-like neurons are also observed (Nieuwenhuys 1967b). Thus, the lamprey may possess components of the main olfactory system, which receives input from the olfactory epithelium and connects to the pallium, as found in other vertebrates. By contrast, differing from many other vertebrates, including tetrapods and lungfishes, the lamprey lacks an accessory olfactory (vomeronasal) system in which neurons in the accessory olfactory bulb receive inputs from the vomeronasal epithelium and connect to the hypothalamus (Gonzalez et al. 2010). However, a recent study suggests that the lamprey olfactory bulb receives axons originating from a specific part of the olfactory epithelium, which is thought to be homologous to the vomeronasal epithelium in gnathostomes. Then, bulbar neurons receiving projections from the presumed vomeronasal epitherium send fibers to the hypothalamus (Chang et al. 2013). This system is thought to be a precursor to the vomeronasal system in gnathostomes. If so, the origin of both the main and accessory olfactory systems may date back to the common ancestor of vertebrates.

The pallium of adult cyclostomes represents a unique morphology. The lamprey medial pallium (primordium hippocampi, phip) displays an unevaginated shape, although other pallial parts represent evagination as seen in many gnathostomes (Nieuwenhuys 1997). In adult hagfishes, the cerebral hemisphere and olfactory bulb are much larger than those of the lamprey. Characteristically, distinct lamination was observed in the hagfish pallium, in which five layers were identified (Nieuwenhuys 1967b). However, identification of the telencephalic subregions is extremely difficult because of a significant reduction of the ventricular systems in addition to the highly differentiated organization of the pallium (Conel 1929).

It has been assumed that the lamprey telencephalon mainly processes olfactory information because it has an enlarged olfactory bulb (OB). In fact, many parts of the lamprey pallium receive extensive connections from the OB. Recent studies have shown that the lamprey pallium receives input not only from the OB, but also from other regions of the brain, such as the thalamus (Northcutt and Wicht 1997). In addition, Ocana and colleagues identified the lamprey pallial region where the projection neurons send axons toward the subnuclei of the basal ganglia, midbrain, and brainstem similar to that of mammalian motor cortex (Fig. 9.8) (Ocana et al. 2015). These studies have suggested that the afferent and efferent neuronal connections of the pallium that are a prerequisite for the subsequent construction of a integrative center are likely to be present in the common ancestor of vertebrates.

9.10.4 Developmental Plan for the Pallium

Pax6 and *Dlx2* homologues of the chondrichthyan shark (*Scyliorhinus torazame*) and cyclostome lamprey (*Lethenteron japonicum*) are expressed exclusively along the dorsoventral axis of the telencephalic primordium (Fig. 9.8) (Sugahara et al. 2013). Moreover, the expression domains of *Emx* cognates in those animals are included in the *Pax6* expression domain in the dorsal telencephalon, as found in osteichthyans (Murakami et al. 2001; Derobert et al. 2002; Tank et al. 2009). This similarity indicates that the pallio–subpallium boundary, and at least two distinct pallial domains, were probably established in the most recent common ancestor of vertebrates (Fig. 9.8).

9.10.5 Developmental Plan for the Subpallium

The vertebrate subpallium can be divided into two major subdivisions, the striatum and pallidum, both of which represent a part of the basal ganglia and are crucial in motor function in the adult brains of gnathostomes (Reiner 2007). During embryogenesis, the striatum and pallidum arise from the lateral and medial ganglionic eminences (LGE and MGE), respectively (Puelles et al. 2000). The MGE produces precursors of inhibitory interneurons that are marked by γ-aminobutyric acid (GABA), and these GABA-positive cells migrate tangentially toward the dorsal pallium during embryonic development (Fig. 9.8) (Marin and Rubenstein 2001). Recent molecular analysis using rodents has shown that the LGE is marked by the expression of *Dlx*, *Gsh1/2*, *Sp8*, and *Isl1*, whereas the MGE is characterized by the expression of *Dlx*, *Nkx2.1*, *Lhx6/7/8*, and *Shh*, and comparative embryology has indicated that the *Dlx-*, *Gsh1/2-*, and *Sp8*-expressing subpallium is commonly observed in tetrapods, teleosts, and lampreys (Fig. 9.8) (Moreno et al. 2009; Sugahara et al. 2011). These results suggest that the LGE might have been present in an evolutionary ancestral (default) stage. A recent comparative study has shown both *Nkx2.1* and *Shh* expression in the ventral telencephalon in the shark (*Scyliorhinus canicula*), which implies the presence of an MGE (presumptive pallidum) (Quintana-Urzainqui et al. 2012). In addition to the gene expression, GABA-positive cells are distributed not only in the subpallium but also in the pallial area of the shark (Carrera et al. 2008). Therefore, the two subpallial domains are likely to have been acquired by the most recent common ancestor of the extant gnathostome lineages. However, the evolutionary origin of the MGE was uncertain because neither *Nkx2.1* nor *Shh* or *Lhx6/7* cognates could be observed in the lamprey subpallium (Fig. 9.7) (Murakami et al. 2001; Sugahara et al. 2011). This absence may indicate that lampreys lack an equivalent region of gnathostome MGE (Murakami et al. 2005). However, recent studies have identified the pallidum in the lamprey subpallium and GABA-immunoreactive cells in the pallium in addition to the subpallium (Stephenson-Jones et al. 2011; Pombal et al. 2011). These studies indicate the presence of MGE derivatives in the lamprey, although the embryonic origin of the pallidum-like region remains unclear. It is also important to clarify whether the GABA-positive/Dlx-positive cells in the lamprey pallium represent interneurons originating from the MGE.

Importantly, two extra lamprey *Nkx2.1* orthologous genes (*Nkx2.1/2.4B* and *Nkx2.1/2.4C*) were recently identified from the *L. japonicum* draft genome sequence. Surprisingly, these genes were expressed in the lamprey subpallium, strongly suggesting the presence of the MGE (Fig. 9.8) (Sugahara et al. 2016). In addition, the hagfish cognate of *Nkx2.1* was also identified and its expression domain was located in a rostral telencephalon, in which the gnathostome MGE would be formed (Sugahara et al. 2016). Taken together, vertebrate telencephalon genoarchitecture is dated back to the latest common ancestor of extant vertebrates (Fig. 9.9).

9.10.6 FGF Signaling from the Anterior Telencephalon

As already mentioned (see 5.2), fibroblast growth factors (FGFs) that are secreted from the ANR, which subsequently develops into a CP, are crucial in the patterning of the vertebrate telencephalon (Fig. 9.8). In lamprey embryogenesis, the lamprey orthologue of *Fgf8* (*LjFgf8/17*) is expressed in the anteroventral region of the telencephalon, which may possibly correspond to the CP of mammals (Sugahara et al. 2011). However, in contrast to gnathostomes, the expression domain is restricted to the ventral side, suggesting that lamprey *Fgf8/17* is expressed only in the presumptive subpallium (Fig. 9.8). In addition, *Sp8*, a downstream target gene in the gnathostome pallium, could not be observed in the lamprey pallium (Fig. 9.8) (Sugahara et al. 2011). These results suggest that the FGF-mediated

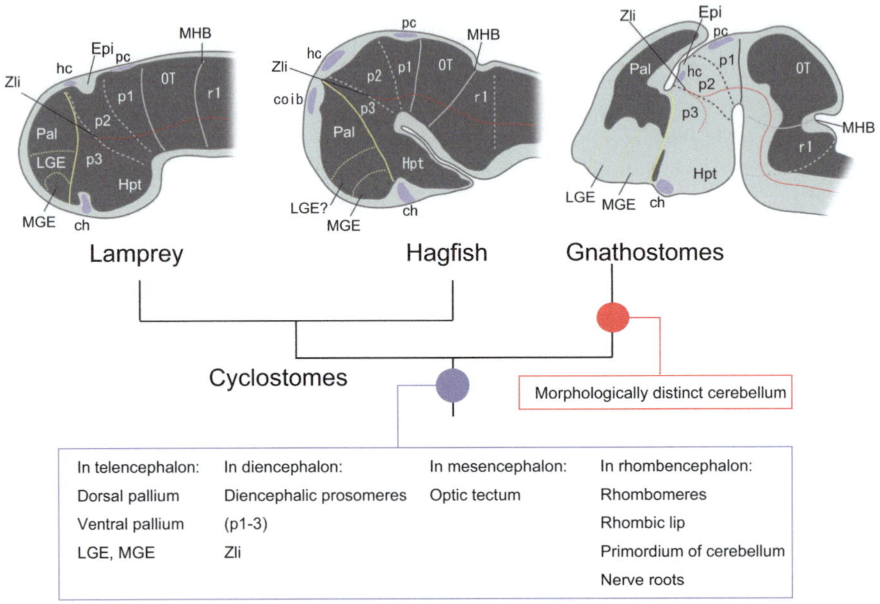

Fig. 9.9 Evolutionary process of the vertebrate brain (After Sugahara et al. 2016). *ch* optic chiasm, *coib* interbulbar commissure, *hc* habenular commissure. See Figs. 9.2, 9.3, 9.4, and 9.5 for abbreviations

pallium patterning mechanism might have been established in the gnathostome lineage, although the lamprey subpallium was patterned by an FGF-based gene network, as occurs in gnathostomes. Thus, to identify the origin of FGF signaling, it is also important to study the development of the hagfish brain.

9.11 Conclusions

The vertebrate brain has been established on a preexisting neuroectoderm as a vertebrate-specific plan. Recent molecular approaches using cyclostomes have demonstrated that the basic arrangement of the neural tube has been conserved through vertebrate evolution under a strict developmental system, in which spatially and temporally regulated gene expression has a crucial role. Some of the combinatory expression of regulatory genes may be inherited from an early stage of animal evolution as a deep homology. By contrast, integrative centers, including the cerebellum, mesencephalon, and telencephalon, are thought to have been established in vertebrates as an evolutionary novelty (Fig. 9.9). The flexibility of the brain developmental program may result in morphological and physiological diversity and enable vertebrates to adapt to complex environments. However, knowledge of the evolutionary process of their brain centers is still fragmentary, although recent comparative approaches have demonstrated additional modification of gene regulatory networks. Further advances in evolutionary developmental biology will provide insight into the origin and evolution of the vertebrate brain.

Acknowledgments We thank Dr. Fumiaki Sugahara, Moe Kawara and Saori Yamakami for technical support and valuable discussions. We thank all past and present members of the YM laboratory for support and constructive discussions. Work in YM's laboratory was supported by RIKEN, Kobe, Japan and the Japan Society for the Promotion of Science (JSPS; grant number 24650178 and 26430018).

References

Abi-Rached L, Gilles A, Shiina T, Pontarotti P, Inoko H (2002) Evidence of en bloc duplication in vertebrate genomes. Nat Genet 31:100–105

Abitua PB, Wagner E, Navarrete IA, Levine M (2012) Identification of a rudimentary neural crest in a non-vertebrate chordate. Nature 492:104–107

Anderson RB, Key B (1999) Novel guidance cues during neuronal pathfinding in the early scaffold of axon tracts in the rostral brain. Development 126:1859–1868

Balanoff AM, Bever GS, Rowe TB, Norell MA (2013) Evolutionary origins of the avian brain. Nature 501:93–96

Barreiro-Iglesias A, Villar-Cheda B, Abalo XM, Anadon R, Rodicio MC (2008) The early scaffold of axon tracts in the brain of a primitive vertebrate, the sea lamprey. Brain Res Bull 75:42–52

Bassham S, Postlethwait JH (2005) The evolutionary history of placodes: a molecular genetic investigation of the larvacean urochordate *Oikopleura dioica*. Development 132:4259–4272

Beard J (1889) On the early development of *Lepidosteus osseus*: preliminary notice. Proc R Soc Lond 46:108–118

Beccari L, Marco-Ferreres R, Bovolenta P (2013) The logic of gene regulatory networks in early vertebrate forebrain patterning. Mech Dev 130:95–111

Bergquist H, Källén B (1953) On the development of neuromeres to migration areas in the vertebrate cerebral tube. Act Anat 18:65–73

Bernhardt RR, Chitnis AB, Lindamer L, Kuwada JY (1990) Identification of spinal neurons in the embryonic and larval zebrafish. J Comp Neurol 302:603–616

Betancur P, Bronner-Fraser M, Sauka-Spengler T (2010) Assembling neural crest regulatory circuits into a gene regulatory network. Annu Rev Cell Dev Biol 26:581–603

Bishop KM, Goudreau G, O'Leary DD (2000) Regulation of area identity in the mammalian neocortex by Emx2 and Pax6. Science 288:344–349

Bishop KM, Rubenstein JL, O'Leary DD (2002) Distinct actions of Emx1, Emx2, and Pax6 in regulating the specification of areas in the developing neocortex. J Neurosci 22:7627–7638

Bone Q (1963) The central nervous system. In: Bodal A, Hinge R (eds) The biology of myxine. Universitetsforlaget, Oslo, pp 50–91

Braun CB (1998) Schreiner organs: a new craniate chemosensory modality in hagfishes. J Comp Neurol 392:135–163

Broccoli V, Boncinelli E, Wurst W (1999) The caudal limit of Otx2 expression positions the isthmic organizer. Nature 401:164–168

Butler AB, Hodos W (2005) Comparative vertebrate neuroanatomy: evolution and adaptation, 2nd edn. Wiley, Hoboken

Butts T, Modrell MS, Baker CV, Wingate RJ (2014) The evolution of the vertebrate cerebellum: absence of a proliferative external granule layer in a non-teleost ray-finned fish. Evol Dev 16:92–100

Carlson BA, Hasan SM, Hollmann M, Miller DB, Harmon LJ, Arnegard ME (2011) Brain evolution triggers increased diversification of electric fishes. Science 332:583–586
Carrera I, Ferreiro-Galve S, Sueiro C, Anadón R, Rodríguez -Moldes I (2008) Tangentially migrating GABAergic cells of subpallial origin invade massively the pallium in developing sharks. Brain Res Bull 75:405–409
Carroll SB (2008) Evo-devo and an expanding evolutionary synthesis: a genetic theory of morphological evolution. Cell 134:25–36
Cecchi C (2002) Emx2: a gene responsible for cortical development, regionalization and area specification. Gene 291:1–9
Chang S, Chung-Davidson YW, Libants SV, Nanlohy KG, Kiupel M, Brown CT, Li W (2013) The sea lamprey has a primordial accessory olfactory system. BMC Evol Biol 13:172
Chaplin N, Tendeng C, Wingate RJ (2010) Absence of an external germinal layer in zebrafish and shark reveals a distinct, anamniote ground plan of cerebellum development. J Neurosci 30:3048–3057
Chitnis AB, Kuwada JY (1990) Axonogenesis in the brain of zebrafish embryos. J Neurosci 10:1892–1905
Clarke JD, Lumsden A (1993) Segmental repetition of neuronal phenotype sets in the chick embryo hindbrain. Development 118:151–162
Coghill GE (1914) Correlated anatomical and physiological studies on the growth of the nervous system of Amphibia. 1. The afferent system of the trunk of *Amblystoma*. J Comp Neurol 24:161–233
Conel JL (1929) The development of the brain of *Bdellostoma stouti*. 1. External growth changes. J Comp Neurol 47:343–403
Cong P, Ma X, Hou X, Edgecombe GD, Strausfeld NJ (2014) Brain structure resolves the segmental affinity of anomalocaridid appendages. Nature 513:538–542
Cooke J, Moens C, Roth L, Durbin L, Shiomi K, Brennan C, Kimmel C, Wilson S, Holder N (2001) Eph signalling functions downstream of Val to regulate cell sorting and boundary formation in the caudal hindbrain. Development 128:571–580
Crow KD, Smith CD, Cheng JF, Wagner GP, Amemiya CT (2012) An independent genome duplication inferred from Hox paralogs in the American paddlefish: a representative basal ray-finned fish and important comparative reference. Genome Biol Evol 4:937–953
Davis SP, Finarelli JA, Coates MI (2012) *Acanthodes* and shark-like conditions in the last common ancestor of modern gnathostomes. Nature 486:247–250
De Robertis EM, Sasai Y (1996) A common plan for dorsoventral patterning in Bilateria. Nature 380:37–40
Denes AS, Jékely G, Steinmetz PR, Raible F, Snyman H, Prud'homme B, Ferrier DE, Balavoine G, Arendt D (2007) Molecular architecture of annelid nerve cord supports common origin of nervous system centralization in Bilateria. Cell 129:277–288
Derobert Y, Plouhinec JL, Sauka-Spengler T, Le Mentec C, Baratte B, Jaillard D, Mazan S (2002) Structure and expression of three Emx genes in the dogfish *Scyliorhinus canicula*: functional and evolutionary implications. Dev Biol 247:390–404
Devine CA, Key B (2008) Robo-Slit interactions regulate longitudinal axon pathfinding in the embryonic vertebrate brain. Dev Biol 313:371–383
Diaz-Regueira S, Anadon R (2000) Calretinin expression in specific neuronal systems in the brain of an advanced teleost, the grey mullet (*Chelon labrosus*). J Comp Neurol 426:81–105
Doldan MJ, Prego B, Holmqvist B, Helvik JV, de Miguel E (2000) Emergence of axonal tracts in the developing brain of the turbot (*Psetta maxima*). Brain Behav Evol 56:300–309
Dupret V, Sanchez S, Goujet D, Tafforeau P, Ahlberg PE (2014) A primitive placoderm sheds light on the origin of the jawed vertebrate face. Nature 507:500–503
Easter SS Jr, Ross LS, Frankfurter A (1993) Initial tract formation in the mouse brain. J Neurosci 13:285–299
Echelard Y, Epstein DJ, St.-Jacques B, Shen L, Mohler J, McMahon JA, McMahon AP (1993) Sonic hedgehog, a member of a family of putative signaling molecules, is implicated in the regulation of CNS polarity. Cell 75:1417–1430

Echevarria D, Vieira C, Gimeno L, Martinez S (2003) Neuroepithelial secondary organizers and cell fate specification in the developing brain. Brain Res Brain Res Rev 43:179–191
Erzurumlu RS, Murakami Y, Rijli FM (2010) Mapping the face in the somatosensory brainstem. Nat Rev Neurosci 11:252–263
Ferran JL, Sanchez-Arrones L, Sandoval JE, Puelles L (2007) A model of early molecular regionalization in the chicken embryonic pretectum. J Comp Neurol 505:379–403
Fritzsch B, Northcutt RG (1993) Cranial and spinal nerve organization in amphioxus and lampreys: evidence for an ancestral craniate pattern. Acta Anat 148:96–109
Gai Z, Donoghue PC, Zhu M, Janvier P, Stampanoni M (2011) Fossil jawless fish from China foreshadows early jawed vertebrate anatomy. Nature 476:324–327
Gaufo GO, Wu S, Capecchi MR (2004) Contribution of Hox genes to the diversity of the hindbrain sensory system. Development 131:1259–1266
Gehring WJ (1996) The master control gene for morphogenesis and evolution of the eye. Genes Cells 1:11–15
Geisen MJ, Di Meglio T, Pasqualetti M, Ducret S, Brunet JF, Chedotal A, Rijli FM (2008) Hox paralog group 2 genes control the migration of mouse pontine neurons through slit-robo signaling. PLoS Biol 6, e142
Gess RW, Coates MI, Rubidge BS (2006) A lamprey from the Devonian period of South Africa. Nature 443:981–984
Gilland E, Baker R (1993) Conservation of neuroepithelial and mesodermal segments in the embryonic vertebrate head. Acta Anat (Basel) 148:110–123
Gonzalez MJ, Yanez J, Anadon R (1999) Afferent and efferent connections of the torus semicircularis in the sea lamprey: an experimental study. Brain Res 826:83–94
Gonzalez A, Morona R, Lopez JM, Moreno N, Northcutt RG (2010) Lungfishes, like tetrapods, possess a vomeronasal system. Front Neuroanat 4
Guillemot F (2007) Spatial and temporal specification of neural fates by transcription factor codes. Development 134:3771–3780
Hanneman E, Trevarrow B, Metcalfe WK, Kimmel CB, Westerfield M (1988) Segmental pattern of development of the hindbrain and spinal cord of the zebrafish embryo. Development 103:49–58
Hartenstein V (1993) Early pattern of neuronal differentiation in the *Xenopus* embryonic brainstem and spinal cord. J Comp Neurol 328:213–231
Heimberg AM, Cowper-Sal-lari R, Semon M, Donoghue PC, Peterson KJ (2010) microRNAs reveal the interrelationships of hagfish, lampreys, and gnathostomes and the nature of the ancestral vertebrate. Proc Natl Acad Sci USA 107:19379–19383
Higashiyama H, Kuratani S (2014) On the maxillary nerve. J Morphol 275:17–38
Hirano M, Guo P, McCurley N, Schorpp M, Das S, Boehm T, Cooper MD (2013) Evolutionary implications of a third lymphocyte lineage in lampreys. Nature 501:435–438
Hocking JC, Hehr CL, Bertolesi GE, Wu JY, McFarlane S (2010) Distinct roles for Robo2 in the regulation of axon and dendrite growth by retinal ganglion cells. Mech Dev 127:36–48
Holland LZ (2013) Evolution of new characters after whole genome duplications: insights from amphioxus. Semin Cell Dev Biol 24:101–109
Holland LZ (2015) The origin and evolution of chordate nervous systems. Philos Trans R Soc Lond B Biol Sci 370(1684):20150048
Holland LZ, Short S (2008) Gene duplication, co-option and recruitment during the origin of the vertebrate brain from the invertebrate chordate brain. Brain Behav Evol 72:91–105
Hughes A (1957) The development of the primary sensory system in *Xenopus laevis* (Daudin). J Anat 91:323–338
Hunt P, Gulisano M, Cook M, Sham MH, Faiella A, Wilkinson D, Boncinelli E, Krumlauf R (1991) A distinct Hox code for the branchial region of the vertebrate head. Nature 353:861–864
Irimia M, Pineiro C, Maeso I, Gomez-Skarmeta JL, Casares F, Garcia-Fernandez J (2010) Conserved developmental expression of Fezf in chordates and Drosophila and the origin of the *Zona Limitans Intrathalamica* (*ZLI*) brain organizer. EvoDevo 1:7
Ishikawa Y, Kage T, Yamamoto N, Yoshimoto M, Yasuda T, Matsumoto A, Maruyama K, Ito H (2004) Axonogenesis in the medaka embryonic brain. J Comp Neurol 476:240–253

Ishikawa Y, Yamamoto N, Yoshimoto M, Ito H (2012) The primary brain vesicles revisited: are the three primary vesicles (forebrain/midbrain/hindbrain) universal in vertebrates? Brain Behav Evol 79:75–83

Iwahori N, Nakamura K, Tsuda A (1996) Neuronal organization of the optic tectum in the hagfish, *Eptatretus burgeri*: a Golgi study. Anat Embryol (Berl) 193:271–279

Jacobson M (1991) Developmental neurobiology. Plenum Press, New York

Janvier P (2002) Early vertebrates. Oxford University Press, Oxford

Jarvis ED, Gunturkun O, Bruce L, Csillag A, Karten H, Kuenzel W, Medina L, Paxinos G, Perkel DJ, Shimizu T, Striedter G, Wild JM, Ball GF, Dugas-Ford J, Durand SE, Hough GE, Husband S, Kubikova L, Lee DW, Mello CV, Powers A, Siang C, Smulders TV, Wada K, White SA, Yamamoto K, Yu J, Reiner A, Butler AB (2005) Avian brains and a new understanding of vertebrate brain evolution. Nat Rev Neurosci 6:151–159

Jarvis ED, Yu J, Rivas MV, Horita H, Feenders G, Whitney O et al (2013) Global view of the functional molecular organization of the avian cerebrum: mirror images and functional columns. J Comp Neurol 521:3614–3665

Jerison HJ (1985) Animal intelligence as encephalization. Philos Trans R Soc Lond B Biol Sci 308:21–35

Jorqensen JM, Lomholt JP, Weber RE, Malte H (eds) (1997) The biology of hagfishes. Chapman & Hall, London/New York

Jorqensen JM, Shichiri M, Genese FA (1998) Morphology of the hagfish inner ear. Acta Zool 79:251–256

Katahira T, Sato T, Sugiyama S, Okafuji T, Araki I, Funahashi J, Nakamura H (2000) Interaction between Otx2 and Gbx2 defines the organizing center for the optic tectum. Mech Dev 91:43–52

Kennedy MC, Rubinson K (1977) Retinal projections in larval, transforming and adult sea lamprey, *Petromyzon marinus*. J Comp Neurol 171:465–479

Kiecker C, Lumsden A (2005) Compartments and their boundaries in vertebrate brain development. Nat Rev Neurosci 6:553–564

Kimmel CB, Sepich DS, Trevarrow B (1988) Development of segmentation in zebrafish. Development 104(suppl):197–207

Kimura S, Hara Y, Pineau T, Fernandez-Salguero P, Fox CH, Ward JM, Gonzalez FJ (1996) The T/ebp null mouse: thyroid-specific enhancer-binding protein is essential for the organogenesis of the thyroid, lung, ventral forebrain, and pituitary. Genes Dev 10:60–69

Kobayashi D, Kobayashi M, Matsumoto K, Ogura T, Nakafuku M, Shimamura K (2002) Early subdivisions in the neural plate define distinct competence for inductive signals. Development 129:83–93

Konishi M (2006) Behavioral guides for sensory neurophysiology. J Comp Physiol A Neuroethol Sens Neural Behav Physiol 192:671–676

Kontges G, Lumsden A (1996) Rhombencephalic neural crest segmentation is preserved throughout craniofacial ontogeny. Development 122:3229–3242

Koyama H, Kishida R, Goris RC, Kusunoki T (1987) Organization of sensory and motor nuclei of the trigeminal nerve in lampreys. J Comp Neurol 264:437–448

Krumlauf R, Hunt P, Sham MH, Whiting J, Nonchev S, Marshall H, Graham A, Papalopulu N, Cook M, Boncinelli E, Wilkinson D (1993) Hox genes: a molecular code for patterning regional diversity in the nervous system and branchial structures. Restor Neurol Neurosci 5:10–12

Kuraku S, Kuratani S (2006) Time scale for cyclostome evolution inferred with a phylogenetic diagnosis of hagfish and lamprey cDNA sequences. Zool Sci 23:1053–1064

Kuratani SC, Eichele G (1993) Rhombomere transplantation repatterns the segmental organization of cranial nerves and reveals cell-autonomous expression of a homeodomain protein. Development 117:105–117

Kuratani S, Horigome N (2000) Developmental morphology of branchiomeric nerves in a cat shark, *Scyliorhinus torazame*, with special reference to rhombomeres, cephalic mesoderm, and distribution patterns of cephalic crest cells. Zool Sci 17:893–909

Kuratani S, Kuraku S, Murakami Y (2002) Lamprey as an evo-devo model: lessons from comparative embryology and molecular phylogenetics. Genesis 34:175–183

Kuratani S, Murakami Y, Nobusada Y, Kusakabe R, Hirano S (2004) Developmental fate of the mandibular mesoderm in the lamprey, Lethenteron japonicum: comparative morphology and development of the gnathostome jaw with special reference to the nature of the trabecula cranii. J Exp Zool B Mol Dev Evol 302:458–468

Kusunoki T, Kadota T, Kishida R (1982) Chemoarchitectonics of the brain stem of the hagfish, Eptatretus burgeri, with special reference to the primordial cerebellum. J Hirnforsch 23:109–119

Lacalli TC (2004) Sensory systems in amphioxus: a window on the ancestral chordate condition. Brain Behav Evol 64:148–162

Lacalli TC, Holland ND, West JE (1994) Landmarks in the anterior central nervous system of amphioxus larvae. Philos Trans R Soc Lond Ser B 344:165–185

Lagadec R, Laguerre L, Menuet A, Amara A, Rocancourt C, Pericard P, Godard BG, Rodicio MC, Rodriguez-Moldes I, Mayeur H, Rougemont Q, Mazan S, Boutet A (2015) The ancestral role of nodal signalling in breaking L/R symmetry in the vertebrate forebrain. Nat Commun 6:6686

Lamborghini JE (1987) Disappearance of Rohon-Beard neurons from the spinal cord of larval *Xenopus* laevis. J Comp Neurol 264:47–55

Langille RM, Hall BK (1988) The organ culture and grafting of lamprey cartilage and teeth. In Vitro Cell Dev Biol 24:1–8

Lannoo MJ, Hawkes R (1997) A search for primitive Purkinje cells: zebrin II expression in sea lampreys (*Petromyzon marinus*). Neurosci Lett 237:53–55

Larsell O (1947) The development of the cerebellum in man in relation to its comparative anatomy. J Comp Neurol 87:85–129

Lee KJ, Jessell TM (1999) The specification of dorsal cell fates in the vertebrate central nervous system. Annu Rev Neurosci 22:261–294

Lee RK, Eaton RC, Zottoli SJ (1993) Segmental arrangement of reticulospinal neurons in the goldfish hindbrain. J Comp Neurol 329:539–556

Lichtneckert R, Reichert H (2005) Insights into the urbilaterian brain: conserved genetic patterning mechanisms in insect and vertebrate brain development. Heredity (Edinb) 94:465–477

Logan C, Wizenmann A, Drescher U, Monschau B, Bonhoeffer F, Lumsden A (1996) Rostral optic tectum acquires caudal characteristics following ectopic engrailed expression. Curr Biol 6:1006–1014

Lopez-Bendito G, Flames N, Ma L, Fouquet C, Di Meglio T, Chedotal A, Tessier-Lavigne M, Marin O (2007) Robo1 and Robo2 cooperate to control the guidance of major axonal tracts in the mammalian forebrain. J Neurosci 27:3395–3407

Lumsden A, Keynes R (1989) Segmental patterns of neuronal development in the chick hindbrain. Nature 337:424–428

MacDonnell MF (1980) Mesencephalic trigeminal nucleus in sharks. A light microscopic study. Brain Behav Evol 17:152–163

Maklad A, Reed C, Johnson NS, Fritzsch B (2014) Anatomy of the lamprey ear: morphological evidence for occurrence of horizontal semicircular ducts in the labyrinth of *Petromyzon marinus*. J Anat 224:432–446

Mallamaci A, Stoykova A (2006) Gene networks controlling early cerebral cortex arealization. Eur J Neurosci 23:847–856

Marin F, Puelles L (1995) Morphological fate of rhombomeres in quail/chick chimeras: a segmental analysis of hindbrain nuclei. Eur J Neurosci 7:1714–1738

Marin O, Rubenstein JLK (2001) A long, remarkable journey: tangential migration in the telencephalon. Nat Rev Neurosci 2:780–790

Matsunaga E, Katahira T, Nakamura H (2002) Role of Lmx1b and Wnt1 in mesencephalon and metencephalon development. Development 129:5269–5277

Mazet F, Hutt JA, Milloz J, Millard J, Graham A, Shimeld SM (2005) Molecular evidence from *Ciona intestinalis* for the evolutionary origin of vertebrate sensory placodes. Dev Biol 282:494–508

McCabe KL, Bronner-Fraser M (2009) Molecular and tissue interactions governing induction of cranial ectodermal placodes. Dev Biol 332:189–195

McCauley DW, Bronner-Fraser M (2003) Neural crest contributions to the lamprey head. Development 130:2317–2327

McCoy VE, Saupe EE, Lamsdell JC, Tarhan LG, McMahon S, Lidgard S, Mayer P, Whalen CD, Soriano C, Finney L, Vogt S, Clark EG, Anderson RP, Petermann H, Locatelli ER, Briggs DE (2016) The ‘Tully monster’ is a vertebrate. Nature 552:496–499

McLysaght A, Hokamp K, Wolfe KH (2002) Extensive genomic duplication during early chordate evolution. Nat Genet 31:200–204

Medina L, Abellan A (2009) Development and evolution of the pallium. Semin Cell Dev Biol 20:698–711

Medina L, Reiner A (2000) Do birds possess homologues of mammalian primary visual, somatosensory and motor cortices? Trends Neurosci 23:1–12

Mehta TK, Ravi V, Yamasaki S, Lee AP, Lian MM, Tay BH, Tohari S, Yanai S, Tay A, Brenner S, Venkatesh B (2013) Evidence for at least six Hox clusters in the Japanese lamprey (*Lethenteron japonicum*). Proc Natl Acad Sci USA 110:16044–16049

Mendelson B (1986) Development of reticulospinal neurons of the zebrafish II. Early axonal outgrowth and cell body position. J Comp Neurol 251:172–184

Metcalfe WK, Mendelson B, Kimmel CB (1986) Segmental homologies among reticulospinal neurons in the hindbrain of the zebrafish larva. J Comp Neurol 251:147–159

Meulemans D, Bronner-Fraser M (2004) Gene-regulatory interactions in neural crest evolution and development. Dev Cell 7:291–299

Meyer A, Van de Peer Y (2005) From 2R to 3R: evidence for a fish-specific genome duplication (FSGD). Bioessays 27:937–945

Milan FJ, Puelles L (2000) Patterns of calretinin, calbindin, and tyrosine-hydroxylase expression are consistent with the prosomeric map of the frog diencephalon. J Comp Neurol 419:96–121

Milet C, Monsoro-Burq AH (2012) Neural crest induction at the neural plate border in vertebrates. Dev Biol 366:22–33

Millet S, Campbell K, Epstein DJ, Losos K, Harris E, Joyner AL (1999) A role for Gbx2 in repression of Otx2 and positioning the mid/hindbrain organizer. Nature 401:161–164

Miyamoto N, Wada H (2013) Hemichordate neurulation and the origin of the neural tube. Nat Commun 4:2713

Miyata T, Suga H (2001) Divergence pattern of animal gene families and relationship with the Cambrian explosion. Bioessays 23:1018–1027

Modrell MS, Hockman D, Uy B, Buckley D, Sauka-Spengler T, Bronner ME, Baker CV (2014) A fate-map for cranial sensory ganglia in the sea lamprey. Dev Biol 385:405–416

Moreno N, Gonzalez A, Retaux S (2009) Development and evolution of the subpallium. Semin Cell Dev Biol 20:735–743

Morris SC, Caron JB (2014) A primitive fish from the Cambrian of North America. Nature 512(7515):419–422

Murakami Y, Ogasawara M, Sugahara F, Hirano S, Satoh N, Kuratani S (2001) Identification and expression of the lamprey Pax6 gene: evolutionary origin of the segmented brain of vertebrates. Development 128:3521–3531

Murakami Y, Pasqualetti M, Takio Y, Hirano S, Rijli FM, Kuratani S (2004) Segmental development of reticulospinal and branchiomotor neurons in lamprey: insights into the evolution of the vertebrate hindbrain. Development 131:983–995

Murakami Y, Uchida K, Rijli FM, Kuratani S (2005) Evolution of the brain developmental plan: insights from agnathans. Dev Biol 280:249–259

Neal HV (1896) A summary of studies on the segmentation of the nervous system in *Squalus acanthias*. Anat Anz 12:377–391

Nelson JS (2006) Fishes of the world, fourth edition. Wiley, Hoboken

Nieuwenhuys R (1967a) Comparative anatomy of the cerebellum. Prog Brain Res 25:1–93

Nieuwenhuys R (1967b) Comparative anatomy of olfactory centres and tracts. Prog Brain Res 23:1–64

Nieuwenhuys R (1997) The central nervous system of vertebrates. Nieuwenhuys R, TenDonkelaar HJ, Nicholson C, (eds) Springer, Berlin/Heidelberg

Niimura Y (2009) On the origin and evolution of vertebrate olfactory receptor genes: comparative genome analysis among 23 chordate species. Genome Biol Evol 1:34–44

Nishizawa H, Kishida R, Kadota T, Goris RC (1988) Somatotopic organization of the primary sensory trigeminal neurons in the hagfish, *Eptatretus burgeri*. J Comp Neurol 267:281–295

Noden DM (1991) Vertebrate craniofacial development: the relation between ontogenetic process and morphological outcome. Brain Behav Evol 38:190–225

Northcutt RG, Wicht H (1997) Afferent and efferent connections of the lateral and medial pallia of the silver lamprey. Brain Behav Evol 49:1–19

O'Leary DD, Sahara S (2008) Genetic regulation of arealization of the neocortex. Curr Opin Neurobiol 18:90–100

Ocana FM, Suryanarayana SM, Saitoh K, Kardamakis AA, Capantini L, Robertson B, Grillner S (2015) The lamprey pallium provides a blueprint of the mammalian motor projections from cortex. Curr Biol 25:413–423

Ohno S (1970) Evolution by gene duplication. Springer, Berlin/New York

Oisi Y, Ota KG, Kuraku S, Fujimoto S, Kuratani S (2013) Craniofacial development of hagfishes and the evolution of vertebrates. Nature 493:175–180

Orr H (1887) Contribution to the embryology of the lizard. J Morphol 1:311–372

Osorio J, Mazan S, Retaux S (2005) Organisation of the lamprey (*Lampetra fluviatilis*) embryonic brain: insights from LIM-homeodomain, Pax and hedgehog genes. Dev Biol 288:100–112

Osorio J, Megias M, Pombal MA, Retaux S (2006) Dynamic expression of the LIM-homeodomain gene Lhx15 through larval brain development of the sea lamprey (*Petromyzon marinus*). Gene Expr Patterns 6:873–878

Ota KG, Kuratani S (2006) The history of scientific endeavors towards understanding hagfish embryology. Zool Sci 23:403–418

Ota KG, Kuraku S, Kuratani S (2007) Hagfish embryology with reference to the evolution of the neural crest. Nature 446:672–675

Ota KG, Fujimoto S, Oisi Y, Kuratani S (2011) Identification of vertebra-like elements and their possible differentiation from sclerotomes in the hagfish. Nat Commun 2:373

Oury F, Murakami Y, Renaud JS, Pasqualetti M, Charnay P, Ren SY, Rijli FM (2006) Hoxa2- and rhombomere-dependent development of the mouse facial somatosensory map. Science 313:1408–1413

Pani AM, Mullarkey EE, Aronowicz J, Assimacopoulos S, Grove EA, Lowe CJ (2012) Ancient deuterostome origins of vertebrate brain signalling centres. Nature 483:289–294

Parker HJ, Bronner ME, Krumlauf R (2014) A Hox regulatory network of hindbrain segmentation is conserved to the base of vertebrates. Nature 514(7523):490–493

Passamaneck YJ, Di Gregorio A (2005) *Ciona intestinalis*: chordate development made simple. Dev Dyn 233:1–19

Pierani A, Wassef M (2009) Cerebral cortex development: from progenitors patterning to neocortical size during evolution. Dev Growth Differ 51:325–342

Plump AS, Erskine L, Sabatier C, Brose K, Epstein CJ, Goodman CS, Mason CA, Tessier-Lavigne M (2002) Slit1 and Slit2 cooperate to prevent premature midline crossing of retinal axons in the mouse visual system. Neuron 33:219–232

Pombal MA, Puelles L (1999) Prosomeric map of the lamprey forebrain based on calretinin immunocytochemistry, Nissl stain, and ancillary markers. J Comp Neurol 414:391–422

Pombal MA, Marin O, Gonzalez A (2001) Distribution of choline acetyltransferase-immunoreactive structures in the lamprey brain. J Comp Neurol 431:105–126

Pombal MA, Megias M, Bardet SM, Puelles L (2009) New and old thoughts on the segmental organization of the forebrain in lampreys. Brain Behav Evol 74:7–19

Pombal MA, Alvarez-Otero R, Perez-Fernandez J, Solveira C, Megias M (2011) Development and organization of the lamprey telencephalon with special reference to the GABAergic system. Front Neuroanat 5:20

Pose-Mendez S, Candal E, Adrio F, Rodriguez-Moldes I (2014) Development of the cerebellar afferent system in the shark *Scyliorhinus canicula*: insights into the basal organization of precerebellar nuclei in gnathostomes. J Comp Neurol 522:131–168

Puelles L, Rubenstein JL (1993) Expression patterns of homeobox and other putative regulatory genes in the embryonic mouse forebrain suggest a neuromeric organization. Trends Neurosci 16:472–479

Puelles L, Rubenstein JL (2003) Forebrain gene expression domains and the evolving prosomeric model. Trends Neurosci 26:469–476

Puelles L, Rubenstein JL (2015) A new scenario of hypothalamic organization: rationale of new hypotheses introduced in the updated prosomeric model. Front Neuroanat 9:27

Puelles L, Kuwana E, Puelles E, Bulfone A, Shimamura K, Keleher J, Smiga S, Rubenstein JL (2000) Pallial and subpallial derivatives in the embryonic chick and mouse telencephalon, traced by the expression of the genes Dlx-2, Emx-1, Nkx-2.1, Pax-6, and Tbr-1. J Comp Neurol 424:409–438

Quintana-Urzainqui I, Sueiro C, Carrera I, Ferreiro-Galve S, Santos-Duran G, Pose-Mendez S, Mazan S, Candal E, Rodriguez-Moldes I (2012) Contributions of developmental studies in the dogfish *Scyliorhinus canicula* to the brain anatomy of elasmobranchs: insights on the basal ganglia. Brain Behav Evol 80:127–141

Redies C, Ast M, Nakagawa S, Takeichi M, Martinez-de-la-Torre M, Puelles L (2000) Morphologic fate of diencephalic prosomeres and their subdivisions revealed by mapping cadherin expression. J Comp Neurol 421:481–514

Reiner A (2007) The evolution of the basal ganglia in mammals and other vertebrates. In: Kaas JH (ed) Evolution of nervous systems, vol 3. Academic, Amsterdam/Boston, pp 397–411

Ricano-Cornejo I, Altick AL, Garcia-Pena CM, Nural HF, Echevarria D, Miquelajauregui A, Mastick GS, Varela-Echavarria A (2011) Slit-Robo signals regulate pioneer axon pathfinding of the tract of the postoptic commissure in the mammalian forebrain. J Neurosci Res 89:1531–1541

Rijli FM, Gavalas A, Chambon P (1998) Segmentation and specification in the branchial region of the head: the role of the Hox selector genes. Int J Dev Biol 42:393–401

Roelink H, Porter JA, Chiang C, Tanabe Y, Chang DT, Beachy PA, Jessell TM (1995) Floor plate and motor neuron induction by different concentrations of the amino-terminal cleavage product of sonic hedgehog autoproteolysis. Cell 81:445–455

Rohr KB, Barth KA, Varga ZM, Wilson SW (2001) The nodal pathway acts upstream of hedgehog signaling to specify ventral telencephalic identity. Neuron 29:341–351

Ross LS, Parrett T, Easter SS Jr (1992) Axonogenesis and morphogenesis in the embryonic zebrafish brain. J Neurosci 12:467–482

Rothberg JM, Jacobs JR, Goodman CS, Artavanis-Tsakonas S (1990) slit: an extracellular protein necessary for development of midline glia and commissural axon pathways contains both EGF and LRR domains. Genes Dev 4:2169–2187

Rowe TB, Macrini TE, Luo ZX (2011) Fossil evidence on origin of the mammalian brain. Science 332:955–957

Rubenstein JL, Martinez S, Shimamura K, Puelles L (1994) The embryonic vertebrate forebrain: the prosomeric model. Science 266:578–580

Rubenstein JL, Shimamura K, Martinez S, Puelles L (1998) Regionalization of the prosencephalic neural plate. Annu Rev Neurosci 21:445–477

Sanes DH (2006) Development of the nervous system. 2nd ed., Sanes DH, Reh TA, Harris W, (eds) Academic Press, Oxford

Schilling TF, Knight RD (2001) Origins of anteroposterior patterning and Hox gene regulation during chordate evolution. Philos Trans R Soc Lond B Biol Sci 356:1599–1613

Schlosser G (2006) Induction and specification of cranial placodes. Dev Biol 294:303–351

Schneider-Maunoury S, Seitanidou T, Charnay P, Lumsden A (1997) Segmental and neuronal architecture of the hindbrain of Krox-20 mouse mutants. Development 124:1215–1226

Schubert M, Holland ND, Laudet V, Holland LZ (2006) A retinoic acid-Hox hierarchy controls both anterior/posterior patterning and neuronal specification in the developing central nervous system of the cephalochordate amphioxus. Dev Biol 296:190–202

Shigetani Y, Funahashi JI, Nakamura H (1997) En-2 regulates the expression of the ligands for Eph type tyrosine kinases in chick embryonic tectum. Neurosci Res 27:211–217

Shimamura K, Hartigan DJ, Martinez S, Puelles L, Rubenstein JL (1995) Longitudinal organization of the anterior neural plate and neural tube. Development 121:3923–3933
Shu T, Richards LJ (2001) Cortical axon guidance by the glial wedge during the development of the corpus callosum. J Neurosci 21:2749–2758
Shu T, Sundaresan V, McCarthy MM, Richards LJ (2003) Slit2 guides both precrossing and postcrossing callosal axons at the midline *in vivo*. J Neurosci 23:8176–8184
Shubin N, Tabin C, Carroll S (1997) Fossils, genes and the evolution of animal limbs. Nature 388:639–648
Sidow A (1996) Gen(om)e duplications in the evolution of early vertebrates. Curr Opin Genet Dev 6:715–722
Stephenson-Jones M, Samuelsson E, Ericsson J, Robertson B, Grillner S (2011) Evolutionary conservation of the basal ganglia as a common vertebrate mechanism for action selection. Curr Biol 21:1081–1091
Stolfi A, Ryan K, Meinertzhagen IA, Christiaen L (2015) Migratory neuronal progenitors arise from the neural plate borders in tunicates. Nature 527:371–374
Strausfeld NJ, Hirth F (2013) Deep homology of arthropod central complex and vertebrate basal ganglia. Science 340:157–161
Streit A (2007) The preplacodal region: an ectodermal domain with multipotential progenitors that contribute to sense organs and cranial sensory ganglia. Int J Dev Biol 51:447–461
Striedter GF (2005) Principles of brain evolution. Sinauer, Sunderland
Sugahara F, Aota S, Kuraku S, Murakami Y, Takio-Ogawa Y, Hirano S, Kuratani S (2011) Involvement of Hedgehog and FGF signalling in the lamprey telencephalon: evolution of regionalization and dorsoventral patterning of the vertebrate forebrain. Development 138:1217–1226
Sugahara F, Murakami Y, Adachi N, Kuratani S (2013) Evolution of the regionalization and patterning of the vertebrate telencephalon: what can we learn from cyclostomes? Curr Opin Genet Dev 23:475–483
Sugahara F, Pascual-Anaya J, Oisi Y, Kuraku S, Aota S, Adachi N, Takagi W, Hirai T, Sato N, Murakami Y, Kuratani S (2016) Evidence from cyclostomes for complex regionalization of the ancestral vertebrate brain. Nature 531:97–100
Suzuki M, Morita H, Ueno N (2012) Molecular mechanisms of cell shape changes that contribute to vertebrate neural tube closure. Dev Growth Differ 54:266–276
Suzuki DG, Murakami Y, Yamazaki Y, Wada H (2015a) Expression patterns of Eph genes in the "dual visual development" of the lamprey and their significance in the evolution of vision in vertebrates. Evol Dev 17:139–147
Suzuki DG, Murakami Y, Escriva H, Wada H (2015b) A comparative examination of neural circuit and brain patterning between the lamprey and amphioxus reveals the evolutionary origin of the vertebrate visual center. J Comp Neurol 523:251–261
Takio Y, Kuraku S, Murakami Y, Pasqualetti M, Rijli FM, Narita Y, Kuratani S, Kusakabe R (2007) Hox gene expression patterns in *Lethenteron japonicum* embryos: insights into the evolution of the vertebrate Hox code. Dev Biol 308:606–620
Tanaka G, Hou X, Ma X, Edgecombe GD, Strausfeld NJ (2013) Chelicerate neural ground pattern in a Cambrian great appendage arthropod. Nature 502:364–367
Tank EM, Dekker RG, Beauchamp K, Wilson KA, Boehmke AE, Langeland JA (2009) Patterns and consequences of vertebrate Emx gene duplications. Evol Dev 11:343–353
Theveneau E, Mayor R (2012) Neural crest delamination and migration: from epithelium-to-mesenchyme transition to collective cell migration. Dev Biol 366:34–54
Tomer R, Denes AS, Tessmar-Raible K, Arendt D (2010) Profiling by image registration reveals common origin of annelid mushroom bodies and vertebrate pallium. Cell 142:800–809
Toresson H, Martinez-Barbera JP, Bardsley A, Caubit X, Krauss S (1998) Conservation of BF-1 expression in amphioxus and zebrafish suggests evolutionary ancestry of anterior cell types that contribute to the vertebrate telencephalon. Dev Genes Evol 208:431–439

Tosa Y, Tsukano K, Itoyama T, Fukagawa M, Nii Y, Ishikawa R, Suzuki KT, Fukui M, Kawaguchi M, Murakami Y (2015) Involvement of Slit-Robo signaling in the development of the posterior commissure and concomitant swimming behavior in *Xenopus laevis*. Zool Lett 1:28

Vaage S (1969) The segmentation of the primitive neural tube in chick embryos (*Gallus domesticus*). Ergeb Anat Entw Gesch 4:1–88

Venkatesh TV, Holland ND, Holland LZ, Su MT, Bodmer R (1999) Sequence and developmental expression of amphioxus AmphiNk2-1: insights into the evolutionary origin of the vertebrate thyroid gland and forebrain. Dev Genes Evol 209:254–259

von Baer K (1828) Über die Entwickelungsgeschichte der Thiere. Königsberg

Voogd J, Glickstein M (1998) The anatomy of the cerebellum. Trends Neurosci 21:370–375

Wicht H, Lacalli TC (2005) The nervous system of amphioxus: structure, development, and evolutionary significance. Can J Zool 83:122–150

Wicht H, Northcutt RG (1992) The forebrain of the Pacific hagfish: a cladistic reconstruction of the ancestral craniate forebrain. Brain Behav Evol 40:25–64

Wilkinson DG, Bhatt S, Chavrier P, Bravo R, Charnay P (1989) Segment-specific expression of a zinc-finger gene in the developing nervous system of the mouse. Nature 337:461–464

Wilson SW, Rubenstein JL (2000) Induction and dorsoventral patterning of the telencephalon. Neuron 28:641–651

Witkovsky P, Roberts BL (1975) The light microscopical structure of the mesencephalic nucleus of the fifth nerve in the selachian brain. Proc R Soc Lond B Biol Sci 190:457–471

Witmer LM, Ridgely RC (2009) New insights into the brain, braincase, and ear region of tyrannosaurs (Dinosauria, Theropoda), with implications for sensory organization and behavior. Anat Rec 292:1266–1296

Witmer LM, Chatterjee S, Franzosa J, Rowe T (2003) Neuroanatomy of flying reptiles and implications for flight, posture and behaviour. Nature 425:950–953

Wullimann MF, Mueller T (2004) Teleostean and mammalian forebrains contrasted: evidence from genes to behavior. J Comp Neurol 475:143–162

Wullimann MF, Puelles L (1999) Postembryonic neural proliferation in the zebrafish forebrain and its relationship to prosomeric domains. Anat Embryol 199:329–348

Yajima H, Suzuki M, Ochi H, Ikeda K, Sato S, Yamamura K, Ogino H, Ueno N, Kawakami K (2014) Six1 is a key regulator of the developmental and evolutionary architecture of sensory neurons in craniates. BMC Biol 12:40

Yamada T, Placzek M, Tanaka H, Dodd J, Jessell TM (1991) Control of cell pattern in the developing nervous system: polarizing activity of the floor plate and notochord. Cell 64:635–647

Yamamoto N, Kato T, Okada Y, Somiya H (2010) Somatosensory nucleus in the torus semicircularis of cyprinid teleosts. J Comp Neurol 518:2475–2502

Yu JK, Holland ND, Holland LZ (2002) An amphioxus winged helix/forkhead gene, AmphiFoxD: insights into vertebrate neural crest evolution. Dev Dyn 225:289–297

Chapter 10
Adaptive Radiation and Vertebrate Brain Diversity: Cases of Teleosts

Naoyuki Yamamoto

Abstract There are numerous species of teleosts in the world. In their widespread distribution in divergent environments, different teleost species have employed variable strategies for survival. The diversity in niches and habits is reflected in the morphology of teleost brains, although all the major brain parts are present that are common to other vertebrates. Species differences are particularly remarkable in the external morphology of sensory brain regions. Generally speaking, the brain part bulges, which is involved in the processing of the most "important" sensory modality for that species. Thus, one can imagine, to a certain extent, the lifestyle of a teleost species, just by looking at its external brain morphology. However, there are also cases of amazing diversity, the reasons for which remain unknown. The diversity of teleost brain morphology is introduced in this chapter, presenting selected cases among the virtually infinite variety. In addition to the intra-teleostean diversity just mentioned, there are features in brain organization that are specific to teleosts (and related non-teleostean actinopterygians) but are lacking in other vertebrates, which is also demonstrated in this chapter. Diversity can be also found in such teleost-specific structures.

Keywords Teleosts • Brain • Morphology • External appearance • Diversity • Niche • Evolution

Abbreviations

a axon
alln anterior lateral line nerve
ALMs major spinal accessory lobes
ALm minor spinal accessory lobe
CC cerebellar corpus (*corpus cerebelli*)
cf climbing fiber

N. Yamamoto (✉)
Laboratory of Fish Biology, Graduate School of Bioagricultural Sciences, Nagoya University, Furo-cho, Chikusa-ku, Nagoya 464-8601, Japan
e-mail: nyama@agr.nagoya-u.ac.jp

S. Shigeno et al. (eds.), *Brain Evolution by Design*, Diversity and Commonality in Animals, DOI 10.1007/978-4-431-56469-0_10

CrC cerebellar crest (*crista cerebellaris*)
d dendrite
DI diencephalon
EG granular eminence (*eminentia granualaris*)
ELLL electrosensory lateral line lobe
Eur eurydendroid cell
g granule cell
GCL granule cell layer
GL glomerular layer
GrL granular layer
L lateral granular cell part of TL
l lateral fibrous part of TL
LI inferior lobe (part of diencephalon)
LVII facial lobe
LIX glossopharyngeal lobe
LX vagal lobe
M medial granular cell part of TL
m medial fibrous part of TL
MCL mitral cell layer
ME mesencephalon
ML molecular layer
MO medulla oblongata
nI olfactory nerve
nII optic nerve
nIII oculomotor nerve
nIV trochlear nerve
nV trigeminal nerve
nVII facial nerve
nVIII octaval nerve
nIX glossopharyngeal nerve
nX vagal nerve
OB olfactory bulb
OE olfactory epithelium
ot olfactory tract
OT optic tectum (part of mesencephalon)
otr optic tract
P pons
pc posterior commissure
PCL Purkinje cell layer
pf parallel fiber
Pit pituitary
plln posterior lateral line nerve
Pur Purkinje cell
Pyr pyramidal cell
SAC *stratum album centrale*

SC spinal cord
SFGS *stratum fibrosum et griseum superficiale*
SGC *stratum griseum centrale*
SM *stratum marginale*
SO *stratum opticum*
SPV *stratum periventriculare*
TE telencephalon
TL *torus longitudinalis*
TS *torus semicircularis*
VC cerebellar valvula (*valvula cerebelli*)
VCl lateral lobe of VC
VCm medial lobe of VC

10.1 Diversity of Teleosts

More than 25,000 extant species constitute the teleosts, making this lineage the largest among vertebrates. Their habitats range from the Equator to the polar regions, from the deep sea and seashores to lakes and rivers. Some species even dwell in the darkness of caves. Living in the same area, some species prefer rocky bottoms, others live on sandy bottoms, and still others stay on seaweed beds. Also, there are nocturnal as well as diurnal species. Feeding habits also differ: there are carnivorous, herbivorous, and omnivorous species. Adapting to the environmental features of their habitats with specific lifestyles, the external morphology of teleosts show a fascinating variety (Nelson 2006). Interestingly, the diversity of niches and habits is also reflected in the morphology of brains in teleosts, as exemplified by Kotrschal et al. (1998) regarding the external morphology. In this chapter, the diversity of brains is described and discussed at the levels of external morphology as well as histology (cytoarchitecture).

10.2 Teleost Brains Are Composed of Brain Parts Common to Other Vertebrates

Before jumping into the jungle of diversity in teleost brain morphology, the organization of the brains in teleosts is briefly explained to facilitate a better understanding of subsequent sections. The brains of teleosts are composed of the telencephalon (cerebrum), diencephalon, mesencephalon (midbrain), pons (pontine region), cerebellum, and medulla oblongata, the latter continuing to the spinal cord (as shown in Fig. 10.1). Figure 10.1 also illustrates cranial nerves, for convenience of subsequent sections, where different cranial nerves are mentioned to explain brain centers. Although it is frequently supposed that the brains of teleosts lack the pons,

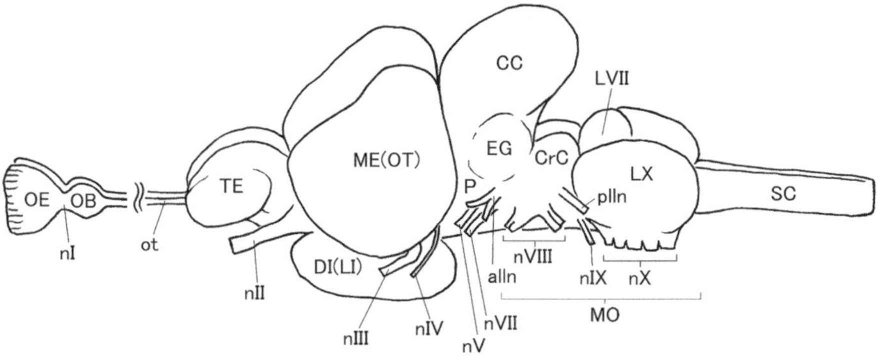

Fig. 10.1 Dorsolateral view of the brain and spinal cord of the common carp *Cyprinus carpio*, showing major parts of the brain and cranial nerves. The sixth cranial nerve or abducens nerve, which emerges from the ventral aspect of the rhombencephalon, is not visible. Rostral is to the *left*. See list for abbreviations (Modified from Yamamoto 2005)

this is not the case. This popular misconception stems from the fact that teleosts do not possess clearly identifiable "pontine nuclei" (although a probable homologue is present: Yang et al. 2004), well-known nuclei in the mammalian pons that relay cortical inputs to the cerebellum. However, the brain region corresponding to the pons of mammals is clearly present, that is, the rhombencephalic region ventral to the cerebellum. Thus, all the brain parts present in other vertebrates can be found in the brains of teleosts.

10.3 Diversity in the Teleost Brains

10.3.1 Optic Tectum

The optic tectum of teleosts is the largest visual center, receiving direct retinal inputs via the optic nerve (or axons of retinal ganglion cells), and occupies the dorsal part of the mesencephalon, usually along its entire rostrocaudal and mediolateral extent. The thread-sail filefish, *Stephanolepis cirrhifer*, is a diurnal species living mainly in groups on sandy bottoms of shallow water. The fish feed on crustaceans, polychaetes, jellyfish, and bivalves. The fish blow away sand to find polychaetes and bivalves, most likely dependent on vision. This species also relies on vision for communication. They change color patterns on the body as well as the position of the first, spiny dorsal fin ray, according to the social relationship to other individuals (dominant, surrender, frightened, and so on) (Ito et al. 2007). This communication is clearly dependent on vision, because changes in color patterns and the position of the ray cannot be detected by other sensory modalities. The filefish is known to anglers as a notorious stealer of bait. They stand in front of the bait to see it

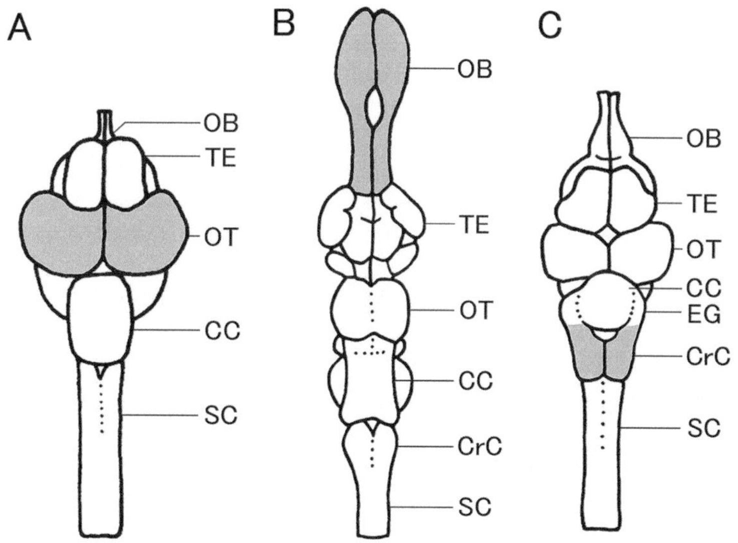

Fig. 10.2 Dorsal views of the brains and spinal cords of the thread-sail filefish *Stephanolepis cirrhifer* (**a**), the kidako moray *Gymnothorax kidako* (**b**), and the silver croaker *Pennahia argentata* (**c**). Hypertrophied parts of the brain are drawn in *gray*. Rostral is to the *top*. See list for abbreviations (Modified from Ito et al. 2007)

binocularly and frequently eat it before the angler becomes aware of this humiliating fact. Thus, the life of the thread-sail filefish is heavily dependent on vision. The optic tectum is huge in this species (Fig. 10.2a). In a nocturnal species, the kidako moray *Gymnothorax kidako* that is referred to in more detail in the next section, the optic tectum is quite small (Fig. 10.2b).

10.3.2 Olfactory Bulb

The olfactory bulb is the primary center for olfaction. It receives axons of olfactory receptor neurons (or the olfactory nerve) and sends fibers mainly to the telencephalon through the olfactory tract (Fig. 10.1). It is present just caudal to the olfactory epithelium in some species such as the common carp (Fig. 10.1) and catfishes (short olfactory nerves and long olfactory tracts, called the “stalked” olfactory bulb), although rostrally adjacent to the telencephalon and away from the olfactory epithelium in other species such as the thread-sail filefish (Fig. 10.2a: long olfactory nerves and short olfactory tracts, called the “sessile” olfactory bulb). These two types of configuration, therefore, provide another case of diversity in teleost brain morphology. The stalked type appears sporadically in the phylogenetic tree: Osteoglossiformes (that includes arowanas), Ostariophysi (that includes catfishes and cyprinids), and Gadiformes (that includes cod). No clear relationship with the lifestyle or niche is recognized so far as the author has explored, and no further survey on the two morphological types of the bulb is made in this chapter. Significant species differences, however, are clearly recognized in the size of the olfactory bulb. For example, the olfactory bulb of the thread-sail filefish is quite small (Fig. 10.2a), whereas the bulb of the kidako moray is extremely hypertrophied and is larger than the optic tectum as observed dorsally (Fig. 10.2b). The kidako moray lives in reefs, staying in rock caves in the daytime. In the nighttime they come out from the caves to feed on cephalopods (in particular, octopus), crustaceans, and fish. The smells of their prey should provide important cues for hunting them under the darkness of the night, and the olfactory epithelium is very large, accompanying many olfactory lamellae. Interestingly, the cytoarchitecture of the olfactory bulb of the moray is bizarre. The olfactory bulb of most teleost species shows a concentric laminar organization (Fig. 10.3a, goldfish) with small granule cells in the center, similar to other vertebrates. However, the olfactory bulb of the moray appears to be composed of many subregions, and groups of granule cell aggregates are scattered in the bulb (Fig. 10.3b). Also, glomeruli are diffusely distributed within the bulb without forming a layer (Fig. 10.3b). This unusual cytoarchitecture of the bulb in the kidako moray might be advantageous in analyzing a large amount of olfactory information.

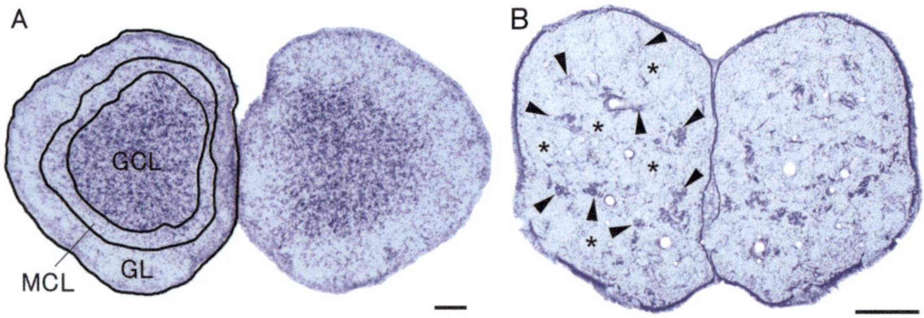

Fig. 10.3 Transverse sections showing the cytoarchitecture of the olfactory bulb of the goldfish *Carassius auratus* (**a**: Nissl staining) and the kidako moray *Gymnothorax kidako* (**b**: Nissl staining). Note concentric laminar organization of the olfactory bulb of goldfish, with the granule cell layer in the *center*. In the moray, however, granule cells form smaller aggregates (*arrowheads*) that are scattered in the bulb. Similarly, the glomeruli (*asterisks*) are scattered within the bulb without forming a layer. *Bars* 100 μm (**a**); 500 μm (**b**)

10.3.3 Cerebellar Crest

The cerebellar crest (*crista cerebellaris* in Latin) is a superficial fibrous layer associated with the octavolateral (auditory, vestibular, and lateral line) nuclei in the rostral medulla oblongata. The crest is composed of fibers that originate from granule cells in the eminentia granularis, a component of the cerebellum in teleosts (see 10.4.2). Ventrally adjacent to the crest, the primary lateral line nucleus is present laterally, and the primary and secondary auditory nuclei are present medially. The primary lateral line nucleus receives information about water movement from the lateral line organs, and the primary auditory nucleus receives fibers that arise from otolith organs in the inner ear; teleosts can detect underwater sounds with otolith organs, although they do not possess externally visible ear auricles and middle ears. Some neurons in the primary lateral line and auditory nuclei extend dendrites into the cerebellar crest to receive input from fibers from the eminentia granularis. The secondary auditory nucleus is situated rostral to the primary nucleus and receives fibers from the primary auditory nucleus. Deep to the three nuclei just enumerated are primary centers receiving sense of balance (vestibular sense) from semicircular canals and otolith organs (i.e., otolith organs serve for both hearing and vestibular sense). The silver croaker *Pennahia argentata* is a species with a nocturnal trend that prefers strong, turbid tidal currents. In the nighttime they come close to shore to hunt for polychaetes and crustaceans. They have to maintain their balance for hunting under turbulence in the water's edge in the dark. Lateral line information and vestibular inputs should be important under such circumstances. It should be also noted that this species makes sounds, or is a sonic fish. When we grasp the fish, they grunt by vibrating their swim bladder with sonic motor muscles. They likely use the sounds for communication in their ordinary life. Related to the presumed sound communication, one of the otoliths of the croaker is extremely large, which may reflect the high sensitivity of this species to sounds. As enumerated, octavolateral information is of great importance for the silver croaker. Sensory nuclei responsible for the processing of octavolateral information become well developed, and the cerebellar crest overlying the nuclei becomes swollen (Fig. 10.2c).

10.3.4 Primary Gustatory Centers

Taste buds are present on the tongue, pharynx, and upper larynx in humans. Therefore, we can taste only those items that are put into our mouth. However, taste buds are also present on the face and body surface in some species of teleosts, which means that those species can taste potential food before they eat it, a feature advantageous for their life under water. Cranial nerves that carry gustatory (taste) sense are the facial (VII), glossopharyngeal (IX), and vagal (X) nerves, as in mammals where the three cranial nerves project to the rostral part of the nucleus of the solitary tract in the medulla oblongata. In teleosts the facial nerve conveys gustatory information from the rostral oral cavity as well as the body surface when extra-oropharyngeal taste buds are present. The glossopharyngeal nerve carries gustatory information from the caudal oral cavity and rostralmost pharynx. The vagal nerve carries gustatory inputs from the caudal pharynx. The three cranial nerves terminate in the dorsal part of the caudal medulla oblongata, basically according to the numerical order of the nerves from rostral to caudal. In some species, such as the thread-sail filefish and the rainbow trout *Oncorhynchus mykiss*, the terminal zones of the three nerves cannot be distinguished as observed externally (Figs. 10.2a and 10.4a) (Meek and Nieuwenhuys 1998) or by observations of nonexperimental brain sections. In such a case the terminal zones of the three nerves, which form together a rostrocaudally elongate primary gustatory column, may be collectively called the nucleus of the solitary tract after the mammalian terminology (Meek and Nieuwenhuys 1998). The rainbow trout find their prey (smaller fish and insects) mainly by relying on vision, and the contribution of the gustatory sense does not appear quite significant; this is evident considering that we can fish for trout with lures that do not emit tasty substances. The morphology of the primary gustatory centers in teleosts with well-developed taste sensory systems is quite different from that of the rainbow trout and thread-sail filefish. For example, the caudalmost region of the primary gustatory zone forms a bilateral pair of huge bulges in the goldfish (Fig. 10.4b). The region receives vagally mediated taste information and is hence called the vagal lobe. The goldfish frequently suck the bottom of the water body in search of food. Connected to this behavior, the goldfish

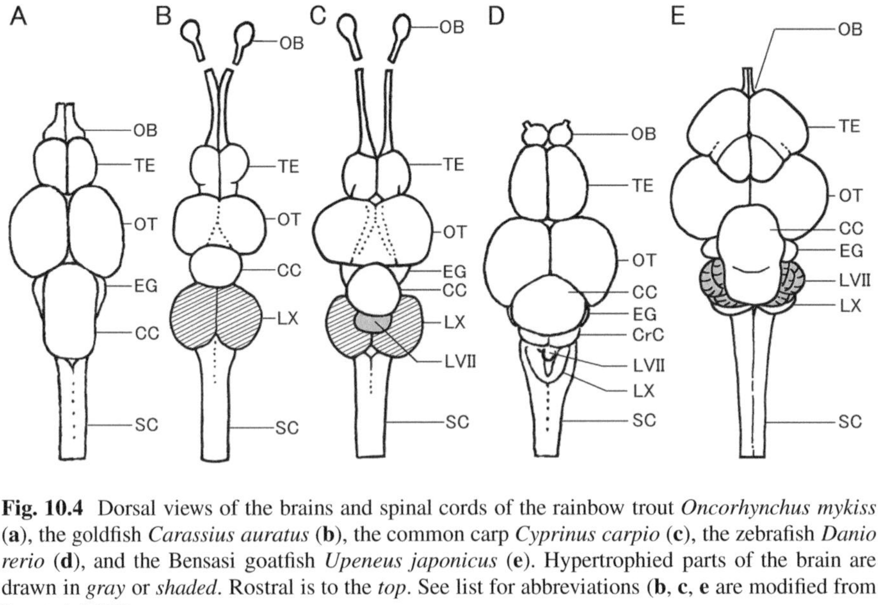

Fig. 10.4 Dorsal views of the brains and spinal cords of the rainbow trout *Oncorhynchus mykiss* (**a**), the goldfish *Carassius auratus* (**b**), the common carp *Cyprinus carpio* (**c**), the zebrafish *Danio rerio* (**d**), and the Bensasi goatfish *Upeneus japonicus* (**e**). Hypertrophied parts of the brain are drawn in *gray* or *shaded*. Rostral is to the *top*. See list for abbreviations (**b**, **c**, **e** are modified from Ito et al. 2007)

possess a specialized muscular organ in the roof of the pharynx, called the palatal organ. There are many taste buds on the surface of the palatal organ and gill rakers facing the organ. Edible particles, which are detected by the taste buds, are sandwiched between the gill rakers and the palatal organ (the region of the organ touching the edible particle protrudes) but sand or mud is not. The latter is washed back out of the mouth, and the goldfish then swallow the edible particles that remain in the pharyngeal cavity (Finger 2008). Imagine that you put into your mouth 50 plastic balls 1 mm in diameter together with 50 candies of 1-mm diameter and then spit out the plastic balls, leaving the candies in your mouth. I do not think you can anticipate a happy ending with a lot of tasty candies and without any plastic balls in your mouth. The goldfish can do such sorting. To determine exact positions of edible particles and protrude tiny portions of the palatal organ that are exactly facing the particles should require much information processing and hence many neurons, leading to the hypertrophied "vagal lobe." The goldfish also possess taste buds on the body surface, and hence a corresponding, relatively large primary gustatory region is present that is called the facial lobe (Fig. 10.5a). This bulge, however, is not easily visible externally because gigantic vagal lobes cover it dorsally (Figs. 10.4b and 10.5a). The glossopharyngeal lobe is also present in the goldfish. The glossopharyngeal nerve innervates relatively small regions of the pharyngeal wall and gill rakers, and thus is thin in comparison with the facial or vagal nerves. Therefore, the glossopharyngeal lobe is much smaller than the facial and vagal lobes and is not identifiable by external observations, hidden between the facial and vagal lobes. It can be identified as observed with transverse sections (Fig. 10.5a). The common carp *Cyprinus carpio* also possess large vagal lobes, because this cyprinid species also uses the palatal organ for sorting food from inedible items (Fig. 10.4c). Being different from the closely related goldfish, however, the common carp have a huge facial lobe (Fig. 10.4c), which is related to the presence of barbels on the snout of the common carp, whereas the goldfish have no barbel. There are many taste buds on the surface of the barbels, leading to an increased amount of information that has to be processed in the facial lobe of the carp. The hypertrophied facial lobe is visible externally in the common carp, in spite of the presence of large vagal lobes. *Pseudogobio esocinus* (Japanese common name: kamatsuka) and *Misgurunus anguillicaudatus* (pond loach), two species of cyprinids equipped with barbels around the mouth, also possess a huge facial lobe (Tuge et al. 1968). The vagal lobe is also huge in these cyprinid species. Enlarged vagal lobes as enumerated here are cases from cyprinids. Therefore, one might suspect that this is a character common to all cyprinids (phylogenetically conserved trait) rather than reflecting the niche and behavior of each species. This assumption is not correct. The vagal lobe of the zebrafish *Danio rerio*, another species of cyprinids, is not as pronounced as those in the goldfish or common carp and does not bulge out (Fig. 10.4d). The zebrafish feed mainly on planktonic items in the water column (Spence et al. 2008), differing from those cyprinid species with the huge vagal lobe enumerated here that prefer the bottom of the water and feed from the substratum. A still different case of diversity in the primary taste centers can be found in the Bensasi goatfish *Upeneus japonicus*. The goatfish possess a pair of long

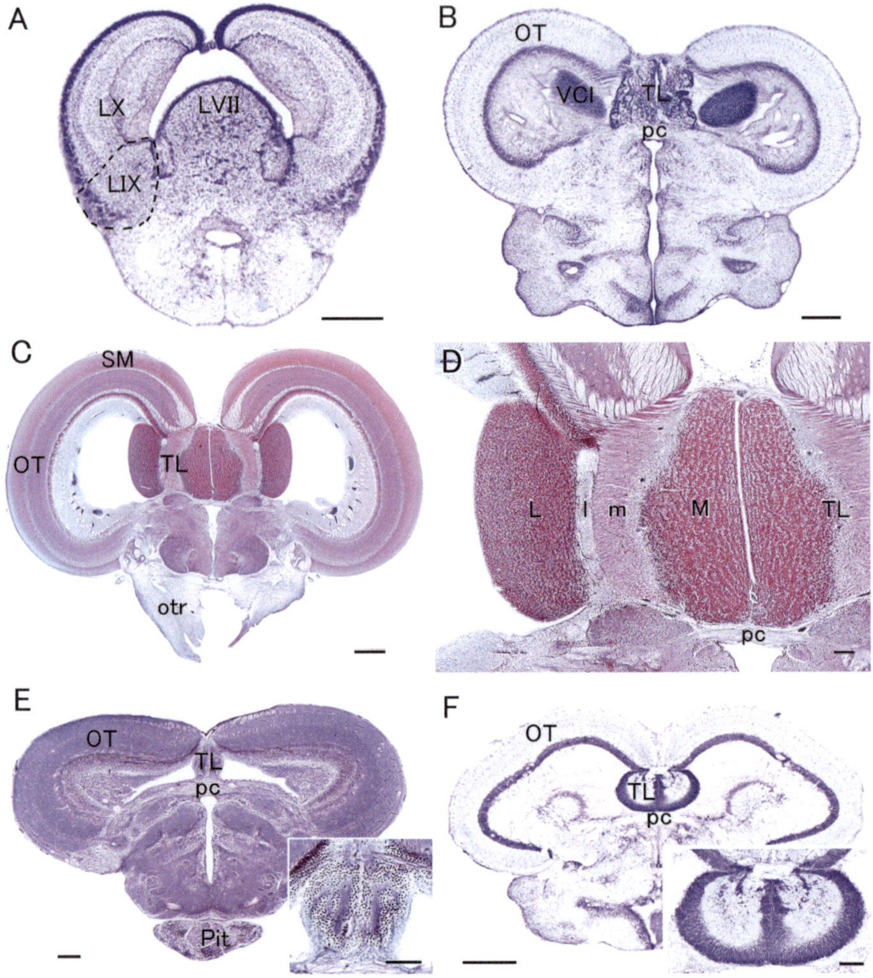

Fig. 10.5 Transverse sections through the primary gustatory centers of the goldfish *Carassius auratus* (**a**: Nissl staining), the *torus longitudinalis* (*TL*) of the goldfish (**b**: Nissl staining), the TL of the north Pacific squirrelfish *Sargocentron spinosissimum* (**c–d**: Bodian staining), the TL of the Japanese ricefish (medaka) *Oryzias latipes* (**e**: Bodian staining), and the TL of the yellowfin goby *Acanthogobius flavimanus* (**f**: Nissl staining). Note that the size and cytoarchitecture of the TL vary substantially in different species of teleosts. See list for other abbreviations. *Bars* 500 μm (**a**, **b**, **c**, **f**); 100 μm (**d**, **e**, inset of **f**); 50 μm (inset of **e**)

barbels extending from the lower jaw. Being different from barbels in most other fishes, the barbels of the goatfish are mobile. The fish swim close to the bottom and vigorously move the barbels in different directions to touch the bottom. The barbels are probes to search for shrimps and sand shrimps and thus are equipped with numerous taste buds (Kiyohara et al. 2002). Gustatory information detected by the barbels is mediated to the facial lobe, because the taste buds are present on the external body surface. The facial lobe of the goatfish is hence large. In the case of the goatfish, however, the facial lobe is not only large but also wrinkled as is our cerebral cortex (Fig. 10.4e). The reason why the facial lobe is wrinkled remains unclear but this phenotype may be related to the fact that the barbels of the goatfish are mobile, in contrast to those of other species, including the common carp. The facial lobe of the goatfish is composed of small and large neurons, and together with fibers these cell types form layers in the facial lobe (Ito and Yoshimoto 1991). That is, the facial lobe of the goatfish is a sheet-like structure whereas the facial lobe in other teleosts such as the goldfish is a mass of neurons (see Fig. 10.5a). The goatfish have to know the direction of the barbel when a shrimp touches it to calculate the precise position of the shrimp that should be targeted as a prey. This action would require computation and integration of sensory and motor-related information, and the laminar organization of neurons may be necessary to perform such a task. To increase the power of information processing, such a sheet-like structure has to enlarge the area rather than the mass. To be housed in the cranial cavity that should have some upper limit in volume, the surface of facial lobe in the goatfish has become wrinkled. The vagal lobe is also laminated in the goldfish (Fig. 10.5a) and common carp, probably because the lobe also processes sensory as well as motor-related (control on the protrusion of palatal organ) information in these species as well. The vagal lobes are not wrinkled in these cyprinids, however, perhaps because the cranial cavity still has room to house the unfolded vagal lobes of these species.

10.3.5 Spinal Cord

The spinal cord is not a part of the brain. However, spectacular cases of diversity, which can be also found in this part of the central nervous system, are introduced in this section. The northern sea robin *Prionotus carolinus* "walk" on sandy bottoms using modified pectoral fins as if they were limbs. There are three pairs of free fin rays lacking the membranous fin between them, and they move independently from each other and from the ordinary part of the pectoral fin with membranes. The three pairs of free pectoral fin rays are used for locomotion, while the remaining, ordinary portions of the pectoral fins are used for swimming and threatening (by expanding the fin; it is colorful). Although "walking" on pectoral fins is bizarre enough for a fish, this is not the whole story. The free fin rays are equipped with specialized chemosensory cells called solitary chemosensory cells (Whitear 1971). Solitary chemosensory cells are similar to single taste cells in taste buds. Multiple taste cells are present in single taste buds. However, solitary chemosensory cells are "solitary"; they do not form groups. In rodents, solitary chemosensory cells can be found in the lumen of the digestive tract (Sbarbati and Osculati 2003) and on the walls of the nasal cavity (in the respiratory epithelium, not in the olfactory epithelium involved in olfaction) (Finger et al. 2003). However, they can be found also on the body surface in teleosts. While "walking," the free spiny fin rays are used to detect prey (shrimps, crabs, and small fish); that is, the sea robin is not walking just for fun but is utilizing this behavior in a serious search for prey. Three conspicuous pairs of bulges, called major spinal accessory lobes, are found on the dorsal aspect of the rostral spinal cord of the sea robin (Fig. 10.6a) (Finger 2000). They correspond to the terminal zones of three spinal nerves innervating the three free rays. Another pair of less pronounced bulges, called minor spinal accessory lobes, is also found on the dorsal aspect of the rostral spinal cord; they receive inputs from the ordinary part of the pectoral fins. The spinal cords of teleosts are usually very long, occupying much of the rostrocaudal extent of the vertebral canal. However, the spinal cord of the long-spined porcupine fish *Diodon holocanthus* is extremely short (Fig. 10.6b). Surprisingly, it is shorter than the brain. There are a number of species in Tetraodontiformes (puffers and filefishes) with such a short spinal cord (e.g., ocean sunfish *Mola mola*). However, other species of Tetraodontiformes (e.g., the thread-sail filefish and blunthead puffer *Sphoeroides pachygaster*) possess an ordinary, long spinal cord (Figs. 10.2a and 10.6c). The reason why the spinal cord has to be (and can be) so short in some species of Tetraodontiformes is not known.

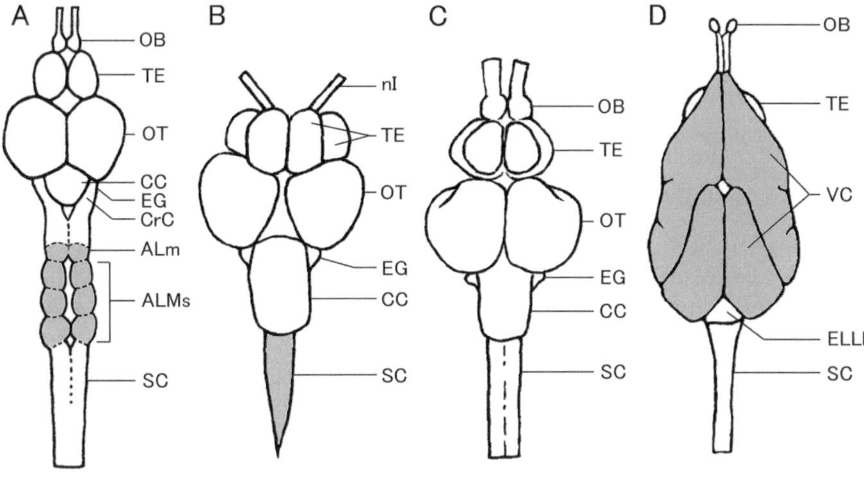

Fig. 10.6 Dorsal views of the brains and spinal cords of the northern sea robin *Prionotus carolinus* (**a**), the long-spined porcupine fish *Diodon holocanthus* (**b**), the blunthead puffer *Sphoeroides pachygaster* (**c**), and the elephant-nose fish *Gnathonemus petersii* (**d**). The olfactory bulb (*OB*) is not visible in the longspined porcupine fish (**b**) because it is hidden below the large telencephalon (*TE*). The olfactory nerve (*nI*) can be seen. The OB of the elephant-nose fish is of stalked type (**d**); this species belongs to Osteoglossiformes. It should be also noted that only the lateral margins of the telencephalon (*TE*) and a posterior portion of the electrosensory lateral line lobe (*ELLL*) are visible dorsally, while the rest of the brain is covered completely by the hypertrophied cerebellar valvula (*VC*) in the elephantnose fish (**d**). Hypertrophied and underdeveloped parts of the brain or spinal cord are drawn in *gray*. Rostral is to the *top*. See list for other abbreviations (**a** is modified from Ito et al. 2007)

10.4 Brain Regions Specific to Teleosts and Their Diversity

As mentioned in Sect. 10.2, the brains of teleosts possess all the major parts common to other vertebrates. However, there are a few brain regions that are not present in other vertebrates. Such teleost-specific structures are be introduced in this section, together with cases of diversity found in those structures.

10.4.1 Torus Longitudinalis

A pair of columnar structures runs longitudinally and ventrally adjacent to the medial margin of the optic tectum (goldfish: Fig. 10.5b), called the *tori longitudinales* (sing. *torus longitudinalis*). The torus can be found in most actinopterygian lineages: teleosts, the bowfin (*Amia calva*), Lepisosteiformes (gars), and Acipensiformes (sturgeons and paddlefishes). The *torus longitudinalis*, however, is not present in Polypteryformes. [In this chapter Polypteryformes is regarded as an actinopterygian lineage, although some researchers consider that Polypteryformes is closely related to sarcopterygians (lungfishes and coelacanths) rather than Actinopterygii.] The *torus longitudinalis* is composed of small cells similar to the granule cells of the cerebellum. In fact, the torus is a component of a neural circuitry that is similar to that seen in the cerebellum. In the cerebellum, axons of granule cells run though the molecular layer parallel to each other and make synaptic contacts with the spiny dendrites of Purkinje cells (Fig. 10.7). The axons of toral cells run through the most superficial layer of the optic tectum, the *stratum marginale* or marginal layer (Fig. 10.5c), which is lacking in the optic tectum of vertebrates without the *torus longitudinalis*. Similarly to cerebellar parallel fibers, toral fibers run parallel to each other through the layer and are called parallel fibers (or marginal fibers). The *stratum marginale* also contains dendrites that originate from tectal neurons situated in a deeper layer (*stratum fibrosum et griseum superficiale*: superficial layer of fibers and gray matter). These tectal cells are called pyramidal cells. Their superficial dendrites are spiny and show a fan-shaped branching pattern within the *stratum marginale*, similar to those of the Purkinje cells in the cerebellum. These dendrites make synaptic contacts with toral fibers, a configuration reminiscent of parallel fiber–Purkinje cell synapses in the cerebellum (Fig. 10.7). In contrast to the Purkinje cells, however, pyramidal cells of teleosts also possess a basal dendrite that extends into deeper tectal layers (Xue et al. 2003). The axon of pyramidal cells forms terminals in a small dimension of a deep tectal layer. The target cell type of pyramidal cell axons remains unclear. Gibbs and Northmore (1996) proposed that the *torus longitudinalis* is involved in the processing of luminance information. Details of the functions of the torus, however, await further study. The size of the *torus longitudialis* varies substantially in different species. For example, the torus of the north Pacific squirrelfish *Sargocentron spinosissimum* is amazingly large (Fig. 10.5c), but the *torus longitudinalis* of the Japanese ricefish (medaka)

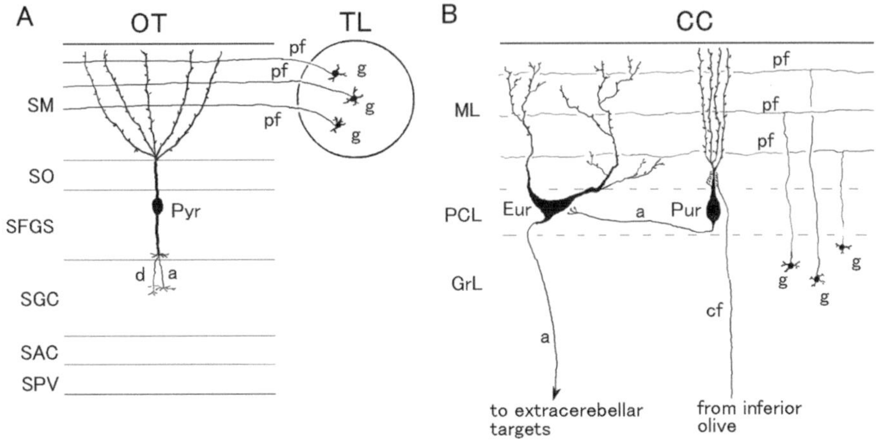

Fig. 10.7 Schematic illustrations showing neural circuits of the *torus longitudinalis* (*TL*)-optic tectum (*OT*) (**a**) and the cerebellar corpus (*CC*) (**b**). (**a**) Granule cells (*g*) of the TL send axons to the *stratum marginale* (*SM*) of the OT. These axons run parallel to each other and are called parallel fibers (*pf*) or marginal fibers. Parallel fibers make synaptic contacts with spiny superficial dendrites of pyramidal cells (*Pyr*). The target tectal cell type of pyramidal cells remains unknown. (**b**) Axons of the granule cells of the cerebellum reach the molecular layer (*ML*), where they run parallel to each other (hence called parallel fibers) to make synaptic contacts with spiny dendrites of Purkinje cells (*Pur*). Dendrites of Purkinje cells spread in the plane perpendicular to parallel fibers and thus two-dimensional spread of the dendrites cannot be appreciated in this illustration. Purkinje cells receive another input from the inferior olive, which is carried by climbing fibers (*cf*). Climbing fibers of teleosts form busy axon terminal arbors around the proximal dendrite (Xue et al. 2008), differing from those of mammals that "climb" superficially along the Purkinje cell dendrite. Purkinje cells send an axon to eurydendroid cells (*Eur*). Eurydendroid cells are output neurons of the cerebellum in teleosts and hence correspond to neurons of cerebellar nuclei of mammals. For other abbreviations, see the list

Oryzias latipes is quite small (Fig. 10.5e). The cytoarchitecture of the torus also show species differences. In the north Pacific squirrelfish the torus can be subdivided into medial and lateral parts, each of which is composed of a granular cell zone and a fibrous zone (Fig. 10.5d). Such subdivisions are not recognized in the goldfish, the Japanese ricefish, or the yellowfin goby *Acanthogobius flavimanus* (Fig. 10.5b, e, f). In the yellowfin goby, a thick shell of small cells enclose the core with scattered cells (Fig. 10.5f), a cytoarchitecture dissimilar to those in the north Pacific squirrelfish, goldfish, or Japanese ricefish. Thus, the size and cytoarchitecture of the *torus longitudinalis* show quite a bit of variation in teleosts. The reason for such diversity, however, remains an open question.

10.4.2 Cerebellar Valvula

The cerebellum of mammals is composed of the hemisphere, vermis, and flocculo-nodular lobe (Steward 2000). The cerebellar hemisphere receives cortical inputs relayed by the pontine nuclei. The vermis mainly receives ascending spinal inputs. The flocculo-nodular lobe is related to equilibrium and control of eye movement and receives fibers of the vestibular nerve. The cerebellum of teleosts is composed of the cerebellar corpus, cerebellar valvula, granular eminence, and caudal lobe. The cerebellar corpus receives inputs from the spinal cord (Finger 2000) and telencephalic inputs relayed by the lateral valvular nucleus (Yang et al. 2004) and hence appears to include components comparable to the hemisphere and vermis of mammals. The granular eminence is a protrusion that is situated caudolaterally adjacent to the corpus. It receives sense of balance through primary afferents of the octaval nerve and lateral line inputs via the lateral line nerve (Yamamoto and Ito 2005; Noro et al. 2007) and projects to the molecular layer of caudal lobe (and also to the cerebellar crest covering the primary lateral line nucleus, and the cerebellar crest overlying primary and secondary auditory nuclei, as mentioned earlier). The caudal lobe is the caudalmost region of the cerebellum, which is present caudally adjacent to the cerebellar corpus. The granular eminence and caudal lobe receiving fibers from the eminence may be comparable, as a unit, to the flocculo-nodular lobe of mammals. Although correspondences of teleost cerebellar regions to mammalian cerebellar regions need to be verified by further studies, those teleost and mammalian cerebellar regions just mentioned are common in that they are present dorsal to the pons. The remaining cerebellar region in teleosts, or the cerebellar valvula, is not situated dorsal to the pons. It protrudes rostrally, from the cerebellar corpus into the mesencephalic ventricle. The cerebellar valvula is present in all actinopterygians except Polypteryformes. It is not found in non-actinopterygian vertebrates. The cerebellar valvula receives descending telencephalic inputs relayed by the lateral valvular nucleus (Yang et al. 2004) but does not appear to receive spinal inputs; a number of previous studies on spinal connections did not report such a connection. The functional significance of the valvula is not well known, although a previous study reported that the common carp sank down onto the bottom after the ablation of the valvula (Ito and Kishida 1978). The cerebellar valvula also shows species differences. In the Japanese ricefish the valvula is small without further divisions (Fig. 10.8a). However, it is very large in the goldfish and is composed of medial and lateral lobes (Fig. 10.8b); the former continues caudally toward the cerebellar corpus. We can see an incredible case of diversity in the cerebellar valvula. The valvula of the elephant nose fish *Gnathonemus petersii* (Mormyridae) reaches an enormous dimension to cover almost completely the other parts of the brain (Fig. 10.6d). This is similar to the situation of us humans, where the cerebral cortex covers the other parts of the brain almost entirely (except the cerebellum), as seen in a dorsal view. The elephant nose fish is a so-called weakly electric fish that emits weak electric currents to survey the environment and also to communicate with other individuals. The fish also possess well-developed electrosensory lateral

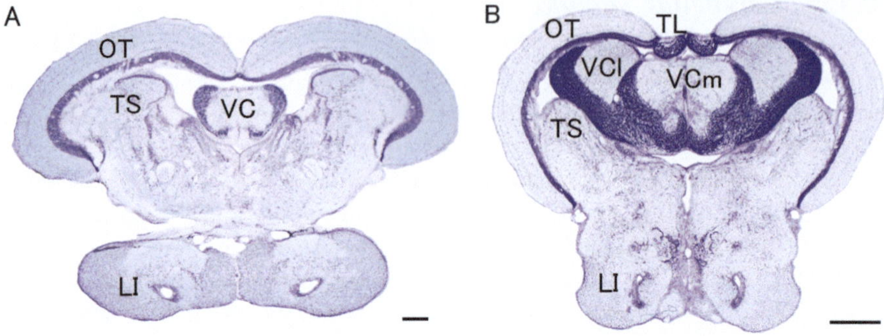

Fig. 10.8 Transverse sections showing the cerebellar valvula (*VC*) of the Japanese ricefish *Oryzias latipes* (**a**: Nissl staining) and the goldfish *Carassius auratus* (**b**: Nissl staining). Note that the VC of the goldfish is very large and composed of the lateral (*VCl*) and medial (*VCm*) lobes, whereas the VC of the ricefish is quite small and not subdivided into lobes. See list for other abbreviations. *Bars* 100 μm (**a**); 500 μm (**b**)

line systems (Fig. 10.6d): the primary sensory structure, the electrosensory lateral line lobe, is huge but only a part of it is visible, hidden by the large cerebellar valvula. This amazing hypertrophy may be related to electrosensory processing or control of electric discharges. However, other weakly electric fish do not possess such a large valvula [e.g., the banded knifefish *Gymnotus carapo* (Gymonotidae)] . The actual cause for this fantastic case of diversity, therefore, remains to be studied further.

10.5 Motor Zones

In general, the cerebellum of vertebrates is involved in motor control in that it processes information that ultimately regulates proper activity patterns of motor neurons. Cerebellar neurons, however, do not directly innervate the muscles or motor neurons (Steward 2000; Meek and Nieuwenhuys 1998). Diversity can be found in the size of cerebellum in teleosts (e.g., compare the northern sea robin and long-spined porcupine fish: Fig. 10.6a, b). However, diversity is not evident regarding the brain regions where motor neurons are present in teleosts. For example, bulges are not seen along the ventral aspect of the spinal cord, where motor neurons are embedded. *Lethotremus awae* (Japanese common name: dango-uo) is a small lampfish 3–4 cm in length. This species adheres to rocks or seaweeds with a sucker or the modified ventral fins. The author investigated the ventral aspect of spinal cord of this species in search of a bulge associated with the motor control of the sucker. Such an enlargement, however, was not found. The Japanese flyingfish *Cypselurus agoo* possesses a pair of huge pectoral fins that are used while gliding. Similarly to the negative finding in *Lethotremus awae*, no bulges are appreciated on the ventral aspect of the spinal cord in the flyingfish (Tuge et al. 1968). In short, no distinct bulges or lobes such as those seen in the sensory regions as described here are present on regions that contain motor neurons, so far as the author is aware. In the spinal cord of mammals slightly thickened regions are recognized at cervical and lumbar levels (cervical and lumbar enlargements) (Steward 2000). These enlargements are thought to reflect the presence of more sensory and motor neurons required for limbs in comparison with the trunk, where the skins are less sensitive to mechanical stimuli and less elaborate muscular systems are present. That is, the enlargements probably emerge from combinatorial effects of sensory and motor requirements. Thus, motor zones per se do not form evident bulges, although sensory zones sometimes result in drastic enlargements. The reason for this difference between the sensory zones and motor zones is not really clear but may be explained as follows. When numerous sensory cells are devoted for greater precision and higher sensitivity of a sensory organ, many sensory neurons have to work in the central nervous system. If sensory inputs from many receptor cells converge on much fewer sensory neurons, this immediately results in a great loss in resolution and/or sensitivity; expending many cells for the sensory organ becomes in vain. In the case of motor neurons, they receive inputs from premotor neurons that configure patterns of motor neuron firing. That is, much of the computation necessary for appropriate control of muscles has already been done before the input reaches the motor neurons; motor neurons just perform the final integration of afferent inputs. Also, motor neurons innervate multiple muscle fibers via axonal branching. Thus, the performance of complex behavior may not require a large number of motor neurons. Simulations with computer models may provide a more quantitative ground for the intuitive speculations mentioned here.

10.6 Developmental Processes that Produce Bulges and Environmental Effects on Brain Morphology

Little is known about the ontogenetic processes leading to the bulged brain regions. Sensory systems are mainly considered herein, because extreme cases of hypertrophy are found mostly in sensory regions. The situations may differ among different sensory modalities. For example, in the case of olfaction, the receptor cells themselves are neurons having an axon that reaches the primary center, or the olfactory bulb. The cause for the enlarged olfactory bulb as seen in the kidako moray may be the increased number of sensory cells; some changes in the gene network regulating the number of olfactory receptor cells may cause the larger size of the olfactory bulb that receives olfactory information. The olfactory bulb is not present when the olfactory placode (anlage of the olfactory epithelium) forms. The olfactory bulb differentiates subsequent to the arrival of the olfactory nerve (fascicle of olfactory receptor axons) onto the developing telencephalon (Honkanen and Ekström 1991). Thus, it is possible that more olfactory nerve axons induce a larger number of telencephalic cells that are recruited into the developing olfactory bulb, although it may not be ruled out that the number of future bulbar neurons is determined on the side of the brain. It is also possible that both factors are involved. Similar arguments may also apply to other sensory modalities where receptor cells are neurons (e.g., free nerve endings in the skin). The story may be different regarding senses that are detected and sent to the central nervous system by two cells. For example, in the case of taste, gustatory stimuli are detected by taste cells in the taste buds. Ganglion cells of the facial, glossopharyngeal, or vagal nerves (primary sensory neurons) receive taste information on the peripheral process and send action potentials to primary centers (i.e., facial, glossopharyngeal, and vagal lobes), through their centrally directed axons. In this case, the cause for enlarged gustatory centers can lie in the sensory receptor cells, primary sensory neurons, or the secondary sensory neurons in the brain. The increase in the number of receptor cells may affect the number of primary afferents reaching the receptor organs, and in turn an increased number of primary afferents results in the production of more secondary sensory neurons in the brain. Flow of influence in the reverse direction, perhaps via trophic factors, might be also present. It is also possible that an increased number of primary sensory neurons results in larger gustatory centers. The amazing diversity of teleost brains, as enumerated in this chapter, provides model cases to study the mechanisms that lead to hypertrophied or poorly developed sensory systems and associated brain regions. The key to solve the question is the mechanism that governs the number of cell divisions made by stem cells during early phases of ontogeny. Differences in the size of brain parts seen in different medaka strains reared under the same conditions suggest that genetic differences underlie this issue (Ishikawa et al. 1999). Also importantly, the body as well as sensory and nervous systems of teleosts continues to grow in adulthood, and the mechanism that determines the number of newly recruited neurons should also have an important role in shaping the brain of teleosts. In fact, the size of brain parts is under the influences of the environment (Eifert et al. 2015). Therefore, the mechanism of the emergence of diversity in brain morphology in teleosts should be considered from the aspects of both genetic and environmental factors.

Acknowledgments The author appreciates Prof. Masami Yoshimoto (University of Tokyo Health Sciences) and Dr. Yuji Ishikawa (National Institute of Radiological Sciences) for a continuing collaboration that forms the basis of the topics introduced in this chapter. The author expresses particular thanks to Prof. Emeritus Hironobu Ito (Nippon Medical School) for guiding me to the fascinating world of fish brain morphology and instructing me about the enormous diversity.

References

Eifert C, Farnworth M, Schultz-Mirbach T, Riesch R, Bierbach D, Klaus S, Wurster A, Tobler M, Streit B, Indy JR, Arias-Rodriguez L, Plath M (2015) Brain size variation in extremophile fish: local adaptation versus phenotypic plasticity. J Zool 295:143–153

Finger TE (2000) Ascending spinal systems in the fish *Prionotus carolinus*. J Comp Neurol 422:106–122

Finger TE (2008) Sorting food from stones: the vagal taste systems in goldfish, *Carassius auratus*. J Comp Physiol A 194:135–143
Finger TE, Böttger B, Hansen A, Anderson KT, Alimohammadi H, Silver W (2003) Solitary chemosensory cells in the nasal cavity serve as sentinels of respiration. Proc Natl Acad Sci USA 100:8981–8986
Gibbs MA, Northmore DPM (1996) The role of torus longitudinalis in equilibrium orientation measured with the dorsal light reflex. Brain Behav Evol 48:115–120
Honkanen T, Ekström P (1991) An immunocytochemical study of the development of the olfactory system in the three-spined stickleback (*Gasterosteus aculeatus* L., Teleostei). Anat Embryol 184:469–477
Ishikawa Y, Yoshimoto M, Yamamoto N, Ito H (1999) Different brain morphologies from different genotypes in a single teleost species, the medaka (*Oryzias latipes*). Brain Behav Evol 53:2–9
Ito H, Kishida R (1978) Afferent and efferent fiber connections of the carp torus longitudinalis. J Comp Neurol 181:465–476
Ito H, Yoshimoto M (1991) Nervous system. In: Itazawa Y, Hanyu I (eds) Fish physiology. Koseisha-Koseikaku, Tokyo, pp 363–402 (in Japanese)
Ito H, Ishikawa Y, Yoshimoto M, Yamamoto N (2007) Diversity of brain morphology in teleosts: brain and ecological niche. Brain Behav Evol 69:76–86
Kiyohara S, Sakata Y, Yoshitomi T, Tsukahara J (2002) The "goatee" of goatfish: innervation of taste buds in the barbels and their representation in the brain. Proc Biol Sci 269:1773–1780
Kotrschal K, van Staaden MJ, Huber R (1998) Fish brains: evolution and environmental relationships. Rev Fish Biol Fish 8:373–408
Meek J, Nieuwenhuys R (1998) Holosteans and teleosts. In: Nieuwenhuys R, Ten Donkelaar HJ, Nicholson C (eds) The central nervous system of vertebrates, vol 2. Springer, Berlin, pp 759–937
Nelson JS (2006) Fishes of the world, 4th edn. Wiley, Hoboken
Noro S, Yamamoto N, Ishikawa Y, Ito H, Ijiri K (2007) Studies on the morphology of the inner ear and semicircular canal end organ projections of *ha*, a medaka behavior mutant. Fish Biol J Medaka 11:31–41
Sbarbati A, Osculati F (2003) Solitary chemosensory cells in mammals? Cells Tissues Organs 175:51–55
Spence R, Gerlach G, Lawrence C, Smith C (2008) The behaviour and ecology of the zebrafish, *Danio rerio*. Biol Rev 83:13–34
Steward O (2000) Functional neuroscience. Springer, New York
Tuge H, Uchihashi K, Shimamura H (1968) An atlas of the brains of fishes of Japan. Tsukiji-Shokan, Tokyo
Whitear M (1971) Cell specialization and sensory function in the fish epidermis. J Zool (Lond) 163:237–264
Xue H-G, Yamamoto N, Yang C-Y, Kerem G, Yoshimoto M, Imura K, Ito H (2003) Fiber connections of the torus longitudinalis and optic tectum in holocentrid teleosts. J Comp Neurol 462:194–212
Xue H-G, Yang C-Y, Yamamoto N (2008) Afferent sources to the inferior olive and distribution of the olivocerebellar climbing fibers in cyprinids. J Comp Neurol 507:1409–1427
Yamamoto N (2005) Nervous system. In: Taniuchi T et al (eds) Encyclopedia of fish science. Asakura Shoten, Tokyo, pp 132–147 (in Japanese)
Yamamoto N, Ito H (2005) Fiber connections of the central nucleus of semicircular torus in cyprinids. J Comp Neurol 491:186–211
Yang C-Y, Yoshimoto M, Xue H-G, Yamamoto N, Imura K, Sawai N, Ishikawa Y, Ito H (2004) Fiber connections of the lateral valvular nucleus in a percomorph teleost, tilapia (*Oreochromis niloticus*). J Comp Neurol 474:209–226

Chapter 11
Molecular Profiling Reveals Insight into Avian Brain Organization and Functional Columnar Commonalities with Mammals

Kazuhiro Wada, Chun-Chun Chen, and Erich D. Jarvis

Abstract The avian cerebrum has pallial functions similar to those of the mammalian cortex. Although the avian pallium is organized as nuclear structures, and the mammalian cortex as layers, the avian pallium supports cognitive abilities similar to those of many mammals. We recently presented a global view of the pallial organization of birds, based on quantitative analyses of constitutively expressed or behaviorally regulated genes in different pallial cell populations (Jarvis et al. J Comp Neurol 521:3614–3665, 2013; Chen et al. J Comp Neurol 521:3666–3701, 2013). Here we present a shortened synopsis of these articles. The findings of the constitutively expressed genes and known neural connectivity reveal four major cell populations: (1) a primary sensory input population, (2) a secondary intrapallial population, (3) a tertiary intrapallial population, and (4) a quaternary output population. These populations have greater similarities to cell layers of the mammalian cortex than to the amygdala. The patterns of behaviorally regulated genes revealed functional columns of activation across boundaries of these cell populations, reminiscent of columns through layers of the mammalian cortex. Each neural cell population contributes portions to columns that control different sensory or motor systems. These findings influence hypotheses on homologies of the avian pallium with other vertebrates.

Keywords Forebrain • Neural activity • Motor behavior • Primary sensory • Neurotransmitter receptors • Pallium • Cortex • Striatum • Pallidum • Basal ganglia

---

K. Wada (✉)
Faculty of Science, Department of Biological Sciences, Hokkaido University, Sapporo, Hokkaido 060-0810, Japan
e-mail: wada@sci.hokudai.ac.jp

C.-C. Chen • E.D. Jarvis (✉)
Department of Neurobiology, Duke University Medical Center, Box 3209, Durham, NC 27710, USA

Howard Hughes Medical Institute, Chevy Chase, MD 20815, USA
e-mail: jarvis@neuro.duke.edu

S. Shigeno et al. (eds.), *Brain Evolution by Design*, Diversity and Commonality in Animals, DOI 10.1007/978-4-431-56469-0_11

11.1 Introduction

In English, the term "birdbrain" is sometimes used to mean "stupid," which is untrue in the field of animal behavior and neuroscience. Many birds have cognitive proficiencies that are quite sophisticated. For example, as reviewed by Jarvis et al. (2005), scrub jays (*Aphelocoma coerulescens*) show episodic memory, the ability to recall autobiographical events that happened at a specific time or place (Clayton and Dickinson 1998). New Caledonian crows (*Corvus moneduloides*) make tools and use them to retrieve food, and are thought to pass this knowledge on to other crows through social learning (Weir et al. 2002; Hunt and Gray 2003). Pigeons (*Columba livia*) can memorize up to 725 different visual patterns (Vonfersen and Delius 1989), learn to categorize objects as 'human-made' versus 'natural' (Lubow 1974), and discriminate cubistic and impressionistic styles of painting (Watanabe et al. 1995). Parrots, hummingbirds, and oscine songbirds possess the rare trait of vocal learning that is a prerequisite in humans for spoken language (Jarvis 2004).

These cognitive abilities are mainly controlled by the telencephalons. The avian telencephalon organization was recently reclassified into two major regions that contain at least seven major subdivisions: pallium (containing hyperpallium, mesopallium, nidopallium, arcopallium, and hippocampus) and subpallium (containing striatum and pallidum) (Reiner et al. 2004b; Jarvis et al. 2005) (Fig. 11.1). In neuroanatomy, the pallium is the cortical region of the telencephalon and the subpallium is equivalent to the basal ganglia. Although the avian pallium supports cognitive abilities similar to those of many mammals, their pallial organization is different. The mammalian pallium, that is, the cortex, is laminar whereas the avian pallium structure is nuclear. In contrast, the avian subpallial subdivisions, striatum and pallidum, are more conserved with their mammalian counterparts in developmental origin, connectivity, cell types, and cell organization (Medina and Abellan 2009; Butler et al. 2011). Because of the less conserved pallium organization, our understanding of the organization of the avian pallium and its cellular homologies with mammals has still been controversial (Jarvis et al. 2005).

To address this issue, we recently performed a quantitative analyses of telencephalic expression profiles of constitutive and activity-responsive genes (52 total) in the adult and a subset of these in the embryonic avian brain to decipher the molecular and functional relationships between different avian telencephalic cell populations (Chen et al. 2013; Jarvis et al. 2013). The 52 genes were selected on the basis of their specific expression patterns in particular sectors of the telencephalon, their relationships with mammalian brain structures, or their functions. Our assumption was that similar brain areas should express similar gene sets to achieve similar functions. Therefore, we chose genes with a wide range of functions and cellular locations, from inside the nucleus to the extracellular space, including neurotransmitter/neuromodulator receptors, protein ligands, transcription factors, enzymes, and a diverse set of membrane and cytoplasmic genes. Among these,

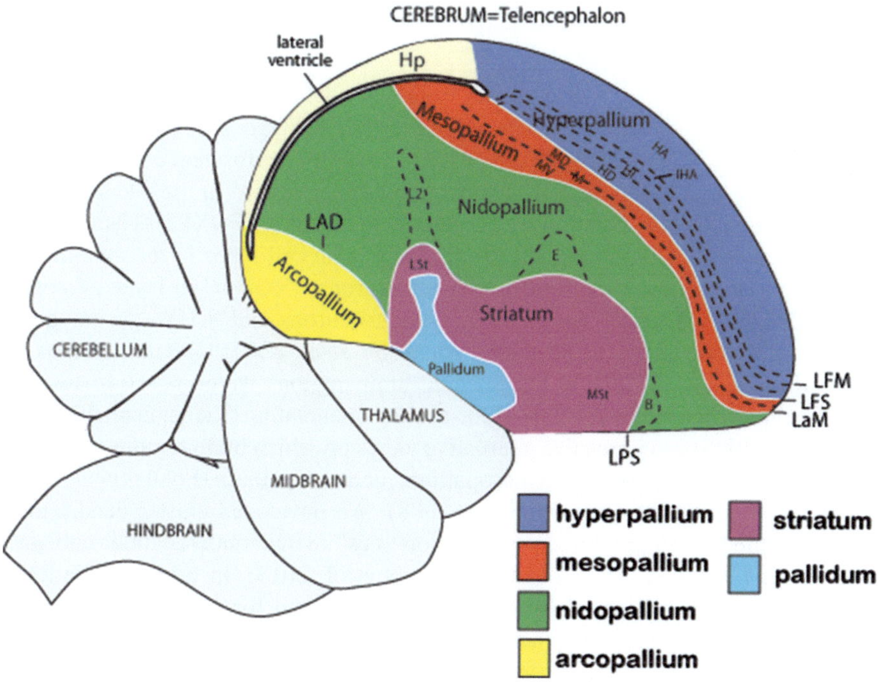

Fig. 11.1 Modern view of avian cerebral organization. Modern (2004–2005) consensus view of avian brain relationships according to the conclusions of the Avian Brain Nomenclature Forum (Reiner et al. 2004a; Jarvis et al. 2005) (Figure panels from Jarvis et al. (2013) used with permission)

we included six genes (BDNF, EGR1, C-FOS, C-JUN, DUSP1, and ARC) that are activity regulated in the brain by sensory and motor behaviors (Wada et al. 2006; Jarvis et al. 2013).

Based on the totality of the findings, we proposed a new view of avian pallium organization as four major cell populations, instead of seven, that wrap above and below the lateral ventricle instead of being different populations below and above. Some of these cell populations have similarities with molecular profiles similar to the different layers of mammalian cortex, as was recently further supported with high-throughput gene expression profiling between avian and primate brains (Pfenning et al. 2014). Furthermore, the patterns of behaviorally regulated genes revealed functional columns of activation across the boundaries of these cell populations, reminiscent of columns through layers of the mammalian cortex. Here we present a synopsis of those findings, including a shortened and updated text of the study. For additional information on the evolving views of avian brain organization, see (Montiel and Molnar 2013).

11.2 Six Major Cerebral Expression Domains

The basal expression levels of 50 genes were quantified in 20 telencephalic regions and 3 cerebellum layers as outgroup regions in the zebra finch (Jarvis et al. 2013), which is a songbird species widely used in neuroscience. The specific genes were chosen based on their distinct expression patterns in the telencephalon, their diverse molecular functions (e.g., glutamate receptors, FOXP2) (Haesler et al. 2004; Wada et al. 2004), and their use to distinguish cell types in the mammalian brain (e.g., "dopamine receptors" in striatal neurons, "ROR-b" in layer IV cortex neurons, "ER81" in layer V cortex and amygdala neurons, and "LHX9" in amygdala neurons) (Molnar and Cheung 2006; Watakabe et al. 2007; Garcia-Lopez et al. 2008; Kubikova et al. 2010). The specific brain regions were chosen to test the relationships proposed in the 2004–2005 brain nomenclature (Riener et al. 2004a, b; Jarvis et al. 2005) and to resolve alternative views on which brain regions constitute the newly defined hyperpallium, mesopallium, and arcopallium (Puelles et al. 2000; Yamamoto et al. 2005; Feenders et al. 2008). We developed cluster dendrogram analyses called "brain phylo-gene expression trees" to quantitatively infer unbiased relationships between cell populations (Jarvis et al. 2013). In brief, first, mRNA signals detected by in situ hybridization were digitized for each brain region of interest. Following this, the values of signal intensity were normalized to a scalar range from 0 to1 to reduce experimental artifacts and adjust discretization (this normalized value is shown as a heatmap in Fig. 11.2). The expression values for each brain region were converted into a vector "x_i" and similarity scores between all pairs of vectors were calculated using either Distance–Correlation or Euclidian–Distance. The Distance–Correlation or Euclidean–Distance similarity values between all pairs of brain regions were used to generate a hierarchical cluster tree. For this expression analysis, all genes used have the same impact on clustering of the phylo-gene expression tree. Phylo-gene expression tree analyses grouped the 20 telencephalic regions into at least six major expression clusters at high similarity (left color-coded in Fig. 11.2). These six expression clusters were as follows.:

1. 1^0-pallium: The primary sensory input fields L2, entopallium (E), and basorostralis (B) and hyperpallium intercalatum (IH) at 0.8 correlation, which together we refer to as primary pallium [Figs. 11.2 (orange), 11.3a].
2. 2^0-pallium: The nidopallium (N) and hyperpallium (H; apicale part) minus primary sensory input fields (L2, E, B, and IH) at 0.86–0.99 correlation, which together we refer to as secondary pallium [Figs. 11.2 (green), 11.3b].
3. 3^0-pallium: The dorsal mesopallium (MD; hyperpallium densocellulare in the 2004–2005 view) and ventral mesopallium (MV; mesopallium in the 2004–2005 view) with near identity at 0.99, which together we refer to as tertiary pallium [Figs. 11.2 (red), 11.3c].
4. 4^0-pallium: The arcopallium (A) regions, at 0.75, which together we refer to as quaternary pallium [Figs. 11.2 (yellow), 11.3d].

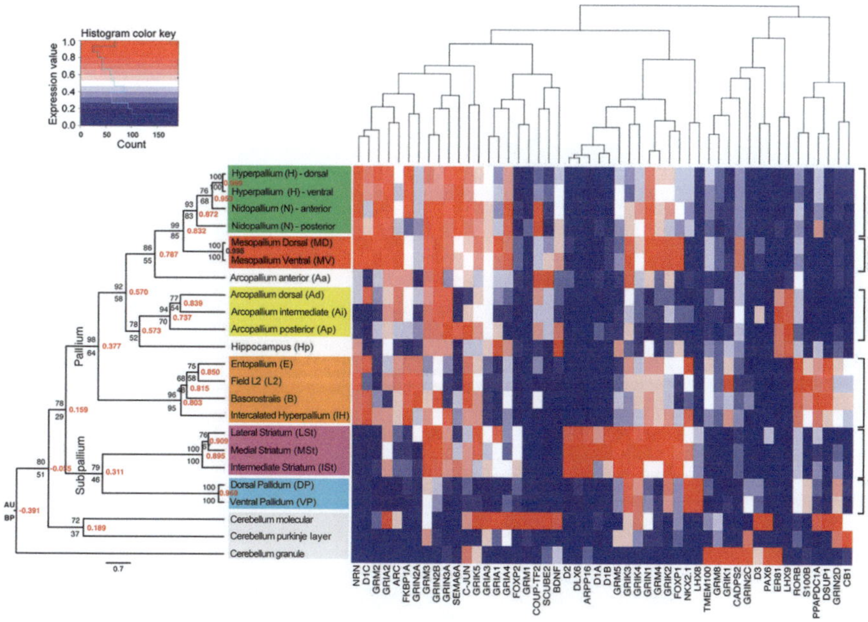

Fig. 11.2 Brain phylo-gene expression tree. Tree (*left*) and gene expression heatmap (*right*) show molecular relationships of 23 brain regions of the zebra finch based on 50 genes. High similarity is considered at 0.7–0.99 correlation coefficient (*red* values inside nodes), 70–100 % bootstrap probability (*number above branches*), and 93–100 approximate unbiased probability supports (*number below branches*). The six major telencephalic subdivisions revealed by the tree are color-coded (names of brain regions). The tree was generated with Distance-Correlation on normalized gene expression data, followed by Biedl's ordering of leaves according to similarity of gene expression vectors. *Far right* is the global numbered pallial and subpallial terminologies based on this tree and known connectivity. The gene expression heatmap shows relative expression levels for each gene scaled between 0 and 1 (*red*, higher than the average for that region relative to other regions; *blue*, lower than the average). Above the heatmap is the tree relationship of the genes based on brain expression (Figure reproduced from Jarvis et al. (2013) with permission)

5. The lateral and medial striatum plus the intermediate striatum (intrapeduncular nucleus in the classical view) at 0.9 [Figs. 11.2 (purple), 11.3e].
6. The dorsal and ventral pallidum at 0.97 [Figs. 11.2 (turquoise), 11.3f].

The higher correlation value (at 0.9) among the subpallial regions (i.e., striatum and pallium) indicates consistent results of distinct expression compared with those among the pallial regions, as described in previous studies (Reiner et al. 2004a; Jarvis et al. 2005).

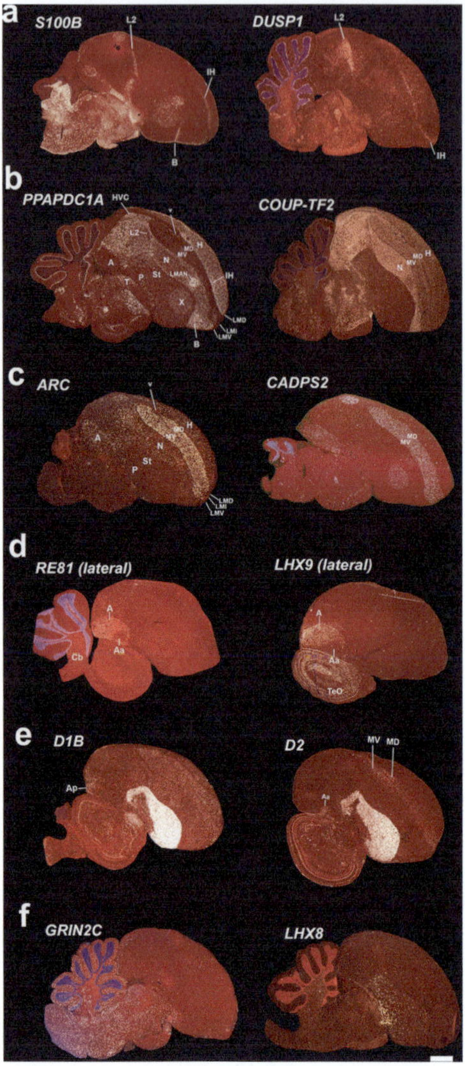

Fig. 11.3 Examples of six major cerebral expressions. (**a**) Examples of intercalated pallium (1^0-pallium)-enriched genes, S100 calcium-binding protein, and dual specificity phosphatase 1 (DUSP1). (**b**) Examples of hyperpallium or nidopallium (2^0-pallium)-enriched genes, phosphatidic acid phosphatase 2 domain containing 1A (*PPAPDC1A*), and chicken ovalbumin upstream promoter transcription factor 2 (*COUP-TF2*) (**c**) Examples of mesopallium (3^0-pallium)-enriched genes, activity-regulated cytoskeleton-associated gene (*ARC*), and calcium-dependent secretion activator 2 (*CADPS2*). (**d**) Examples of arcopallium (4^0-pallium)-enriched genes, Ets-related 81 (*ER81*) transcription factor, and LIM home domain 9 (*LHX9*). (**e**) Examples of striatum-enriched genes, dopamine 1B (*D1B*) receptor, and dopamine 2 (*D2*) receptor. (**f**) Examples of pallidum-enriched genes, glutamate receptor ionotropic NMDA subunit 2C (*GRIN2C*), and LIM homeobox 8 (*LHX8*) transcription factor.Images are in situ hybridizations with cDNAs of the associated genes in zebra finch brain tissue. Gene expression mRNA signal is *white silver grains*; general cellular stain is cresyl violet (*red*) (Figure reconstructed from panels in Jarvis et al. (2013), used with permission.) *Bar* 1 mm

11.3 Functional Columns of Brain Activation in the Avian Brain

The basal expression patterns of the genes studied were confined to lamina-defined subdivision boundaries. However, the induced expression patterns of the activity-dependent genes (EGR1, C-FOS, C-JUN, ARC, BDNF, and DUSP1) were not confined to the lamina-defined boundaries (Mello et al. 1992; Jarvis and Nottebohm 1997; Wada et al. 2006; Feenders et al. 2008; Horita et al. 2010, 2012; Jarvis et al. 2013). These genes were upregulated in subsets of cell types in different brain regions when the animals processed specific sensory stimuli or performed repeated motor behaviors. Thus, the genes were used to map physiological activation of different cell types within functionally connected neural systems (Feenders et al. 2008; Horita et al. 2010, 2012). EGR1, C-FOS, C-JUN, and ARC are all inducible in pallial and striatal cells except the primary sensory neurons (particularly for EGR1; Mello and Clayton 1995; Wada et al. 2006; Feenders et al. 2008). DUSP1 is mainly inducible in primary sensory neurons of the telencephalon and thalamus (Horita et al. 2010). BDNF is mainly inducible in pallial cells (Wada et al. 2006). By examination of the profiles of these genes from prior studies with our modified view of avian brain organization, it was found that the avian brain shows semi-"columnar" patterns of activation across specific combinations of brain subdivisions. Four such columns are as follows.

11.3.1 Auditory Column

When songbirds hear playbacks of song, while sitting in the dark in sound isolation chambers, and do not vocalize in response, a column of adjacent brain regions show early growth response (EGR)1 and dual-specificity phosphatase (DUSP)1 activation with boundaries that transverse brain subdivisions (Fig. 11.4a, b, vs. c, d) (Mello et al. 1992; Velho et al. 2005; Feenders et al. 2008; Horita et al. 2010). The EGR1-activated portion of the column consisted of a caudal part of the ventral mesopallium (often called the caudal mesopallium, CM), the subadjacent caudal nidopallium (comprising what has been called L1, L3, and NCM), and a subadjacent part of caudal striatum (called CSt). The DUSP1-activated portion was the primary sensory L2 cells of nidopallium intercalatum (IN; combined population of L2, E, and B) (Fig. 11.4d). The only major auditory-activated telencephalic region separate from this column was the RA (robust arcopallium) cup in the arcopallium, adjacent to the RA song nucleus (Mello and Clayton 1994; Feenders et al. 2008).

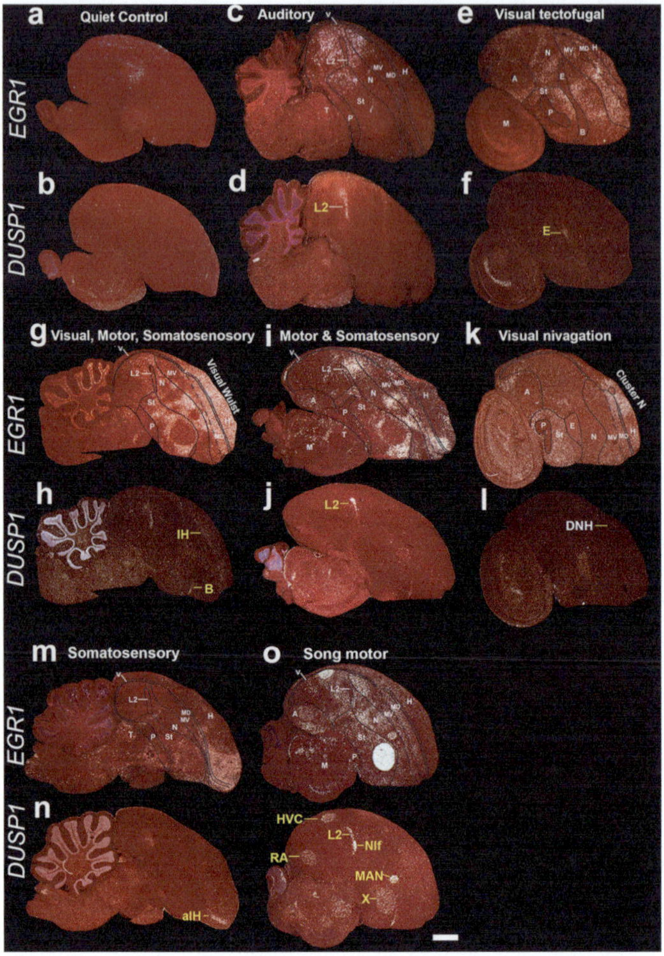

Fig. 11.4 Functional columns revealed by activity-dependent gene expression in the songbird brain. (**a**, **b**) Example of basal expression in an awake zebra finch, sitting still in the dark. (**c**, **d**) Example of hearing-induced gene expression in a zebra finch that heard playbacks of three different conspecific songs for 30 min. (**e**, **f**) Example of light-induced gene expression in the visual pathway of the ventral pallium from a zebra finch sitting still and stimulated with daylight for 1 h after an overnight period of darkness. (**g**, **h**) Example of induced gene expression in three columns: the visual pathway of the dorsal pallium, the adjacent anterior somatosensory pathway of the dorsal pallium, and a motor-activated region of the ventral pallium surrounding the song nuclei in a zebra finch male that hopped around a cylindrical cage for 30 min with lights on. (**i**, **j**) Example of hopping-induced gene expression in the dorsal pallium somatosensory pathway and the ventral pallium motor regions surrounding the song nuclei in a deaf male zebra finch that hopped in a rotating wheel with lights off. (**k**, **l**) Example of dim-light, magnetic vision-induced gene expression in a light-dependent magnetic compass sensing column (Cluster N) of the dorsal pallium found in night-migrating garden warblers. (**m**, **n**) Example of hopping-induced gene expression in a similarly treated animal, but in a more medial part of the dorsal pallial somatosensory pathway. (**o**, **p**) Example of singing-induced gene expression in song nuclei of a male zebra finch that sang for 30 min and made some hopping movements between singing bouts. *Bar* 1 mm (Figure reproduced from Jarvis et al. (2013) with permission)

11.3.2 Visual Columns

When birds are stimulated with light after an overnight period of darkness, two clusters of activation are seen in known visual pathways (Feenders et al. 2008; Hara et al. 2009; Horita et al. 2010). The first cluster consists of a patchy column that includes EGR1 activation in a portion of MV near the E, the subadjacent N near E, the subadjacent striatum ventral to E, and DUSP1 activation in E of the IN (Fig. 11.4e, f). The second and larger column is in the Wulst and consists of EGR1 activation in the central part of MD, in the overlying central part of the H, and DUSP1 in the central part of IH (Fig. 11.4g, h); the anterior end of this column is not visual, but includes somatosensory parts of H, IH, and MD, activated during movement (Fig. 11.4i, j) (Feenders et al. 2008). The caudal end of this cluster in migratory songbirds has a dim-light magnetic sensing-activated column called cluster N, also consisting of portions of H, IH, and MD involved in processing light-dependent magnetic compass information (Fig. 11.4k, l) (Mouritsen et al. 2005; Liedvogel et al. 2007; Zapka et al. 2009; Zapka et al. 2010). A difference from the ventral pallial visual columns is that with the dorsal columns we could not find not striatal regions of activation associated with them.

11.3.3 Somatosensory and Motor Columns

When birds hop, walk, or fly (particularly while deaf and in the dark to eliminate auditory and visual activation), activated columns of brain regions are found in the two known somatosensory pathways and an apparent motor pathway (Feenders et al. 2008; Horita et al. 2010). The somatosensory pathway columns include (1) the MV and N adjacent to B (in anterior IN), B itself (for DUSP1), and the adjacent striatum; and (2) the anterior MD and H adjacent to anterior IH, and anterior IH itself (for DUSP1; Fig. 11.4m, n). The motor-activated column includes a proportion of the anterior MV, the subadjacent anterior N, and the subadjacent anterior striatum, all surrounding anterior song nuclei (MO, MAN, and Area X, respectively) in song learning species (Fig. 11.4g, i). Another movement-activated column includes the posterior-lateral MV, the adjacent N, and interestingly a part of L2 of the IN located within the auditory column; that is, these parts of MV, N, and L2 have both motor and auditory activation (Feenders et al. 2008). Similar to the auditory pathway, there was a movement-activated region in the arcopallium, the lateral intermediate arcopallium, that was not part of a column (Feenders et al. 2008).

11.3.4 Song Nuclei

When song-learning birds sing, singing-driven gene expression occurs in seven analogous song nuclei. These song nuclei include the mesopallium oval nucleus (MO) that according to our modified view is located in MV, the underlying lateral magnocellular nucleus nucleus (LMAN) in the anterior nidopallium, and the underlying area X in the anterior striatum, forming a column within a column (Fig. 11.5a, b). The four posterior song nuclei in songbirds do not show a columnar organization, although we note that the interfacial nucleus (NIf) in the nidopallium meets up with avalanche (Av) in MV at the lamina border (best seen with C-FOS; Fig. 11.5b). In contrast, in hummingbirds and parrots the HVC-like and RA-like song nuclei are adjacent to each other in a semi-columnar organization

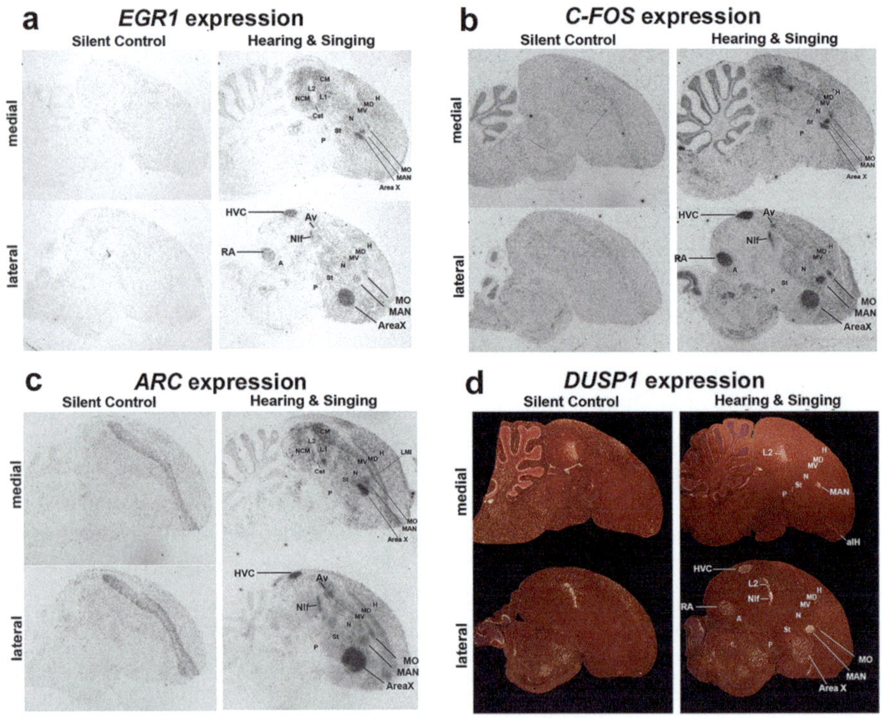

Fig. 11.5 Singing-driven neural activity-dependent gene expression. (**a**) EGR1. (**b**) C-Fos. (**c**) ARC. (**d**) DUSP1. *Left* images in each panel: Medial and lateral sections from a quiet control, sitting still animal, showing basal neural activity-dependent gene expression. *Right* images in each panel: Medial and lateral sections from a singing animal (for 30 min), perched, without much hopping, showing the highest levels of singing-driven increased gene expression in song nuclei (*lines*), and hearing-driven increased expression in the auditory pathway (NCM, L1, L3, CM, CSt) from hearing itself sing. (**a–c**) Images from X-ray film exposure. (**d**) Images from emulsion-dipped slides (Figure panels reproduced from Jarvis et al. (2013) with permission)

at the arcopallium lamina border (Jarvis and Mello 2000; Jarvis et al. 2000). The activated patterns conform to the differential expression differences already present at baseline. For example, EGR1 and C-FOS are induced in all nonprimary sensory pallial regions, and they are induced in all song nuclei in the corresponding brain subdivisions (Fig. 11.5a, b). C-JUN is low in the mesopallial regions at baseline, and it is not induced in the mesopallial song nuclei (Av and MO) by singing (Wada et al. 2006). ARC is high in the mesopallium at baseline, and it is induced in the mesopallial song nuclei by singing (Fig. 11.5c). BDNF is not expressed in the striatum at baseline, and it is not induced in area X by singing (Wada et al. 2006). The only exception is DUSP1, which is not induced in higher order (nonprimary) connected neurons in the forebrain except in song nuclei by singing (Fig. 11.5d) (Horita et al. 2012).

11.4 Conclusions

The Jarvis et al. (2013) study contains the highest number of genes and the most quantitative for comparative in situ expression profiling of avian brain subdivisions to date. Along with the companion study by Chen et al. (2013) on the developmental profiles of some of these genes, the key new elucidated points are (1) the expression patterns of most genes in the avian telencephalon can be grouped into one of six expression populations, four pallial and two subpallial; (2) the pallial populations form a partial mirror image profile of each other above and below the ventricle and the associated lamina through the mesopallium; (3) these populations form first as layers of cells in the dorsal or ventral pallium that later during development wrap around the ventricle space and mesopallium to form the mirror image profiles; and (4) three of the pallial populations (IN + IH; nidopallium + hyperpallium; mesopallium) function in a columnar-like organization for specific sensory or motor systems (Fig. 11.6a).

11.4.1 Functional Column Organization of the Avian Cerebrum

The activity-dependent gene expression results suggest that the avian pallium also has a functional columnar organization akin to the mammalian pallium. The finding by Jarvis et al. (2013) is accordance with those of previous studies (Karten 1997; Medina and Reiner 2000; Wang et al. 2010). However, we emphasized this concept of "functional column" on the finding of a semi-mirror pallial organization. In mammals, the columns span the six cortical layers or pallial amygdala regions for particular sensory or motor systems, with topographic projections from the columnar layer 5 cells to the striatum (Swanson 2000). In the avian brain, the functional adjacent columns wrap from ventral pallial (auditory, visual, somatic motor,

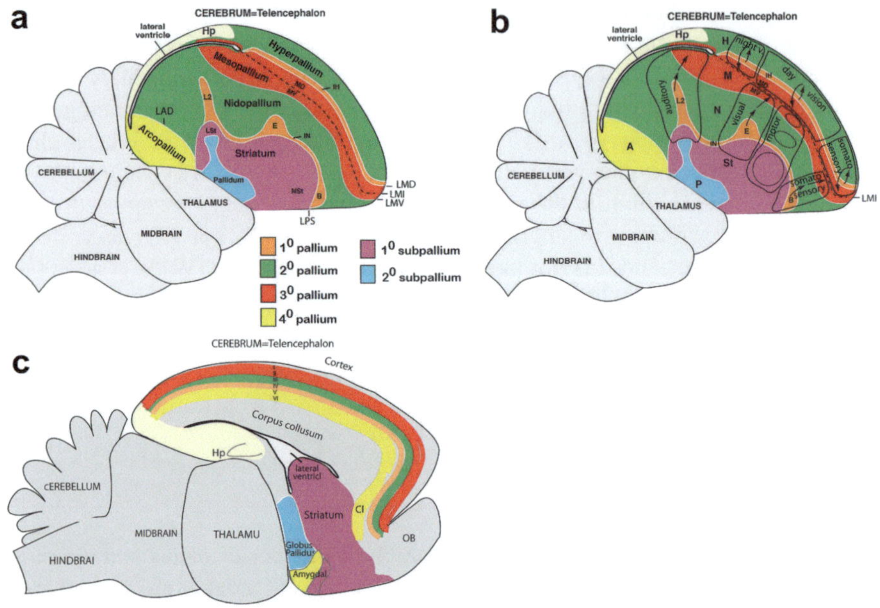

Fig. 11.6 General model of avian brain organization proposed by Jarvis et al. 2013. (**a**) Drawing in sagittal view with subdivision shapes based on songbirds. (**b**) Same drawing as in (**a**) with outlines of different brain systems that show columnar activation of immediate early genes (IEGs). *Arrows* show known connectivity. (**c**) Color-coded scheme of rodent brains according to the field hypothesis of homology with the avian brain proposed in our study (Jarvis et al. 2013). For all images, *solid white lines* are lamina (relatively cell-sparse zones) that separate subdivisions; *dashed lines* divide regions within a subdivision, whether or not a lamina is present

somatosensory) to dorsal pallial (somatosensory, visual) regions, each containing a primary (IN or IH), secondary (N or H), and tertiary (MV or MD) pallial population (Fig. 11.6b); when seen in this view the two avian somatosensory pathways are near each other in the anterior forebrain. The avian ventral pallial columns also include topographically positioned striatal-activated regions. We do not yet have an immediate early gene (IEG) that is activated in pallidal cells to determine if it too forms topographically organized columns with the striatum and pallium. The pallidal cells intermingled in the anterior striatum of birds (Kuenzel et al. 2011) presumably make up the pallidal component of the somatic motor and vocal motor columns. We do not know if there are striatal, pallidal, and arcopallial components to the dorsal pallial columns, although this is likely, because the hyperpallium, similar to the nidopallium, sends robust projections to the striatum (Veenman et al. 1995). Based on these findings, Jarvis et al. (2013) suggested that a minimal column system to process sensory information in the avian telencephalon is to incorporate adjacent sectors of 1^0-pallium (IN or IH), 2^0-pallium (N or H), and 3^0-pallium (MV or MD), and the striatum, with feedback to the brainstem primarily via the 4^0-pallium (arcopallium). A minimal system to produce behavior is to incorporate adjacent sectors of the 2^0-pallium, 3^0-pallium, striatum, and projections to brainstem/spinal cord motor nuclei via 4^0-pallium.

11.4.2 Brain Homologies Between Birds and Mammals

There are still two major controversial hypotheses on specific homologies of different cell populations between the avian and mammalian pallium. (1) The *nuclear-to-layer* hypothesis posits that the different nuclear subdivisions of the avian telencephalon below the lateral ventricle, called the dorsal ridge (DVR), contain cell types that are homologous to layers I–IV of the mammalian cortex. (2) The *nuclear-to-claustrum/amygdala* hypothesis posits that the DVR subdivisions are, in contrast, homologous to the mammalian amygdala and claustrum. Although we believe that the reviewed findings impact the two competing hypotheses on avian and mammalian pallial homologies, we caution against using the reviewed findings alone to equate the "homology" of the avian columns with the mammalian six-layered cortex columns because it is also possible that each vertebrate group develops a similar functional organization by different mechanisms. The mammalian cortical columns extend across layers of cells that predominantly arrive in their locations by radial migration from the same sector of the ventricle zone during development, whereas the avian pallial columns extend across larger clusters of cells that may predominantly arrive in their location by tangential migration from different sectors of the ventricle zone (Medina and Abellan 2009). The pallial portions of the mammalian amygdala and claustrum develop by diverse mechanisms, including both local radial and long-distance tangential migration of cells from the dorsal pallium, the thalamus, and preoptic area (Carney et al. 2006; Hirata et al. 2009; Soma et al. 2009; Garcia-Moreno et al. 2010; Bupesh et al. 2011). Recent studies of columnar circuit formation in mammals have revealed that the columnar neural connection across layers within the mammalian cortex is accomplished by preferential synaptic connections among clonally related neurons, which are originally generated by a single neural progenitor (Yu et al. 2009; Ohtsuki et al. 2012; Gao et al. 2013). In contrast, the developmental mechanism of avian columnar circuit formation is still almost unidentified. At least, avian species possess a genetically controlled developmental program for the chronological generation of the layer-specific neural subtypes (Suzuki et al. 2012).

Nevertheless, regardless of how the pallial and subpallial cells arrive to their final destinations and how the layer-specific subtype neurons connect with other cell populations to develop functionally specialized columnar circuits in mammals and birds, the final outcome is similar: columns across different cell populations organized as layers (cortex in mammals) or thick nuclear slabs (pallium in birds and claustrum/amygdala in mammals) (Fig. 11.6c). In a previous study performed by our group, we analyzed expression profiles of more than 7000 homologous genes between avian and primate genomes of the avian and primate brains, respectively, using oligo-microarrays by focusing on vocal learning-related regions (Pfenning et al. 2014). In that study, we found that the molecular profile of RA, a song nucleus located in the arcopallium, was more similar to that of layer 5 cells of the primate primary motor cortex than to that of any other brain region; HVC, a nidopallium song nucleus, was more similar to layers 2 and 3 of the cortex; and basal ganglia Area X was most similar to parts of the anterior putamen/caudate. There was a negative correlation between RA and the amygdala and claustrum. These findings support the cortex-layered hypothesis of avian and mammalian homologies; thus, the model is proposed in Fig. 11.6a, b. However, we emphasize that our hypotheses need to be further tested with additional comparative high-throughput gene expression profiles between mammals, birds, and non-avian reptiles as well as comparative analyses at other biological levels. Although further molecular profiling investigations should be performed to better resolve forebrain homologies (Belgard et al. 2013; Zhang et al. 2014), we believe that two criteria must be met. (1) One needs to know the global expression profile within each species, including cortical, amygdala, and claustrum patterns, in mammals and the major subdivisions described here for birds. (2) One needs to compare multiple genes across species to guard against making broad conclusions on the basis of one or a few genes that could show variations within a brain subdivision.

References

Belgard TG, Montiel JF, Wang WZ, Garcia-Moreno F, Margulies EH, Ponting CP, Molnar Z (2013) Adult pallium transcriptomes surprise in not reflecting predicted homologies across diverse chicken and mouse pallial sectors. Proc Natl Acad Sci USA 110:13150–13155. doi:10.1073/pnas.1307444110

Bupesh M, Legaz I, Abellan A, Medina L (2011) Multiple telencephalic and extratelencephalic embryonic domains contribute neurons to the medial extended amygdala. J Comp Neurol 519:1505–1525. doi:10.1002/cne.22581

Butler AB, Reiner A, Karten HJ (2011) Evolution of the amniote pallium and the origins of mammalian neocortex. Ann N Y Acad Sci 1225:14–27. doi:10.1111/j.1749-6632.2011.06006.x

Carney RS, Alfonso TB, Cohen D et al (2006) Cell migration along the lateral cortical stream to the developing basal telencephalic limbic system. J Neurosci 26:11562–11574. doi:10.1523/JNEUROSCI.3092-06.2006

Chen CC, Winkler CM, Pfenning AR, Jarvis ED (2013) Molecular profiling of the developing avian telencephalon: regional timing and brain subdivision continuities. J Comp Neurol. doi:10.1002/cne.23406

Clayton NS, Dickinson A (1998) Episodic-like memory during cache recovery by scrub jays. Nature (Lond) 395:272–274. doi:10.1038/26216

Feenders G, Liedvogel M, Rivas M et al (2008) Molecular mapping of movement-associated areas in the avian brain: a motor theory for vocal learning origin. PLoS One 3, e1768. doi:10.1371/journal.pone.0001768

Gao P, Sultan KT, Zhang XJ, Shi SH (2013) Lineage-dependent circuit assembly in the neocortex. Development (Camb) 140:2645–2655. doi:10.1242/dev.087668

Garcia-Lopez M, Abellan A, Legaz I, Rubenstein JL, Puelles L, Medina L (2008) Histogenetic compartments of the mouse centromedial and extended amygdala based on gene expression patterns during development. J Comp Neurol 506:46–74. doi:10.1002/cne.21524

Garcia-Moreno F, Pedraza M, Di Giovannantonio LG, Di Salvio M, Lopez-Mascaraque L, Simeone A, De Carlos JA (2010) A neuronal migratory pathway crossing from diencephalon to telencephalon populates amygdala nuclei. Nat Neurosci 13:680–689. doi:10.1038/nn.2556

Haesler S, Wada K, Nshdejan A, Morrisey EE, Lints T, Jarvis ED, Scharff C (2004) FoxP2 expression in avian vocal learners and non-learners. J Neurosci 24:3164–3175. doi:10.1523/JNEUROSCI.4369-03.2004

Hara E, Kubikova L, Hessler NA, Jarvis ED (2009) Assessing visual requirements for social context-dependent activation of the songbird song system. Proc Biol Sci 276:279–289. doi:10.1098/rspb.2008.1138

Hirata T, Li P, Lanuza GM, Cocas LA, Huntsman MM, Corbin JG (2009) Identification of distinct telencephalic progenitor pools for neuronal diversity in the amygdala. Nat Neurosci 12:141–149. doi:10.1038/nn.2241

Horita H, Wada K, Rivas MV, Hara E, Jarvis ED (2010) The dusp1 immediate early gene is regulated by natural stimuli predominantly in sensory input neurons. J Comp Neurol 518:2873–2901. doi:10.1002/cne.22370

Horita H, Kobayashi M, Liu WC, Oka K, Jarvis ED, Wada K (2012) Specialized motor-driven dusp1 expression in the song systems of multiple lineages of vocal learning birds. PLoS One 7, e42173. doi:10.1371/journal.pone.0042173

Hunt GR, Gray RD (2003) Diversification and cumulative evolution in New Caledonian crow tool manufacture. Proc Biol Sci 270:867–874. doi:10.1098/rspb.2002.2302

Jarvis ED (2004) Learned birdsong and the neurobiology of human language. Ann N Y Acad Sci 1016:749–777. doi:10.1196/annals.1298.038

Jarvis ED, Mello CV (2000) Molecular mapping of brain areas involved in parrot vocal communication. J Comp Neurol 419:1–31

Jarvis ED, Nottebohm F (1997) Motor-driven gene expression. Proc Natl Acad Sci USA 94:4097–4102

Jarvis ED, Ribeiro S, da Silva ML, Ventura D, Vielliard J, Mello CV (2000) Behaviourally driven gene expression reveals song nuclei in hummingbird brain. Nature (Lond) 406:628–632. doi:10.1038/35020570

Jarvis ED, Gunturkun O, Bruce L et al (2005) Avian brains and a new understanding of vertebrate brain evolution. Nat Rev Neurosci 6:151–159. doi:10.1038/nrn1606

Jarvis ED, Yu J, Rivas MV et al (2013) A global view of the functional molecular organization of the avian cerebrum: mirror images and functional columns. J Comp Neurol. doi:10.1002/cne.23404

Karten HJ (1997) Evolutionary developmental biology meets the brain: the origins of mammalian cortex. Proc Natl Acad Sci USA 94:2800–2804

Kubikova L, Wada K, Jarvis ED (2010) Dopamine receptors in a songbird brain. J Comp Neurol 518:741–769. doi:10.1002/cne.22255

Kuenzel WJ, Medina L, Csillag A, Perkel DJ, Reiner A (2011) The avian subpallium: new insights into structural and functional subdivisions occupying the lateral subpallial wall and their embryological origins. Brain Res 1424:67–101. doi:10.1016/j.brainres.2011.09.037

Liedvogel M, Feenders G, Wada K, Troje NF, Jarvis ED, Mouritsen H (2007) Lateralized activation of Cluster N in the brains of migratory songbirds. Eur J Neurosci 25:1166–1173. doi:10.1111/j.1460-9568.2007.05350.x

Lubow RE (1974) High-order concept formation in the pigeon. J Exp Anal Behav 21:475–483

Medina L, Abellan A (2009) Development and evolution of the pallium. Semin Cell Dev Biol 20:698–711. doi:10.1016/j.semcdb.2009.04.008

Medina L, Reiner A (2000) Do birds possess homologues of mammalian primary visual, somatosensory and motor cortices? Trends Neurosci 23:1–12

Mello CV, Clayton DF (1994) Song-induced ZENK gene expression in auditory pathways of songbird brain and its relation to the song control system. J Neurosci 14:6652–6666

Mello CV, Clayton DF (1995) Differential induction of the ZENK gene in the avian forebrain and song control circuit after metrazole-induced depolarization. J Neurobiol 26:145–161. doi:10.1002/neu.480260112

Mello CV, Vicario DS, Clayton DF (1992) Song presentation induces gene expression in the songbird forebrain. Proc Natl Acad Sci USA 89:6818–6822

Molnar Z, Cheung AF (2006) Towards the classification of subpopulations of layer V pyramidal projection neurons. Neurosci Res 55:105–115. doi:10.1016/j.neures.2006.02.008

Montiel JF, Molnar Z (2013) The impact of gene expression analysis on evolving views of avian brain organization. J Comp Neurol 521:3604–3613. doi:10.1002/cne.23403

Mouritsen H, Feenders G, Liedvogel M, Wada K, Jarvis ED (2005) Night-vision brain area in migratory songbirds. Proc Natl Acad Sci USA 102:8339–8344. doi:10.1073/pnas.0409575102

Ohtsuki G, Nishiyama M, Yoshida T, Murakami T, Histed M, Lois C, Ohki K (2012) Similarity of visual selectivity among clonally related neurons in visual cortex. Neuron 75:65–72. doi:10.1016/j.neuron.2012.05.023

Pfenning AR, Hara E, Whitney O et al (2014) Convergent transcriptional specializations in the brains of humans and song-learning birds. Science 346:1256846. doi:10.1126/science.1256846

Puelles L, Kuwana E, Puelles E et al (2000) Pallial and subpallial derivatives in the embryonic chick and mouse telencephalon, traced by the expression of the genes Dlx-2, Emx-1, Nkx-2.1, Pax-6, and Tbr-1. J Comp Neurol 424:409–438

Reiner A, Perkel DJ, Bruce LL et al (2004a) Revised nomenclature for avian telencephalon and some related brainstem nuclei. J Comp Neurol 473:377–414. doi:10.1002/cne.20118

Reiner A, Perkel DJ, Bruce LL et al (2004b) The avian brain nomenclature forum: terminology for a new century in comparative neuroanatomy. J Comp Neurol 473:E1–E6. doi:10.1002/cne.20119

Soma M, Aizawa H, Ito Y et al (2009) Development of the mouse amygdala as revealed by enhanced green fluorescent protein gene transfer by means of in utero electroporation. J Comp Neurol 513:113–128. doi:10.1002/cne.21945

Suzuki IK, Kawasaki T, Gojobori T, Hirata T (2012) The temporal sequence of the mammalian neocortical neurogenetic program drives mediolateral pattern in the chick pallium. Dev Cell 22:863–870. doi:10.1016/j.devcel.2012.01.004

Swanson LW (2000) Cerebral hemisphere regulation of motivated behavior. Brain Res 886:113–164

Veenman CL, Wild JM, Reiner A (1995) Organization of the avian "corticostriatal" projection system: a retrograde and anterograde pathway tracing study in pigeons. J Comp Neurol 354:87–126. doi:10.1002/cne.903540108

Velho TA, Pinaud R, Rodrigues PV, Mello CV (2005) Co-induction of activity-dependent genes in songbirds. Eur J Neurosci 22:1667–1678. doi:10.1111/j.1460-9568.2005.04369.x

Vonfersen L, Delius JD (1989) Long-term retention of many visual-patterns by pigeons. Ethology 82:141–155

Wada K, Sakaguchi H, Jarvis ED, Hagiwara M (2004) Differential expression of glutamate receptors in avian neural pathways for learned vocalization. J Comp Neurol 476:44–64. doi:10.1002/cne.20201

Wada K, Howard JT, McConnell P et al (2006) A molecular neuroethological approach for identifying and characterizing a cascade of behaviorally regulated genes. Proc Natl Acad Sci USA 103:15212–15217. doi:10.1073/pnas.0607098103

Wang Y, Brzozowska-Prechtl A, Karten HJ (2010) Laminar and columnar auditory cortex in avian brain. Proc Natl Acad Sci USA 107:12676–12681. doi:10.1073/pnas.1006645107

Watakabe A, Ichinohe N, Ohsawa S, Hashikawa T, Komatsu Y, Rockland KS, Yamamori T (2007) Comparative analysis of layer-specific genes in mammalian neocortex. Cereb Cortex 17:1918–1933. doi:10.1093/cercor/bhl102

Watanabe S, Sakamoto J, Wakita M (1995) Pigeons' discrimination of paintings by Monet and Picasso. J Exp Anal Behav 63:165–174

Weir AA, Chappell J, Kacelnik A (2002) Shaping of hooks in New Caledonian crows. Science 297:981. doi:10.1126/science.1073433

Yamamoto K, Sun Z, Wang HB, Reiner A (2005) Subpallial amygdala and nucleus taeniae in birds resemble extended amygdala and medial amygdala in mammals in their expression of markers of regional identity. Brain Res Bull 66:341–347. doi:10.1016/j.brainresbull.2005.02.016

Yu YC, Bultje RS, Wang X, Shi SH (2009) Specific synapses develop preferentially among sister excitatory neurons in the neocortex. Nature (Lond) 458:501–504. doi:10.1038/nature07722
Zapka M, Heyers D, Hein CM et al (2009) Visual but not trigeminal mediation of magnetic compass information in a migratory bird. Nature (Lond) 461:1274–1277. doi:10.1038/nature08528
Zapka M, Heyers D, Liedvogel M, Jarvis ED, Mouritsen H (2010) Night-time neuronal activation of Cluster N in a day- and night-migrating songbird. Eur J Neurosci 32:619–624. doi:10.1111/j.1460-9568.2010.07311.x
Zhang G, Li C, Li Q et al (2014) Comparative genomics reveals insights into avian genome evolution and adaptation. Science 346:1311–1320. doi:10.1126/science.1251385

Chapter 12
The Neocortex and Dorsal Ventricular Ridge: Functional Convergence and Underlying Developmental Mechanisms

Wataru Yamashita and Tadashi Nomura

Abstract Extensive radiation of extant amniotes could have been achieved by the innovation of several unique characteristics in the body plans of their ancestors. In particular, distinct brain regions were enlarged independently to acquire similar functional properties in different amniote lineages. The neocortex and dorsal ventricular ridge (DVR) are a typical case of such parallel brain evolution in mammalian and reptilian lineages. Although these structures have distinct developmental origins, striking functional similarities in the neocortex and DVR have led to long-lasting arguments regarding their evolutionary development from ancestral amniotes. Here, we introduce morphological, neuroanatomical, and developmental aspects of the convergent and divergent features of the neocortex and DVR in amniotes. Furthermore, we discuss possible genetic changes that provided these remarkable brain structures, with special interest in the role of the Pax6 gene, an essential regulator of neural stem/progenitor cell dynamics. Comparative functional analyses of the regulatory genes required for pallial development will provide significant insights into the evolutionary origin of the hallmarks of mammalian and reptilian brains.

Keywords Neocortex • DVR • Sensory inputs • Neural stem/progenitor cells • Pax6

W. Yamashita
Developmental Neurobiology, Kyoto Prefectural University of Medicine, INAMORI Memorial Building, 1-5 Shimogamo-hangi cho, Sakyo-ku, Kyoto, 606-8023 Japan

T. Nomura (✉)
Developmental Neurobiology, Kyoto Prefectural University of Medicine, INAMORI Memorial Building, 1-5 Shimogamo-hangi cho, Sakyo-ku, Kyoto, 606-8023 Japan

Japan Science and Technology Agency (JST), PRESTO, 4-1-8 Honcho, Kawaguchi, Saitama, 332-0012 Japan
e-mail: tadnom@koto.kpu-m.ac.jp

S. Shigeno et al. (eds.), *Brain Evolution by Design*, Diversity and Commonality in Animals, DOI 10.1007/978-4-431-56469-0_12

12.1 Introduction

Accumulating paleontological evidence has suggested that the first appearance of tetrapods on land occurred during the Permian period, which was approximately 300 million years ago (MYA) (Carroll 1988; Modesto et al. 2015; Benton et al. 1988). In addition to the radiation of ancestral amphibians, primitive amniotes that adapted to a terrestrial lifestyle could have evolved during the late Permian (Romer 1957). The early stages of amniotes are sometimes called "reptiliomorpha" and include a large number of ancestral tetrapods with unique characteristics, although their phylogenetic positions remain under debate (Laurin 2004). By contrast, fossil evidence and molecular phylogenetics have suggested that ancestral amniotes rapidly radiated and diverged into distinct lineages, which are classified by anatomical differences such as the number of openings in the lateral side of the skull (Carroll 1988). These lineages include synapsid and diapsid, which were the ancestry of extant mammals and reptiles, respectively (Ruta et al. 2003; Sanchez-Villagra 2010; Kemp 2007).

Recent X-ray tomographic analyses of fossil endocasts have provided a fascinating scenario of the brain evolution in which the volume of the forebrain was rapidly increased in mammalian lineages (Rowe et al. 2011). Brains might have become larger as the result of an increased dependency on specific sensory modalities, such as olfactory, somatosensory, and auditory information (Rowe 1996; Rowe et al. 2011). In extant mammals, multiple types of sensory information are integrated into the dorsal part of the telencephalon, which is called the neocortex. Indeed, one of the distinguishing characteristics of the mammalian brain is the enlarged neocortex with tangential expansion of its surface area and a six-layered laminar structure composed of multiple neuronal subtypes (Nieuwenhuys 1994).

By contrast, diapsid lineages established distinct styles of forebrains. In extant reptiles, multiple sensory modalities are unified and processed in the dorsal ventricular ridge (DVR), which is a large, bulge-like structure that protrudes in the lateral side of the ventricular zone (Ulinski 1983; Bulter and Hodos 2005; Nieuwenhuys et al. 1998). Thus, the mammalian neocortex and reptilian DVR are comparable brain regions regarding functional commonalities. However, how these unique structures evolved in parallel in distantly related amniote lineages and what types of genetic modifications contributed to the creation of such different forebrain designs remain unknown. In this chapter, we introduce the mammalian neocortex and reptilian DVR as a typical example of the convergent evolution of amniote brain architectures. First, we provide evidence regarding the neuroanatomical similarities of the neocortex and DVR with respect to sensory inputs. Later in the chapter, possible molecular mechanisms that provided parallel evolution of the neocortex and DVR are discussed, focusing on the role of Pax6 as a master regulatory gene of brain development in vertebrates.

12.2 Morphological Diversity of the Amniote Telencephalon

The dorsal part of the vertebrate telencephalon, designated the pallium, is recognized in all vertebrate telencephali (Bulter and Hodos 2005; Holmgren 1922, 1925). The pallium is classified into four subdivisions, including the medial, dorsal, lateral, and ventral pallium, based on their anatomical positions and transcription factor expression patterns (Puelles et al. 2000). These pallial subdivisions are evolutionarily conserved in all vertebrates, whereas mature derivatives of each pallial region show considerable morphological diversity among phyla and species (Puelles and Medina 2002). In mammals, the dorsal pallium gives rise to the neocortex, which is characterized by tangential expansion of six-layered laminar sheets (Fig. 12.1a, c). This neocortical landmark is commonly equipped in all extant mammalian species, including eutheria, marsupials, and monotremes, suggesting that the neocortical structure could have already evolved in common ancestor(s) of modern mammalian lineages (Molnár et al. 2006).

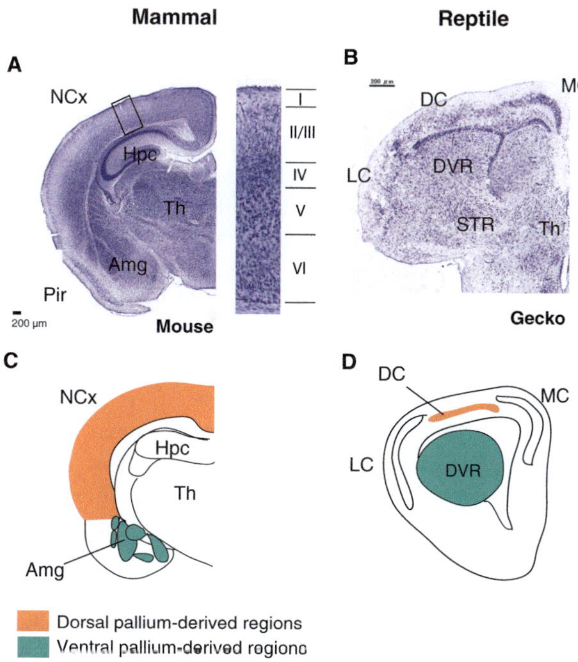

Fig. 12.1 Comparison of mammalian and reptilian telencephali. Coronal sections of neonatal mouse (**a**) and gecko (**b**) telencephali with Nissl staining. Note the enlarged neocortex (*NCx*) and dorsal ventricular ridge (*DVR*) in the mouse and gecko pallia, respectively. The mammalian neocortex is consists of a six-layered neuronal structure. (**c**, **d**) Schematic illustration of mouse (**c**) and gecko (**d**) telencephalic structures. *Hpc* hippocampus, *Amg* amygdala, *Th* thalamus, *MC* medial cortex, *DC* dorsal cortex, *LC* lateral cortex, *Pir* piriform cortex

Extant nonmammalian amniotes consist of two phylogenetic groups, lepidosaurs and archosaurs (Pyron et al. 2013; Sterling 2011). The former group includes snakes, lizards, and related species, and the latter includes crocodiles and birds. Species of *Sphenodon* (tuatara), which are considered to retain the ancient characteristics of amniotes, are also classified in the lepidosaurs (Jones and Cree 2012). Recent molecular phylogenetic analyses have demonstrated that turtles should be classified in the out group of archosaurs, although their external morphology and skeletons could have been extremely modified (Wang et al. 2013; Nagashima et al. 2013; Field et al. 2014). In this chapter, we use the term "reptiles" to refer to amniote groups that include lepidosaurs and archosaurs but that exclude modern birds, which might be acceptable for general readers.

Modern reptiles and birds have unique telencephalon architectures, some of which are not fully comparable to the mammalian counterparts. The reptilian cortex, which is thought to be a homologous region of the mammalian neocortex, exhibits a three-layered laminar organization (Ulinski 1990; Nomura et al. 2013b). Although cortical cellular arrangements show subtle variations among reptilian species, the middle layer (layer II), in which neuronal cell bodies are densely accumulated between the upper (layer I) and lower layer (layer III), is discernible in the dorsal pallium of all reptiles. In addition to the cortex, the DVR is a remarkable structure in the reptilian telencephalon, which is derived from the embryonic ventral pallium, and is characterized as "*the stippled mass that extends from the ventrolateral wall of the cerebral hemisphere into the lateral ventricle*" (Ulinski 1983) (Fig. 12.1b, d). The reptilian DVR is subdivided into the anterior and posterior compartments (ADVR and PDVR), which are distinguished by anatomical structures and by sensory projection patterns. Generally, the ADVR receives visual, auditory and somatosensory information and the PDVR receives olfactory information (Ulinski 1983; Bulter and Hodos 2005).

The avian telencephalon demonstrates extensively specified morphology. The dorsal part of the telencephalon gives rise to a tissue slab, called the hyperpallium or Wulst, which is composed of nuclear organizations rather than horizontally oriented laminar structures (Jarvis et al. 2005; Medina and Reiner 2000). However, a large stratified Wulst, which could be a derived architecture as a result of evolutionary adaptation, has been described in some avian species, such as parrots and owls (Striedter 2005). Similar to reptiles, the most obvious region in the avian telencephalon is also the DVR, which consists of several neuronal compartments, such as the mesopallium and nidopallium. As indicated by their names, all these compartments are derived from the embryonic pallium, although their precise origins remain under debate (Medina et al. 2013).

12.3 Functional Similarities of the Neocortex and DVR

Remarkable similarities regarding neuronal circuits can be found among the mammalian neocortex, reptilian/avian cortex/Wulst, and DVR. The mammalian neocortex receives several types of sensory information, such as somatosensory, visual, and auditory inputs through specific thalamic nuclei (Nieuwenhuys 1994; Nieuwenhuys et al. 1998). These sensory pathways maintain their topographic order within the forebrain, and project to the specific area of the neocortex, namely, the primary somatosensory, visual, and auditory cortex. Similarly, the reptilian cortex and DVR (ADVR) also receive multiple ascending sensory inputs from the thalamus (Ulinski 1983). Each sensory afferent terminates in distinct regions of the reptilian dorsal cortex and DVR, as in the mammalian neocortex. Notably, topographic arrangement of somatosensory, motor and visual areas in the dorsal pallium is highly conserved among amniotes, suggesting that common developmental mechanisms control the sensory projection pattern in the dorsal pallium (Medina and Reiner 2000). However, molecular mechanisms underlying the establishment of the sensory areas in the developing reptilian and avian DVR remain elusive. A recent study revealed that the expression of EphA4 and Slit2 in the developing turtle thalamus was similar to those in the mouse. Thus, basic molecular mechanisms that control thalamic axon guidance are shared in mammals and reptilian/avian lineages (Tosa et al. 2015).

Based on the similarities in sensory connection patterns, Karten claimed that the nuclear-specific neurons in the reptilian/avian pallia correspond to the layer-specific neurons in the mammalian neocortex (Karten 1969). This "equivalent cell hypothesis" proposes that homologues of the mammalian neocortical neurons are widely distributed in the reptilian/avian pallia, including the dorsal cortex and DVR. Concomitantly, recent studies have revealed that gene expression patterns associated with neuronal connections are highly conserved among mammalian and nonmammalian pallia. For example, layer IV neurons in the mammalian neocortex receive thalamic inputs, and these neurons are characterized by the expression of *Rorβ* and *Eag2* (Dugas-Ford et al. 2012; Hevner et al. 2003). Similar to mammals, orthologous genes of *Rorβ* and *Eag2* are expressed in thalamo-recipient neurons in the dorsal cortex and DVR in turtles (Dugas-Ford et al. 2012). However, subsequent comprehensive transcriptome analyses have demonstrated that gene expression patterns in the adult mouse neocortex are not comparable to the gene expression patterns of the adult chicken pallium (Belgard et al. 2013; Jarvis et al. 2013). Thus, a certain degree of similarity in gene expression patterns between distantly related species (e.g., mouse and chick) could be the results of functional convergence. Indeed, reptilian DVR and avian mesopallium/nidopallium are derived from the ventral pallium, which shares a developmental origin with the amygdala, claustrum, and endopiriform complex, not the neocortex, in the developing mammalian pallium (Puelles et al. 2000; Puelles and Medina 2002; Fernandez et al. 1998) (Fig. 12.1d).

12.4 Developmental Mechanisms Caused the Convergent Evolution of the Neocortex and DVR

One fundamental question is how the neocortex and DVR have acquired similar functions, although these structures have distinct developmental origins. Of note, significant similarities in neuronal connections, particularly regarding the sensory inputs from the dorsal thalamus, have been observed between the mammalian neocortex and reptilian/avian DVR.

The thalamic afferent pathway includes the collothalamic and lemnothalamic pathways: the former is relayed through the midbrain, although the latter is *not* mediated by the midbrain. These two pathways are well conserved among vertebrates, although each neural pathway projects to distinct pallial targets. For example, in mammals, the collothalamic visual pathway is relayed through several thalamic nuclei including posterior nuclear groups, which target the visual cortex and lateral nucleus of the amygdala (Bulter and Hodos 2005) (Fig. 12.2a). A similar pathway exists in nonmammalian amniotes, such as turtles and birds (Fig. 12.2b). In these animals, collothalamic visual pathways are mediated by the optic tectum and thalamic nuclei, which project to the visual region of the ADVR in reptiles and to the ectopallium in birds (Ulinski 1983). In contrast, lemnothalamic visual pathways are relayed through the dorsal geniculate nucleus in all amniotes, which target the visual cortex in mammals, dorsal cortex in reptiles, and visual Wulst in birds (Nieuwenhuys et al. 1998; Bulter and Hodos 2005; Ulinski 1990) (Fig. 12.2a, b).

Extensive similarities in the patterns of sensory inputs in distinct amniote lineages suggest that these characteristics are derived from common ancestor(s). If

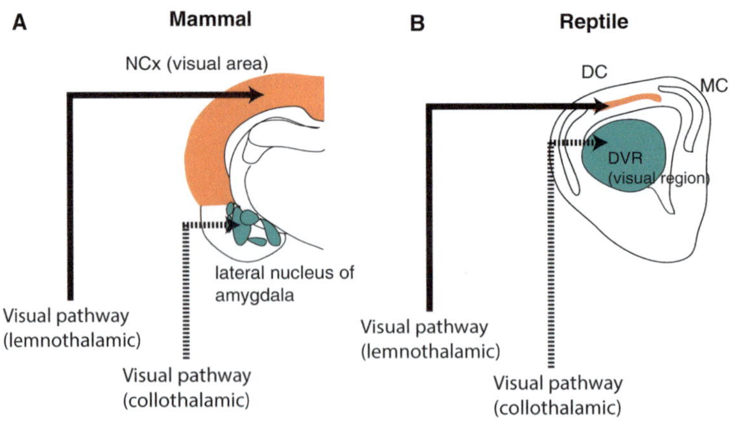

Fig. 12.2 Illustration of common sensory pathways in mammals and reptiles. Target regions of lemnothalamic and collothalamic visual pathways in the pallium are well conserved in amniotes. The regions with the *same colors* are thought to be homologous (This illustration was based on Bulter and Hodos (2005) with slight modifications)

this is the case, how did the striking architectural differences of pallial subdomains evolve during amniote evolution? Expansion and/or protrusion of the dorsal and ventral pallia give rise to the neocortex and DVR, as a result of differences in the patterns of neural stem cell proliferation, differentiation, and migration. Recent studies have clarified that morphological diversities are provided by small but significant modifications in genetic programs, such as changes in the spatiotemporal patterns of core-regulatory gene networks during embryogenesis (Galant and Carroll 2002; Ronshaugen et al. 2002; Chan et al. 2010; Guerreiro et al. 2013). Accumulating evidence has indicated that neurogenesis regulation has a key role in the disproportional expansion of the human neocortex (Borrell and Calegari 2014; Charvet et al. 2011). In the next sections, we discuss a possible molecular mechanism that generated the enlarged neocortex and DVR by focusing on the role of the transcription factor, Pax6.

12.5 Phenocopies of the Reptilian Cortex and DVR in Pax6 Mutant Mouse Brains

Pax6 is a member of the Pax family of transcription factors. The structure of the Pax6 gene is highly conserved among a wide variety of organisms including invertebrates and vertebrates (Quiring et al. 1994; Wang et al. 2010). *Pax* family genes are characterized by the presence of two unique DNA-binding domains called paired domains (PDs) and homeodomains (HDs), and each domain can recognize specific DNA sequences in either a cooperative or independent manner (Jun and Desplan 1996; Lang et al. 2007). The *Pax* gene was identified as being responsible for mouse and human eye development (Hill et al. 1991; Quiring et al. 1994; Halder et al. 1995). In the central nervous systems of vertebrates, *Pax6* is expressed in the forebrain, hindbrain, and spinal cord in a specific spatiotemporal manner (Simpson and Price 2002). The role of Pax6 in cortical development has been studied in mouse embryos (Georgala et al. 2011).

Several lines of evidence have suggested that alterations in *Pax6*-dependent genetic programs underlie the diversification of pallial morphology during amniote evolution (Molnár and Butler 2002; Aboitiz and Zamorano 2013). In the *Pax6* mutant mouse, the size and thickness of the neocortex is severely reduced (Quinn et al. 2007) whereas the ventrolateral portion of the pallium is abnormally enlarged (Jones et al. 2002). Therefore, the cytoarchitecture of the *Pax6* mutant mouse pallium is reminiscent of the reptilian and avian pallia: a small and thin dorsal cortex and a DVR-like protrusion at the lateroventral region (Molnár and Butler 2002). Because the expression of Pax6 is highly conserved in all amniote pallia

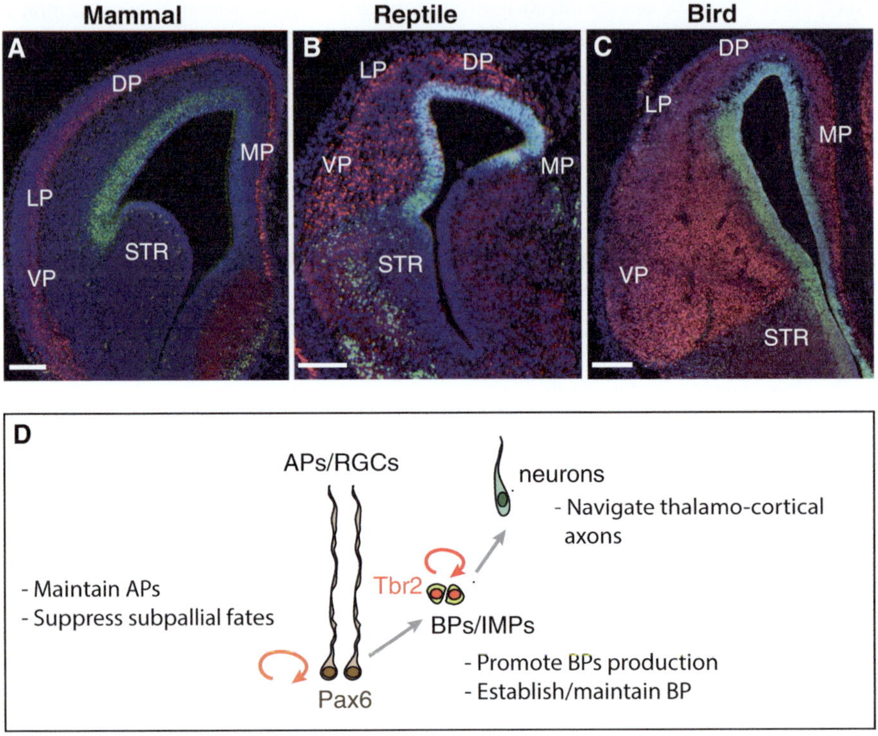

Fig. 12.3 The expression pattern of Pax6 among amniotes. Coronal sections of the telencephali of mouse (E13.5, **a**), Madagascar ground gecko (19 d.p.o., **b**) and chick (HH stage27, **c**). Sections are immunostained with anti-Pax6 antibody (*green*) and anti-Tbr1 antibody (*red*). Pax6 is strongly expressed in the pallial neural stem/progenitor cells, which produce Tbr1-positive excitatory neurons. (**d**) Schematic illustration of how Pax6 functions in neurogenesis. *MP* medial pallium, *DP* dorsal pallium, *LP* lateral pallium, *VP* ventral pallium, *STR* striatum. *Bars* 100 μm

(Fig. 12.3a–c), the phenotype of the *Pax6* mutant mouse is not attributed to the same genetic programs as those of reptilian/avian brains. However, altered cellular dynamics in the *Pax6* mutant mouse brain may resemble those in reptilian/avian brains, although this hypothesis has not been argued in detail.

12.6 Phenotypic Similarities Between the Pax6 Mutant Mouse Neocortex and Reptilian/Avian Dorsal Pallia

Among the many regulatory functions in cortical development, Pax6 promotes the expression of many transcription factors and cell-cycle genes to modulate the self-renewal and neuronal differentiation of the neural stem cells (NSCs) (Fig. 12.3d).

A significant reduction in the volume of the *Pax6* mutant mouse cortex is caused by impairments in NSC maintenance and in neural differentiation. The developing mammalian neocortex contains several types of neural stem/progenitor cells with distinct morphological and molecular features (Stancik et al. 2010; Fietz and Huttner 2011). In the pallia of all amniotes, Pax6 is expressed in the apical progenitor cells (or radial glial cells) that are localized in the pallial ventricular zone. During early stages of cortical development, apical progenitors undergo symmetrical division to increase the progenitor pool. As corticogenesis proceeds, apical progenitors produce postmitotic neurons or basal progenitors by asymmetrical division (Tabata and Nakajima 2003; Noctor et al. 2004; Haubensak et al. 2004; Miyata et al. 2004; Hevner et al. 2006). In the *Pax6* mutant mouse neocortex, apical progenitors are not adequately maintained (Estivill-Torrus et al. 2002; Quinn et al. 2007; Asami et al. 2011; Tuoc et al. 2009), resulting in the precocious depletion of neural stem/progenitor cells. In addition, radial fibers of apical progenitors in the *Pax6* mutant mouse pallium do not extend in a straight manner, probably because of the alteration of basal lamina components (Caric et al. 1997; Gotz et al. 1998).

Recent studies have shown the unique characteristics of apical progenitors in reptilian and avian dorsal pallia (Nomura et al. 2013a, 2014; Suzuki et al. 2012; Charvet 2010). In the developing gecko pallium, the rates of proliferation and differentiation of apical progenitors are significantly lower than are those in the mammalian neocortex. These characteristics of apical progenitors in the gecko are the result of prolonged cell-cycle lengths and lower frequencies of asymmetrical division, resulting in a lower production of cortical neurons during a limited neurogenic period (Nomura et al. 2013a). Apical progenitors in the avian dorsal pallium also exhibit a restricted mitotic potential (Suzuki et al. 2012; Charvet 2010). Furthermore, a wavy extension of radial fibers was evident in the developing reptilian and avian dorsal pallia, as in the case of the *Pax6* mutant mouse neocortex.

In the developing mammalian neocortex, proliferating basal progenitors are abundant in the subventricular zone, and these progenitors undergo symmetrical neuron production divisions and contribute to the expansion of cortical neuronal numbers (Sessa et al. 2008; Martinez-Cerdeno et al. 2006; Nonaka-Kinoshita et al. 2013). By contrast, a few subventricular mitoses have been reported in the reptilian and avian dorsal pallia (Cheung et al. 2007; Suzuki et al. 2012). Tbr2, a T-box transcription factor, is responsible for the generation and maintenance of basal progenitors (Sessa et al. 2008). Although Tbr2-positive cells are present in the basal side of the ventricular zone in reptiles, these cells do not exhibit proliferative activity (Nomura et al. 2013a). Of note, the number of Tbr2-positive basal progenitors is severely reduced in the *Pax6* mutant mouse neocortex (Quinn et al. 2007), suggesting that Pax6 promotes the transition from apical progenitors to basal progenitors.

12.7 The Role of Pax6 in the Development of the Ventral Pallium

In mammals, the ventral pallium develops as a source of unique neuronal subpopulations. At early stages of mouse cortical development, a subtype of Cajal–Retzius (CR) cells is generated from the progenitors in the ventral pallium, along with the cortical hem and the septum, and these CR cells migrate to the surface of the neocortex (Bielle et al. 2005; Takiguchi-Hayashi et al. 2004; Yoshida et al. 2006). CR cells are essential in the inside-out pattern of corticogenesis via the secreted molecule Reelin (Kirischuk et al. 2014; Sekine et al. 2014; Frotscher 1998). Recent studies have demonstrated that CR cells regulate rostral-caudal patterning of the neocortex (Griveau et al. 2010). Subsequent to CR cells, a subpopulation of excitatory glutamatergic neurons differentiates from the ventral pallium and migrates tangentially to the developing neocortex (Teissier et al. 2010). Although the majority of these neurons undergo apoptosis during corticogenesis, this transient neuronal population enhances the production of the later-born cortical neurons (Teissier et al. 2012). Another neuronal population derived from the ventral pallium migrates toward the ventrolateral part of the telencephalon and gives rise to the amygdala, claustrum, and endopiriform nucleus (Hicks and D'Amato 1968). This migratory route is called the lateral cortical stream (Carney et al. 2006). In mice, most of these neuronal populations are derived from *Dbx1*-positive progenitors in the ventral pallium (Bielle et al. 2005; Teissier et al. 2010; Hirata et al. 2009). The loss of *Pax6* gene functions disrupts the dorsoventral patterning of the telencephalon, and ventral pallial markers including *Dbx1* are not expressed in the *Pax6* mutant mice (Assimacopoulos et al. 2003; Hirata et al. 2002). Furthermore, the majority of neurons derived from the ventral pallium accumulate at the junction between the cortex and striatum (Jones et al. 2002). Consequently, a portion of amygdala and associated nuclear structures are not developed at the lateroventral portion of the Pax6 mutant telencephalon (Tole et al. 2005).

Migration patterns of ventral pallium-derived neurons in reptiles and birds largely differ from those in mammals. In reptiles and birds, the majority of neurons that differentiate at the ventral pallium migrate toward the pial surface in an outside-in manner: early-born neurons position at the surface area whereas later-born neurons occupy the deep area (Tsai et al. 1981; Goffinet et al. 1986). In the reptilian DVR, an obvious laminar structure is not evident, but neuronal clusters are present (Ulinski 1983). As discussed in the previous section, thalamic projection contributes to an increase in the volume of the DVR. Notably, the thalamocortical projection in the *Pax6* mouse mutant stalls at the border between the cortex and striatum and intermingles with abnormally accumulated neurons at the lateroventral portion of the telencephalon (Jones et al. 2002). Compared to mammals, the numbers of Reelin-positive CR cells in the reptilian and avian pallia are much fewer (Nomura et al. 2008; Cabrera-Socorro et al. 2007; Tissir et al. 2003; Bernier et al. 2000), possibly because of differences in the potential of Reelin expression or in the generation of CR cells among amniotes. Indeed, *Dbx1* expression at the ventral pallium is specifically absent in the developing chick telencephalon, and Dbx1 overexpression induces ectopic Reelin-positive cells in chick embryos (Bielle et al. 2005; Nomura et al. 2008). These data suggest that *Pax6*-dependent regulation of *Dbx1* expression had been acquired in mammalian lineages, resulting in the amplification of CR cells and in the establishment of a unique migratory stream of neurons to form the amygdaloid complex. The expression of Dbx1 in the developing reptilian pallium remains to be elucidated.

12.8 A Possible Scenario Regarding Changes in Pax6 Functions and the Emergence of the Neocortex and DVR

Evolutionary developmental biologists have hypothesized that a few but significant changes in core-regulatory factors of developmental processes resulted in considerable anatomical differences (King and Wilson 1975; Rakic 1995; Jacob 1977). In this context, subtle alterations in the gene regulatory networks under the control of the Pax6 gene may have provided species-specific progenitor dynamics in the developing amniote pallium. Consistent with this possibility, a previous report clarified the differential regulation of pluripotent and neural genes by Pax6 between mouse and human (Zhang et al. 2010). This differential regulation might be caused by heterotopic or heterochronic changes in the expression patterns of Pax6 or related genes or to alterations in the *cis*-regulatory elements of downstream genes (Molnár and Butler 2002; Aboitiz and Zamorano 2013). In the developing mouse neocortex, the expression level of Pax6 shows a rostrolateral-high and caudomedial-low gradient. In contrast, the gradient in the expression of Pax6 is not obviously detected in the reptilian/avian pallium (Fig. 12.3a–c). A previous study indicated that neural progenitors in the Pax6-low regions (include the medial and dorsal pallium) have higher proliferative capacity than those in the Pax6-high regions (lateral and ventral pallum) because of the dose-dependent negative regulation of Pax6 in cell-cycle progression (Mi et al. 2013; Sansom et al. 2009). Thus, it is possible that the small volume of the dorsal pallium in reptiles/birds might be the result of the relatively higher expression level of Pax6 in the dorsal pallium compared to mammals. In contrast, Aboitiz (Aboitiz and Zamorano 2013) hypothesized upregulation of Pax6 in the ventral pallium of reptiles/birds might contribute to the expansion of the DVR, which could be attributed to positive regulation of Pax6 in cell proliferation. Indeed, Pax6 not only suppress the expression of Cdk6 (Mi et al. 2013) but also positively regulates other cell-cycle regulators such as cyclin D2 (Sansom et al. 2009). A possible scenario that integrates these controversial theories is that the role of Pax6 is context dependent, and the heterometric changes of Pax6 contributed to the increase/decrease of different pallial regions, although we need to have more experimental data about the dose-dependent function of Pax6 in different amniote species.

We summarized putative changes in Pax6-dependent regulation that are relevant to the creation of the neocortex and DVR (Table 12.1). Notably, the tangential expansion and six-layered laminar structure of the mammalian neocortex could have evolved by a massive increase in apical progenitor mitosis, together with the amplification of basal progenitors: these developmental events are tightly controlled by Pax6-dependent genetic networks in mammals. By contrast, the emergence of DVR structures in reptilian lineages might be caused by other changes in Pax6-dependent regulation of neural stem/progenitor cells, resulting in the expansion of apical progenitors and in the massive production of excitatory neurons in the reptilian ventral pallium. In addition to the control of progenitor proliferation, species-specific patterns of cell migration in the dorsal and ventral pallium might also be accomplished by changes in Pax6 functions during evolution. Along with the inside-out patterns of excitatory neuron migration in the dorsal pallium, the emergence of the lateral cortical stream is a critical step for making mammalian-specific telencephalic structures. Recent studies have reported differential expression of guidance molecules for axon elongation and neuronal migration in developing mammalian and nonmammalian telencephali (Bielle et al. 2011; Tosa et al. 2015). Thus, changes in the expression patterns, levels and/or timing of these molecules might have contributed to the establishment of novel migratory patterns of dorsal and ventral pallium in mammalian and nonmammalian lineages.

12.9 Conclusion and Perspective

In addition to anatomical, physiological, and behavioral differences, the neocortex and DVR have remarkable characteristics to classify amniotes into two major groups that descended from ancestral synapsids and diapsids. Although these brain architectures might have developed independently in ancient amniote lineages, genetic and environmental constraints have provided amazing functional similarities. Comparative neuroanatomical studies have identified that some of these commonalities are based on the conserved neural circuits, which possibly derived from common ancestral characteristics. Elaboration of the neocortex and DVR as a higher cognitive center might have resulted in successful adaptive radiation of mammals and reptiles.

Currently, much evidence has been accumulating regarding the developmental regulation of the mammalian neocortex and underlying molecular mechanisms. However, we do not have sufficient information regarding the lineage- and species-specific functions of regulatory genes such as Pax6 in different contexts of brain development. Although most genes and their expression patterns during embryogenesis are highly conserved among amniotes, no evidence has indicated that orthologous genes have identical functions among amniote lineages because unique brain designs must underlie species-specific developmental constraints. Thus, we

Table 12.1 Possible changes in *Pax6*-dependent cortical developmental programs in mammals and reptiles

	Mammals	Reptiles	Putative downstream genes[a]
Dorsal pallium	**Tangential expansion of the surface area**	**Lesser expansion of the surface area**	Neurogenin 2
	Enhanced proliferation of apical progenitors	Slower proliferation of the apical progenitors	Tbr2
	Amplification of basal progenitors	No or minimum amplification of basal progenitors	Cdk6
	Amplification of outer radial glial cells[b]		Dmrta1
	Navigation of thalamic axons		Hes family genes
	Establishment of internal capsule		Wnt family genes
			Eph/ephrins
			Slit/Robo family genes
Ventral pallium	**Generation of *Dbx1*-positive cell populations**	**Formation of the DVR**	Dbx1
	Production of Cajal–Retzius cells	Enhanced proliferation of apical progenitors	Neurogenin 2
	Production of transient excitatory neurons	Amplification of basal progenitors[c]	Tbr2
	Production of amygdala neurons	**Establishment of nuclear structures**	Cdk6
	Establishment of the lateral cortical stream	Outside-in patterns of cortico/palliogenesis	Dmrta1
	Establishment of claustrum/amygdala complex	**Navigation of thalamic axons**	Hes
			Wnt
			Reelin/Dab1
			Eph/ephrins
			Slit/Robo family genes

[a]References are cited in the text
[b]Massively amplified in gyrencephalic mammals
[c]Massively amplified in avian lineages

could not simply extrapolate the molecular logic of the mammalian neocortex to understand nonmammalian brain development. Manipulations of orthologous gene functions in various animals are powerful approaches to unveil the genetic mechanisms of the evolution of brain diversity. Recent advancements in molecular genetics such as RNA interference, genome editing, and comprehensive RNA sequencing technologies will provide new insights into comparative analyses of orthologous gene functions. Identifying regulatory systems that constitute species-specific developmental constraints is essential to understanding the origin and evolution of the neocortex and DVR as landmarks of amniotes.

Acknowledgments We thank Drs. Katsuhiko Ono and Hitoshi Gotoh for providing critical comments and suggestions for the research and Ms. Misato Kawami and Mr. Kazuhiro Arimura for technical support. This work was supported by Grant-in-Aid for Challenging Exploratory Research (#24657158), Scientific Research on Innovative Areas (The Empathetic Systems, #26118510) and PRESTO, JST.

References

Aboitiz F, Zamorano F (2013) Neural progenitors, patterning and ecology in neocortical origins. Front Neuroanat 7:38. doi:10.3389/fnana.2013.00038

Asami M, Pilz GA, Ninkovic J, Godinho L, Schroeder T, Huttner WB, Gotz M (2011) The role of Pax6 in regulating the orientation and mode of cell division of progenitors in the mouse cerebral cortex. Development 138(23):5067–5078. doi:10.1242/dev.074591

Assimacopoulos S, Grove EA, Ragsdale CW (2003) Identification of a Pax6-dependent epidermal growth factor family signaling source at the lateral edge of the embryonic cerebral cortex. J Neurosci 23(16):6399–6403

Belgard TG, Montiel JF, Wang WZ, Garcia-Moreno F, Margulies EH, Ponting CP, Molnár Z (2013) Adult pallium transcriptomes surprise in not reflecting predicted homologies across diverse chicken and mouse pallial sectors. Proc Natl Acad Sci USA 110(32):13150–13155. doi:10.1073/pnas.1307444110

Benton M, Gauthier J, Kluge A, Rowe TB (1988) The early evolution of the Amniota. In: Benton MJ (ed) The phylogeny and classification of the tetrapods. Clarendon, Oxford, pp 103–155

Bernier B, Bar I, D'Arcangelo G, Curran T, Goffinet AM (2000) Reelin mRNA expression during embryonic brain development in the chick. J Comp Neurol 422(3):448–463

Bielle F, Griveau A, Narboux-Neme N, Vigneau S, Sigrist M, Arber S, Wassef M, Pierani A (2005) Multiple origins of Cajal–Retzius cells at the borders of the developing pallium. Nat Neurosci 8(8):1002–1012. doi:10.1038/nn1511

Bielle F, Marcos-Mondejar P, Keita M, Mailhes C, Verney C, Nguyen Ba-Charvet K, Tessier-Lavigne M, Lopez-Bendito G, Garel S (2011) Slit2 activity in the migration of guidepost neurons shapes thalamic projections during development and evolution. Neuron 69(6):1085–1098. doi:10.1016/j.neuron.2011.02.026

Borrell V, Calegari F (2014) Mechanisms of brain evolution: regulation of neural progenitor cell diversity and cell cycle length. Neurosci Res 86:14–24. doi:10.1016/j.neures.2014.04.004

Bulter A, Hodos W (2005) Comparative vertebrate neuroanatomy: evolution and adaptation. Wiley, Hoboken

Cabrera-Socorro A, Hernandez-Acosta NC, Gonzalez-Gomez M, Meyer G (2007) Comparative aspects of p73 and Reelin expression in Cajal-Retzius cells and the cortical hem in lizard, mouse and human. Brain Res 1132(1):59–70. doi:10.1016/j.brainres.2006.11.015

Caric D, Gooday D, Hill RE, McConnell SK, Price DJ (1997) Determination of the migratory capacity of embryonic cortical cells lacking the transcription factor Pax-6. Development 124(24):5087–5096

Carney RS, Alfonso TB, Cohen D, Dai H, Nery S, Stoica B, Slotkin J, Bregman BS, Fishell G, Corbin JG (2006) Cell migration along the lateral cortical stream to the developing basal telencephalic limbic system. J Neurosci 26(45):11562–11574. doi:10.1523/JNEUROSCI.3092-06.2006

Carroll RL (1988) Vertebrate paleontology and evolution. Freeman, New York

Chan YF, Marks ME, Jones FC, Villarreal G Jr, Shapiro MD, Brady SD, Southwick AM, Absher DM, Grimwood J, Schmutz J, Myers RM, Petrov D, Jonsson B, Schluter D, Bell MA, Kingsley DM (2010) Adaptive evolution of pelvic reduction in sticklebacks by recurrent deletion of a Pitx1 enhancer. Science 327(5963):302–305. doi:10.1126/science.1182213

Charvet CJ (2010) A reduced progenitor pool population accounts for the rudimentary appearance of the septum, medial pallium and dorsal pallium in birds. Brain Behav Evol 76(3-4):289–300. doi:10.1159/000322102

Charvet CJ, Striedter GF, Finlay BL (2011) Evo-devo and brain scaling: candidate developmental mechanisms for variation and constancy in vertebrate brain evolution. Brain Behav Evol 78(3):248–257. doi:10.1159/000329851

Cheung AF, Pollen AA, Tavare A, DeProto J, Molnár Z (2007) Comparative aspects of cortical neurogenesis in vertebrates. J Anat 211(2):164–176. doi:10.1111/j.1469-7580.2007.00769.x

Dugas-Ford J, Rowell JJ, Ragsdale CW (2012) Cell-type homologies and the origins of the neocortex. Proc Natl Acad Sci USA 109(42):16974–16979. doi:10.1073/pnas.1204773109

Estivill-Torrus G, Pearson H, van Heyningen V, Price DJ, Rashbass P (2002) Pax6 is required to regulate the cell cycle and the rate of progression from symmetrical to asymmetrical division in mammalian cortical progenitors. Development 129(2):455–466

Fernandez AS, Pieau C, Reperant J, Boncinelli E, Wassef M (1998) Expression of the Emx-1 and Dlx-1 homeobox genes define three molecularly distinct domains in the telencephalon of mouse, chick, turtle and frog embryos: implications for the evolution of telencephalic subdivisions in amniotes. Development 125(11):2099–2111

Field DJ, Gauthier JA, King BL, Pisani D, Lyson TR, Peterson KJ (2014) Toward consilience in reptile phylogeny: miRNAs support an archosaur, not lepidosaur, affinity for turtles. Evol Dev 16(4):189–196. doi:10.1111/ede.12081

Fietz SA, Huttner WB (2011) Cortical progenitor expansion, self-renewal and neurogenesis-a polarized perspective. Curr Opin Neurobiol 21(1):23–35. doi:10.1016/j.conb.2010.10.002

Frotscher M (1998) Cajal–Retzius cells, Reelin, and the formation of layers. Curr Opin Neurobiol 8(5):570–575

Galant R, Carroll SB (2002) Evolution of a transcriptional repression domain in an insect Hox protein. Nature 415(6874):910–913. doi:10.1038/nature717

Georgala PA, Carr CB, Price DJ (2011) The role of Pax6 in forebrain development. Dev Neurobiol 71(8):690–709. doi:10.1002/dneu.20895

Goffinet AM, Daumerie C, Langerwerf B, Pieau C (1986) Neurogenesis in reptilian cortical structures: ^{3}H-thymidine autoradiographic analysis. J Comp Neurol 243(1):106–116. doi:10.1002/cne.902430109

Gotz M, Stoykova A, Gruss P (1998) Pax6 controls radial glia differentiation in the cerebral cortex. Neuron 21(5):1031–1044

Griveau A, Borello U, Causeret F, Tissir F, Boggetto N, Karaz S, Pierani A (2010) A novel role for Dbx1-derived Cajal–Retzius cells in early regionalization of the cerebral cortical neuroepithelium. PLoS Biol 8(7), e1000440. doi:10.1371/journal.pbio.1000440

Guerreiro I, Nunes A, Woltering JM, Casaca A, Novoa A, Vinagre T, Hunter ME, Duboule D, Mallo M (2013) Role of a polymorphism in a Hox/Pax-responsive enhancer in the evolution of the vertebrate spine. Proc Natl Acad Sci USA 110(26):10682–10686. doi:10.1073/pnas.1300592110

Halder G, Callaerts P, Gehring WJ (1995) Induction of ectopic eyes by targeted expression of the eyeless gene in *Drosophila*. Science 267(5205):1788–1792

Haubensak W, Attardo A, Denk W, Huttner WB (2004) Neurons arise in the basal neuroepithelium of the early mammalian telencephalon: a major site of neurogenesis. Proc Natl Acad Sci USA 101(9):3196–3201. doi:10.1073/pnas.0308600100

Hevner RF, Daza RA, Rubenstein JL, Stunnenberg H, Olavarria JF, Englund C (2003) Beyond laminar fate: toward a molecular classification of cortical projection/pyramidal neurons. Dev Neurosci 25(24):139–151. doi:72263

Hevner RF, Hodge RD, Daza RA, Englund C (2006) Transcription factors in glutamatergic neurogenesis: conserved programs in neocortex, cerebellum, and adult hippocampus. Neurosci Res 55(3):223–233. doi:10.1016/j.neures.2006.03.004

Hicks SP, D'Amato CJ (1968) Cell migrations to the isocortex in the rat. Anat Rec 160(3):619–634. doi:10.1002/ar.1091600311

Hill RE, Favor J, Hogan BL, Ton CC, Saunders GF, Hanson IM, Prosser J, Jordan T, Hastie ND, van Heyningen V (1991) Mouse small eye results from mutations in a paired-like homeobox-containing gene. Nature 354(6354):522–525. doi:10.1038/354522a0

Hirata T, Nomura T, Takagi Y, Sato Y, Tomioka N, Fujisawa H, Osumi N (2002) Mosaic development of the olfactory cortex with Pax6-dependent and -independent components. Brain Res Dev Brain Res 136(1):17–26

Hirata T, Li P, Lanuza GM, Cocas LA, Huntsman MM, Corbin JG (2009) Identification of distinct telencephalic progenitor pools for neuronal diversity in the amygdala. Nat Neurosci 12(2):141–149. doi:10.1038/nn.2241

Holmgren N (1922) Points of view concerning forebrain morphology in lower vertebrates. J Comp Neurol 34:391–459

Holmgren N (1925) Points of view concerning forebrain morphology in higher vertebrates. Acta Zool (Stockh) 6:413–477

Jacob F (1977) Evolution and tinkering. Science 196(4295):1161–1166

Jarvis ED, Gunturkun O, Bruce L, Csillag A, Karten H, Kuenzel W, Medina L, Paxinos G, Perkel DJ, Shimizu T, Striedter G, Wild JM, Ball GF, Dugas-Ford J, Durand SE, Hough GE, Husband S, Kubikova L, Lee DW, Mello CV, Powers A, Siang C, Smulders TV, Wada K, White SA, Yamamoto K, Yu J, Reiner A, Butler AB (2005) Avian brains and a new understanding of vertebrate brain evolution. Nat Rev Neurosci 6(2):151–159. doi:10.1038/nrn1606

Jarvis ED, Yu J, Rivas MV, Horita H, Feenders G, Whitney O, Jarvis SC, Jarvis ER, Kubikova L, Puck AE, Siang-Bakshi C, Martin S, McElroy M, Hara E, Howard J, Pfenning A, Mouritsen H, Chen CC, Wada K (2013) Global view of the functional molecular organization of the avian cerebrum: mirror images and functional columns. J Comp Neurol 521(16):3614–3665. doi:10.1002/cne.23404

Jones ME, Cree A (2012) Tuatara. Curr Biol 22(23):R986–R987. doi:10.1016/j.cub.2012.10.049

Jones L, Lopez-Bendito G, Gruss P, Stoykova A, Molnár Z (2002) Pax6 is required for the normal development of the forebrain axonal connections. Development 129(21):5041–5052

Jun S, Desplan C (1996) Cooperative interactions between paired domain and homeodomain. Development 122:2639–2650

Karten H (1969) The organization of the avian telencephalon and some speculations on the phylogeny of the amniote telencephalon. Ann NY Acad Sci 167:164–179

Kemp TS (2007) The origin of higher taxa: macroevolutionary processes, and the case of the mammals. Acta Zool (Stockh) 88:3–22

King MC, Wilson AC (1975) Evolution at two levels in humans and chimpanzees. Science 188(4184):107–116

Kirischuk S, Luhmann HJ, Kilb W (2014) Cajal-Retzius cells: update on structural and functional properties of these mystic neurons that bridged the 20th century. Neuroscience 275:33–46. doi:10.1016/j.neuroscience.2014.06.009

Lang D, Powell SK, Plummer RS, Young KP, Ruggeri BA (2007) PAX genes: roles in development, pathophysiology, and cancer. Biochem Pharmacol 73(1):1–14. doi:10.1016/j.bcp.2006.06.024

Laurin M (2004) The evolution of body size, Cope's rule and the origin of amniotes. Syst Biol 53(4):594–622. doi:10.1080/10635150490445706

Martinez-Cerdeno V, Noctor SC, Kriegstein AR (2006) The role of intermediate progenitor cells in the evolutionary expansion of the cerebral cortex. Cereb Cortex 16((suppl 1)):i152–i161. doi:10.1093/cercor/bhk017

Medina L, Reiner A (2000) Do birds possess homologues of mammalian primary visual, somatosensory and motor cortices? Trends Neurosci 23(1):1–12

Medina L, Abellan A, Desfilis E (2013) A never-ending search for the evolutionary origin of the neocortex: rethinking the homology concept. Brain Behav Evol 81(3):150–153. doi:10.1159/000348282

Mi D, Carr CB, Georgala PA, Huang YT, Manuel MN, Jeanes E, Niisato E, Sansom SN, Livesey FJ, Theil T, Hasenpusch-Theil K, Simpson TI, Mason JO, Price DJ (2013) Pax6 exerts regional control of cortical progenitor proliferation via direct repression of Cdk6 and hypophosphorylation of pRb. Neuron 78(2):269–284. doi:10.1016/j.neuron.2013.02.012

Miyata T, Kawaguchi A, Saito K, Kawano M, Muto T, Ogawa M (2004) Asymmetric production of surface-dividing and non-surface-dividing cortical progenitor cells. Development 131(13):3133–3145. doi:10.1242/dev.01173

Modesto SP, Scott DM, MacDougall MJ, Sues HD, Evans DC, Reisz RR (2015) The oldest parareptile and the early diversification of reptiles. Proc R Soc B Biol Sci 282(1801):1–9. doi:10.1098/rspb.2014.1912

Molnár Z, Butler AB (2002) The corticostriatal junction: a crucial region for forebrain development and evolution. BioEssays 24(6):530–541. doi:10.1002/bies.10100

Molnár Z, Metin C, Stoykova A, Tarabykin V, Price DJ, Francis F, Meyer G, Dehay C, Kennedy H (2006) Comparative aspects of cerebral cortical development. Eur J Neurosci 23(4):921–934. doi:10.1111/j.1460-9568.2006.04611.x

Nagashima H, Hirasawa T, Sugahara F, Takechi M, Usuda R, Sato N, Kuratani S (2013) Origin of the unique morphology of the shoulder girdle in turtles. J Anat 223(6):547–556. doi:10.1111/joa.12116

Nieuwenhuys R (1994) The neocortex. An overview of its evolutionary development, structural organization and synaptology. Anat Embryol 190(4):307–337

Nieuwenhuys R, Ten Donkelaar HJ, Nicholson C (1998) The central nervous system of vertebrates. Springer, Berlin

Noctor SC, Martinez-Cerdeno V, Ivic L, Kriegstein AR (2004) Cortical neurons arise in symmetric and asymmetric division zones and migrate through specific phases. Nat Neurosci 7(2):136–144. doi:10.1038/nn1172

Nomura T, Takahashi M, Hara Y, Osumi N (2008) Patterns of neurogenesis and amplitude of Reelin expression are essential for making a mammalian-type cortex. PLoS One 3(1), e1454. doi:10.1371/journal.pone.0001454

Nomura T, Gotoh H, Ono K (2013a) Changes in the regulation of cortical neurogenesis contribute to encephalization during amniote brain evolution. Nat Commun 4:2206. doi:10.1038/ncomms3206

Nomura T, Kawaguchi M, Ono K, Murakami Y (2013b) Reptiles: a new model for brain evo-devo research. J Exp Zool B Mol Dev Evol 320(2):57–73. doi:10.1002/jez.b.22484

Nomura T, Murakami Y, Gotoh H, Ono K (2014) Reconstruction of ancestral brains: exploring the evolutionary process of encephalization in amniotes. Neurosci Res 86:25–36. doi:10.1016/j.neures.2014.03.004

Nonaka-Kinoshita M, Reillo I, Artegiani B, Martinez-Martinez MA, Nelson M, Borrell V, Calegari F (2013) Regulation of cerebral cortex size and folding by expansion of basal progenitors. EMBO J 32(13):1817–1828. doi:10.1038/emboj.2013.96

Puelles L, Medina L (2002) Field homology as a way to reconcile genetic and developmental variability with adult homology. Brain Res Bull 57(3-4):243–255

Puelles L, Kuwana E, Puelles E, Bulfone A, Shimamura K, Keleher J, Smiga S, Rubenstein JL (2000) Pallial and subpallial derivatives in the embryonic chick and mouse telencephalon, traced by the expression of the genes Dlx-2, Emx-1, Nkx-2.1, Pax-6, and Tbr-1. J Comp Neurol 424(3):409–438

Pyron RA, Burbrink FT, Wiens JJ (2013) A phylogeny and revised classification of Squamata, including 4161 species of lizards and snakes. BMC Evol Biol 13:93. doi:10.1186/1471-2148-13-93

Quinn JC, Molinek M, Martynoga BS, Zaki PA, Faedo A, Bulfone A, Hevner RF, West JD, Price DJ (2007) Pax6 controls cerebral cortical cell number by regulating exit from the cell cycle and specifies cortical cell identity by a cell autonomous mechanism. Dev Biol 302(1):50–65. doi:10.1016/j.ydbio.2006.08.035

Quiring R, Walldorf U, Kloter U, Gehring WJ (1994) Homology of the eyeless gene of *Drosophila* to the Small eye gene in mice and Aniridia in humans. Science 265(5173):785–789

Rakic P (1995) A small step for the cell, a giant leap for mankind: a hypothesis of neocortical expansion during evolution. Trends Neurosci 18(9):383–388

Romer AS (1957) Origin of the amniote egg. Sci Mon 85(57-63)

Ronshaugen M, McGinnis N, McGinnis W (2002) Hox protein mutation and macroevolution of the insect body plan. Nature 415(6874):914–917. doi:10.1038/nature716

Rowe T (1996) Coevolution of the mammalian middle ear and neocortex. Science 273(5275):651–654

Rowe TB, Macrini TE, Luo ZX (2011) Fossil evidence on origin of the mammalian brain. Science 332(6032):955–957. doi:10.1126/science.1203117

Ruta M, Coates MI, Quicke DL (2003) Early tetrapod relationships revisited. Biol Rev Camb Philos Soc 78(2):251–345

Sanchez-Villagra MR (2010) Developmental palaeontology in synapsids: the fossil record of ontogeny in mammals and their closest relatives. Proc R Soc B Biol Sci 277(1685):1139–1147. doi:10.1098/rspb.2009.2005

Sansom SN, Griffiths DS, Faedo A, Kleinjan DJ, Ruan Y, Smith J, van Heyningen V, Rubenstein JL, Livesey FJ (2009) The level of the transcription factor Pax6 is essential for controlling the balance between neural stem cell self-renewal and neurogenesis. PLoS Genet 5(6), e1000511. doi:10.1371/journal.pgen.1000511

Sekine K, Kubo K, Nakajima K (2014) How does Reelin control neuronal migration and layer formation in the developing mammalian neocortex? Neurosci Res 86:50–58. doi:10.1016/j.neures.2014.06.004

Sessa A, Mao CA, Hadjantonakis AK, Klein WH, Broccoli V (2008) Tbr2 directs conversion of radial glia into basal precursors and guides neuronal amplification by indirect neurogenesis in the developing neocortex. Neuron 60(1):56–69. doi:10.1016/j.neuron.2008.09.028

Simpson TI, Price DJ (2002) Pax6: a pleiotropic player in development. BioEssays 24(11):1041–1051. doi:10.1002/bies.10174

Stancik EK, Navarro-Quiroga I, Sellke R, Haydar TF (2010) Heterogeneity in ventricular zone neural precursors contributes to neuronal fate diversity in the postnatal neocortex. J Neurosci 30(20):7028–7036. doi:10.1523/JNEUROSCI.6131-09.2010

Sterling JN (2011) The early evolution of archosaurs: relationships and the origin of major clades. Bull Am Mus Nat Hist 352:1–292

Striedter GF (2005) Principle of brain evolution. Sinauer, Sunderland

Suzuki IK, Kawasaki T, Gojobori T, Hirata T (2012) The temporal sequence of the mammalian neocortical neurogenetic program drives mediolateral pattern in the chick pallium. Dev Cell 22(4):863–870. doi:10.1016/j.devcel.2012.01.004

Tabata H, Nakajima K (2003) Multipolar migration: the third mode of radial neuronal migration in the developing cerebral cortex. J Neurosci 23(31):9996–10001

Takiguchi-Hayashi K, Sekiguchi M, Ashigaki S, Takamatsu M, Hasegawa H, Suzuki-Migishima R, Yokoyama M, Nakanishi S, Tanabe Y (2004) Generation of reelin-positive marginal zone cells from the caudomedial wall of telencephalic vesicles. J Neurosci 24(9):2286–2295. doi:10.1523/JNEUROSCI.4671-03.2004

Teissier A, Griveau A, Vigier L, Piolot T, Borello U, Pierani A (2010) A novel transient glutamatergic population migrating from the pallial-subpallial boundary contributes to neocortical development. J Neurosci 30(31):10563–10574. doi:10.1523/JNEUROSCI.0776-10.2010

Teissier A, Waclaw RR, Griveau A, Campbell K, Pierani A (2012) Tangentially migrating transient glutamatergic neurons control neurogenesis and maintenance of cerebral cortical progenitor pools. Cereb Cortex 22(2):403–416. doi:10.1093/cercor/bhr122

Tissir F, Lambert De Rouvroit C, Sire JY, Meyer G, Goffinet AM (2003) Reelin expression during embryonic brain development in *Crocodylus niloticus*. J Comp Neurol 457(3):250–262. doi:10.1002/cne.10573

Tole S, Remedios R, Saha B, Stoykova A (2005) Selective requirement of Pax6, but not Emx2, in the specification and development of several nuclei of the amygdaloid complex. J Neurosci 25(10):2753–2760. doi:10.1523/JNEUROSCI.3014-04.2005

Tosa Y, Hirao A, Matsubara I, Kawaguchi M, Fukui M, Kuratani S, Murakami Y (2015) Development of the thalamo-dorsal ventricular ridge tract in the Chinese soft-shelled turtle, *Pelodiscus sinensis*. Dev Growth Differ 57(1):40–57. doi:10.1111/dgd.12186

Tsai HM, Garber BB, Larramendi LM (1981) ^{3}H-Thymidine autoradiographic analysis of telencephalic histogenesis in the chick embryo: I. Neuronal birthdates of telencephalic compartments in situ. J Comp Neurol 198(2):275–292. doi:10.1002/cne.901980207

Tuoc TC, Radyushkin K, Tonchev AB, Pinon MC, Ashery-Padan R, Molnár Z, Davidoff MS, Stoykova A (2009) Selective cortical layering abnormalities and behavioral deficits in cortex-specific Pax6 knock-out mice. J Neurosci 29(26):8335–8349. doi:10.1523/JNEUROSCI.5669-08.2009

Ulinski PS (1983) Dorsal ventricular ridge: a treatise on forebrain organization in reptiles and birds. Wiley, New York

Ulinski PS (1990) The cerebral cortex of reptiles. In: Cerebral cortex, vol 8A. Plenum Press, New York, pp 139–215

Wang W, Zhong J, Wang YQ (2010) Comparative genomic analysis reveals the evolutionary conservation of Pax gene family. Genes Genet Syst 85(3):193–206

Wang Z, Pascual-Anaya J, Zadissa A, Li W, Niimura Y, Huang Z, Li C, White S, Xiong Z, Fang D, Wang B, Ming Y, Chen Y, Zheng Y, Kuraku S, Pignatelli M, Herrero J, Beal K, Nozawa M, Li Q, Wang J, Zhang H, Yu L, Shigenobu S, Wang J, Liu J, Flicek P, Searle S, Wang J, Kuratani S, Yin Y, Aken B, Zhang G, Irie N (2013) The draft genomes of soft-shell turtle and green sea turtle yield insights into the development and evolution of the turtle-specific body plan. Nat Genet 45(6):701–706. doi:10.1038/ng.2615

Yoshida M, Assimacopoulos S, Jones KR, Grove EA (2006) Massive loss of Cajal–Retzius cells does not disrupt neocortical layer order. Development 133(3):537–545. doi:10.1242/dev.02209

Zhang X, Huang CT, Chen J, Pankratz MT, Xi J, Li J, Yang Y, Lavaute TM, Li XJ, Ayala M, Bondarenko GI, Du ZW, Jin Y, Golos TG, Zhang SC (2010) Pax6 is a human neuroectoderm cell fate determinant. Cell Stem Cell 7(1):90–100. doi:10.1016/j.stem.2010.04.017

Chapter 13
Molecular Investigations of the Structure and Development of the Brain of Carnivores

Yohei Shinmyo, Tomohisa Toda, Kosuke Masuda, Yoshio Hoshiba, Haruka Ebisu, Naoyuki Matsumoto, and Hiroshi Kawasaki

Abstract The brains of higher mammals such as carnivores and primates contain developed structures that are not found in the brains of mice. Revealing the physiological importance, developmental mechanisms, and evolution of these structures using carnivores and primates would greatly contribute to our understanding of the human brain and its diseases. Although the anatomical and physiological characteristics of the brains of carnivores and primates have been intensively investigated, molecular investigations are still limited. Recently, genetic techniques that can be applied to carnivores and primates have been explored, and molecules whose expression patterns were related to their developed brain structures were reported. To investigate the functional importance of these molecules, rapid and efficient genetic manipulation methods were established. Here we review recent advances in molecular investigations of the development and evolution of the brains of higher mammals, mainly focusing on the ferret (*Mustela putorius furo*).

Keywords Marmoset • Ferret • Cerebral cortex • Gyrus • Outer subventricular zone • Outer radial glia • In utero electroporation

13.1 Introduction

Genetically modified mice have been widely used for investigating the molecular mechanisms underlying the function and development of the brain and the pathophysiology of brain diseases. Recently, higher mammals such as carnivores and primates have attracted more attention from researchers in neuroscience because

---

Y. Shinmyo • T. Toda • K. Masuda • Y. Hoshiba • H. Ebisu • N. Matsumoto
Graduate School of Medical Sciences, Kanazawa University, Takara-machi 13-1, Kanazawa, Ishikawa 920-8640, Japan

H. Kawasaki, M.D., Ph.D. (✉)
Graduate School of Medical Sciences, Kanazawa University, Takara-machi 13-1, Kanazawa, Ishikawa 920-8640, Japan

Brain/Liver Interface Medicine Research Center, Kanazawa University, Ishikawa 920-8640, Japan
e-mail: hiroshi-kawasaki@umin.ac.jp

S. Shigeno et al. (eds.), *Brain Evolution by Design*, Diversity and Commonality in Animals, DOI 10.1007/978-4-431-56469-0_13

the brains of carnivores and primates contain developed brain structures that mice do not seem to have. These brain structures include ocular dominance columns (ODCs) in the visual cortex, the magnocellular (M) and parvocellular (P) pathways in the visual system, and the gyri and the outer subventricular zone (OSVZ) of the cerebral cortex. Because these structures have been proposed to be important for higher brain functions, determining the physiological importance, developmental mechanisms, and evolution of these structures using carnivores and primates would lead to our understanding of the human brain and its diseases, which are often difficult to investigate using mice. Although the anatomical and physiological properties have been intensively investigated, the molecular mechanisms of the formation, function, pathophysiology, and evolution of these structures are still unclear, not only because these structures are only poorly developed in mice, but also because genetic manipulation techniques that can be used for carnivores and primates were not readily available until recently.

To overcome these limitations, genetic manipulation techniques that can be applied to carnivores and primates have been explored. Pioneering studies have reported successful application of virus vectors to make transgenic primates such as monkeys and marmosets (Chan et al. 2001; Lois et al. 2002; Sasaki et al. 2009). The injection of a lentiviral vector into marmoset eggs led to transgenic marmosets that expressed transgenes in several organs (Sasaki et al. 2009). Importantly, germline transmission of the transgene was demonstrated, and transgenic offspring developed normally (Sasaki et al. 2009). Although the creation of transgenic marmosets would provide novel animal models for various kinds of human diseases (Okano et al. 2012), developing new transgenic animals takes time and effort and requires special animal facilities. Therefore, it was desirable to establish a rapid and simple genetic manipulation technique for carnivores and primates. To achieve the expression of transgenes in the cerebral cortex of higher mammals, we applied in utero electroporation to ferrets (*Mustela putorius furo*) (Fig. 13.1) and succeeded in expressing GFP in the cerebral cortex, as described next (Kawasaki et al. 2012, 2013). The ferret, similar to the weasel, badger, and skunk, belongs to Mustelidae,

Fig. 13.1 An adult ferret (*Mustela putorius furo*). The ferret has an average length of about 50 cm and weighs about 1–2 kg

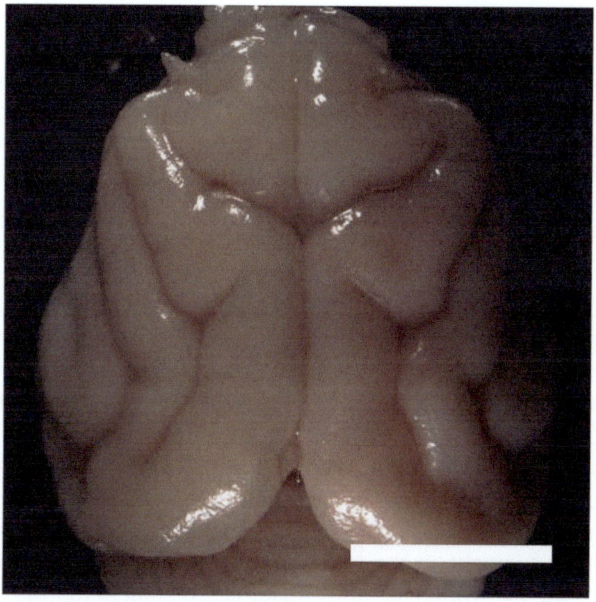

Fig. 13.2 Dorsal view of the ferret brain. Cortical gyri and sulci are clearly visible. *Bar* 1 cm (Adapted from Kawasaki et al. 2012)

which is a family of carnivorous mammals. They have an average length of about 50 cm and weight of about 1–2 kg. Ferrets have a long history as animal model subjects because they have developed brain structures that mice do not have (Fig. 13.2). Combining transgenic marmosets and electroporated ferrets would greatly facilitate our understanding of the development and evolution of the brain of higher mammals. In this chapter, we summarize recent advances in the molecular understanding of the brains of higher mammals, especially focusing on ferrets.

13.2 The Ferret as a Model Animal for Investigating the Cerebral Cortex

One of the most prominent features of the brains of higher mammals is the formation of folds in the cerebral cortex (i.e., cortical gyrus) (Fig. 13.2). Humans, monkeys, and ferrets have gyrencephalic brains (i.e., brains with cortical folds), whereas the brains of rodents are often lissencephalic (i.e., lacking cortical folds). It has been proposed that the increased number of cortical neurons resulted in expansion and folding of the cerebral cortex during evolution. It is therefore important to investigate the mechanisms regulating the proliferation, migration, and differentiation of neural progenitors in the cerebral cortex during development in higher mammals (Borrell and Reillo 2012; Dehay and Kennedy 2007; Rakic 2009;

Fietz and Huttner 2011; Lui et al. 2011; Hevner and Haydar 2012; Molnar and Clowry 2012).

During development, cortical neurons are produced from radial glial cells (RG cells, also known as apical progenitors/apical RG cells/ventricular RG cells). RG cells are epithelial stem cells in the ventricular zone (VZ) that are located along the cerebral ventricles and have apical fibers and basal fibers (Malatesta et al. 2000; Miyata et al. 2001; Noctor et al. 2001). RG cells undergo multiple rounds of asymmetrical cell divisions and produce intermediate progenitor cells (IP cells/basal progenitors). IP cells migrate into the subventricular zone (SVZ) and further proliferate to increase neuronal number (Haubensak et al. 2004; Noctor et al. 2004). One of the important features of the developing cerebral cortex of carnivores and primates is the appearance of the large SVZ, which has an inner region (ISVZ) and an outer region (OSVZ), often divided by a thin layer of fibers called the inner fiber layer (IFL) (Smart et al. 2002; Zecevic et al. 2005). Recent pioneering studies uncovered a novel class of progenitor cells in the OSVZ, OSVZ radial glial cells (oRG cells/outer RG cells/basal RG cells/intermediate RG cells/translocating RG cells) (Fietz et al. 2010; Hansen et al. 2010; Reillo et al. 2011). In contrast to RG cells in the VZ, oRG cells in the OSVZ are unipolar and have a basal fiber that ascends toward the pia without an apical fiber descending to the ventricle. Because a major underlying cause of the expansion and gyrification of the cerebral cortex during evolution seems to be the increase in population size of neural progenitors in the OSVZ, which is a specialized germinal zone characteristic of higher mammals, it would be intriguing to investigate the mechanisms underlying the proliferation, migration, and differentiation of neural progenitors in the OSVZ.

Ferrets seem to be a good option for investigating such mechanisms, given that genetic manipulation using ferrets has become feasible (Borrell 2010; Kawasaki et al. 2012, 2013; Reillo et al. 2011; Nonaka-Kinoshita et al. 2013; Pilz et al. 2013). Recently, lists of genes expressed in the VZ and SVZ of mice and in various regions of the cerebral cortex in monkeys were reported (Ayoub et al. 2011; Bernard et al. 2012). Manipulating these genes in ferrets should lead to identifying the mechanisms underlying the formation and evolution of the gyrus and the OSVZ in the cerebral cortex. Recent studies also identified oRG-like progenitors in mice and marmosets. These animals also appear useful for examining the developmental mechanisms of oRG cells (Garcia-Moreno et al. 2012; Wang et al. 2011; Shitamukai et al. 2011; Kelava et al. 2011). Discovering molecules whose expression patterns correspond to brain structures unique to carnivores and primates will facilitate our understanding of the development, function, pathophysiology, and evolution of the brain. Because a recent report identified four distinct morphologies of oRG cells in macaques (Betizeau et al. 2013), it is intriguing to examine the morphological diversity of oRG cells in ferrets.

Neuronal migration has also been examined using the ferret cerebral cortex (Anderson et al. 2002; O'Rourke et al. 1992, 1995, 1997; Borrell et al. 2006). Time-lapse confocal microscopic analyses using cultured cortical slices demonstrated that most cells migrated along radial fibers, whereas a subset of cells migrated orthogonal to radial fibers (O'Rourke et al. 1992). In vivo DiI focal injection showed that labeled cells migrated in all directions and over long distances (O'Rourke et al. 1997). These results suggest that cortical cells migrate not only radially but also non-radially during development, which may lead to tangential dispersion.

13.3 The Ferret as a Model Animal for Investigating the Visual Cortex

Using the developed visual system in carnivores and primates, important concepts about the intrinsic and extrinsic regulatory factors responsible for brain development have been intensively investigated (Chalupa and Werner 2003). Anatomical properties of ocular dominance columns (ODCs) in the visual cortex and retinogeniculate projections in the lateral geniculate nucleus (LGN), physiological descriptions of cortical responses, and the dendritic morphology of neurons in the visual cortex have been investigated (Callaway and Katz 1993; Usrey et al. 2003; Yu et al. 2005; Crowley and Katz 1999, 2000; Kawasaki et al. 2004; Huberman et al. 2005; Iwai et al. 2013; Matsui et al. 2013; Horch and Katz 2002). A pioneering study by Hubel and Wiesel initially demonstrated ODCs in the primary visual cortex (V1) of cats in the early 1960s (Fig. 13.3) (Hubel and Wiesel 1962). Cortical neurons in V1 respond differentially to one of the two eyes, and those neurons with similar eye preference are segregated into cortical columns in V1. These columns contain LGN axons derived from right and left eye-specific layers of the LGN, and LGN axons segregate to form alternating stripes in cortical layer 4 of V1 (Fig. 13.3). ODCs can be revealed with electrophysiological techniques and transneuronal tracers such as tritiated amino acids. Although it has been proposed that ODCs are involved in binocular vision, it is still unclear whether ODCs are functionally important or are

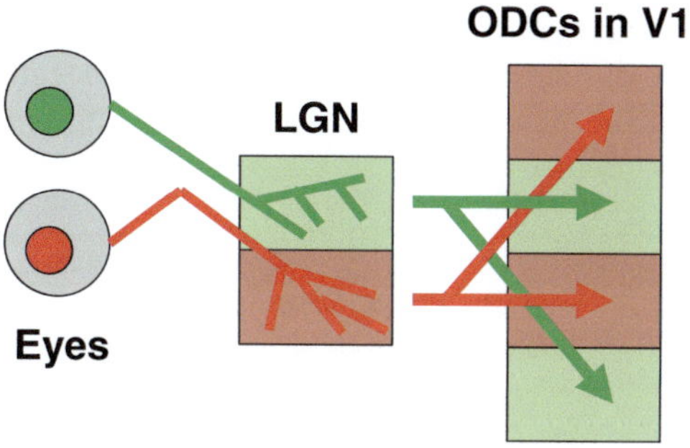

Fig. 13.3 Schematic representation of ocular dominance columns (ODCs) in the primary visual cortex (V1). Note that cortical neurons in one ODC are activated preferentially by one of the two eyes. *Green* and *red* in the lateral geniculate nucleus (LGN) and V1 represent the areas that are preferentially activated by visual inputs to the green eye and the red eye, respectively

just a by-product. Notably, ODCs in V1 are observed in both primates and carnivores including ferrets, but not in mice.

ODCs in V1 of primates and carnivores have been widely used for investigating the mechanisms of developmental plasticity. Monocular eyelid closure during the first few months of life (i.e., the critical period) reduced the number of neurons activated by the closed eye and increased the number of neurons activated by the intact open eye (Wiesel and Hubel 1963, 1965a, b). The investigations of the effects of monocular deprivation demonstrated that preexisting neuronal connections are substantially modified by an activity-dependent competitive process mediated by the NMDA receptor. Interestingly, visualization of geniculocortical axons by direct tracer injections into an eye-specific layer of the LGN in early postnatal ferrets revealed segregated ODCs before the critical period (Crowley and Katz 2000). These findings distinguished between the innate mechanisms that determine the initial formation of cortical columnar architecture and the experience-dependent, competition-based refinement responsible for their later modification during the critical period in V1.

13.4 The Ferret as a Model Animal for Investigating Projection Patterns of Retinal Axons

The visual information detected by the retina is transferred to the LGN via the retinogeniculate axons of retinal ganglion cells (RGCs). These retinogeniculate axons contain several distinct information properties including eye specificity, ON/OFF responses, and magnocellular/parvocellular/koniocellular (M/P/K) responses.

Using eye-specific projections, the mechanisms underlying the activity-dependent refinement of connections have been investigated. In adult mammals, RGC axons from the two eyes are segregated into eye-specific regions in the LGN (Fig. 13.4). Early in development, however, when retinogeniculate projections are initially formed, RGC axons from the two eyes are intermingled and are not

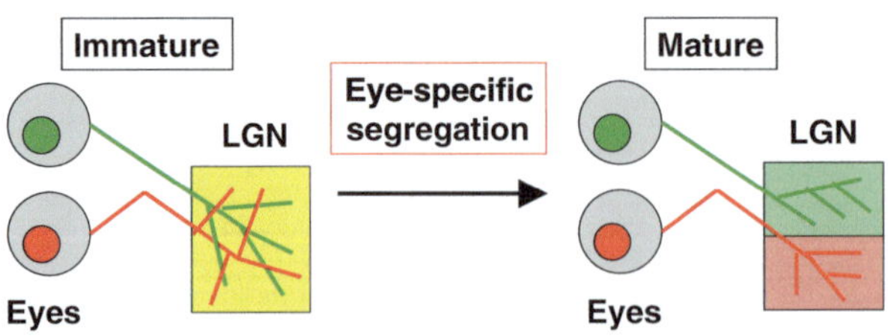

Fig. 13.4 A schematic representation of eye-specific segregation of retinogeniculate axons in the LGN. Retinal ganglion cell (RGC) axons from the two eyes are initially intermingled in the LGN (*left, yellow*) and then are segregated into eye-specific regions (*right, green and red*) during development

segregated in the LGN (Fig. 13.4). After this initial formation, the refinement of retinogeniculate projections proceeds to make distinct eye-specific regions during development (eye-specific segregation). The mechanisms of eye-specific segregation have been examined, and eye-specific segregation was found to require spontaneous retinal activity called retinal waves. Monocular blockade of retinal waves prevented normal eye-specific segregation in the LGN (Penn et al. 1998). Retinogeniculate projections from the intact retina were greatly expanded in the LGN, whereas those from the inhibited retina were substantially decreased. These findings indicate that spontaneous neuronal activity leads to highly stereotyped patterns of projections preceding visual experience (Penn et al. 1998; Huberman et al. 2003). It has been proposed that the firing patterns of retinal waves provide the appropriate spatial and temporal information to guide the refinement of retinogeniculate projections (Meister et al. 1991). Retinal waves are observed during the same period in which eye-specific segregation proceeds in the LGN and subside gradually as retinogeniculate projections stabilize (Wong et al. 1993). Although ferrets have been widely used for investigating these mechanisms, mice are becoming another choice for examining the mechanisms of eye-specific segregation during development (Chen and Regehr 2000; Demas et al. 2006; Upton et al. 1999; Hayakawa and Kawasaki 2010; Iwai and Kawasaki 2009).

In contrast to eye-specific projections, the ON/OFF pathways and the M/P/K pathways are more developed in the visual system in ferrets than in mice. For example, ON cells and OFF cells, which receive inputs from ON-center RGCs and OFF-center RGCs, are spatially mixed in the mouse LGN, whereas they are distributed in distinct ON and OFF sublaminae in the ferret LGN (Stryker and Zahs 1983). Using ON and OFF sublaminae in the ferret LGN, the mechanisms underlying the formation of ON and OFF sublaminae have been examined. Pharmacological experiments demonstrated that the NMDA receptor was required for the formation of ON and OFF sublaminae (Hahm et al. 1991). Because the formation of ON and OFF sublaminae was disrupted by the nitric oxide (NO) synthase inhibitor L-NoArg and the soluble guanylyl cyclase (sGC) inhibitor ODQ, NO and sGC seemed to be downstream mediators of the NMDA receptor (Cramer et al. 1996; Leamey et al. 2001). It seemed likely that calcium influx through the NMDA receptor activates NO synthase, and activation of sGC leads to the formation of ON and OFF sublaminae in the LGN during development. It remains unclear whether NO and sGC work presynaptically or postsynaptically. It should be noted that the NMDA receptor is crucial for the ON/OFF segregation of retinogeniculate axons in the ferret LGN, although they are dispensable for eye-specific segregation of retinogeniculate axons (Smetters et al. 1994). It seems likely that distinct mechanisms underlie different types of segregation of retinogeniculate axons in the ferret LGN.

13.5 The Ferret as a Model Animal for Investigating the Parallel Visual Pathways

The magnocellular/parvocellular/koniocellular parallel visual pathways are also more prominent in the visual system of higher mammals than in that of mice. Visual information detected by the retina is conveyed to the LGN and then to V1 along the parallel visual pathways. The parallel visual pathways are composed of three pathways with distinct anatomical and physiological properties. These three pathways are known as the X, Y, and W pathways in carnivores and as the parvocellular (P), magnocellular (M), and koniocellular (K) pathways in primates (DeYoe and Van Essen 1988; Sherman and Spear 1982; Livingstone and Hubel 1987; Felleman and Van Essen 1991; Maunsell 1992; Hendry and Reid 2000; Sherman and Guillery 2004; Jones 2007; Nassi and Callaway 2009; Wassle 2004). It is believed that these parallel visual pathways contribute to visual perception differently. The visual ability of monkeys was investigated after selectively damaging either the M or P layers of the LGN by injecting pharmacological reagents. Selective damage to the M layers had little effect on visual acuity or color vision, but strongly reduced the ability to recognize moving stimuli. In contrast, damage to the P layers did not show obvious effects on motion perception but markedly impaired visual acuity and color perception. These results suggest that the visual information conveyed by the P pathway is related to the detailed analysis of the shape, size, and color of objects, whereas the M pathway is mainly concerned with information about the movement of objects.

To elucidate the molecular mechanisms underlying the formation of the developed brain structures unique to higher mammals, it is important to identify molecules whose expression patterns correspond to these structures. Several laboratories have identified such molecules using carnivores and primates (Bernard et al. 2012; Kawasaki et al. 2004; Iwai et al. 2013; Yamamori 2011; Johnson et al. 2009; Murray et al. 2008; Mashiko et al. 2012). For example, we recently found that the Forkhead transcription factor FoxP2 was selectively expressed in X cells of the adult ferret dLGN and in the P layers of the adult monkey LGN (Iwai et al. 2013). One of the long-standing questions about the evolution of the visual system is the relationship between M and P cells in primates and Y and X cells in carnivores (Fig. 13.5). It has been suggested that X and Y cells in carnivores are comparable to

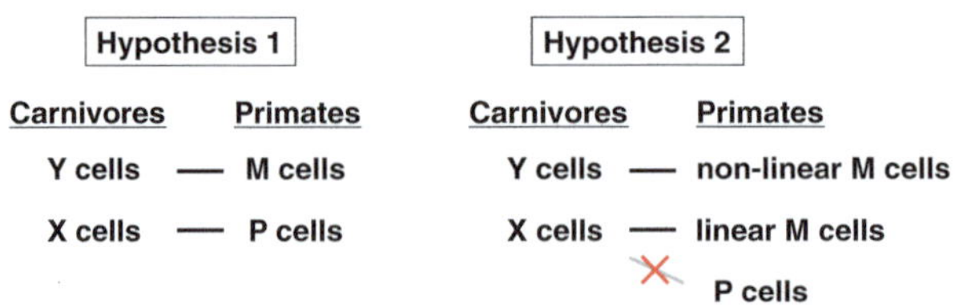

Fig. 13.5 Two hypotheses about the relationship between M and P cells in primates and Y and X cells in carnivores. The expression pattern of FoxP2 suggests that hypothesis 1 is correct

P and M cells in primates, respectively. An alternative view was that X and Y cells are homologous to linear and nonlinear M cells, respectively, and that P cells are unique to primates (Kaplan 2004). Because our findings demonstrated that FoxP2 was selectively expressed in X cells in the ferret LGN and also in the P layers in the monkey LGN, these findings provide new evidence for a homology between X cells of ferrets and P cells of monkeys (Iwai et al. 2013) (Fig. 13.5). Because FoxP2 is the first transcription factor found to be selectively expressed in one of the three pathways, further investigations of FoxP2 would be important for detailing the mechanisms underlying the development and evolution of the M/P/K parallel visual pathways. It would be important to investigate the roles of FoxP2 in the development of X cells of the LGN in carnivores and the P layers of the LGN in primates because the molecular mechanisms underlying the formation of the parallel visual pathways during development are still unknown. Detailed analyses of the visual functions of the KE family, whose members have a mutated *FOXP2* gene and developmental speech-language abnormalities (Lai et al. 2001), could also aid in uncovering the roles of FoxP2 in the development of visual system. Another application of FoxP2 would be the characterization of the FoxP2 promoter, which should be useful for selectively expressing genes of interest such as GFP, Kir2.1, and NaChBac in the P pathway of the LGN. Expressing such genes would uncover the precise neuronal circuits and functional roles of the P pathway. A deeper understanding of the parallel visual pathways in higher mammals will not only provide information about visual recognition in humans but will also contribute to our understanding of the development and evolution of the visual system and the general mechanisms by which the brain integrates sensory information derived from the external world.

13.6 Other Studies Using the Ferret as a Model Animal

Remodeling of axonal projection patterns has also been examined using the visual system of ferrets (Sur et al. 1988; Roe et al. 1992, 1993; Angelucci et al. 1997). Using neonatal surgical manipulations, RGC axons, which usually innervate the LGN, can be induced to innervate the auditory thalamus (Sur et al. 1988). Importantly, neuronal activity in response to visual stimuli can be recorded in the primary auditory cortex (Sur et al. 1988). Similar to neurons in the normal primary visual cortex, those in the primary auditory cortex of manipulated animals show orientation and direction selectivity and have simple and complex receptive fields (Roe et al. 1992). Among the X, Y, and W types of retinal ganglion cells of the ferret retina, W cells are mainly responsible for this remodeling (Roe et al. 1993). Interestingly, eye-specific projection patterns found in the LGN of the visual system are recapitulated in the auditory thalamus of manipulated ferrets (Angelucci et al. 1997). These findings indicate that functional neuronal circuits are reconstituted in the auditory thalamus of manipulated ferrets.

13.7 Genetic Manipulation of Ferrets Using In Utero Electroporation

Although transgenic marmosets provided a new avenue to examine the pathophysiological mechanisms of human diseases (Okano et al. 2012), it was also desirable to establish rapid and simple genetic manipulation methods for carnivores and primates. In utero electroporation is well known to be useful for expressing genes of interest in the living rodent brain (Ako et al. 2011; Saito and Nakatsuji 2001; Tabata and Nakajima 2001; Fukuchi-Shimogori and Grove 2001; Sehara et al. 2010). Because successful application of in utero electroporation in carnivores and primates had not been achieved, we developed a rapid and efficient procedure of in utero electroporation for the ferret brain (Fig. 13.6) (Kawasaki et al. 2012, 2013). Using our procedure, electroporated ferret babies can be obtained within a couple of days after in utero electroporation is performed. Expression of transgenes becomes detectable in the embryo and is observed at least 2 months after birth. Transgenes can be expressed in both superficial and deep neurons in the cerebral cortex, depending on when in utero electroporation is performed. In utero electroporation performed at E31 and E37 results in transgene expression in superficial and deep cortical neurons, respectively. Our electroporation procedure is useful for transfecting not only postmitotic neurons but also neural progenitors, which include radial glial cells (RG cells) in the ventricular zone (VZ), outer radial glial cells (oRG cells) in the outer subventricular zone (OSVZ), and intermediate progenitor cells (IP cells) (Kawasaki et al. 2012, 2013). Because the OSVZ of the cerebral cortex is especially prominent in gyrencephalic mammals during development, it would be

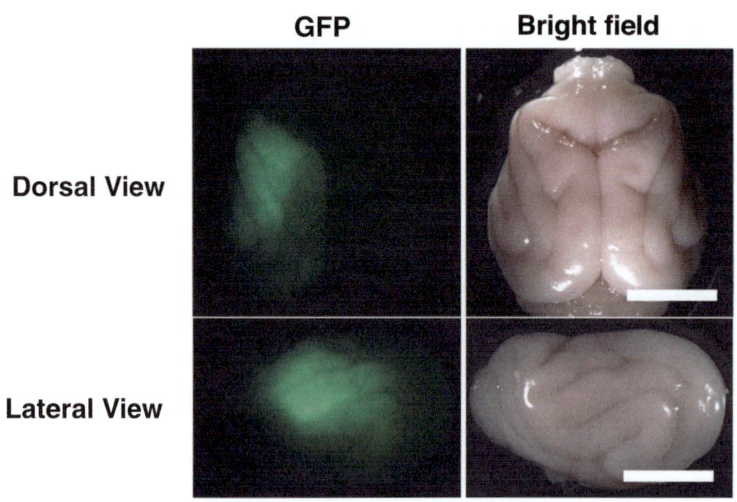

Fig. 13.6 Green fluorescent protein (GFP) in the ferret brain introduced by using in utero electroporation. *Bars* 1 cm (Adapted from Kawasaki et al. 2012)

exciting to examine the mechanisms of the formation of the OSVZ and the roles of the OSVZ in the formation of the gyrencephalic cortex. These experiments may lead to our understanding of the mechanisms of the formation of gyri during evolution.

In utero electroporation is useful not only for investigating molecular mechanisms but also for identifying novel neuronal circuitry (Sehara et al. 2010, 2012). Development of the cerebral cortex in higher mammals such as monkeys and humans is characterized by the appearance of the inner fiber layer (IFL), a fiber layer located between the ISVZ and the OSVZ (Smart et al. 2002; Zecevic et al. 2005). However, a previous study using ferrets failed to identify an obvious fiber layer corresponding to the IFL of the primate cerebral cortex (Martinez-Cerdeno et al. 2012). In addition, it was unclear from which neurons the IFL is derived (Molnar and Clowry 2012). Interestingly, when we expressed GFP in layer 2/3 neurons of the ferret cerebral cortex using in utero electroporation, we found GFP-positive fibers located in the inner OSVZ (Fig. 13.7). This finding suggests that ferrets do also have a fiber layer corresponding to the IFL in the primate cerebral cortex (Kawasaki et al. 2013). This finding also suggests that layer 2/3 neurons in the cerebral cortex are responsible for the IFL, at least partially, in ferrets (Fig. 13.7) (Kawasaki et al. 2013). One attractive hypothesis would be that an increase in the number of layer 2/3 neurons during evolution led to the formation of the thick bundle comprising the IFL in the cerebral cortex of higher mammals.

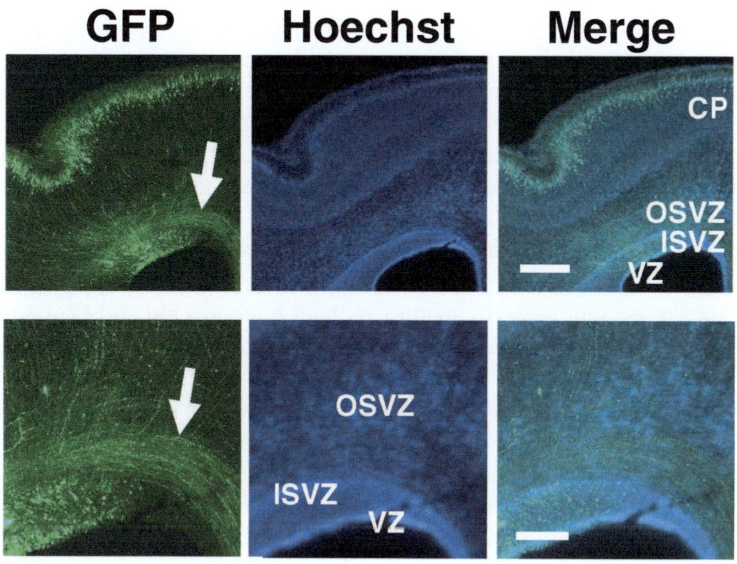

Fig. 13.7 Inner fiber layer (IFL)-like fibers in the developing ferret cerebral cortex. When GFP was expressed in layer 2/3 neurons using in utero electroporation, GFP-positive fibers were found in the inner region of the outer subventricular zone (OSVZ) (*arrows*). *Upper panels* are low-magnification images; *lower panels* are high-magnification images. *CP* cortical plate. *Bars* 500 μm (*upper*); 200 μm (*lower*) (Adapted from Kawasaki et al. 2013)

In utero electroporation has several important characteristics. First, it is easy to perform co-transfection, and as a result, multiple genes can be expressed simultaneously. When in utero electroporation is performed using a mixture of GFP and mCherry expression plasmids, most GFP-positive neurons are also mCherry positive in the ferret cerebral cortex, suggesting that co-transfection efficiencies are reasonably high. Second, the location of the transfected area can be modified by adjusting the direction of the electrodes and by changing the age at which in utero electroporation is performed. Based on the results obtained using rodents (Borrell et al. 2005; Kataoka and Shimogori 2008; Soma et al. 2009; Nakahira and Yuasa 2005; Hatanaka et al. 2004; Garcia-Frigola et al. 2007), it seems plausible that genetic manipulation using in utero electroporation is applicable to other brain regions such as the hippocampus, the thalamus, the retina, and the amygdala in ferrets. Third, because larger plasmids can be used for in utero electroporation, cell type-specific promoters would be valuable for regulating the distribution patterns of transgenes in the electroporated ferret brain. For example, we recently combined in utero electroporation and the Thy1S promoter for labeling neurons sparsely in mice (Ako et al. 2011). Finally, it does not take much time and effort to express genes of interest using in utero electroporation. It takes only a couple of days to make transfected ferrets. Because in utero electroporation is useful not only for rodents but also for carnivores, it seems possible that it would also be useful for other higher mammals such as primates.

Besides in utero electroporation, postnatal electroporation was reported to be a feasible method for expressing genes into the cerebral cortex of ferrets (Borrell 2010). When postnatal electroporation was used, transfected neurons were mostly distributed in superficial layer 2/3 in the ferret cerebral cortex. This distribution seems to occur because most cortical neurons had already moved into the cortical plate from the ventricular zone when postnatal electroporation was performed. Combining postnatal and in utero electroporation, excitatory neurons in most of the cerebral cortex of ferrets can be manipulated.

Although both species are candidates for research into the brains of higher mammals, ferrets and marmosets differ in a number of important ways. Ferrets and marmosets become sexually mature at about 8 months old and 14 months old, respectively, and average lifespans are 5–10 years in ferrets and 10–15 years in marmosets. The gestation period of ferrets is shorter than that of marmosets (ferrets, 42 days; marmosets, 150 days). About eight babies are often born from one pregnant ferret mother, while one or two are born from a pregnant marmoset. This larger number of babies per pregnant mother is helpful for obtaining a sufficient number of experimental samples. Therefore, the ferret should be an interesting option for exploring developed brain structures in higher mammals.

13.8 Conclusions

Here we have summarized recent advances in the molecular understanding of the development and evolution of the brain of higher mammals, mainly focusing on ferrets. Given that one of the ultimate goals of neuroscience research is to understand the human brain, molecular investigations of the brain of carnivores and primates are of great importance. Because a rapid and simple genetic manipulation is now feasible using ferrets, the ferret should be an important option for neuroscience researchers.

Acknowledgments We apologize to the numerous authors whose valuable papers we were not able to cite because of limitations. We thank the Kawasaki Laboratory members for their contributions.

References

Ako R, Wakimoto M, Ebisu H, Tanno K, Hira R, Kasai H, Matsuzaki M, Kawasaki H (2011) Simultaneous visualization of multiple neuronal properties with single-cell resolution in the living rodent brain. Mol Cell Neurosci 48(3):246–257

Anderson SA, Kaznowski CE, Horn C, Rubenstein JL, McConnell SK (2002) Distinct origins of neocortical projection neurons and interneurons in vivo. Cereb Cortex 12(7):702–709

Angelucci A, Clasca F, Bricolo E, Cramer KS, Sur M (1997) Experimentally induced retinal projections to the ferret auditory thalamus: development of clustered eye-specific patterns in a novel target. J Neurosci 17(6):2040–2055

Ayoub AE, Oh S, Xie Y, Leng J, Cotney J, Dominguez MH, Noonan JP, Rakic P (2011) Transcriptional programs in transient embryonic zones of the cerebral cortex defined by high-resolution mRNA sequencing. Proc Natl Acad Sci USA 108(36):14950–14955

Bernard A, Lubbers LS, Tanis KQ, Luo R, Podtelezhnikov AA, Finney EM, McWhorter MME, Serikawa K, Lemon T, Morgan R, Copeland C, Smith K, Cullen V, Davis-Turak J, Lee C-K, Sunkin SM, Loboda AP, Levine DM, Stone DJ, Hawrylycz MJ, Roberts CJ, Jones AR, Geschwind DH, Lein ES (2012) Transcriptional architecture of the primate neocortex. Neuron 73:1083–1099

Betizeau M, Cortay V, Patti D, Pfister S, Gautier E, Bellemin-Menard A, Afanassieff M, Huissoud C, Douglas RJ, Kennedy H, Dehay C (2013) Precursor diversity and complexity of lineage relationships in the outer subventricular zone of the primate. Neuron 80(2):442–457

Borrell V (2010) In vivo gene delivery to the postnatal ferret cerebral cortex by DNA electroporation. J Neurosci Methods 186(2):186–195

Borrell V, Reillo I (2012) Emerging roles of neural stem cells in cerebral cortex development and evolution. Dev Neurobiol 72(7):955–971

Borrell V, Yoshimura Y, Callaway EM (2005) Targeted gene delivery to telencephalic inhibitory neurons by directional in utero electroporation. J Neurosci Methods 143(2):151–158

Borrell V, Kaspar BK, Gage FH, Callaway EM (2006) In vivo evidence for radial migration of neurons by long-distance somal translocation in the developing ferret visual cortex. Cereb Cortex 16(11):1571–1583

Callaway EM, Katz LC (1993) Photostimulation using caged glutamate reveals functional circuitry in living brain slices. Proc Natl Acad Sci USA 90(16):7661–7665

Chalupa LM, Werner JS (2003) The visual neuroscience. MIT Press, Cambridge

Chan AW, Chong KY, Martinovich C, Simerly C, Schatten G (2001) Transgenic monkeys produced by retroviral gene transfer into mature oocytes. Science 291(5502):309–312
Chen C, Regehr WG (2000) Developmental remodeling of the retinogeniculate synapse. Neuron 28(3):955–966
Cramer KS, Angelucci A, Hahm JO, Bogdanov MB, Sur M (1996) A role for nitric oxide in the development of the ferret retinogeniculate projection. J Neurosci 16(24):7995–8004
Crowley JC, Katz LC (1999) Development of ocular dominance columns in the absence of retinal input. Nat Neurosci 2(12):1125–1130
Crowley JC, Katz LC (2000) Early development of ocular dominance columns. Science 290(5495):1321–1324
Dehay C, Kennedy H (2007) Cell-cycle control and cortical development. Nat Rev Neurosci 8(6):438–450
Demas J, Sagdullaev BT, Green E, Jaubert-Miazza L, McCall MA, Gregg RG, Wong RO, Guido W (2006) Failure to maintain eye-specific segregation in nob, a mutant with abnormally patterned retinal activity. Neuron 50(2):247–259
DeYoe EA, Van Essen DC (1988) Concurrent processing streams in monkey visual cortex. Trends Neurosci 11(5):219–226
Felleman DJ, Van Essen DC (1991) Distributed hierarchical processing in the primate cerebral cortex. Cereb Cortex 1(1):1–47
Fietz SA, Huttner WB (2011) Cortical progenitor expansion, self-renewal and neurogenesis: a polarized perspective. Curr Opin Neurobiol 21(1):23–35
Fietz SA, Kelava I, Vogt J, Wilsch-Brauninger M, Stenzel D, Fish JL, Corbeil D, Riehn A, Distler W, Nitsch R, Huttner WB (2010) OSVZ progenitors of human and ferret neocortex are epithelial-like and expand by integrin signaling. Nat Neurosci 13(6):690–699
Fukuchi-Shimogori T, Grove EA (2001) Neocortex patterning by the secreted signaling molecule FGF8. Science 294(5544):1071–1074
Garcia-Frigola C, Carreres MI, Vegar C, Herrera E (2007) Gene delivery into mouse retinal ganglion cells by in utero electroporation. BMC Dev Biol 7:103
Garcia-Moreno F, Vasistha NA, Trevia N, Bourne JA, Molnar Z (2012) Compartmentalization of cerebral cortical germinal zones in a lissencephalic primate and gyrencephalic rodent. Cereb Cortex 22(2):482–492
Hahm JO, Langdon RB, Sur M (1991) Disruption of retinogeniculate afferent segregation by antagonists to NMDA receptors. Nature 351(6327):568–570
Hansen DV, Lui JH, Parker PR, Kriegstein AR (2010) Neurogenic radial glia in the outer subventricular zone of human neocortex. Nature 464(7288):554–561
Hatanaka Y, Hisanaga S, Heizmann CW, Murakami F (2004) Distinct migratory behavior of early- and late-born neurons derived from the cortical ventricular zone. J Comp Neurol 479(1):1–14
Haubensak W, Attardo A, Denk W, Huttner WB (2004) Neurons arise in the basal neuroepithelium of the early mammalian telencephalon: a major site of neurogenesis. Proc Natl Acad Sci USA 101(9):3196–3201
Hayakawa I, Kawasaki H (2010) Rearrangement of retinogeniculate projection patterns after eye-specific segregation in mice. PLoS One 5(6), e11001
Hendry SH, Reid RC (2000) The koniocellular pathway in primate vision. Annu Rev Neurosci 23:127–153
Hevner RF, Haydar TF (2012) The (not necessarily) convoluted role of basal radial glia in cortical neurogenesis. Cereb Cortex 22(2):465–468
Horch HW, Katz LC (2002) BDNF release from single cells elicits local dendritic growth in nearby neurons. Nat Neurosci 5(11):1177–1184
Hubel DH, Wiesel TN (1962) Receptive fields, binocular interaction and functional architecture in the cat's visual cortex. J Physiol 160:106–154
Huberman AD, Wang GY, Liets LC, Collins OA, Chapman B, Chalupa LM (2003) Eye-specific retinogeniculate segregation independent of normal neuronal activity. Science 300(5621):994–998

Huberman AD, Murray KD, Warland DK, Feldheim DA, Chapman B (2005) Ephrin-As mediate targeting of eye-specific projections to the lateral geniculate nucleus. Nat Neurosci 8(8):1013–1021

Iwai L, Kawasaki H (2009) Molecular development of the lateral geniculate nucleus in the absence of retinal waves during the time of retinal axon eye-specific segregation. Neuroscience 159(4):1326–1337

Iwai L, Ohashi Y, van der List D, Usrey WM, Miyashita Y, Kawasaki H (2013) FoxP2 is a parvocellular-specific transcription factor in the visual thalamus of monkeys and ferrets. Cereb Cortex 23:2204–2212

Johnson MB, Kawasawa YI, Mason CE, Krsnik Z, Coppola G, Bogdanovic D, Geschwind DH, Mane SM, State MW, Sestan N (2009) Functional and evolutionary insights into human brain development through global transcriptome analysis. Neuron 62(4):494–509

Jones EG (2007) The thalamus, 2nd edn. Cambridge University Press, Cambridge

Kaplan E (2004) The M, P, and K pathways of the primate visual system. In: Chalupa LM, Werner JS (eds) The visual neurosciences, vol 1. MIT Press, Cambridge, pp 481–493

Kataoka A, Shimogori T (2008) Fgf8 controls regional identity in the developing thalamus. Development 135(17):2873–2881

Kawasaki H, Crowley JC, Livesey FJ, Katz LC (2004) Molecular organization of the ferret visual thalamus. J Neurosci 24(44):9962–9970

Kawasaki H, Iwai L, Tanno K (2012) Rapid and efficient genetic manipulation of gyrencephalic carnivores using in utero electroporation. Mol Brain 5:24

Kawasaki H, Toda T, Tanno K (2013) In vivo genetic manipulation of cortical progenitors in gyrencephalic carnivores using in utero electroporation. Biol Open 2(1):95–100

Kelava I, Reillo I, Murayama AY, Kalinka AT, Stenzel D, Tomancak P, Matsuzaki F, Lebrand C, Sasaki E, Schwamborn JC, Okano H, Huttner WB, Borrell V (2011) Abundant occurrence of basal radial glia in the subventricular zone of embryonic neocortex of a lissencephalic primate, the common marmoset *Callithrix jacchus*. Cereb Cortex 22(2):469–481

Lai CS, Fisher SE, Hurst JA, Vargha-Khadem F, Monaco AP (2001) A forkhead-domain gene is mutated in a severe speech and language disorder. Nature 413(6855):519–523

Leamey CA, Ho-Pao CL, Sur M (2001) Disruption of retinogeniculate pattern formation by inhibition of soluble guanylyl cyclase. J Neurosci 21(11):3871–3880

Livingstone MS, Hubel DH (1987) Psychophysical evidence for separate channels for the perception of form, color, movement, and depth. J Neurosci 7(11):3416–3468

Lois C, Hong EJ, Pease S, Brown EJ, Baltimore D (2002) Germline transmission and tissue-specific expression of transgenes delivered by lentiviral vectors. Science 295(5556):868–872

Lui JH, Hansen DV, Kriegstein AR (2011) Development and evolution of the human neocortex. Cell 146(1):18–36

Malatesta P, Hartfuss E, Gotz M (2000) Isolation of radial glial cells by fluorescent-activated cell sorting reveals a neuronal lineage. Development 127(24):5253–5263

Martinez-Cerdeno V, Cunningham CL, Camacho J, Antczak JL, Prakash AN, Cziep ME, Walker AI, Noctor SC (2012) Comparative analysis of the subventricular zone in rat, ferret and macaque: evidence for an outer subventricular zone in rodents. PLoS One 7(1), e30178

Mashiko H, Yoshida AC, Kikuchi SS, Niimi K, Takahashi E, Aruga J, Okano H, Shimogori T (2012) Comparative anatomy of marmoset and mouse cortex from genomic expression. J Neurosci 32(15):5039–5053

Matsui A, Tran M, Yoshida AC, Kikuchi SS, Mami U, Ogawa M, Shimogori T (2013) BTBD3 controls dendrite orientation toward active axons in mammalian neocortex. Science 342(6162):1114–1118

Maunsell JH (1992) Functional visual streams. Curr Opin Neurobiol 2(4):506–510

Meister M, Wong RO, Baylor DA, Shatz CJ (1991) Synchronous bursts of action potentials in ganglion cells of the developing mammalian retina. Science 252(5008):939–943

Miyata T, Kawaguchi A, Okano H, Ogawa M (2001) Asymmetric inheritance of radial glial fibers by cortical neurons. Neuron 31(5):727–741

Molnar Z, Clowry G (2012) Cerebral cortical development in rodents and primates. Prog Brain Res 195:45–70

Murray KD, Rubin CM, Jones EG, Chalupa LM (2008) Molecular correlates of laminar differences in the macaque dorsal lateral geniculate nucleus. J Neurosci 28(46):12010–12022

Nakahira E, Yuasa S (2005) Neuronal generation, migration, and differentiation in the mouse hippocampal primoridium as revealed by enhanced green fluorescent protein gene transfer by means of in utero electroporation. J Comp Neurol 483(3):329–340

Nassi JJ, Callaway EM (2009) Parallel processing strategies of the primate visual system. Nat Rev Neurosci 10(5):360–372

Noctor SC, Flint AC, Weissman TA, Dammerman RS, Kriegstein AR (2001) Neurons derived from radial glial cells establish radial units in neocortex. Nature 409(6821):714–720

Noctor SC, Martinez-Cerdeno V, Ivic L, Kriegstein AR (2004) Cortical neurons arise in symmetric and asymmetric division zones and migrate through specific phases. Nat Neurosci 7(2):136–144

Nonaka-Kinoshita M, Reillo I, Artegiani B, Martinez-Martinez MA, Nelson M, Borrell V, Calegari F (2013) Regulation of cerebral cortex size and folding by expansion of basal progenitors. EMBO J 32(13):1817–1828

O'Rourke NA, Dailey ME, Smith SJ, McConnell SK (1992) Diverse migratory pathways in the developing cerebral cortex. Science 258(5080):299–302

O'Rourke NA, Sullivan DP, Kaznowski CE, Jacobs AA, McConnell SK (1995) Tangential migration of neurons in the developing cerebral cortex. Development 121(7):2165–2176

O'Rourke NA, Chenn A, McConnell SK (1997) Postmitotic neurons migrate tangentially in the cortical ventricular zone. Development 124(5):997–1005

Okano H, Hikishima K, Iriki A, Sasaki E (2012) The common marmoset as a novel animal model system for biomedical and neuroscience research applications. Semin Fetal Neonatal Med 17(6):336–340

Penn AA, Riquelme PA, Feller MB, Shatz CJ (1998) Competition in retinogeniculate patterning driven by spontaneous activity. Science 279(5359):2108–2112

Pilz GA, Shitamukai A, Reillo I, Pacary E, Schwausch J, Stahl R, Ninkovic J, Snippert HJ, Clevers H, Godinho L, Guillemot F, Borrell V, Matsuzaki F, Gotz M (2013) Amplification of progenitors in the mammalian telencephalon includes a new radial glial cell type. Nat Commun 4:2125

Rakic P (2009) Evolution of the neocortex: a perspective from developmental biology. Nat Rev Neurosci 10(10):724–735

Reillo I, de Juan Romero C, Garcia-Cabezas MA, Borrell V (2011) A role for intermediate radial glia in the tangential expansion of the mammalian cerebral cortex. Cereb Cortex 21(7):1674–1694

Roe AW, Pallas SL, Kwon YH, Sur M (1992) Visual projections routed to the auditory pathway in ferrets: receptive fields of visual neurons in primary auditory cortex. J Neurosci 12(9):3651–3664

Roe AW, Garraghty PE, Esguerra M, Sur M (1993) Experimentally induced visual projections to the auditory thalamus in ferrets: evidence for a W cell pathway. J Comp Neurol 334(2):263–280

Saito T, Nakatsuji N (2001) Efficient gene transfer into the embryonic mouse brain using in vivo electroporation. Dev Biol 240(1):237–246

Sasaki E, Suemizu H, Shimada A, Hanazawa K, Oiwa R, Kamioka M, Tomioka I, Sotomaru Y, Hirakawa R, Eto T, Shiozawa S, Maeda T, Ito M, Ito R, Kito C, Yagihashi C, Kawai K, Miyoshi H, Tanioka Y, Tamaoki N, Habu S, Okano H, Nomura T (2009) Generation of transgenic non-human primates with germline transmission. Nature 459(7246):523–527

Sehara K, Toda T, Iwai L, Wakimoto M, Tanno K, Matsubayashi Y, Kawasaki H (2010) Whisker-related axonal patterns and plasticity of layer 2/3 neurons in the mouse barrel cortex. J Neurosci 30(8):3082–3092

Sehara K, Wakimoto M, Ako R, Kawasaki H (2012) Distinct developmental principles underlie the formation of ipsilateral and contralateral whisker-related axonal patterns of layer 2/3 neurons in the barrel cortex. Neuroscience 226:289–304

Sherman SM, Guillery RW (2004) Thalamus. In: Shepherd GM (ed) The synaptic organization of the brain, 5th edn. Oxford University Press, New York, pp 311–359

Sherman SM, Spear PD (1982) Organization of visual pathways in normal and visually deprived cats. Physiol Rev 62(2):738–855

Shitamukai A, Konno D, Matsuzaki F (2011) Oblique radial glial divisions in the developing mouse neocortex induce self-renewing progenitors outside the germinal zone that resemble primate outer subventricular zone progenitors. J Neurosci 31(10):3683–3695

Smart IH, Dehay C, Giroud P, Berland M, Kennedy H (2002) Unique morphological features of the proliferative zones and postmitotic compartments of the neural epithelium giving rise to striate and extrastriate cortex in the monkey. Cereb Cortex 12(1):37–53

Smetters DK, Hahm J, Sur M (1994) An *N*-methyl-D-aspartate receptor antagonist does not prevent eye-specific segregation in the ferret retinogeniculate pathway. Brain Res 658(1–2):168–178

Soma M, Aizawa H, Ito Y, Maekawa M, Osumi N, Nakahira E, Okamoto H, Tanaka K, Yuasa S (2009) Development of the mouse amygdala as revealed by enhanced green fluorescent protein gene transfer by means of in utero electroporation. J Comp Neurol 513(1):113–128

Stryker MP, Zahs KR (1983) On and off sublaminae in the lateral geniculate nucleus of the ferret. J Neurosci 3(10):1943–1951

Sur M, Garraghty PE, Roe AW (1988) Experimentally induced visual projections into auditory thalamus and cortex. Science 242(4884):1437–1441

Tabata H, Nakajima K (2001) Efficient in utero gene transfer system to the developing mouse brain using electroporation: visualization of neuronal migration in the developing cortex. Neuroscience 103(4):865–872

Upton AL, Salichon N, Lebrand C, Ravary A, Blakely R, Seif I, Gaspar P (1999) Excess of serotonin (5-HT) alters the segregation of ispilateral and contralateral retinal projections in monoamine oxidase A knock-out mice: possible role of 5-HT uptake in retinal ganglion cells during development. J Neurosci 19(16):7007–7024

Usrey WM, Sceniak MP, Chapman B (2003) Receptive fields and response properties of neurons in layer 4 of ferret visual cortex. J Neurophysiol 89(2):1003–1015

Wang X, Tsai JW, LaMonica B, Kriegstein AR (2011) A new subtype of progenitor cell in the mouse embryonic neocortex. Nat Neurosci 14(5):555–561

Wassle H (2004) Parallel processing in the mammalian retina. Nat Rev Neurosci 5(10):747–757

Wiesel TN, Hubel DH (1963) Single-cell responses in striate cortex of kittens deprived of vision in one eye. J Neurophysiol 26:1003–1017

Wiesel TN, Hubel DH (1965a) Comparison of the effects of unilateral and bilateral eye closure on cortical unit responses in kittens. J Neurophysiol 28(6):1029–1040

Wiesel TN, Hubel DH (1965b) Extent of recovery from the effects of visual deprivation in kittens. J Neurophysiol 28(6):1060–1072

Wong RO, Meister M, Shatz CJ (1993) Transient period of correlated bursting activity during development of the mammalian retina. Neuron 11(5):923–938

Yamamori T (2011) Selective gene expression in regions of primate neocortex: implications for cortical specialization. Prog Neurobiol 94(3):201–222

Yu H, Farley BJ, Jin DZ, Sur M (2005) The coordinated mapping of visual space and response features in visual cortex. Neuron 47(2):267–280

Zecevic N, Chen Y, Filipovic R (2005) Contributions of cortical subventricular zone to the development of the human cerebral cortex. J Comp Neurol 491(2):109–122

Chapter 14
Evolution of the Mammalian Brain with a Focus on the Whale Olfactory Bulb

Takushi Kishida

Abstract Mammals possess relatively larger brains compared to body size, and there are many studies focusing on the evolutionary changes of the relative brain size in mammals. A recent study showed that increased resolution in olfaction drove the enlargement of mammalian brains. However, the olfactory bulbs are degenerated among highly encephalized mammals, primates and whales. Several species of whales possess functional olfactory bulbs, but their olfactory bulbs lack a specific area known to induce innate avoidance behavior against odors of predators and spoiled foods. In this chapter, evolutionary changes of the encephalization quotient among mammals and the degeneration processes of olfactory bulbs among whales are discussed from paleontological, anatomical, and genomic points of view.

Keywords Cetacea • Encephalization quotient • Glomeruli • Mysticeti • Olfaction • Primates

14.1 Origin of the Mammalian Brain and the Evolutionary Changes of the Olfactory Receptor Genes in Basal Mammals

14.1.1 Increased Resolution in Olfaction Drove the Enlargement of Mammalian Brains

Mammals are a class of vertebrates that originated in or before the Early Jurassic and were derived from extinct Cynodontia (Rowe et al. 2008, 2011; Luo 2007; O'Leary et al. 2013; Rowe 1988; Rowe and Gauthier 1992). Mammals possess larger brains relative to body size, mainly because of the enlarged neocortex (isocortex), which is unique to mammalian brains. Especially in humans, 90 % of the cerebral cortex is neocortex (Noback et al. 2005). As a matter of course, many studies on the evolution of mammalian brains have been focused on the origin of the neocortex

T. Kishida (✉)
Wildlife Research Center, Kyoto University, Tanaka Sekiden-cho, Sakyo, Kyoto 606-8203, Japan
e-mail: taku.kishida@gmail.com

S. Shigeno et al. (eds.), *Brain Evolution by Design*, Diversity and Commonality in Animals, DOI 10.1007/978-4-431-56469-0_14

and the increases in brain size (Jerison 1973, 1988; Rowe et al. 2011; Montiel et al. 2011, 2013; Aboitiz 2011; Aboitiz and Zamorano 2013; Aboitiz and Montiel 2007). Encephalization quotient (EQ) is often used to compare the relative brain size to body size across different groups at many taxonomic levels. EQ value is generally calculated as follows: EQ = (brain weight)/0.12 (body weight)$^{2/3}$ (Jerison 1973)

Fossils are the keys to understanding the evolutionary changes of relative brain size. Figure 14.1 shows how EQ values have increased from basal cynodonts to modern mammals. The EQs are 0.16–0.23 in cynodonts, but increase with the emergence of Mammaliaformes, a clade including modern mammals and their closest extinct relatives (Rowe 1988). The EQ of *Morganucodon*, a basal mammaliaform, is 0.32, nearly 50 % larger than in basal cynodonts (Fig. 14.1). Although no clear evidences of the possession of a neocortex has been found in a *Morganucodon* endocranial cast, a skeleton of its close relative *Castorocauda* preserves the oldest evidence of the differentiation of the neocortex (Ji et al. 2006; Rowe et al. 2011), suggesting that the mammalian neocortex originated at this point in evolution (Fig. 14.1). However, the larger EQ in *Morganucodon* is seen mainly because of its enlarged olfactory bulbs (OBs) rather than the acquisition of a neocortex (Rowe et al. 2011). Rowe et al. (2011) pointed that relative brain size expanded to mammalian levels in two evolutionary pulses, and that the initial pulse was the enlargement of OBs in basal mammaliaforms, probably driven by increased resolution in olfaction. The second pulse was shown in *Hadrocodium*, a mammaliaform closely related to modern mammals. The EQ

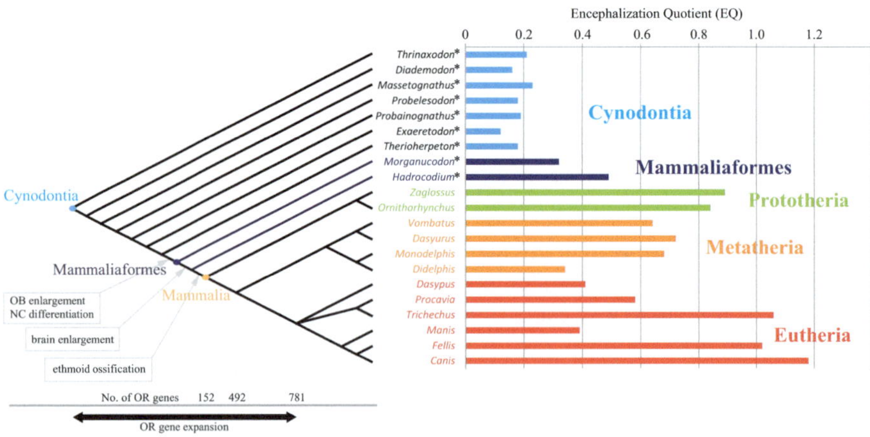

Fig. 14.1 A schematic phylogenetic relationships of basal cynodonts, mammaliaformes, and three major groups of modern mammals (Estimated encephalization quotients (EQ) of each species were taken from Rowe et al. 2011. *Asterisks* indicate extinct species. The estimated numbers of ancestral olfactory receptor (OR) genes are shown under the phylogenetic tree from Niimura and Nei 2007 and Niimura et al. 2014. *OB* olfactory bulb, *NC* neocortex)

of *Hadrocodium* is measured as 0.49 (Fig. 14.1). At this point, EQ reached the mammalian levels.

14.1.2 The Mammalian Olfactory System: From Receptors to Glomeruli

Mammals detect environmental volatile odorants using repertoires of olfactory receptors (ORs), which are encoded by intronless *OR* genes that constitute one of the largest multigene families (Nei et al. 2008). All mammalian *OR* genes can be classified into two monophyletic subfamilies, class I and class II, based on sequence similarities (Niimura and Nei 2006). The numbers of *OR* genes vary greatly among mammals, from approximately 2000 in African elephants (Niimura et al. 2014) to only 12 in bottlenose dolphins (Kishida et al. 2015) (Fig. 14.2). These ORs detect

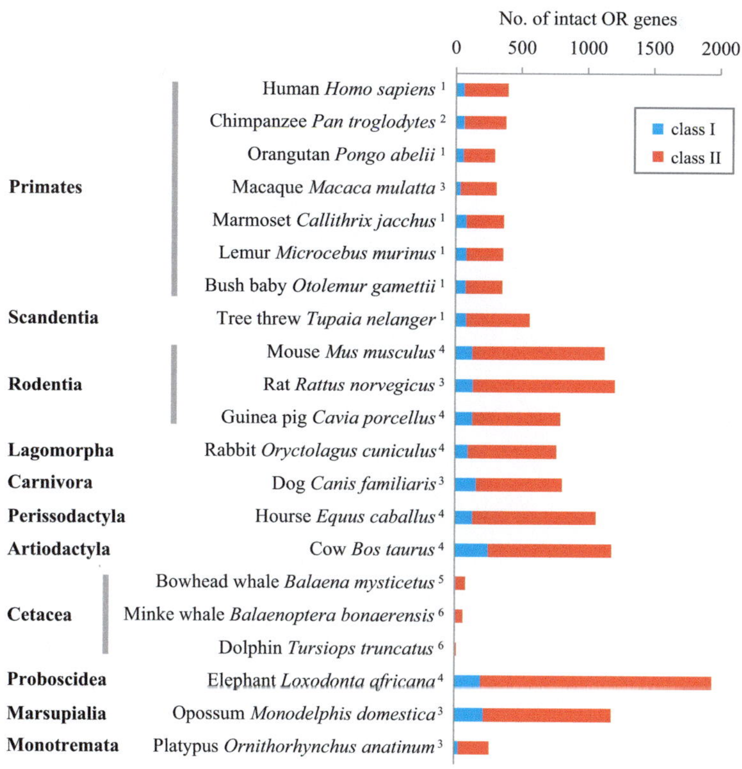

Fig. 14.2 Numbers of intact OR genes in various mammalian species. *Blue bars* numbers of class I OR genes; *red bars* numbers of class II OR genes (From Matsui et al. 2010 (1), Go and Niimura 2008(2), Niimura and Nei 2007 (3), Niimura et al. 2014 (4), Kishida et al., 2015b (5), Kishida et al. 2015 (6))

odorants in a combinatorial manner: a single OR may detect multiple odorants and a single odorant may be detected by multiple ORs (Malnic et al. 1999).

The mammalian olfactory system has long been studied using mice and rats as model organisms. Olfactory sensory neurons (OSNs) are located in the olfactory epithelium of the nasal cavity, and each OSN expresses only one OR gene (one neuron–one receptor rule) (Serizawa et al. 2003, 2004). The OSNs project to the glomeruli, which are located near the surface of the OB (Fig. 14.3), and each glomerulus is projected by ONSs that express the same OR (one glomerulus–one receptor rule) (Mombaerts et al. 1996). It is reported that any one OR is typically represented by two glomeruli (Ressler et al. 1994; Vassar et al. 1994; Mombaerts et al. 1996), which indicates that the number of glomeruli in an OB is approximately twice that of the number of *OR* genes in its genome.

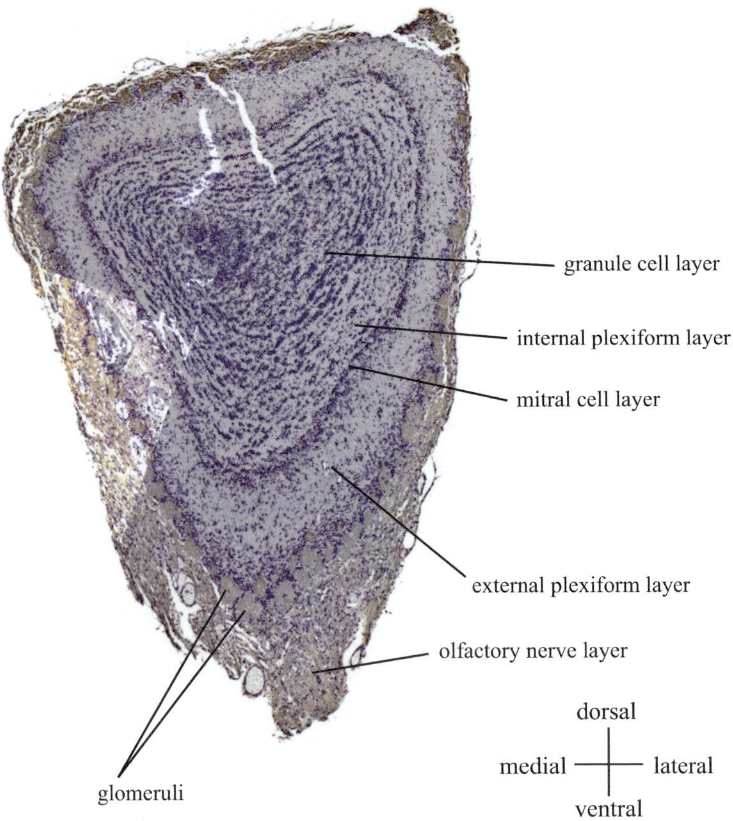

Fig. 14.3 Coronal section of an OB of *Macaca mulatta*, stained with anti-OMP (olfactory marker protein) antibody (Santa Cruz Biotechnology, cat sc-2023) and counterstained with thionin, showing layers

14.1.3 Inconsistency Between the Enlargement of Olfactory Bulbs and the Increase of Olfactory Receptor Genes

As shown, olfactory signals are initially produced by ORs, and the number of glomeruli in an OB is highly related to the number of *OR* genes. Therefore, from the evolutionary context, it is expected that the size of OBs and the number of ORs should be increased coordinately. Rowe et al. (2011) showed that the OBs of mammaliaforms were as large as that of modern mammals. Certainly, it is indicated that large-scale duplications of *OR* genes had been occurred in the mammalian lineage before the Prototheria–Eutheria split (Kishida 2008). However, the number of *OR* genes possessed by the last common ancestor (LCA) of all modern mammals, estimated by the reconciled tree method (Goodman et al. 1979; Page and Charleston 1997) in which the topology of a gene tree is reconciled with that of a species tree, is 152 (Fig. 14.1), much less than that of modern terrestrial mammals (Fig. 14.2). The number of *OR* genes had increased gradually in the eutherian lineage, and that of the LCA of all modern eutherians reached the modern eutherian level (Fig. 14.1). These findings suggest that the initial pulse of the brain enlargement, the enlargement of OBs, did not correlate to the enlargement of *OR* gene repertoires. Thus, the initial odor signals had not been increased when the basal mammaliaforms emerged. It is thus a mystery why the OBs had been enlarged at that point in evolution.

14.2 The Largest Brain: Evolution of the Primate and Cetacean Brains

Among modern mammals, it is widely known that primates (humans, apes, monkeys, prosimians) and cetaceans (whales, dolphins, porpoises) have larger brains compared to body mass. Especially, the EQ of our own species is highest among all mammals (Fig. 14.1; Table 14.1). However, the second highest EQ is not found in primates, but in cetaceans (Table 14.1) (Marino 1998), and it has been debated why several species of cetaceans have such larger brains (Marino 2007; Marino et al. 2008; Manger 2006).

From the evolutionary aspect, primates show a strong trend for directional increase in relative brain size (Montgomery et al. 2010). Increases in brain size are almost ubiquitous across the primate phylogenetic tree in a background of body mass evolution that shows no significant trend to increase through time (Montgomery et al. 2010). However, the evolutionary pattern in cetaceans contrasts with that in primates. Cetaceans is an order of mammals that originated in the early Eocene and which was derived from Artiodactyls (Thewissen et al. 2009). All living cetaceans are classified into two suborders, Mysticeti (baleen whales) and Odontoceti (toothed whales), and both of these are fully aquatic (Fig. 14.4). As shown in Fig. 14.4, both body mass and brain mass had increased during the basal cetacean evolution before the basilosaurids–modern cetaceans split, and

Table 14.1 Encephalization quotients (EQs) and body weights of primates and cetaceans

		Species	EQ	Body weight (g)
Primates		*Callithrix jacchus*	1.66	233
		Callithrix geoffroyi	1.89	793
		Cebuella pygmaea	1.42	120
		Cebus capucinus	2.63	71
		Saimiri oerstedii	2.38	753
		Macaca mulatta	2.15	6,652
		Papio anubis	1.76	30,000
		Cercopithecus aethiops	2.04	4,324
		Colobus badius	1.72	7,000
		Hylobates lar	2.54	5,664
		Pongo pygmaeus	1.77	63,730
		Gorilla gorilla	1.63	114,654
		Pan troglodytes	2.34	48,893
		Homo sapiens	7.06	57,333
Cetartiodactyla	**Odontoceti**	*Physeter macrocephalus*	0.58	35,833,330
		Kogia breviceps	1.78	305,000
		Kogia sima	1.63	168,500
		Mesoplodon mirus	1.97	929,500
		Mesoplodon europaeus	2.11	732,500
		Mesoplodon densirostris	1.39	767,000
		Ziphius cavirostris	0.92	2,273,000
		Delphinapterus leucas	2.24	636,000
		Monodon monoceros	1.76	1,578,330
		Lipotes vexillifer	2.17	82,000
		Inia geoffrensis	2.51	90,830
		Platanista gangetica	1.55	59,630
		Pontoporia blainvillei	1.67	34,890
		Phocoena phocoena	2.59	61,100
		Phocoenoides dalli	3.54	86,830
		Tursiops truncatus	4.14	209,530
		Sotalia fluviatilis	4.56	42,240
		Lagenorhynchus obliquidens	4.55	91,050
		Delphinus delphis	4.26	60,170
		Grampus griseus	4.01	328,000
		Globicephala melaena	2.39	943,200
		Orcinus orca	2.57	1,955,450
	Mysticeti	*Balaenoptera physalus*	0.49	38,421,500
		Balaenoptera musculus	0.21	50,904,000
		Megaptera novaeangliae	0.44	39,295,000

(continued)

Table 14.1 (continued)

		Species	EQ	Body weight (g)
	Artiodactyla	*Sus scrofa*	0.58	124,640
		Giraffa camelopardalis	1.20	318,850
		Cervus elaphus	0.92	141,890
		Tragulus javanicus	0.12	32,370
		Hyemoschus aquaticus	0.14	56,230

Source: Hof et al. (2005) and Marino (1998)

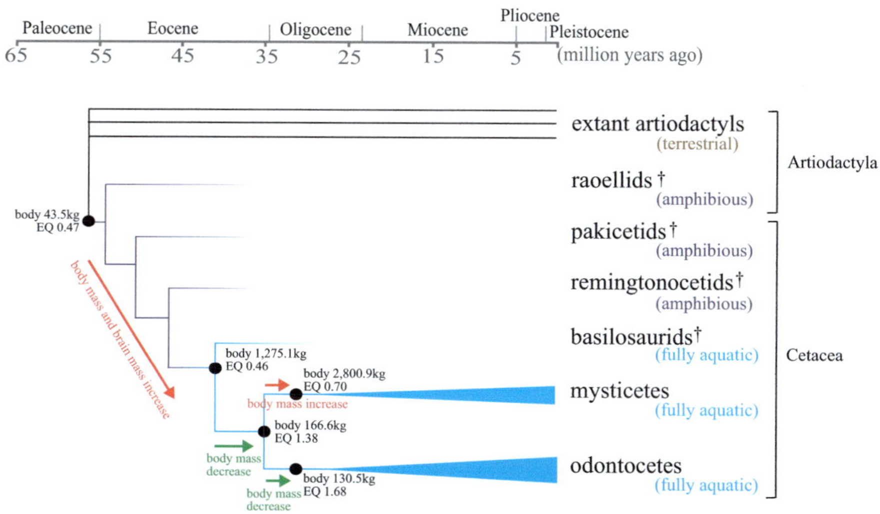

Fig. 14.4 Modified phylogenetic tree of cetaceans inferred from Thewissen et al. and Uhen (Thewissen et al. 2007, 2009; Uhen 2007). Extinct groups indicated by *daggers*. *Purple branches* amphibious lifestyle, *blue branches* fully aquatic. Estimated body weights and EQ values in several ancestral nodes were taken from Montgomery et al. (2013)

the EQ of the LCA of basilosaurids and modern cetaceans is estimated to have been 0.46. Body mass had decreased profoundly in the modern cetacean lineage between the basilosaurids–modern cetaceans split and the mysticetes–odontocetes split, resulting in the increase of the EQ value. In the cetacean lineages, evolutionary changes of body mass have had more important roles for the changes of EQ values compared to that of brain mass (Fig. 14.4) (Montgomery et al. 2013). Actually, EQs of bigger whales (sperm whales, *Physeter macrocephalus*, and mysticetes) are relatively small, and especially, the biggest blue whales (*Balaenoptera musculus*) show the smallest EQ (0.21) among all cetaceans (Table 14.1).

Interestingly, both primates and cetaceans have reduced or lost the sense of smell and OBs (Table 14.2) (Stephan et al. 1981; Ridgway 1988; Dehnhardt 2002; Pihlström 2008), and both these orders possess smaller numbers of *OR* genes (Fig. 14.2). There is a traditional view that olfaction is a 'primitive' sense, although

Table 14.2 Relative size of olfactory bulb (OB) to the whole brain

Species		Relative OB size	References
Primates	*Homo sapiens*	0.01 %	Stephan et al. (1981)
	Pan troglodytes	0.06 %	Stephan et al. (1981)
	Colobus badius	0.07 %	Stephan et al. (1981)
	Papio anubis	0.14 %	Stephan et al. (1981)
	Macaca mulatta	0.09 %	Stephan et al. (1981)
	Callithrix jacchus	0.30 %	Stephan et al. (1981)
	Saguinus oedipus	0.19 %	Stephan et al. (1981)
	Galago demidovii	2.46 %	Stephan et al. (1981)
	Indri indri	0.44 %	Stephan et al. (1981)
	Lemur fulvus	0.89 %	Stephan et al. (1981)
Cetacea	*Balaena mysticetus*	0.13 %	Thewissen et al. (2011)
	Modern odontocetes	0.00 %	
Afrosoricida	*Tenrec ecaudatus*	12.22 %	Stephan et al. (1981)
	Oryzorictes talpoides	8.09 %	Stephan et al. (1981)
Soricomorpha	*Solenodon paradoxus*	9.96 %	Stephan et al. (1981)
	Sorex araneus	7.85 %	Stephan et al. (1981)
Erinaceomorpha	*Erinaceus europaeus*	9.91 %	Stephan et al. (1981)
Scandentia	*Tupaia glis*	4.11 %	Stephan et al. (1981)

this is not true because OR genes are found only in the chordate species, suggesting that the *OR* genes appeared relatively recently (Niimura 2009). In my opinion, it is partly because the sense of smell has been reduced in these highly encephalized mammals.

14.3 Evolution of the Degeneration of Olfactory Bulbs in Cetaceans

In this section, recent studies of the evolution and function of cetacean OBs are discussed. Until recently, it had been widely been considered that cetaceans either have a minimal sense of smell or lack it altogether (Dehnhardt 2002; Pihlström 2008). Indeed, all living odontocetes have no nervous system structures that mediate olfaction, such as OBs, olfactory tract, and cranial nerve I (Oelschläger 2008). However, highly degenerated but fully equipped olfactory system and OBs are found in at least some species of mysticetes (Fig. 14.5) (Thewissen et al. 2011). The numbers of OR genes are higher in mysticetes than in odontocetes (Fig. 14.2). The evolutionary changes of the synonymous and nonsynonymous change ratios of an olfactory-involved gene [*olfactory marker protein* (*OMP*) gene] suggest that mysticetes use olfaction in foraging (Kishida and Thewissen 2012).

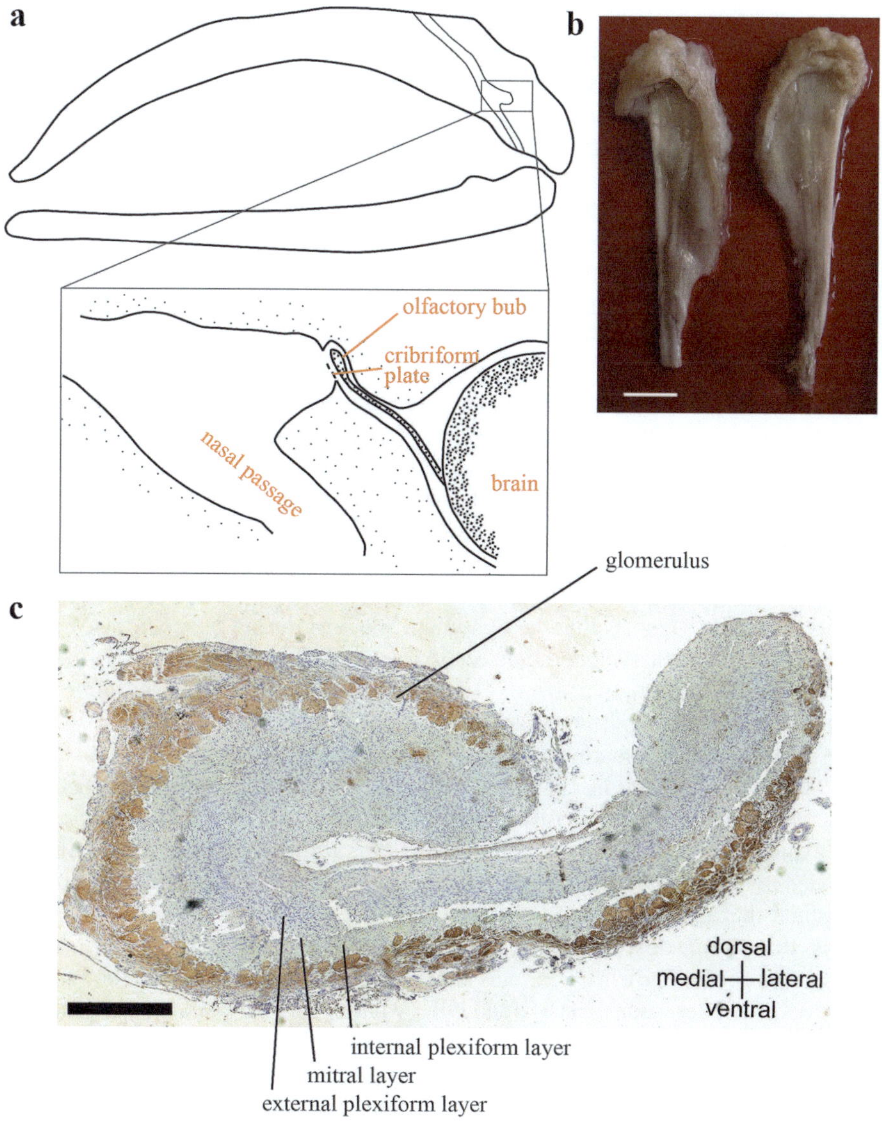

Fig. 14.5 Bowhead whale (*Balaena mysticetus*, Mysticeti) olfactory bulbs (Modified after Thewissen et al. (2011) and Kishida et al., submitted). (**a**) Diagram of location of olfactory bulb in the bowhead whale skull. (**b**) Dorsal view of left and right olfactory bulbs of a bowhead whale. *Bar* 10 mm. (**c**) Coronal section of right olfactory bulb of a bowhead whale. Glomeruli were stained with DAB using anti-OMP antibody; the whole tissue was counterstained with thionin. *Bar* 1 mm

Generally, glomeruli are distributed in both dorsal and ventral sides of OBs among mammals (e.g., Fig. 14.3). However, it was found that dorsal glomeruli are absent or nearly absent in the bowhead whale OB (Fig. 14.5). The glomerular layer of the OB can be divided into two nonoverlapping areas, a dorsal domain (D domain) and a ventral domain (V domain), based on the expression patterns of domain-specific marker genes; the D domain is defined by the expression of the *OMACS* gene and the V domain is defined by the expression of the *OCAM* gene (Kobayakawa et al. 2007; Imai and Sakano 2007; Oka et al. 2003; Yoshihara et al. 1997). It was reported that although the *OCAM* gene is highly conserved widely among cetacean species, the *OMACS* gene have become functionless pseudogenes in a mysticete, the Antarctic minke whale (*Balaena bonaerensis*) (Kishida et al. 2015). Comparative genomics revealed that the mysticete *OMACS* gene turned into a pseudogene before the Odontoceti–Mysticeti split (Kishida et al. 2015). Regarding OR–glomeruli projection, most OSNs expressing class I ORs project specifically to the D domain (Tsuboi et al. 2006), whereas OSNs expressing class II ORs project to both D and V domains (Miyamichi et al. 2005). As shown in Fig. 14.2, cetaceans possess very small numbers of class I ORs compared to their class II ORs. These findings strongly suggest that the D domain had been lost from the mysticete OBs. The OB communicates with the nasal cavity via the clibriform plate (Fig. 14.5a), which fossilizes. Fossils of basal amphibious cetacean pakicetids (Fig. 14.4) show that a part of their cribriform plate faces dorsally, but this could not be found in the fossils of remingtonocetids (Fig. 14.4) (Kishida et al. 2015). This observation suggests that the D domain was lost during the course of Eocene between the pakicetids–modern cetaceans split and the remingtonocetids–modern cetaceans split. The relative size of OBs is also reduced in Primates (Table 14.2), but glomeruli are present in both dorsal and ventral sides of their OB (Fig. 14.3).

D domain-ablated mutant mice (ΔD mice) were generated by Kobayakawa et al. (2007), who reported that the ΔD mice fail to show innate avoidance behavior against predator odors and spoiled smells, and that the D domain is sufficient for such avoidance behavior. It is not obvious that this study using mice can be directly extended to other mammals, but Kishida et al. (2015) discussed that it is reasonable to assume that the olfactory capability of myeticetes resembles that of ΔD mice, that is, that mysticetes lack innate avoidance behavior against odors of predators and spoiled foods. Terrestrial animals cannot prey on fully aquatic whales, and the predators of whales, such as sharks and killer whales, cannot be detected by smelling in air. In addition, differing from the nares of other mammals, the nares of whales are not located at the tip of their snout (Fig. 14.5a), and the nasal passage of whales is not connected directly to their oral cavity, indicating that it is difficult for whales to rely on olfaction to judge whether something they are about to swallow is edible. Further studies will test this assumption.

The number of glomeruli in a bowhead whale OB is estimated to be more than 4000, whereas bowhead whales have only 80 OR genes (Kishida et al., submitted), but this is inconsistent with the widely assumed theory that the ratio of OR to glomeruli is approximately 1:2 (see Sect. 14.1). The same tendency was also reported in humans. Humans have much larger numbers of glomeruli in their OB (5600 on average) compared to the number of their ORs (396; Fig. 14.2) (Maresh et al. 2008). These findings show the conceptual limits of using rodents as model organisms for understanding the initial coding of odor information among mammals.

14.4 Conclusions

In this chapter, I showed the evolution of mammal brains mainly based on quantitative analyses. As shown in Sect. 14.1, encephalization of the mammalian brain was led by the encephalization of OBs. However, interestingly, the OBs are profoundly degenerated or lost in two highly encephalized mammalian orders, the primates and cetaceans. Details of the degenerated OBs possessed by living mysticetes are shown in Sect. 14.3. The mysticete OB has changed drastically not only its relative size but also in its shape and function, indicating that brains may easily be changed, not only quantitatively but qualitatively. Qualitative analyses of the mammalian brains including fossil specimens are required to understand the origin and evolution of brains among the mammals.

Acknowledgments I thank the editors, especially Yasunori Murakami, for inviting me to contribute to this volume. This work was financially supported in part by MEXT KAKENHI (grant no. 24770075).

References

Aboitiz F (2011) Genetic and developmental homology in amniote brains. Toward conciliating radical views of brain evolution. Brain Res Bull 84(2):125–136. doi:10.1016/j.brainresbull.2010.12.003

Aboitiz F, Montiel JF (2007) Origin and evolution of the vertebrate telencephalon, with special reference to the mammalian neocortex. Springer, Berlin/New York

Aboitiz F, Zamorano F (2013) Neural progenitors, patterning and ecology in neocortical origins. Front Neuroanat 7:38. doi:10.3389/fnana.2013.00038

Dehnhardt G (2002) Sensory systems. In: Hoelzel R (ed) Marine mammal biology. Blackwell, Oxford, pp 116–141

Go Y, Niimura Y (2008) Similar numbers but different repertoires of olfactory receptor genes in humans and chimpanzees. Mol Biol Evol 25(9):1897–1907. doi:10.1093/molbev/msn135

Goodman M, Czelusniak J, Moore GW, Romero-Herrera AE, Matsuda G (1979) Fitting the gene lineage into its species lineage, a parsimony strategy illustrated by cladograms constructed from globin sequences. Syst Biol 28(2):132–163. doi:10.1093/sysbio/28.2.132

Hof PR, Chanis R, Marino L (2005) Cortical complexity in cetacean brains. Anat Rec A Discov Mol Cell Evol Biol 287A(1):1142–1152. doi:10.1002/ar.a.20258

Imai T, Sakano H (2007) Roles of odorant receptors in projecting axons in the mouse olfactory system. Curr Opin Neurobiol 17(5):507–515. doi:10.1016/j.conb.2007.08.002

Jerison HJ (1973) Evolution of the brain and intelligence. Academic, New York

Jerison H (1988) Brain size. In: Comparative neuroscience and neurobiology. Readings from the encyclopedia of neuroscience. Birkhäuser, Boston, pp 15–17. doi:10.1007/978-1-4899-6776-3_7

Ji Q, Luo Z-X, Yuan C-X, Tabrum AR (2006) A swimming mammaliaform from the Middle Jurassic and ecomorphological diversification of early mammals. Science 311(5764):1123–1127. doi:10.1126/science.1123026

Kishida T (2008) Pattern of the divergence of olfactory receptor genes during tetrapod evolution. PLoS One 3(6), e2385. doi:10.1371/journal.pone.0002385

Kishida T, Thewissen JGM (2012) Evolutionary changes of the importance of olfaction in cetaceans based on the olfactory marker protein gene. Gene 492(2):349–353. doi:10.1016/j.gene.2011.11.013

Kishida T, Thewissen JGM, Hayakawa T, Imai H, Agata K (2015) Aquatic adaptation and the evolution of smell and taste in whales. Zool Lett 1:9. doi: 10.1186/s40851-014-0002-z

Kishida T, Thewissen JGM, Usip S, Suydam RS, George JC (2015b) Organization and distribution of glomeruli in the bowhead whale olfactory bulb. PeerJ 3:e897. doi:10.7717/peerj.897

Kobayakawa K, Kobayakawa R, Matsumoto H, Oka Y, Imai T, Ikawa M, Okabe M, Ikeda T, Itohara S, Kikusui T, Mori K, Sakano H (2007) Innate versus learned odour processing in the mouse olfactory bulb. Nature 450(7169):503–508. doi:10.1038/nature06281

Luo Z-X (2007) Transformation and diversification in early mammal evolution. Nature 450(7172):1011–1019

Malnic B, Hirono J, Sato T, Buck LB (1999) Combinatorial receptor codes for odors. Cell 96(5):713–723. doi:10.1016/S0092-8674(00)80581-4

Manger PR (2006) An examination of cetacean brain structure with a novel hypothesis correlating thermogenesis to the evolution of a big brain. Biol Rev 81(2):293–338. doi:10.1017/S1464793106007019

Maresh A, Rodriguez Gil D, Whitman MC, Greer CA (2008) Principles of glomerular organization in the human olfactory bulb: implications for odor processing. PLoS One 3(7), e2640. doi:10.1371/journal.pone.0002640

Marino L (1998) A comparison of encephalization between odontocete cetaceans and anthropoid primates. Brain Behav Evol 51(4):230–238

Marino L (2007) Cetacean brains: how aquatic are they? Anat Rec 290(6):694–700. doi:10.1002/ar.20530

Marino L, Butti C, Connor RC, Fordyce RE, Herman LM, Hof PR, Lefebvre L, Lusseau D, McCowan B, Nimchinsky EA, Pack AA, Reidenberg JS, Reiss D, Rendell L, Uhen MD, Van der Gucht E, Whitehead H (2008) A claim in search of evidence: reply to Manger's thermogenesis hypothesis of cetacean brain structure. Biol Rev 83(4):417–440. doi:10.1111/j.1469-185X.2008.00049.x

Matsui A, Go Y, Niimura Y (2010) Degeneration of olfactory receptor gene repertories in primates: no direct link to full trichromatic vision. Mol Biol Evol 27(5):1192–1200. doi:10.1093/molbev/msq003

Miyamichi K, Serizawa S, Kimura HM, Sakano H (2005) Continuous and overlapping expression domains of odorant receptor genes in the olfactory epithelium determine the dorsal/ventral positioning of glomeruli in the olfactory bulb. J Neurosci 25(14):3586–3592. doi:10.1523/jneurosci.0324-05.2005

Mombaerts P, Wang F, Dulac C, Chao SK, Nemes A, Mendelsohn M, Edmondson J, Axel R (1996) Visualizing an olfactory sensory map. Cell 87(4):675–686

Montgomery SH, Capellini I, Barton RA, Mundy NI (2010) Reconstructing the ups and downs of primate brain evolution: implications for adaptive hypotheses and *Homo floresiensis*. BMC Biol 8:9. doi:10.1186/1741-7007-8-9

Montgomery SH, Geisler JH, McGowen MR, Fox C, Marino L, Gatesy J (2013) The evolutionary history of cetacean brain and body size. Evolution 67(11):3339–3353. doi:10.1111/evo.12197

Montiel JF, Wang WZ, Oeschger FM, Hoerder-Suabedissen A, Tung WL, García-Moreno F, Holm IE, Villalón A, Molnár Z (2011) Hypothesis on the dual origin of the mammalian subplate. Front Neuroanat 5:25. doi:10.3389/fnana.2011.00025

Montiel JF, Kaune H, Maliqueo M (2013) Maternal-fetal unit interactions and eutherian neocortical development and evolution. Front Neuroanat 7:22. doi:10.3389/fnana.2013.00022

Nei M, Niimura Y, Nozawa M (2008) The evolution of animal chemosensory receptor gene repertoires: roles of chance and necessity. Nat Rev Genet 9(12):951–963. doi:10.1038/nrg2480

Niimura Y (2009) Evolutionary dynamics of olfactory receptor genes in chordates: interaction between environments and genomic contents. Hum Genomics 4(2):107–118. doi:U3T40206KH4144N5 [pii]

Niimura Y, Nei M (2006) Evolutionary dynamics of olfactory and other chemosensory receptor genes in vertebrates. J Hum Genet 51(6):505–517. doi:10.1007/s10038-006-0391-8

Niimura Y, Nei M (2007) Extensive gains and losses of olfactory receptor genes in mammalian evolution. PLoS One 2(1), e708. doi:10.1371/journal.pone.0000708

Niimura Y, Matsui A, Touhara K (2014) Extreme expansion of the olfactory receptor gene repertoire in African elephants and evolutionary dynamics of orthologous gene groups in 13 placental mammals. Genome Res 24:1485–1496. doi:10.1101/gr.169532.113

Noback CR, Strominger NL, Demarest RJ, Ruggiero DA (2005) The human nervous system: structure and function, 6th edn. Humana Press, Totowa

O'Leary MA, Bloch JI, Flynn JJ, Gaudin TJ, Giallombardo A, Giannini NP, Goldberg SL, Kraatz BP, Luo ZX, Meng J, Ni X, Novacek MJ, Perini FA, Randall ZS, Rougier GW, Sargis EJ, Silcox MT, Simmons NB, Spaulding M, Velazco PM, Weksler M, Wible JR, Cirranello AL (2013) The placental mammal ancestor and the post–K-Pg radiation of placentals. Science 339(6120):662–667. doi:10.1126/science.1229237

Oelschläger HHA (2008) The dolphin brain—a challenge for synthetic neurobiology. Brain Res Bull 75(2-4):450–459. doi:10.1016/j.brainresbull.2007.10.051

Oka Y, Kobayakawa K, Nishizumi H, Miyamichi K, Hirose S, Tsuboi A, Sakano H (2003) O-MACS, a novel member of the medium-chain acyl-CoA synthetase family, specifically expressed in the olfactory epithelium in a zone-specific manner. Eur J Biochem 270(9):1995–2004. doi:10.1046/j.1432-1033.2003.03571.x

Page RDM, Charleston MA (1997) From gene to organismal phylogeny: reconciled trees and the gene tree/species tree problem. Mol Phylogenet Evol 7(2):231–240. doi:10.1006/mpev.1996.0390

Pihlström H (2008) Comparative anatomy and physiology of chemical senses in aquatic mammals. In: Thewissen JGM, Nummela S (eds) Sensory evolution on the threshold: adaptations in secondarily aquatic vertebrates. University of California Press, Berkeley, pp 95–109

Ressler KJ, Sullivan SL, Buck LB (1994) Information coding in the olfactory system: evidence for a stereotyped and highly organized epitope map in the olfactory bulb. Cell 79(7):1245–1255. doi:10.1016/0092-8674(94)90015-9

Ridgway S (1988) The cetacean central nervous system. In: Comparative neuroscience and neurobiology. Readings from the encyclopedia of neuroscience. Birkhäuser, Boston, pp 20–25. doi:10.1007/978-1-4899-6776-3_9

Rowe T (1988) Definition, diagnosis, and origin of Mammalia. J Vertebr Paleontol 8(3):241–264. doi:10.1080/02724634.1988.10011708

Rowe T, Gauthier J (1992) Ancestry, paleontology, and definition of the name Mammalia. Syst Biol 41(3):372–378

Rowe T, Rich TH, Vickers-Rich P, Springer M, Woodburne MO (2008) The oldest platypus and its bearing on divergence timing of the platypus and echidna clades. Proc Natl Acad Sci USA 105(4):1238–1242. doi:10.1073/pnas.0706385105

Rowe TB, Macrini TE, Luo ZX (2011) Fossil evidence on origin of the mammalian brain. Science 332(6032):955–957. doi:10.1126/science.1203117

Serizawa S, Miyamichi K, Nakatani H, Suzuki M, Saito M, Yoshihara Y, Sakano H (2003) Negative feedback regulation ensures the one receptor–one olfactory neuron rule in mouse. Science 302(5653):2088–2094. doi:10.1126/science.1089122

Serizawa S, Miyamichi K, Sakano H (2004) One neuron–one receptor rule in the mouse olfactory system. Trends Genet 20(12):648–653. doi:10.1016/j.tig.2004.09.006

Stephan H, Frahm H, Baron G (1981) New and revised data on volumes of brain structures in insectivores and primates. Folia Primatol 35(1):1–29

Thewissen JGM, Cooper LN, Clementz MT, Bajpai S, Tiwari BN (2007) Whales originated from aquatic artiodactyls in the Eocene epoch of India. Nature 450(7173):1190–1194. doi:10.1038/nature06343

Thewissen JGM, Cooper LN, George JC, Bajpai S (2009) From land to water: the origin of whales, dolphins, and porpoises. Evol Educ Outreach 2(2):272–288. doi:10.1007/s12052-009-0135-2

Thewissen JGM, George J, Rosa C, Kishida T (2011) Olfaction and brain size in the bowhead whale (*Balaena mysticetus*). Mar Mamm Sci 27(2):282–294. doi:10.1111/j.1748-7692.2010.00406.x

Tsuboi A, Miyazaki T, Imai T, Sakano H (2006) Olfactory sensory neurons expressing class I odorant receptors converge their axons on an antero-dorsal domain of the olfactory bulb in the mouse. Eur J Neurosci 23(6):1436–1444. doi:10.1111/j.1460-9568.2006.04675.x

Uhen MD (2007) Evolution of marine mammals: back to the sea after 300 million years. Anat Rec 290(6):514–522. doi:10.1002/ar.20545

Vassar R, Chao SK, Sitcheran R, Nuñez JM, Vosshall LB, Axel R (1994) Topographic organization of sensory projections to the olfactory bulb. Cell 79(6):981–991. doi:10.1016/0092-8674(94)90029-9

Yoshihara Y, Kawasaki M, Tamada A, Fujita H, Hayashi H, Kagamiyama H, Mori K (1997) OCAM: a new member of the neural cell adhesion molecule family related to zone-to-zone projection of olfactory and vomeronasal axons. J Neurosci 17(15):5830–5842

Chapter 15
The Evolution and Function of Sleep

Yu Hayashi and Chih-Yao Liu

Abstract Sleep is a common physiological state appearing in the everyday life of humans and other animals. In humans, sleep occupies approximately one third of our whole lifetime. People have thus kept asking the question of why we sleep. Sleep deprivation in rats results in lethality, indicating its essential roles (Rechtschaffen A, Bergmann BM, Sleep 25:18–24, 2002; Rechtschaffen A, Bergmann BM, Everson CA, Kushida CA, Gilliland MA, Sleep 12:68–87, 1989). From the aspect of evolution, sleep or sleep-like states are conserved across diverse animal species, implying an existent function for fulfilling a common purpose that may benefit the survival of animals. Up to now, however, the function and mechanism of sleep are still largely unknown. Recently, simple genetic animal models including fruit flies (*Drosophila melanogaster*), roundworms (*Caenorhabditis elegans*), and zebrafish (*Danio rerio*) have been actively studied to reveal the evolutionarily conserved components of sleep, which may lead to solving the fundamental question about the evolutionary origin of sleep (Hendricks JC, Finn SM, Panckeri KA, Chavkin J, Williams JA, Sehgal A, Pack AI, Neuron 25:129–138, 2000; Raizen DM, Zimmerman JE, Maycock MH, Ta UD, You YJ, Sundaram MV, Pack AI, Nature 451:569–572, 2008; Shaw PJ, Cirelli C, Greenspan RJ, Tononi G, Science 287:1834–1837, 2000; Singh K, Ju JY, Walsh MB, DiIorio MA, Hart AC, Sleep 37:1439–1451, 2014; Zhdanova IV, Wang SY, Leclair OU, Danilova NP, Brain Res 903:263–268, 2001). In addition, with the development of new techniques such as two-photon microscopy, optogenetics, and pharmacogenetics, researchers have obtained more ability to observe and manipulate neurons or their activity. Partly owing to the breakthrough of such new tools, researchers have found some evidence suggesting that sleep serves several functions including memory consolidation, clearance of brain metabolites,

Y. Hayashi (✉)
International Institute for Integrative Sleep Medicine (WPI-IIIS), University of Tsukuba, 1-1-1 Tennodai, Tsukuba, Ibaraki, 305-8575 Japan

Japan Science and Technology Agency (JST), PRESTO, 4-1-8 Honcho, Kawaguchi, Saitama, 332-0012 Japan
e-mail: hayashi.yu.fp@u.tsukuba.ac.jp

C.-Y. Liu
International Institute for Integrative Sleep Medicine (WPI-IIIS), University of Tsukuba, 1-1-1 Tennodai, Tsukuba, Ibaraki, 305-8575 Japan

S. Shigeno et al. (eds.), *Brain Evolution by Design*, Diversity and Commonality in Animals, DOI 10.1007/978-4-431-56469-0_15

spine remodeling, and brain development (Bushey D, Tononi G, Cirelli C, Science 332:1576–1581, 2011; Donlea JM, Thimgan MS, Suzuki Y, Gottschalk L, Shaw PJ, Science 332:1571–1576, 2011; Kayser MS, Yue Z, Sehgal A, Science 344:269–274, 2014; Rasch B, Buchel C, Gais S, Born J, Science 315:1426–1429, 2007; Rechtschaffen A, Bergmann BM, Everson CA, Kushida CA, Gilliland MA, Sleep 12:68–87, 1989; Xie L, Kang H, Xu Q, Chen MJ, Liao Y, Thiyagarajan M, O'Donnell J, Christensen DJ, Nicholson C, Iliff JJ, et al. Science 342:373–377, 2013; Yang G, Lai CS, Cichon J, Ma L, Li W, Gan WB, Science 344:1173–1178, 2014). These studies have shown the relationship between sleep and other biological processes in different animals, and it further brings us to the question of whether the function of sleep is only for one purpose or is for multiple purposes. Up to now, our knowledge about sleep seems to be merely the tip of the iceberg. Further research is needed to understand the general function of sleep across species.

Here, we first introduce general criteria for sleep, which allows its definition in animals other than mammals (Sect. 15.1). Then we introduce REM sleep and non-REM sleep, which are the two major sleep stages of mammalian and avian sleep (Sect. 15.2), and introduce studies and hypotheses related to how they evolved (Sect. 15.3). Next, we briefly introduce sleep in aquatic mammals, which have made a unique change from their ancestral mammals to adapt to their lifestyle (Sect. 15.4). Then we introduce the current progress in studies using simple genetic animal models, namely, zebrafish, fruit flies, and roundworms (Sect. 15.5 and Sect. 15.6). Finally, we compare the suggested functions of sleep between mammals and invertebrate animals (Sect. 15.7).

Keywords Sleep • Evolution • REM sleep • Slow wave • Mammal • Reptile • Dolphin • Unihemispheric sleep • Zebrafish (*Danio rerio*) • Roundworm (*Caenorhabditis elegans*) • Fruit fly (*Drosophila melanogaster*)

15.1 Definition of Sleep

Sleep in mammals and birds is a relatively evident state. In these animals, electroencephalograms (EEG) and electromyograms (EMG) can be used to easily distinguish sleep and wakefulness. However, criteria based on EEG and EMG are often difficult to apply to other animals. Therefore, researchers have proposed several features of sleep to define it in species across the animal kingdom. These features include loss of locomotion, distinct postures, enhanced arousal thresholds to environmental signals, rapid reversibility, preference for specific environmental spots, and homeostatic rebound to sleep deprivation (Campbell and Tobler 1984; Rial et al. 2010). Based on these definitions, sleep has been characterized in fruit flies (*Drosophila melanogaster*), roundworms (*Caenorhabditis elegans*), zebrafish (*Danio rerio*), and various other animal species (Hendricks et al. 2000; Raizen et al. 2008; Shaw et al. 2000; Singh et al. 2014; Zhdanova et al. 2001) (Fig. 15.3). These criteria also differentiate sleep from other quiescent states within mammals. For example, hibernation in some mammals is distinct from sleep in that reversibility to the active state is not rapid. Furthermore, the studies applying these definitions to simple animal models led to findings that several genetic components related to sleep are conserved across different species, including the epidermal growth factor (EGF) signaling pathway (Foltenyi et al. 2007; Kramer et al. 2001; Kushikata et al. 1998; Van Buskirk and Sternberg 2007), protein kinase G (PKG) (Langmesser et al. 2009; Raizen et al. 2008), the cyclic adenosine monophosphate signaling pathway (Graves et al. 2003; Hendricks et al. 2001; Raizen et al. 2008), the dopaminergic pathway (Kume et al. 2005; Singh et al. 2014; Wisor et al. 2001), the histaminergic pathway (Haas et al. 2008; Monnier et al. 1967; Nicholson et al. 1985; Oh et al. 2013; Renier et al. 2007; Sundvik et al. 2011), etc. (summarized in Table 15.1). However, cautious interpretation is necessary, as these genes or neurotransmitters are involved in a diverse array of cellular events, rather than involved specifically in sleep. Genes involved in circadian rhythm can also have roles independent of circadian rhythm in the regulation of sleep (Franken et al. 2007; Monsalve et al. 2011; Naylor et al. 2000; Shaw et al. 2002; Viola et al. 2007; Wisor et al. 2002). These findings suggest that using the foregoing features to dissect sleep in nonmammalian species is feasible.

15.2 REM Sleep and Non-REM Sleep

Human sleep can be roughly divided into two main stages, rapid eye movement sleep (REM sleep) and non-REM sleep (NREM sleep). NREM sleep is further subdivided into multiple stages. During a night's sleep, a normal person cycles between these stages (Figs. 15.1 and 15.2). Each sleep stage can be distinguished by means of electroencephalography (EEG), recording the electrical activity derived from the brain surface (Fig. 15.1) and electromyography (EMG), recording the electrical activity of skeletal muscle. Each stage has its distinct pattern of EEG and EMG as well as other features. The EEG of a wakeful person mostly shows alpha and beta activity (neural oscillations in the frequency range of 7.5–12.5 Hz and 12.5–30 Hz, respectively) (Fig. 15.1). Beta activity appears when the person is conducting a behavior that requires attention. By contrast, alpha activity appears when the person is in a relaxed and quiet state. After the onset of sleep, a person first enters NREM sleep, which can be further differentiated into three stages N1–N3 (classically four stages). In stage N1, alpha activity gradually decreases and theta activity (4–8 Hz) appears (Figs. 15.1 and 15.2). During stage N2, EEG is characterized by the appearance of sleep spindles [e.g., brief bursts of high-frequency (7–14 Hz) activity) and K-complexes (e.g., one cycle of slow oscillation) (Figs. 15.1 and 15.2). When the person enters stage N3, delta activity (1–4 Hz) or so-called slow wave activity (SWA) starts to dominate the EEG (Figs. 15.1 and 15.2). Classically, stage N3 was further divided into two stages based on the amount of slow waves (Fig. 15.1). Stage N3 is typically called slow wave sleep (SWS), and during this stage, the person is situated in the deepest sleep and is hardest to be wakened (Fig. 15.1). These series of NREM sleep are usually followed by REM

Table 15.1 Neurotransmitters or signaling pathways involved in sleep/wake regulation across animal species

	Fruit fly (*Drosophila melanogaster*)	Roundworm (*Caenorhabditis elegans*)	Zebrafish (*Danio rerio*)	Mammals
Dopamine	Wakefulness ↑	Wakefulness ↑	–	Wakefulness ↑
	Amphetamine feeding causes higher activity. Genetic loss of dopamine transporter reduces sleep (Andretic et al. 2005; Kume et al. 2005)	Genetic loss of dopamine transporter reduces lethargus and loss of dopamine receptor increases quiescence (Singh et al. 2014)		Genetic loss of dopamine transporter in mice causes less NREM sleep and more awake time. (Lu et al. 2006a; Wisor et al. 2001)
Noradrenaline	–	–	–	Wakefulness↑
				Optogenetic activation of locus coeruleus neurons induces wake (Carter et al. 2010)
Histamine	Wakefulness ↑	–	Wakefulness ↑	Wakefulness ↑
	Genetic loss of histidine decarboxylase increases sleep (Oh et al. 2013)		Genetic knockdown of histidine decarboxylase causes less locomotor activity in the light phase (Sundvik et al. 2011)	Histamine causes increased wakefulness through the activation of H1 receptor (Haas et al. 2008)
Serotonin	Sleep ↑	Wakefulness ↑	–	Mixed results
	Genetic loss of serotonin receptor 1A has reduces sleep (Yuan et al. 2006)	Genetic loss of serotonin receptor reduces quiescence (Singh et al. 2014)		Serotonin receptor 1A knockout mice show increased REM sleep (Boutrel et al. 2002). The serotonin precursor 5-HTP administration to mice at light onset increases wakefulness, whereas administration at dark onset increases NREM sleep (Morrow et al. 2008).

Orexin/hypocretin	–	–	Mixed results	Wakefulness ↑ (or perhaps stabilization of each state)
			Orexin/hypocretin overexpression increases locomotor activity (Prober et al. 2006), whereas genetic loss of orexin/hypocretin receptor reduces sleep and injection of orexin/hypocretin reduces locomotor activity (Yokogawa et al. 2007).	Orexin/hypocretin knockout mice show fragmented sleep and a mutation of orexin/hypocretin receptor in dog causes narcolepsy (Chemelli et al. 1999; Lin et al. 1999b)
GABA	Sleep ↑	Sleep ↑	–	NREM sleep ↑
	Genetic inhibition of GABA release causes less sleep (Agosto et al. 2008)	Genetic loss of glutamic acid decarboxylase or $GABA_A$ receptor reduces quiescence (Singh et al. 2014)		Benzodiazepines produce hypnotic effects mainly through $GABA_A$ receptor (Harrison 2007)
cAMP signaling	Wakefulness ↑	Wakefulness ↑	–	Wakefulness ↑
	Higher levels of cAMP reduce sleep, whereas lower levels of cAMP increase sleep (Hendricks et al. 2001)	Higher level of cAMP reduces arousal threshold (Raizen et al. 2008)		Mice with CREB mutation have less NREM sleep (Graves et al. 2003)
EGF signaling	Sleep ↑	Sleep ↑	–	NREM sleep ↑
	Activation of EGFR increases sleep (Foltenyi et al. 2007)	Overexpression of EGF-like ligand increases quiescence (Van Buskirk and Sternberg 2007)		Administration of EGF increases NREM sleep in rats and rabbits, whereas mice with an EGFR mutation have reduced sleep (Kushikata et al. 1998; Obal et al. 1988) (Kramer et al. 2001)

(continued)

Table. 15.1 (continued)

	Fruit fly (*Drosophila melanogaster*)	Roundworm (*Caenorhabditis elegans*)	Zebrafish (*Danio rerio*)	Mammals
PKG signaling	–	Sleep ↑	–	Sleep ↑
		Gain-of-function mutation of cGMP-dependent kinase increases quiescence, whereas loss-of-function mutation reduces quiescence (Raizen et al. 2008)		Mice with mutation of cGMP-dependent kinase type 1 have less REM sleep, low-quality NREM sleep, and disrupted sleep/wake cycle (Langmesser et al. 2009)
Voltage-gated potassium channel	Sleep ↑	Sleep ↑	–	NREM sleep ↑
	Mutant of Shaker potassium channel has less sleep (Cirelli et al. 2005)	Genetic loss of Shaker potassium channel reduces quiescence (Singh et al. 2014)		Kcna2 knockout mice have reduced NREM sleep (Douglas et al. 2007)

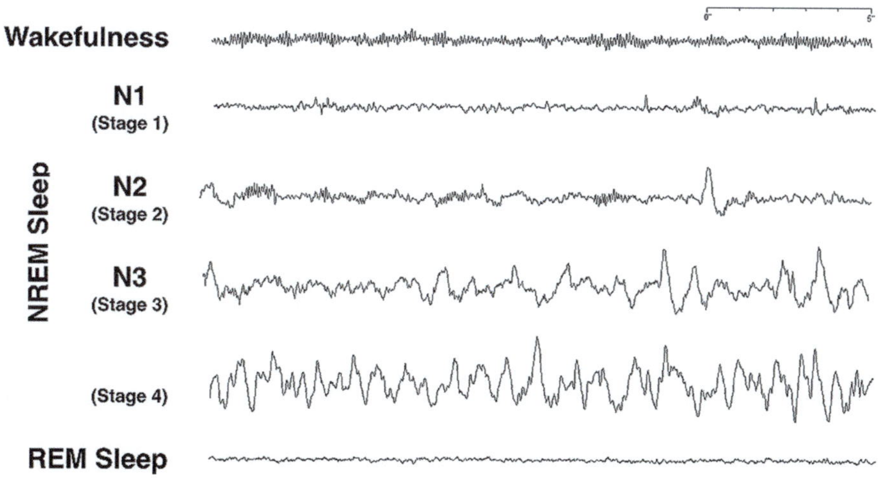

Fig. 15.1 Typical electroencephalogram (*EEG*) patterns of wakefulness and sleep in a normal human adult. Non-rapid eye movement (NREM) sleep was classified according to the current (*N1–N3*) and classical (*1–4*) classification. EEG data were kindly provided by Dr. Makoto Satoh and Ms. Kumiko Shimoyama (International Institute for Integrative Sleep Medicine (WPI-IIIS), University of Tsukuba, Japan)

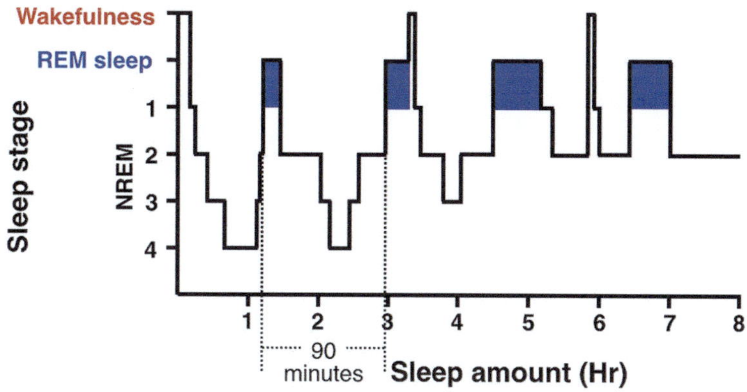

Fig. 15.2 Typical hypnogram of a normal human adult. NREM sleep was classified according to the classical (1–4) classification. Hypnogram data were kindly provided by Dr. Takashi Kanbayashi (Akita University)

sleep, in which beta activity comes up again, along with theta activity (Figs. 15.1 and 15.2). More distinctively, the person during REM sleep shows rapid eye movement, muscular atonia, and loss of homeothermy. Therefore, REM sleep can be easily distinguished from wakefulness by the low signal on EMG, although the EEG patterns look similar. At the end of REM sleep is the timing when the person can be wakened easiest with small disturbance. Afterward, the person goes into stage N1 of NREM sleep again (Fig. 15.2). The total time from the onset of NREM sleep to the end of REM sleep is on average around one and a half hours. A normal person takes about 8 h of sleep a day, which alternates between REM sleep and NREM sleep by 4–5 times.

REM sleep was the latest sleep stage to be discovered, first reported in 1953 by Kleitman and Aserinsky (Aserinsky and Kleitman 1953). Several years later, Kleitman and Dement further demonstrated that REM sleep is closely associated with dreaming (Dement and Kleitman 1957). Soon after, Dement, and subsequently Jouvet's group, discovered a similar state in cats (Dement 1958; Jouvet et al. 1959). REM sleep is also termed paradoxical sleep, as the EEG patterns are similar to wakefulness although the muscle tone is lost. Ever since the discovery of REM sleep in cats, intense pharmacological, physiological, and genetic studies have been carried out by Jouvet's group as well as other groups to elucidate the neural substrate of REM and NREM sleep. These studies led to discoveries that the brainstem pontine tegmental area and adjacent mesencephalic and medullary regions contain neurons that have key roles in generating REM and NREM sleep (Boissard et al. 2002; Crochet et al. 2006; Lu et al. 2006b; Sakai et al. 2001; Saper et al. 2010; Vanni-Mercier et al. 1989).

15.3 Evolution of REM and NREM Sleep

NREM and REM sleep are two distinct sleep states observed in various mammalian and avian species but not in other vertebrates, which has motivated researchers to investigate what evolutionary events occurred during the transitions from ancestral vertebrates to mammals and birds (Fig. 15.3). The key to answering this question could be located in reptiles, which are likely closest to the ancestors of mammals and birds. More precisely, according to phylogenetic analysis, the primitive form of reptiles, cotylosaurs, is regarded as the common ancestor of reptiles, mammals, and birds (Young 1981). Although the sleep/wake patterns of cotylosaurs cannot be assessed, current reptiles may still retain some features inherited from its ancestor. Compared with mammals and birds, current reptiles seem to share more features with cotylosaurs in metabolism and brain architecture, including poikilothermy and simple telencephalic structures. By contrast, mammals and birds are both homeotherms and have advanced telencephalic structures such as the neocortex in mammals and the neostriate in birds. Given that the reptilian EEG is highly dependent on body temperature and information processing are largely different between reptiles and mammals/birds because of the distinct brain architecture, some researchers hypothesize that the transitions in sleep from the ancestral style to mammalian or avian styles may be related to the evolution of homeothermy and telencephalic structures (Nicolau et al. 2000).

Nonetheless, although a host of studies seeking the origin of NREM and REM sleep have been carried out, solid and consistent conclusions cannot be drawn among researchers. Given that, up to now, obvious signs of NREM and REM sleep could only be detected in mammals and birds, there is a possibility that

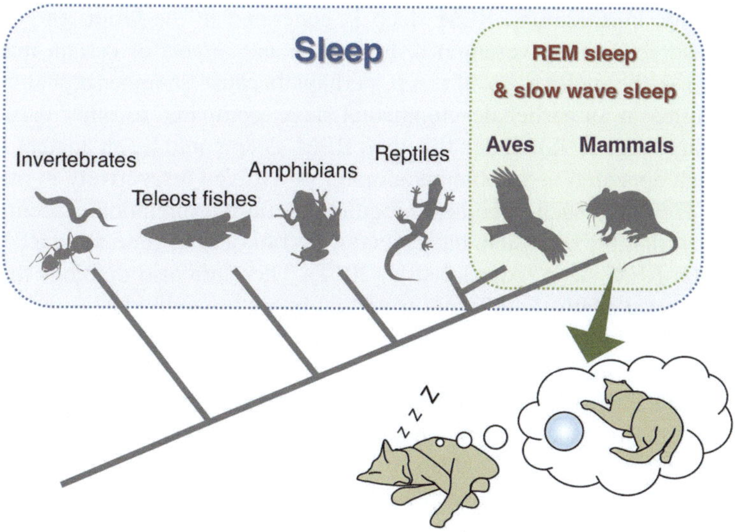

Fig. 15.3 Sleep or sleep-like states are observed across the animal kingdom

these sleep states emerged independently in the respective evolutionary route of mammals and birds. In further addressing this issue, the monotremes, which are egg-laying mammals, have received much attention. The extant monotremes are the echidna and the platypus. During the evolution of mammals, the divergence of the monotreme lineage from other mammalian lineages (the marsupial lineage and the placental lineage) happened prior to the divergence between marsupials and placental mammals. Monotremes are thought to most closely resemble ancestral primitive mammals. Early studies that carefully recorded various measures of sleep/wake in the echidna (*Tachyglossus aculeatus*) suggested that this animal undergoes NREM sleep with slow waves but shows no sign of REM sleep (Allison et al. 1972). This could be interpreted that REM sleep emerged in mammals after the divergence of the monotremes from other mammals, and that NREM sleep closely resembles primitive sleep. When another group simultaneously recorded brainstem neural activity, however, they observed a REM sleep-like pattern, although the EEG detected slow wave activity (Siegel et al. 1996). Thus this putatively primitive form of sleep contains features of both REM and NREM sleep. In contrast to these studies, other researchers reported that sleep in echidna could be clearly differentiated to REM sleep and NREM sleep by cortical EEG, but that REM sleep could only be detected at an appropriate temperature (Nicol et al. 2000). This report is, however, challenged by a view that quiet wakefulness and REM sleep were not sufficiently differentiated. Overall, whether REM sleep and NREM sleep were present in the common ancestors of mammals and bird remains to be solved.

There are also opinions that REM sleep, rather than NREM sleep, closely resembles a state in ancestral animals. During REM sleep, the regulation of body temperature in mammals and birds is reduced, namely, a partial loss of homeothermy. Furthermore, REM sleep is generated in the brainstem, which is highly conserved among vertebrates. In fetuses and infants of certain mammals, REM sleep is the major state of sleep, perhaps because rhombencephalon (hindbrain) matures in an earlier developmental stage, compared to other parts of the brain (Finlay et al. 1998). Taken together, REM sleep could be an ancient form of sleep, which appeared in a common ancestor or evolved respectively in mammals and birds. However, such hypotheses require careful interpretation. Recent studies clearly show that the brainstem has crucial roles not only in generating REM sleep but also non-REM sleep (Anaclet et al. 2014). There are also opinions that sleep in neonates is qualitatively different from that in mature individuals, and that their sleep should be classified as active sleep and quiet sleep, instead of REM sleep and NREM sleep.

Although REM sleep and NREM sleep have not been detected in reptiles, researchers have pointed out that some characteristics of reptilian EEG patterns are reminiscent of that of NREM sleep or REM sleep. While the sleep-like state in reptiles is not accompanied by increased slow wave activity, Rattenborg proposed that the high-voltage spikes observed in reptilian sleep could be a counterpart of hippocampal activity in the mammalian hippocampus during NREM sleep (Hartse 1994; Rattenborg 2006). Rattenborg further discussed that SWA was not detectable in reptiles because of the lack of corticocortical connections in reptilian cortex. This "interconnectivity" hypothesis also explained the avian NREM sleep similarly in term of the extensive interconnectivity in the avian pallium. Rial and colleagues, on the other hand, emphasized the increase of a slow wave-like EEG during wakefulness in reptiles, and proposed that mammalian NREM sleep evolved from reptilian wakefulness. They further proposed that mammalian wakefulness is a novel state that emerged as a result of the development of the cortex (Rial et al. 2007, 2008; Rial et al. 2010). Future studies addressing the effects of neurotransmitters and neuromodulators such as noradrenaline (norepinephrine), histamine, dopamine, and orexin (hypocretin), which are important in the control of the wake–sleep cycle in mammals, in reptiles might be meaningful. As described below, at least histamine and dopamine seem to have a deeply conserved role in promoting wakefulness across the animal kingdom, but the role of orexin (hypocretin) seems more complicated. Another effective approach would be to clarify the molecular identity of various neurons in the brainstem that regulate REM/NREM sleep. As the anatomical features of the brainstem are highly conserved, such studies will allow detailed comparison of sleep-regulating cells among various vertebrate species.

15.4 Sleep in Aquatic Mammals

One might imagine that whales and dolphins, which spend their whole life in water, need somehow to adopt a special form of sleep, or otherwise they will drown or be under a high threat of being attacked by predators while sleeping. Intriguingly, dolphins circumvent such risks by sleeping one hemisphere of the brain at a

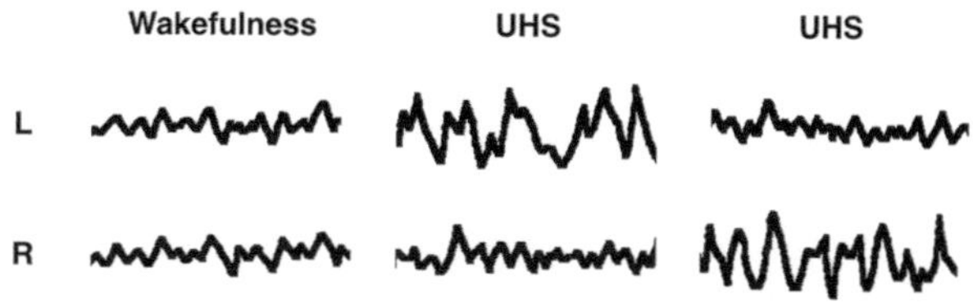

Fig. 15.4 Schematic of EEG patterns during wake and unihemispheric sleep in a dolphin. *UHS* unihemispheric sleep, *L* left cerebral hemisphere, *R* right cerebral hemisphere

time, which is termed unihemispheric sleep. EEG recording from dolphins clearly demonstrated that slow waves were generated in a single hemisphere (Mukhametov et al. 1977) (Fig. 15.4). During unihemispheric sleep, one half of the brain retains a low level of alertness so that dolphins can keep swimming. Another interesting aspect of sleep in dolphins is that REM sleep amount is very low, if any. A general feature of REM sleep in mammals is the loss of muscle tone, which could be dangerous for aquatic mammals. Thus dolphins might have evolved to minimize or even abandon REM sleep.

Unihemispheric sleep is also observed in other marine mammals, such as the fur seal (Mukhametov et al. 1985). In contrast to dolphins, fur seals live both on land and in water. Interestingly, while sleeping on land, fur seals predominantly display bilateral sleep, whereas when sleeping in water, they primarily display unihemispheric sleep. How the brain switches between these two modes of sleep remains to be revealed.

In addition, various bird species also show unihemispheric sleep (reviewed in Rattenborg et al. 2000), which might be beneficial in reducing the risk of predation. Indeed, mallard ducks exhibited increased unihemispheric sleep under higher predation risk, which allowed rapid response to visual stimuli (Rattenborg et al. 1999). It may also aid birds, especially migratory birds, to retain consciousness during a long period of flight, although direct evidence of unihemispheric sleep during flight is lacking (Rattenborg and Martinez-Gonzalez 2014).

These studies in aquatic mammals and birds strongly suggest that, during evolution, vertebrates have created unihemispheric sleep multiple times, independently (convergent evolution). This result might imply the innate nature of vertebrate sleep to be locally regulated. Indeed, in rats, prolonged wakefulness leads to local emergence of slow waves even though, as a whole, the individual appears to be awake (Vyazovskiy et al. 2011). Unihemispheric sleep might be an extreme form of such local sleep.

15.5 Sleep in the Zebrafish (*Danio rerio*)

Along with fruit flies (*Drosophila melanogaster*) and roundworms (*Caenorhabditis elegans*), the zebrafish (*Danio rerio*) is one of the most efficient animal models for genetic analyses. The zebrafish is a teleost fish, which is much closely related to mammals than invertebrate animal models. For example, the design of the brain roughly resembles those of mammals, especially in the brainstem, a key structure regulating sleep/wake. Moreover, many genes, including genes for neuropeptides and their receptors, are conserved. Thus, to address the evolutionary origin of wakefulness and REM/NREM sleep, the roles of zebrafish brain regions, genes, or neuromodulators homologous to those involved in mammalian sleep are of interest.

Zebrafish displays circadian quiescence that basically follows the criteria for sleep described in Sect. 15.1 (Yokogawa et al. 2007; Zhdanova et al. 2001). One notable thing about zebrafish sleep is that it is under the strong influence of light. There are reports that, under constant light conditions for as long as 3 days, sleep was completely suppressed (Yokogawa et al. 2007). This finding suggests that light has stronger effects than circadian influences. Moreover, surprisingly, sleep deprivation by light exposure is not accompanied by a sleep rebound (Yokogawa et al. 2007), which suggests that light influences can even overcome homeostatic responses. It is not clear how this sensitivity to light is adaptive. Perhaps, for zebrafish, falling asleep in a bright place is of extreme danger, compromising them to attacks by predators.

Hints on whether sleep/wake states are conserved across vertebrates might be discussed from the apparently conserved roles of histamine. In mammals, histamine is a well-addressed wake-inducing molecule. Similarly, histamine seems to promote wake in zebrafish. Mutants of the *histidine decarboxylase (hdc)*, a gene required for histamine synthesis, have increased sleep (Sundvik et al. 2011). In addition, H1 histaminergic antagonists increase the amount of sleep (Renier et al. 2007).

Although the conserved roles of histamine support the conservation between mammalian sleep and zebrafish sleep, the case is not so simple for orexin (hypocretin), a neuropeptide important for maintaining mammalian wake (de Lecea et al. 1998; Sakurai et al. 1998). In humans, loss of neurons producing orexin (hypocretin) leads to narcolepsy, a sleep disorder characterized by excessive daytime sleepiness, cataplexy (sudden brief episodes of muscle paralysis that are typically triggered by strong positive emotions), and dream-like hallucinations at sleep onset likely caused by direct transition from wake to REM sleep (Peyron et al. 2000; Thannickal et al. 2000). Similar phenotypes are observed in mice or dogs lacking either orexin (hypocretin), their receptors, or the neurons synthesizing orexin (Chemelli et al. 1999; Hara et al. 2001; Lin et al. 1999a; Willie et al. 2003). On the other hand, rats and mice receiving orexin (hypocretin) administration show increased wakefulness (Hagan et al. 1999; Mieda et al. 2011; Piper et al. 2000). Orexinergic neurons fire most during wakefulness, and become silent upon entering sleep (Lee et al. 2005; Mileykovskiy et al. 2005; Takahashi et al. 2008). Optogenetic activation of these neurons is sufficient to induce wakefulness (Adamantidis et al. 2007). The orexin (hypocretin) receptors are expressed in monoaminergic and cholinergic systems that promote wakefulness. Thus, in mammals, orexin (hypocretin) seems to have a role to maintain wakefulness and suppress REM sleep.

By contrast, zebrafish mutants lacking the orexin (hypocretin) receptor display no obvious phenotypes during the daytime, and increased wakefulness and fragmented sleep during the nighttime (Yokogawa et al. 2007). Even more surprisingly, administration of orexin (hypocretin) induces sleep in a receptor-dependent manner (Yokogawa et al. 2007). Moreover, in contrast to mammals, the receptor expression does not match with the monoaminergic and cholinergic arousal system, although there seem to be some inconsistencies among studies (Mieda et al. 2011; Prober et al. 2006; Yokogawa et al. 2007). All these results suggest a somewhat differential role for orexin (hypocretin) in mammalian and teleost fish wakefulness. However, cautious interpretation is required, as the function of orexin (hypocretin) in mammals is itself complicated. Mice deficient in orexin (hypocretin) signaling, similar to the zebrafish mutants, also display fragmented sleep, and the total sleep amount actually seems unaffected (Mochizuki et al. 2004). Future studies to assess the firing patterns of orexinergic neurons in zebrafish might be meaningful.

In mammals, the catecholamines dopamine and noradrenaline (norepinephrine) also comprise a major component of the arousal circuit. As described next, catecholamines also promote wakefulness in invertebrates. Their roles in zebrafish sleep, however, have not been well studied yet.

15.6 Sleep in Invertebrate Animals

15.6.1 General Introduction

Humans have had tight relationships with silkworms ever since the history of silk culture started in China in the very old days. Thus, the life cycle of this insect is very well characterized. The silkworm larvae undergo four molts before forming a cocoon. Before each molt, they stop feeding and become immobile. In Japan, this behavior was termed "min," which means "sleep". As in this example, people have felt certain commonness between our sleep and particular quiescence states in invertebrate animals. Currently, we do not know very well whether our sleep and the invertebrate sleep-like states are derived from a common evolutionary origin and share similar mechanisms. If they did, it will have at least two important meanings. The first is that sleep would be of an extremely ancient origin, perhaps extending back to ancestral animals with a primitive nervous system. The second is that we can utilize the fruit fly (*Drosophila melanogaster*) and the roundworm (*Caenorhabditis elegans*), two of the most efficient genetic animal models, to understand the molecular bases underlying human sleep. Many researchers have already started to intensively study the molecular mechanisms and functions of sleep in these two animal species, and thus it is expected that we will know in the near future to what extent sleep is conserved across the animal kingdom.

15.6.2 Sleep in the Fruit Fly (Drosophila melanogaster)

Adult fruit flies display a circadian rhythm of locomotor activity, with periods of rest at night. This has been a model extensively studied by circadian biologists, and such studies have had major contributions to elucidating the molecular entity of our circadian clock. Now researchers have gone further into the investigation of the periods of rest displayed at night, to see whether it satisfies the basic criteria for sleep. Rest in fruit flies not only receives circadian influences, but also homeostatic regulation, as demonstrated by a rebound after deprivation by mechanical stimulation or social interaction (Hendricks et al. 2000; Shaw et al. 2000). Arousal threshold was also increased. Thus, this behavior follows the basic criteria for sleep described in Sect. 15.1, and cannot be explained simply as a rest period regulated by the circadian clock. Moreover, similar to human sleep, the rest period was decreased in old flies compared to young flies, and caffeine, which promotes wake in mammals, also efficiently promoted arousal in fruit flies (Hendricks et al. 2000; Wu et al. 2009). In mammals, caffeine acts by antagonizing the receptor for adenosine, which is one of the few known endogenous sleep-inducing substances in mammals. In mice lacking the A2A adenosine receptor, caffeine does not increase wakefulness (Huang et al. 2005). Although the apparently conserved effect of caffeine seems intriguing, arousal effect on fruit flies, however, was independent of the putative adenosine receptor homologue, raising the possibility that the underlying mechanism might be largely different between mammals and fruit flies (Wu et al. 2009).

Catecholamine neurotransmitters, namely dopamine and noradrenaline (norepinephrine) in the vertebrate central nervous system, have important roles in mammalian sleep/wake regulation. Dopamine promotes wake in mammals. Mice lacking the *dopamine transporter* (*DAT*) gene, in which extracellular dopamine is increased, exhibit increased wake and reduced NREM sleep (Wisor et al. 2001). Similarly, in fruit flies, a mutation in the *DAT* gene results in increased activity and reduced rest (Kume et al. 2005). Noradrenaline also enhances wake in mammals. Optogenetic activation of mouse noradrenaline-releasing neurons in the brainstem locus coeruleus promoted sleep-to-wake transitions, whereas optogenetic inhibition reduced wakefulness (Carter et al. 2010). Although invertebrates do not possess noradrenaline, the catecholamine octopamine is generally regarded as the invertebrate counterpart to noradrenaline, both involved in conserved behaviors such as the fight-or-flight response (Haller et al. 1998; Libersat and Pflueger 2004). Similar to the case of mouse noradrenaline, genetic excitation or silencing of octopamine-releasing neurons in fruit flies increased or decreased wakefulness, respectively (Crocker and Sehgal 2008).

The roles of histamine in promoting wakefulness also seem conserved. As with zebrafish (*Danio rerio*), mutants of *hdc* display increased sleep (Oh et al. 2013). As in caffeine, however, the downstream pathway might be different. In mammals, the H1 receptor, which is a G-protein-coupled receptor, conveys the wake-promoting effect of histamine. On the other hand, in fruit flies, a histamine-gated chloride channel seems crucial (Oh et al. 2013).

In mammals, the inhibitory neurotransmitter GABA is involved in promoting sleep. GABAergic neurons in the preoptic hypothalamic area or the brainstem medulla promote NREM sleep (Anaclet et al. 2014; Zhang et al. 2015). In addition, the $GABA_A$ receptor is a major target for treating insomnia in humans, although the responsible brain areas remain to be clarified (Equihua et al. 2013). GABA and the $GABA_A$ receptor are also important for promoting sleep in fruit flies. Genetic hyperpolarization of GABAergic neurons reduced sleep, and a mutation in the *$GABA_A$ receptor* gene that results in increased channel current led to increased sleep (Agosto et al. 2008).

A disadvantage in using fruit flies (or roundworms) as a model for sleep might be the low conservation of neuropeptides. Although orexin (hypocretin) has a crucial role in maintaining wake in mammals, invertebrates lack an obvious orexin (hypocretin) homologue. Instead, in fruit flies, the neuropeptide pigment dispersing factor (PDF), which is not present in vertebrates, is critical in regulating circadian rhythms and maintaining wakefulness (Parisky et al. 2008).

15.6.3 Sleep in the Roundworm (*Caenorhabditis elegans*)

Roundworms take less than a week to mature, and no obvious circadian behavior has been reported. However, recently, this small animal has received much attention as a model for studying sleep. Roundworms enter a quiescent state under certain conditions. The most intensely studied is a state termed lethargus, which is actually similar to the silkworm "min" state. Similar to silkworm larvae, roundworm larvae undergo four molts before becoming adults. And as with the silkworm larvae, before each molt, the larvae become quiescent, and this state is termed lethargus in the case of roundworms (Fig. 15.5). Although lethargus is not a daily event but rather occurs in intervals of 7–12 h, this state satisfies various criteria of sleep, including the display of a homeostatic rebound following "sleep deprivation" by mechanical stimulation and increased arousal threshold (Raizen et al. 2008). The robustness of

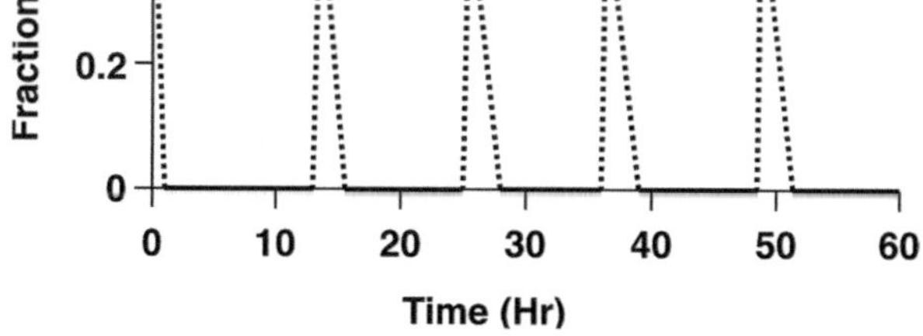

Fig. 15.5 Schematic of the periodic emergence of lethargus during the development of a roundworm (*Caenorhabditis elegans*) larva. *L1–L4* indicate each larval stage. Each lethargus is followed by molting and transition to the next developmental stage

this quiescent state compared to other quiescent states in roundworms is another advantage as a sleep model.

For roundworm lethargus, the lack of a 24-h cycle and its restriction to a particular immature stage largely separates it from sleep in other animal models. However, several studies support that roundworm lethargus and fruit fly sleep are conserved at the molecular level. For example, the roundworm homologue of PDF, PDF-1, negatively regulates lethargus (Choi et al. 2013). In another study, a wide survey was conducted to examine to what extent roundworm lethargus and fruit fly sleep are conserved (Singh et al. 2014). Of 26 genes known to be required for fruit fly sleep, 20 orthologous roundworm genes were examined for their involvement in lethargus. Surprisingly, all 20 genes affected lethargus quantity and arousal thresholds. Moreover, for 18 genes including the *DAT* and $GABA_A$ *receptor* genes, the direction of the effect matched the fruit fly genes. These studies highlight the deep conservation between fruit fly sleep and roundworm lethargus. In the future, a major question will be, of course, whether there is also a deep conservation between these animals' sleep and mammalian sleep.

15.7 Is Sleep Function Conserved Across the Animal Kingdom?

The function of sleep is even less understood than its regulatory mechanisms. Sleep deprivation leads to death in rats, fruit flies (*Drosophila melanogaster*), and roundworms (*Caenorhabditis elegans*) (Driver et al. 2013; Rechtschaffen and Bergmann 2002; Shaw et al. 2002). In all cases, the exact cause of the lethality remains obscure. In mammals, sleep is roughly classified to REM and NREM sleep. Although the function of REM sleep is almost a complete mystery, NREM sleep has been reported to be involved in the secretion of growth hormone, synaptic plasticity, memory consolidation, and clearance of brain metabolites (Chauvette et al. 2012; Marshall et al. 2006; Rasch et al. 2007; Takahashi et al. 1968; Xie et al. 2013; Yang et al. 2014). Despite the fact that slow waves occurring during NREM sleep are a unique feature of mammals and birds, as discussed next, there seem to be some conservation between the roles of our sleep and sleep in invertebrates.

Novel imaging techniques using two photon microscopes have allowed in vivo observation of the mouse brain at the synaptic level: this led to findings that, during mammalian sleep, dendritic spines, which are postsynaptic structures, undergo active remodeling (Maret et al. 2011; Yang and Gan 2012; Yang et al. 2014). Similarly, in fruit flies, sleep was associated with structural changes in multiple brain areas including the mushroom body, which is important for learning and memory (Bushey et al. 2011).

At the behavioral level, NREM sleep is related to memory consolidation. In humans, reactivation of a specific memory by means of odor cues during NREM sleep, but not during wake or REM sleep, improved memory consolidation (Rasch et al. 2007). Enhancing slow wave activity during NREM sleep also improved memory consolidation (Marshall et al. 2006). Sleep also seems to be related to learning and memory in fruit flies. In fruit flies, artificial induction of sleep by stimulating sleep-regulating neurons promoted formation of long-term memories, and this effect was canceled by sleep deprivation (Donlea et al. 2011).

Chronic sleep deprivation in rats results in the disability to regulate the body temperature, skin lesions, high metabolic rate, and weight loss in spite of increased food intake. Within 2–3 weeks, the sleep-deprived rats died (Rechtschaffen and Bergmann 2002; Rechtschaffen et al. 1989). Although this finding indicates the necessity of sleep, the direct cause of lethality remains elusive. A study addressed this issue using fruit flies (Shaw et al. 2002), identifying two strains that are extremely sensitive to sleep deprivation. These strains, carrying a mutation in either the circadian clock gene *cycle*, or the heat shock-induced gene *hsp83*, showed an exaggerated sleep rebound after 3 h of sleep deprivation. Moreover, individuals of these strains started dying after just 10 h of sleep deprivation. These findings might provide hints to the fundamental function of sleep at the molecular level. The *hsp83* gene encodes a chaperone protein, and thus sleep might be required for quality management of certain proteins.

In roundworms, continuous disturbance of lethargus by mechanical stimulation results in lethality (Driver et al. 2013), just as in rats. This finding supports that there is an active role for lethargus. In addition to preparation for molting, in the nervous system, synaptic remodeling and pruning events coincide with the lethargus timing (Hallam and Jin 1998; Hayashi et al. 2009; White et al. 1978), raising the possibility that, as with mammalian NREM sleep and fruit fly sleep, lethargus is a state where neural circuit remodeling is enhanced.

The foregoing sleep deprivation study in rats also suggests that metabolism is tightly connected with sleep, as sleep deprivation caused a higher metabolic rate and enhanced appetite (Rechtschaffen et al. 1989). Indeed, hypothalamic neurons such as the orexinergic neurons and the melatonin-concentrating hormone (MCH)-releasing neurons have dual roles in the regulation of sleep and feeding, supporting their coordinated regulation (Chemelli et al. 1999; Hassani et al. 2009; Jego et al. 2013; Konadhode et al. 2013; Modirrousta et al. 2005; Qu et al. 1996; Sakurai et al. 1998; Szentirmai and Krueger 2006; Tsunematsu et al. 2014; Verret et al. 2003). Starvation suppresses sleep in fruit flies, suggesting that feeding and sleep interact in invertebrates, too (Keene et al. 2010). In humans, the amount of sleep is associated with the regulation of metabolic hormones, body mass index, and obesity (Baumgartner et al. 1993; Taheri et al. 2004; Van Cauter and Knutson 2008). In addition, in humans, sleep deprivation also affected the release of sex hormones including luteinizing hormone (LH), estradiol, and follicle-stimulating hormone (FSH) (Baumgartner et al. 1993). Taken together, these observations suggest that sleep might be an important strategy used to regulate the energy distribution and conservation through the adjustment of metabolism and feeding or reproductive behavior, although more detailed examinations and accurate interpretations are needed (Rechtschaffen et al. 1989).

15.8 Concluding Remarks and Future Prospects

In this review, we featured studies using various vertebrate or invertebrate animal species to clarify the mysteries of the evolution and function of sleep. Recent studies using novel genetic tools or simple animal models have offered substantial advantages in identifying neural circuits or genes that are engaged in sleep regulation. There are, however, fundamental questions yet to be solved. Finally, we propose such important issues together with the future directions for addressing these issues.

Is there a universal molecular pathway that encodes “sleepiness”? Although various neurotransmitters and some signaling molecules have been identified, our understanding of the molecular basis of sleep is still weak compared to, for example, the circadian clock. In the future, intense high-throughput genetic and molecular approaches might lead to the elucidation of the overall picture of the core sleep-regulatory pathway.

Why is sleep essential? Whether there is a general purpose for sleep across animal species still remains unsolved. The discovery of unihemispheric sleep in aquatic mammals and birds and local sleep in rodents may imply that at least some aspects of sleep functions can be fulfilled without the change in state of the whole individual. Recent advances in genetic tools for precise manipulation of neuronal activity, such as optogenetics and pharmacogenetics, have led to the identification of neural circuits critical for sleep. In the future, these tools are also expected to be effective for manipulating sleep/wake states and subsequent analyses of the phenotypic outcome of sleep induction or inhibition.

How did REM and NREM sleep evolve, and how were they beneficial? The individual roles of each sleep state are even less well understood. In the future, the precise identification of neural circuits that play central roles in generating these sleep states will offer substantial advantages. Inhibition or induction of REM sleep by manipulation of such circuits may provide clues to the function of REM sleep. In addition, identification and analyses of homologous neural circuits in animals lacking REM/NREM sleep may provide clues to the evolutionary origin of these sleep states. These approaches will require the integration of multiple approaches, including genetics, physiology, comparative neurology, behavioral studies, and developmental studies in various animal species.

References

Adamantidis AR, Zhang F, Aravanis AM, Deisseroth K, de Lecea L (2007) Neural substrates of awakening probed with optogenetic control of hypocretin neurons. Nature 450:420–424

Agosto J, Choi JC, Parisky KM, Stilwell G, Rosbash M, Griffith LC (2008) Modulation of GABAA receptor desensitization uncouples sleep onset and maintenance in *Drosophila*. Nat Neurosci 11:354–359

Allison T, Van Twyver H, Goff WR (1972) Electrophysiological studies of the echidna, *Tachyglossus aculeatus*. I. Waking and sleep. Arch Ital Biol 110:145–184

Anaclet C, Ferrari L, Arrigoni E, Bass CE, Saper CB, Lu J, Fuller PM (2014) The GABAergic parafacial zone is a medullary slow wave sleep-promoting center. Nat Neurosci 17:1217–1224
Andretic R, van Swinderen B, Greenspan RJ (2005) Dopaminergic modulation of arousal in *Drosophila*. Curr Biol 15:1165–1175
Aserinsky E, Kleitman N (1953) Regularly occurring periods of eye motility, and concomitant phenomena, during sleep. Science 118:273–274
Baumgartner A, Dietzel M, Saletu B, Wolf R, Campos-Barros A, Graf KJ, Kurten I, Mannsmann U (1993) Influence of partial sleep deprivation on the secretion of thyrotropin, thyroid hormones, growth hormone, prolactin, luteinizing hormone, follicle stimulating hormone, and estradiol in healthy young women. Psychiatry Res 48:153–178
Boissard R, Gervasoni D, Schmidt MH, Barbagli B, Fort P, Luppi PH (2002) The rat ponto-medullary network responsible for paradoxical sleep onset and maintenance: a combined microinjection and functional neuroanatomical study. Eur J Neurosci 16:1959–1973
Boutrel B, Monaca C, Hen R, Hamon M, Adrien J (2002) Involvement of 5-HT1A receptors in homeostatic and stress-induced adaptive regulations of paradoxical sleep: studies in 5-HT1A knock-out mice. J Neurosci 22:4686–4692
Bushey D, Tononi G, Cirelli C (2011) Sleep and synaptic homeostasis: structural evidence in *Drosophila*. Science 332:1576–1581
Campbell SS, Tobler I (1984) Animal sleep: a review of sleep duration across phylogeny. Neurosci Biobehav Rev 8:269–300
Carter ME, Yizhar O, Chikahisa S, Nguyen H, Adamantidis A, Nishino S, Deisseroth K, de Lecea L (2010) Tuning arousal with optogenetic modulation of locus coeruleus neurons. Nat Neurosci 13:1526–1533
Chauvette S, Seigneur J, Timofeev I (2012) Sleep oscillations in the thalamocortical system induce long-term neuronal plasticity. Neuron 75:1105–1113
Chemelli RM, Willie JT, Sinton CM, Elmquist JK, Scammell T, Lee C, Richardson JA, Williams SC, Xiong Y, Kisanuki Y et al (1999) Narcolepsy in orexin knockout mice: molecular genetics of sleep regulation. Cell 98:437–451
Choi S, Chatzigeorgiou M, Taylor KP, Schafer WR, Kaplan JM (2013) Analysis of NPR-1 reveals a circuit mechanism for behavioral quiescence in *C. elegans*. Neuron 78:869–880
Cirelli C, Bushey D, Hill S, Huber R, Kreber R, Ganetzky B, Tononi G (2005) Reduced sleep in *Drosophila* Shaker mutants. Nature 434:1087–1092
Crochet S, Onoe H, Sakai K (2006) A potent non-monoaminergic paradoxical sleep inhibitory system: a reverse microdialysis and single-unit recording study. Eur J Neurosci 24:1404–1412
Crocker A, Sehgal A (2008) Octopamine regulates sleep in *Drosophila* through protein kinase A-dependent mechanisms. J Neurosci 28:9377–9385
de Lecea L, Kilduff TS, Peyron C, Gao X, Foye PE, Danielson PE, Fukuhara C, Battenberg EL, Gautvik VT, Bartlett FS 2nd et al (1998) The hypocretins: hypothalamus-specific peptides with neuroexcitatory activity. Proc Natl Acad Sci USA 95:322–327
Dement W (1958) The occurrence of low voltage, fast, electroencephalogram patterns during behavioral sleep in the cat. Electroencephalogr Clin Neurophysiol 10:291–296
Dement W, Kleitman N (1957) Cyclic variations in EEG during sleep and their relation to eye movements, body motility, and dreaming. Electroencephalogr Clin Neurophysiol 9:673–690
Donlea JM, Thimgan MS, Suzuki Y, Gottschalk L, Shaw PJ (2011) Inducing sleep by remote control facilitates memory consolidation in *Drosophila*. Science 332:1571–1576
Douglas CL, Vyazovskiy V, Southard T, Chiu SY, Messing A, Tononi G, Cirelli C (2007) Sleep in Kcna2 knockout mice. BMC Biol 5:42
Driver RJ, Lamb AL, Wyner AJ, Raizen DM (2013) DAF-16/FOXO regulates homeostasis of essential sleep-like behavior during larval transitions in *C. elegans*. Curr Biol 23:501–506
Equihua AC, De La Herran-Arita AK, Drucker-Colin R (2013) Orexin receptor antagonists as therapeutic agents for insomnia. Front Pharmacol 4:163
Finlay BL, Hersman MN, Darlington RB (1998) Patterns of vertebrate neurogenesis and the paths of vertebrate evolution. Brain Behav Evol 52:232–242

Foltenyi K, Greenspan RJ, Newport JW (2007) Activation of EGFR and ERK by rhomboid signaling regulates the consolidation and maintenance of sleep in *Drosophila*. Nat Neurosci 10:1160–1167

Franken P, Thomason R, Heller HC, O'Hara BF (2007) A non-circadian role for clock-genes in sleep homeostasis: a strain comparison. BMC Neurosci 8:87

Graves LA, Hellman K, Veasey S, Blendy JA, Pack AI, Abel T (2003) Genetic evidence for a role of CREB in sustained cortical arousal. J Neurophysiol 90:1152–1159

Haas HL, Sergeeva OA, Selbach O (2008) Histamine in the nervous system. Physiol Rev 88:1183–1241

Hagan JJ, Leslie RA, Patel S, Evans ML, Wattam TA, Holmes S, Benham CD, Taylor SG, Routledge C, Hemmati P et al (1999) Orexin A activates locus coeruleus cell firing and increases arousal in the rat. Proc Natl Acad Sci USA 96:10911–10916

Hallam SJ, Jin Y (1998) lin-14 regulates the timing of synaptic remodelling in *Caenorhabditis elegans*. Nature 395:78–82

Haller J, Makara GB, Kruk MR (1998) Catecholaminergic involvement in the control of aggression: hormones, the peripheral sympathetic, and central noradrenergic systems. Neurosci Biobehav Rev 22:85–97

Hara J, Beuckmann CT, Nambu T, Willie JT, Chemelli RM, Sinton CM, Sugiyama F, Yagami K, Goto K, Yanagisawa M et al (2001) Genetic ablation of orexin neurons in mice results in narcolepsy, hypophagia, and obesity. Neuron 30:345–354

Harrison NL (2007) Mechanisms of sleep induction by GABA(A) receptor agonists. J Clin Psychiatry 68(suppl 5):6–12

Hartse K (1994) Sleep in insects and nonmammalian vertebrates. Principles and practice of sleep medicine, 2nd edn. Saunders, Philadelphia, pp 95–104

Hassani OK, Lee MG, Jones BE (2009) Melanin-concentrating hormone neurons discharge in a reciprocal manner to orexin neurons across the sleep–wake cycle. Proc Natl Acad Sci USA 106:2418–2422

Hayashi Y, Hirotsu T, Iwata R, Kage-Nakadai E, Kunitomo H, Ishihara T, Iino Y, Kubo T (2009) A trophic role for Wnt-Ror kinase signaling during developmental pruning in *Caenorhabditis elegans*. Nat Neurosci 12:981–987

Hendricks JC, Finn SM, Panckeri KA, Chavkin J, Williams JA, Sehgal A, Pack AI (2000) Rest in *Drosophila* is a sleep-like state. Neuron 25:129–138

Hendricks JC, Williams JA, Panckeri K, Kirk D, Tello M, Yin JC, Sehgal A (2001) A non-circadian role for cAMP signaling and CREB activity in *Drosophila* rest homeostasis. Nat Neurosci 4:1108–1115

Huang ZL, Qu WM, Eguchi N, Chen JF, Schwarzschild MA, Fredholm BB, Urade Y, Hayaishi O (2005) Adenosine A2A, but not A1, receptors mediate the arousal effect of caffeine. Nat Neurosci 8:858–859

Jego S, Glasgow SD, Herrera CG, Ekstrand M, Reed SJ, Boyce R, Friedman J, Burdakov D, Adamantidis AR (2013) Optogenetic identification of a rapid eye movement sleep modulatory circuit in the hypothalamus. Nat Neurosci 16:1637–1643

Jouvet M, Michel F, Courjon J (1959) On a stage of rapid cerebral electrical activity in the course of physiological sleep. CR Seances Soc Biol Fil 153:1024–1028

Kayser MS, Yue Z, Sehgal A (2014) A critical period of sleep for development of courtship circuitry and behavior in *Drosophila*. Science 344:269–274

Keene AC, Duboue ER, McDonald DM, Dus M, Suh GS, Waddell S, Blau J (2010) Clock and cycle limit starvation-induced sleep loss in *Drosophila*. Curr Biol 20:1209–1215

Konadhode RR, Pelluru D, Blanco-Centurion C, Zayachkivsky A, Liu M, Uhde T, Glen WB Jr, van den Pol AN, Mulholland PJ, Shiromani PJ (2013) Optogenetic stimulation of MCH neurons increases sleep. J Neurosci 33:10257–10263

Kramer A, Yang FC, Snodgrass P, Li X, Scammell TE, Davis FC, Weitz CJ (2001) Regulation of daily locomotor activity and sleep by hypothalamic EGF receptor signaling. Science 294:2511–2515

Kume K, Kume S, Park SK, Hirsh J, Jackson FR (2005) Dopamine is a regulator of arousal in the fruit fly. J Neurosci 25:7377–7384

Kushikata T, Fang J, Chen Z, Wang Y, Krueger JM (1998) Epidermal growth factor enhances spontaneous sleep in rabbits. Am J Physiol 275:R509–R514

Langmesser S, Franken P, Feil S, Emmenegger Y, Albrecht U, Feil R (2009) cGMP-dependent protein kinase type I is implicated in the regulation of the timing and quality of sleep and wakefulness. PLoS One 4, e4238

Lee MG, Hassani OK, Jones BE (2005) Discharge of identified orexin/hypocretin neurons across the sleep-waking cycle. J Neurosci 25:6716–6720

Libersat F, Pflueger HJ (2004) Monoamines and the orchestration of behavior. Bioscience 54:17–25

Lin L, Faraco J, Li R, Kadotani H, Rogers W, Lin X, Qiu X, de Jong PJ, Nishino S, Mignot E (1999a) The sleep disorder canine narcolepsy is caused by a mutation in the hypocretin (orexin) receptor 2 gene. Cell 98:365–376

Lin L, Faraco J, Li R, Kadotani H, Rogers W, Lin XY, Qiu XH, de Jong PJ, Nishino S, Mignot E (1999b) The sleep disorder canine narcolepsy is caused by a mutation in the hypocretin (orexin) receptor 2 gene. Cell 98:365–376

Lu J, Jhou TC, Saper CB (2006a) Identification of wake-active dopaminergic neurons in the ventral periaqueductal gray matter. J Neurosci 26:193–202

Lu J, Sherman D, Devor M, Saper CB (2006b) A putative flip-flop switch for control of REM sleep. Nature 441:589–594

Maret S, Faraguna U, Nelson AB, Cirelli C, Tononi G (2011) Sleep and waking modulate spine turnover in the adolescent mouse cortex. Nat Neurosci 14:1418–1420

Marshall L, Helgadottir H, Molle M, Born J (2006) Boosting slow oscillations during sleep potentiates memory. Nature 444:610–613

Mieda M, Hasegawa E, Kisanuki YY, Sinton CM, Yanagisawa M, Sakurai T (2011) Differential roles of orexin receptor-1 and −2 in the regulation of non-REM and REM sleep. J Neurosci 31:6518–6526

Mileykovskiy BY, Kiyashchenko LI, Siegel JM (2005) Behavioral correlates of activity in identified hypocretin/orexin neurons. Neuron 46:787–798

Mochizuki T, Crocker A, McCormack S, Yanagisawa M, Sakurai T, Scammell TE (2004) Behavioral state instability in orexin knock-out mice. J Neurosci 24:6291–6300

Modirrousta M, Mainville L, Jones BE (2005) Orexin and MCH neurons express c-Fos differently after sleep deprivation vs. recovery and bear different adrenergic receptors. Eur J Neurosci 21:2807–2816

Monnier M, Fallert M, Battacharya IC (1967) The waking action of histamine. Experientia 23:21–22

Monsalve GC, Van Buskirk C, Frand AR (2011) LIN-42/PERIOD controls cyclical and developmental progression of *C. elegans* molts. Curr Biol 21:2033–2045

Morrow JD, Vikraman S, Imeri L, Opp MR (2008) Effects of serotonergic activation by 5-hydroxytryptophan on sleep and body temperature of C57BL/6J and interleukin-6-deficient mice are dose and time related. Sleep 31:21–33

Mukhametov LM, Supin AY, Polyakova IG (1977) Interhemispheric asymmetry of the electroencephalographic sleep patterns in dolphins. Brain Res 134:581–584

Mukhametov LM, Lyamin OI, Polyakova IG (1985) Interhemispheric asynchrony of the sleep EEG in northern fur seals. Experientia 41:1034–1035

Naylor E, Bergmann BM, Krauski K, Zee PC, Takahashi JS, Vitaterna MH, Turek FW (2000) The circadian clock mutation alters sleep homeostasis in the mouse. J Neurosci 20:8138–8143

Nicholson AN, Pascoe PA, Stone BM (1985) Histaminergic systems and sleep. Studies in man with H1 and H2 antagonists. Neuropharmacology 24:245–250

Nicol SC, Andersen NA, Phillips NH, Berger RJ (2000) The echidna manifests typical characteristics of rapid eye movement sleep. Neurosci Lett 283:49–52

Nicolau MC, Akaarir M, Gamundi A, Gonzalez J, Rial RV (2000) Why we sleep: the evolutionary pathway to the mammalian sleep. Prog Neurobiol 62:379–406

Obal F Jr, Alfoldi P, Cady AB, Johannsen L, Sary G, Krueger JM (1988) Growth hormone-releasing factor enhances sleep in rats and rabbits. Am J Physiol 255:R310–R316
Oh Y, Jang D, Sonn JY, Choe J (2013) Histamine-HisCl1 receptor axis regulates wake-promoting signals in *Drosophila melanogaster*. PLoS One 8, e68269
Parisky KM, Agosto J, Pulver SR, Shang Y, Kuklin E, Hodge JJ, Kang K, Liu X, Garrity PA, Rosbash M et al (2008) PDF cells are a GABA-responsive wake-promoting component of the *Drosophila* sleep circuit. Neuron 60:672–682
Peyron C, Faraco J, Rogers W, Ripley B, Overeem S, Charnay Y, Nevsimalova S, Aldrich M, Reynolds D, Albin R et al (2000) A mutation in a case of early onset narcolepsy and a generalized absence of hypocretin peptides in human narcoleptic brains. Nat Med 6:991–997
Piper DC, Upton N, Smith MI, Hunter AJ (2000) The novel brain neuropeptide, orexin-A, modulates the sleep–wake cycle of rats. Eur J Neurosci 12:726–730
Prober DA, Rihel J, Onah AA, Sung RJ, Schier AF (2006) Hypocretin/orexin overexpression induces an insomnia-like phenotype in zebrafish. J Neurosci 26:13400–13410
Qu DQ, Ludwig DS, Gammeltoft S, Piper M, Pelleymounter MA, Cullen MJ, Mathes WF, Przypek J, Kanarek R, Maratos-Flier E (1996) A role for melanin-concentrating hormone in the central regulation of feeding behaviour. Nature 380:243–247
Raizen DM, Zimmerman JE, Maycock MH, Ta UD, You YJ, Sundaram MV, Pack AI (2008) Lethargus is a *Caenorhabditis elegans* sleep-like state. Nature 451:569–572
Rasch B, Buchel C, Gais S, Born J (2007) Odor cues during slow-wave sleep prompt declarative memory consolidation. Science 315:1426–1429
Rattenborg NC (2006) Evolution of slow-wave sleep and palliopallial connectivity in mammals and birds: a hypothesis. Brain Res Bull 69:20–29
Rattenborg NC, Martinez-Gonzalez D (2014) Avian versus mammalian sleep: the fruits of comparing apples and oranges. Curr Sleep Med Rep 1:55–63
Rattenborg NC, Lima SL, Amlaner CJ (1999) Facultative control of avian unihemispheric sleep under the risk of predation. Behav Brain Res 105:163–172
Rattenborg NC, Amlaner CJ, Lima SL (2000) Behavioral, neurophysiological and evolutionary perspectives on unihemispheric sleep. Neurosci Biobehav Rev 24:817–842
Rechtschaffen A, Bergmann BM (2002) Sleep deprivation in the rat: an update of the 1989 paper. Sleep 25:18–24
Rechtschaffen A, Bergmann BM, Everson CA, Kushida CA, Gilliland MA (1989) Sleep deprivation in the rat: X. Integration and discussion of the findings. Sleep 12:68–87
Renier C, Faraco JH, Bourgin P, Motley T, Bonaventure P, Rosa F, Mignot E (2007) Genomic and functional conservation of sedative-hypnotic targets in the zebrafish. Pharmacogenet Genomics 17:237–253
Rial RV, Nicolau MC, Gamundi A, Akaarir M, Garau C, Aparicio S, Tejada S, Moranta D, Gene L, Esteban S (2007) Comments on evolution of sleep and the palliopallial connectivity in mammals and birds. Brain Res Bull 72:183–186
Rial RV, Akaarir M, Gamundi A, Garau C, Aparicio S, Tejada S, Gene L, Nicolau MC, Esteban S (2008) Wake and sleep hypothalamic regulation in diurnal and nocturnal chronotypes. J Pineal Res 45:225–226
Rial RV, Akaarir M, Gamundi A, Nicolau C, Garau C, Aparicio S, Tejada S, Gene L, Gonzalez J, De Vera LM et al (2010) Evolution of wakefulness, sleep and hibernation: from reptiles to mammals. Neurosci Biobehav Rev 34:1144–1160
Sakai K, Crochet S, Onoe H (2001) Pontine structures and mechanisms involved in the generation of paradoxical (REM) sleep. Arch Ital Biol 139:93–107
Sakurai T, Amemiya A, Ishii M, Matsuzaki I, Chemelli RM, Tanaka H, Williams SC, Richardson JA, Kozlowski GP, Wilson S et al (1998) Orexins and orexin receptors: a family of hypothalamic neuropeptides and G protein-coupled receptors that regulate feeding behavior. Cell 92:573–585
Saper CB, Fuller PM, Pedersen NP, Lu J, Scammell TE (2010) Sleep state switching. Neuron 68:1023–1042
Shaw PJ, Cirelli C, Greenspan RJ, Tononi G (2000) Correlates of sleep and waking in *Drosophila melanogaster*. Science 287:1834–1837

Shaw PJ, Tononi G, Greenspan RJ, Robinson DF (2002) Stress response genes protect against lethal effects of sleep deprivation in *Drosophila*. Nature 417:287–291

Siegel JM, Manger PR, Nienhuis R, Fahringer HM, Pettigrew JD (1996) The echidna *Tachyglossus aculeatus* combines REM and non-REM aspects in a single sleep state: implications for the evolution of sleep. J Neurosci 16:3500–3506

Singh K, Ju JY, Walsh MB, DiIorio MA, Hart AC (2014) Deep conservation of genes required for both *Drosophila melanogaster* and *Caenorhabditis elegans* sleep includes a role for dopaminergic signaling. Sleep 37:1439–1451

Sundvik M, Kudo H, Toivonen P, Rozov S, Chen YC, Panula P (2011) The histaminergic system regulates wakefulness and orexin/hypocretin neuron development via histamine receptor H1 in zebrafish. FASEB J 25:4338–4347

Szentirmai E, Krueger JM (2006) Central administration of neuropeptide Y induces wakefulness in rats. Am J Physiol Regul Integr Comp Physiol 291:R473–R480

Taheri S, Lin L, Austin D, Young T, Mignot E (2004) Short sleep duration is associated with reduced leptin, elevated ghrelin, and increased body mass index. PLoS Med 1, e62

Takahashi Y, Kipnis DM, Daughaday WH (1968) Growth hormone secretion during sleep. J Clin Invest 47:2079–2090

Takahashi K, Lin JS, Sakai K (2008) Neuronal activity of orexin and non-orexin waking-active neurons during wake-sleep states in the mouse. Neuroscience 153:860–870

Thannickal TC, Moore RY, Nienhuis R, Ramanathan L, Gulyani S, Aldrich M, Cornford M, Siegel JM (2000) Reduced number of hypocretin neurons in human narcolepsy. Neuron 27:469–474

Tsunematsu T, Ueno T, Tabuchi S, Inutsuka A, Tanaka KF, Hasuwa H, Kilduff TS, Terao A, Yamanaka A (2014) Optogenetic manipulation of activity and temporally controlled cell-specific ablation reveal a role for MCH neurons in sleep/wake regulation. J Neurosci 34:6896–6909

Van Buskirk C, Sternberg PW (2007) Epidermal growth factor signaling induces behavioral quiescence in *Caenorhabditis elegans*. Nat Neurosci 10:1300–1307

Van Cauter E, Knutson KL (2008) Sleep and the epidemic of obesity in children and adults. Eur J Endocrinol Eur Fed Endocrine Soc 159(suppl 1):S59–S66

Vanni-Mercier G, Sakai K, Lin JS, Jouvet M (1989) Mapping of cholinoceptive brainstem structures responsible for the generation of paradoxical sleep in the cat. Arch Ital Biol 127:133–164

Verret L, Goutagny R, Fort P, Cagnon L, Salvert D, Leger L, Boissard R, Salin P, Peyron C, Luppi PH (2003) A role of melanin-concentrating hormone producing neurons in the central regulation of paradoxical sleep. BMC Neurosci 4:19

Viola AU, Archer SN, James LM, Groeger JA, Lo JC, Skene DJ, von Schantz M, Dijk DJ (2007) PER3 polymorphism predicts sleep structure and waking performance. Curr Biol 17:613–618

Vyazovskiy VV, Olcese U, Hanlon EC, Nir Y, Cirelli C, Tononi G (2011) Local sleep in awake rats. Nature 472:443–447

White JG, Albertson DG, Anness MA (1978) Connectivity changes in a class of motoneurone during the development of a nematode. Nature 271:764–766

Willie JT, Chemelli RM, Sinton CM, Tokita S, Williams SC, Kisanuki YY, Marcus JN, Lee C, Elmquist JK, Kohlmeier KA et al (2003) Distinct narcolepsy syndromes in Orexin receptor-2 and Orexin null mice: molecular genetic dissection of non-REM and REM sleep regulatory processes. Neuron 38:715–730

Wisor JP, Nishino S, Sora I, Uhl GH, Mignot E, Edgar DM (2001) Dopaminergic role in stimulant-induced wakefulness. J Neurosci 21:1787–1794

Wisor JP, O'Hara BF, Terao A, Selby CP, Kilduff TS, Sancar A, Edgar DM, Franken P (2002) A role for cryptochromes in sleep regulation. BMC Neurosci 3:20

Wu MN, Ho K, Crocker A, Yue Z, Koh K, Sehgal A (2009) The effects of caffeine on sleep in *Drosophila* require PKA activity, but not the adenosine receptor. J Neurosci 29:11029–11037

Xie L, Kang H, Xu Q, Chen MJ, Liao Y, Thiyagarajan M, O'Donnell J, Christensen DJ, Nicholson C, Iliff JJ et al (2013) Sleep drives metabolite clearance from the adult brain. Science 342:373–377

Yang G, Gan WB (2012) Sleep contributes to dendritic spine formation and elimination in the developing mouse somatosensory cortex. Dev Neurobiol 72:1391–1398

Yang G, Lai CS, Cichon J, Ma L, Li W, Gan WB (2014) Sleep promotes branch-specific formation of dendritic spines after learning. Science 344:1173–1178

Yokogawa T, Marin W, Faraco J, Pezeron G, Appelbaum L, Zhang J, Rosa F, Mourrain P, Mignot E (2007) Characterization of sleep in zebrafish and insomnia in hypocretin receptor mutants. PLoS Biol 5, e277

Young JZ (1981) The life of vertebrates, 3rd edn. Clarendon, Oxford

Yuan Q, Joiner WJ, Sehgal A (2006) A sleep-promoting role for the *Drosophila* serotonin receptor 1A. Curr Biol 16:1051–1062

Zhang Z, Ferretti V, Guntan I, Moro A, Steinberg EA, Ye Z, Zecharia AY, Yu X, Vyssotski AL, Brickley SG et al (2015) Neuronal ensembles sufficient for recovery sleep and the sedative actions of alpha2 adrenergic agonists. Nat Neurosci 18:553–561

Zhdanova IV, Wang SY, Leclair OU, Danilova NP (2001) Melatonin promotes sleep-like state in zebrafish. Brain Res 903:263–268

Chapter 16
Prefrontal Anatomical Architecture and Top-Down Behavioral Control in Human and Nonhuman Primates

Yosuke Morishima

Abstract Primates, including humans, have great cognitive capability, can adapt to their environments, and have a brain is characterized by a large volume of prefrontal cortex. In this chapter, I provide an overview on how the primate prefrontal cortex differs from that of other species, and I discuss the structural similarities and differences of the prefrontal cortex among primate species. In particular, I discuss how the human prefrontal cortex has unique characteristics among primate species. I also provide an overview of the neural mechanisms of top-down control of visual attention and discuss how cognitive research in human and non-human primates is integrated to understand brain mechanisms. In summary, I will argue that comparative and integrative approaches aid the understanding of the biological basis of human cognition.

Keywords Attention • Comparative anatomy • Executive function • Human brain • Nonhuman primate • Prefrontal cortex • Top-down control

16.1 Introduction

Adaptive behavior requires flexible processing of external information for behavioral guidance according to the demands of environments (Miller and Cohen 2001; Real 1991). Organisms receive external information through sensory organs, but not all information received can be actively processed at the same time. For example, we cannot report two consecutively presented visual stimuli, called a phenomenon of attentional blink (Marois and Ivanoff 2005). The capacity of working memory is limited to five to seven items (Baddeley et al. 1974; Kane and Engle 2002). Therefore, the first step in achieving adaptive behavior is to select relevant information and ignore irrelevant information pertaining to the demand at hand. In the next step, the

---

Y. Morishima (✉)
Division of Systems Neuroscience of Psychopathology, Translational Research Center, University Hospital of Psychiatry, University of Bern, Bolligenstrasse 111, 3000 Bern, Switzerland
e-mail: yosuke.morishima@puk.unibe.ch

S. Shigeno et al. (eds.), *Brain Evolution by Design*, Diversity and Commonality in Animals, DOI 10.1007/978-4-431-56469-0_16

processed information is utilized and evaluated according to the current demand. In the last step, a relevant action is selected and executed. Higher cognitive functions, such as attention, working memory, planning, decision making, memory retrieval, motivated behavior, and social cognition, can be fractionated into these steps that are served by the prefrontal cortex (PFC) (Desimone and Duncan 1995; Fujii et al. 2009; Fuster 1988; Goldman-Rakic 1996; Grabenhorst and Rolls 2011; Iacoboni et al. 2004; Karafin et al. 2004; Miller and Cohen 2001; Passingham 2008; Passingham and Wise 2012; Rudebeck et al. 2006; Rushworth et al. 2007a, b; Sakai 2008; Sallet et al. 2011; Squire et al. 2013; Tsujimoto et al. 2011).

The PFC is a part of the cerebral cortex that is observed only in mammals. Amphibians and lower organisms do not have a PFC (Butler and Hodos 2005; Wilczynski 2009). Birds have a brain area analogous to the mammalian PFC, but the area does not have the columnar structure of the cerebral cortex that is found in mammals (Fuster 1988; Wilczynski 2009). Among mammalian species, primates have a considerably larger PFC compared to other species (Roth and Dicke 2005). Although the function of the PFC has been studied extensively in humans and nonhuman primates, rodents have recently been used to study PFC functions as well.

The goal of this chapter is to discuss human cognition by taking a comparative approach of brain architecture. To this end, in the first part of this chapter, I provide an overview of the anatomical architecture of the PFC in human and nonhuman primates and illustrate the differences from other nonprimate species. Then, I discuss the structural commonality and differences of the PFC among primate species. In the second part of this chapter, I review recent research on visual attention. I chose visual attention in this chapter, because researchers have investigated the neural mechanism of the top-down control of visual attention with human and nonhuman primates, and studies with different species have been nicely integrated to understand human cognition.

16.2 Anatomical Architecture of the PFC in Humans and Nonhuman Primates

16.2.1 The Definition of the PFC

First, I clarify the definition of the PFC. In this chapter, I define the PFC as a part of the cerebral cortex that covers cortical areas in the frontal lobe, except for motor areas, such as the primary motor area, premotor, and supplementary motor area. Therefore, the PFC is different from the frontal lobe; the frontal lobe is separated from the parietal lobe by the central sulcus and separated from the temporal lobe by the lateral sulcus (in humans, the Sylvian fissure). Within the primate PFC, distinct types of cytoarchitecture have been observed: granular, dysgranular, and agranular parts of the PFC (Fuster 1988; Passingham and Wise 2012). These three cytoarchitecture types are differentiated by the cell density of the internal granular layer; that is, layer IV. The granular PFC has an evident layer IV, whereas few neuronal cell bodies are observed in the agranular PFC. The dysgranular PFC shows an immature layer IV (Öngür and Price 2000; Petrides and Pandya 1999, 2007). Because of the subtle differences, the dysgranular PFC was included in Brodmann's classical definition of the PFC (Brodmann 1909). Rodents are recognized to have a PFC, but no evidence has been shown for the existence of the granular PFC in rodents (Öngür and Price 2000; Preuss 1995; Uylings et al. 1990). Brodmann thus did not acknowledge the existence of the PFC in rodents, although the presence of a PFC in rodents is currently being considered (Preuss 1995; Uylings et al. 1990, 2003). Although the granular PFC is observed in a limited number of species, the agranular PFC is observed in lower mammals, such as rodents. Hence, the agranular PFC is considered to be an evolutionarily older PFC and the granular PFC is considered to contribute to the complex cognitive processes of primates.

16.2.2 Subdivision of the PFC in Humans, Macaques, and Marmosets

The PFC is not a homogeneous structure. Hence, the PFC is subdivided into subregions based on macroscopic and microscopic criteria. The cerebral cortex of mammals comprises six laminated layers based on distinct neuronal cell density layers. The patterns of cell densities in the six layers are not uniform across the entire cerebral cortex but are considerably different among different brain areas. Therefore, the cerebral cortex has been parcellated into subregions based on distinct patterns of cytoarchitecture. Brodmann's initial parcellation of human brain areas is still of value (Brodmann 1909), and a similar naming rule has been adopted to label the brain areas of other primate species (Walker 1940). Figure 16.1 illustrates the parcellated brain areas of the PFC in humans, macaque monkeys, and marmoset monkeys (Burman et al. 2006; Burman and Rosa 2009; Öngür and Price 2000; Petrides and Pandya 1999, 2007). As shown in Fig. 16.1, the agranular PFC is observed in the more caudal part of the lateral and medial PFC, whereas the granular PFC is observed in the more rostral part of the lateral and medial PFC. The dysgranular PFC is located in between the two. In addition, the more rostral part of the orbital PFC has a thicker layer IV along with the anterior posterior line (Mackey and Petrides 2010). This rule is uniformly applied to humans, macaque monkeys, and marmoset monkeys. Although such comprehensive cytoarchitecture information is not available for other primate species, this rule could be applied to other primate species. The commonality in the naming rules among species helps to easily relate knowledge obtained from one species to another, in particular between humans and macaque monkeys, as both species have been extensively used to study the neural basis of cognition. The functional similarity of a particular area between human and nonhuman primates is ubiquitously reported. In addition, more

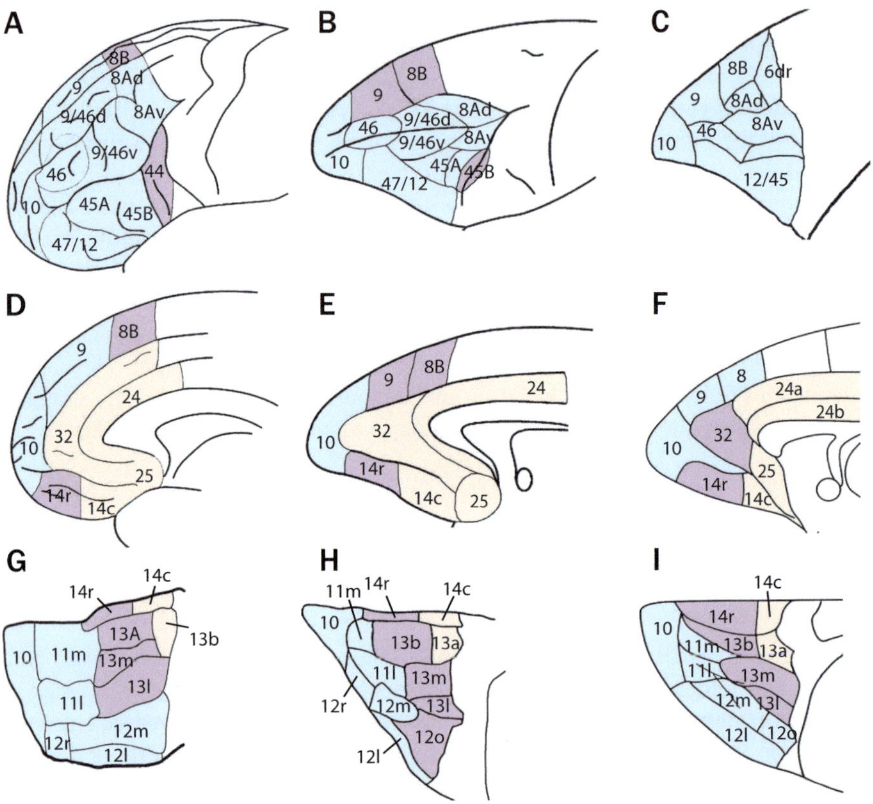

Fig. 16.1 Cytoarchitectural subdivision of the prefrontal cortex (PFC): the subdivisions of the PFC in humans (**a**, **d**, **g**), macaque monkeys (**b**, **e**, **h**), and marmoset monkeys (**c**, **f**, **i**). Each lateral (**a–c**), medial (**d–f**), and orbital (**g–i**) surface is described. Subdivision of the surface of human prefrontal cortex is shown. *Colors* indicate the existence of granular layer IV. The granular, dysgranular, and agranular PFC areas are depicted by *blue*, *purple*, and *yellow*, respectively. As is the convention, the rostral ventral prefrontal area is described as area 47/12 or 45/12 in the lateral surface, whereas the area is labeled as area 12 in the orbital surface (Petrides and Pandya 2002) (Figures are adapted and modified from Petrides and Pandya (1999) (**a**, **b**, **d**, **e**), Öngür and Price (2000) (**g**, **h**), Burman et al. (2006) (**c**), and Burman and Rosa (2009) (**f**, **i**))

recent studies have examined the pattern of anatomical and functional connectivity among brain areas and found certain similarities between human and primate brains (De Schotten et al. 2012; Margulies et al. 2009; Neubert et al. 2014, 2015; Ramnani et al. 2006; Rilling et al. 2008).

The second labeling rule of the PFC that is widely used is based on gross anatomical features, and the PFC is subdivided into five areas: the dorsolateral prefrontal cortex (DLPFC), ventrolateral prefrontal cortex (VLPFC), medial prefrontal cortex (MFPC), orbitofrontal cortex (OFC), and frontopolar cortex (otherwise called the anterior PFC or polar PFC). These definitions are convenient to roughly note the locus of activation, recording, or lesion. However, the nomenclature causes misunderstanding because each subregion covers broad cortical areas. For example, the DLPFC sometimes refers to the core area of executive control function, and BA 9 and BA 46 are involved in the process, but at the same time, the DLPFC in this definition includes BA 8, BA 9, BA 10, and BA 46 (Cieslik et al. 2012). Because of the confusion of the terminology, in particular in human neuroimaging studies, the foci of activation should be carefully confirmed to interpret the results. We sometimes encounter a study that refers to DLPFC activation as the control mechanism, but the activation was observed in the posterior extreme of area 6, which is less likely to be involved in executive control. Hence, the precise definition of brain areas is prerequisite for the interpretation of functional associations.

16.2.3 Evolutionarily Relevant Neural Index for Intelligence

What feature of the brain accounts for our intelligence? In this section, I overview potential neural indices that may account for the interspecies variation of intelligence: Total brain volume, total brain volume scaled by body size, PFC volume scaled by total cortex, and granular PFC size scaled by total cortex (Table 16.1). The brain volume of primates is generally larger than that of most other mammalian species (Roth and Dicke 2005). However, there are many species whose brain is larger than that of humans, such as elephants and whales. Hence, the total brain volume does not represent the uniqueness of primate intelligence, and another measure is needed to relate the structural variation among species with intelligence. If we scale total brain volume by body size, it may potentially account for the difference in intelligence across mammalian species (Fig. 16.2a) (Van Dongen 1998), but that is not yet an ideal measure to account for the association between brain structure and intelligence. In fact, the scaled total brain volume of humans

Table 16.1 Possible evolutionarily relevant neural indices for intelligence

Index	Relevance	Reason	References
Total brain volume	No	Elephants and whales have larger brain than primates	Roth and Dicke (2005)
Brain volume scaled by body size	No	Mice have larger relative brain size than most primates	Van Dongen (1998)
PFC volume scaled by total cerebral cortex	No	Rats have larger relative brain size than primates, including humans	Uylings et al. (1990)
Granular PFC volume scaled by frontal lobe size	Yes	Consistent with interspecies variation of intelligence	Elston (2003, 2006)

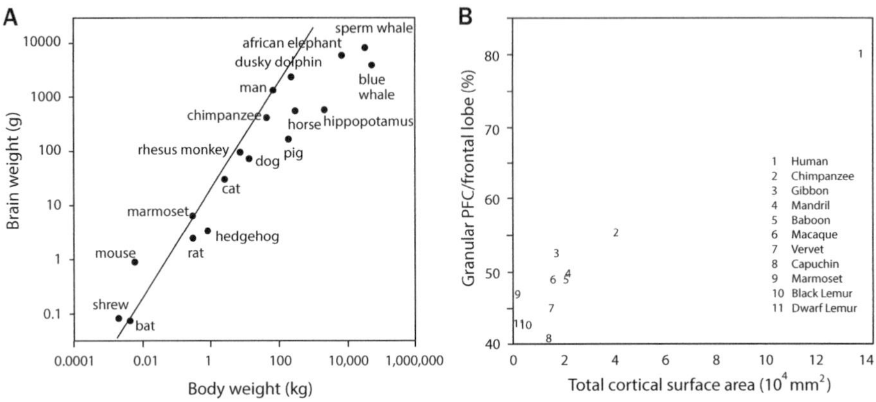

Fig. 16.2 Relative brain and prefrontal size. (**a**) Relationship between the body weight and brain weight in mammalian species. The *oblique line* is drawn through the relative brain size equal to humans. The relative brain weight of marmoset monkeys is equal to humans, and the relative brain weight of the shrew is larger than for humans (Figure adapted and modified from Van Dongen 1998). (**b**) The relationship between the total cortical surface and fraction of the granular PFC in primates. In the human brain, the granular PFC occupies 80 % of the volume of the frontal lobe, whereas the granular PFC occupies between 40 and 60 % of the frontal lobe in other species. Among nonhuman primates, apes have a larger relative granular PFC area compared with other nonhuman primates (Figure adapted and modified from Elston et al. 2006)

is lower than that of mice and is comparable to that of marmosets (Fig. 16.2a). As we are discussing the PFC in this chapter, we consider the size of the PFC across species. If we look at the PFC volume scaled by total cerebral cortex, among primate species, humans have the highest relative PFC size, followed by apes, macaque monkeys, and marmosets (Semendeferi et al. 2002; Uylings et al. 1990). However, the relative PFC size of rats is considerably larger than that of humans (Uylings et al. 1990). Thus, PFC volume scaled by total cerebral cortex is not an appropriate measure to account for the association of brain size measures with primate intelligence. However, the relatively large PFC size in rats may account for the fact that rats have been widely used to study cognition.

Because the existence of granular cell layer IV in the primate PFC is a unique characteristic of the primate PFC, we can look at the size of the granular area of the PFC. The granular PFC is observed mostly in primates and among primate species, and the human granular PFC is extraordinarily large relative to the frontal lobe size (Elston et al. 2006) (Fig. 16.2b). Thus, granular PFC volume scaled by frontal lobe size is a promising neural index for the interspecies variation of intelligence in mammals. The granular part of the PFC is considered to characterize the uniqueness of the primate PFC. In primates, the granular PFC occupies a considerable part of the total cortex (from 8 % in marmosets to 29 % in humans), whereas in *Pteropus* (a bat) and Leporidae (rabbits), the granular area makes up approximately 2 % of the total cortex (Elston 2003). These results suggest that the larger granular PFC reflects the high intelligence of human and nonhuman primates but also results raises a new question: what is the function of the granular layer IV? From the microscopic perspective, neurons in layer IV consist of glutamatergic stellate neurons and GABAergic interneurons (de Almeida and Mengod 2008), which send excitatory and inhibitory inputs to layers II, III, and V. As layer IV receives input from the thalamus (Sherman 2007), layer IV in the granular PFC may integrate information from subcortical structures, local cortical inputs, and cortical inputs from other brain areas mediated via the thalamus. In sum, the existence and size of the granular PFC support the large capacity for processing information and may reflect the high intelligence of humans and nonhuman primates. Studying prefrontal functions of primates is thus important for the understanding not only of human intelligence but also of the biological origins of intelligence.

16.2.4 What Is Special About the Human Brain?

In this section, we consider the unique characteristics of the human PFC compared to that of other primate species. As already discussed, the human PFC has a considerably larger granular PFC (80 %) compared to that of other primates (40–55 %) (Elston et al. 2006). Other than the relative granular PFC size, the white matter volume is disproportionally larger in the human PFC than in that of other primates (Schoenemann et al. 2005). One point of caution when interpreting this study is that it adopted an uncommon definition of the PFC. Because it is hard to examine the cytoarchitecture of the cerebral cortex with MRI, the study specified the PFC as "all portions of the frontal cortex anterior to the genu of the corpus callosum, in a plane perpendicular to the line connecting the anterior and posterior commissures" (Schoenemann et al. 2005). The uncommon definition may create bias to estimate the PFC size across species. In addition to the size of the granular PFC area, the human PFC has area 44. Although area 44 is categorized as the dysgranular PFC, area 44 is observed in neither macaque monkeys nor marmoset monkeys and it is a part of Broca's area, which is crucial for the production and comprehension of language. In humans, the volume of the left areas 44 and 45 is larger than that of right areas 44 and 45 (Amunts et al. 1999), which apparently reflects the dominance of language processing in the left side of the brain. Surprisingly, this asymmetry of the posterior part of the inferior frontal gyrus is observed not only in humans but also in great ape species (Cantalupo and Hopkins 2001), suggesting that the ape's brain emerged with the anatomical capacity to process complex language systems. Other studies of the size of area 10 in humans and apes included that of Semendeferi et al. (2001), which found that the volume of area 10 relative to the total brain volume in humans is double that in other apes. Area 10 is located in the most rostral part of the PFC and is situated at the top of the prefrontal hierarchy (Badre and D'Esposito 2007; Botvinick 2008; Koechlin and Summerfield 2007). In fact, area 10 is involved in more complex processes of executive function, such as complex planning and decision making, metacognition, and lie telling (Burgess et al. 2007; Fleming et al. 2010; Karim et al. 2010; Passingham and Wise 2012; Ramnani and Owen 2004;

Stuss and Knight 2013; Tsujimoto et al. 2011), and, in particular, the larger gray matter volume in area 10 is associated with better metacognitive ability (Fleming et al. 2010). Therefore, a large area 10 in the human PFC may reflect the capacity of human intelligence.

At a microscopic level, several unique characteristics of the human PFC have been reported. First, Elston and colleagues studied the number of dendritic spines in seven primate species including humans (Elston et al. 2006). They showed that the number of spines is significantly larger in the human granular PFC than that in the granular PFC of other primates, although this difference was not observed in visual areas V1 and V2. Because spines are the locus for the reception of neural inputs to pyramidal neurons, these results suggested that human prefrontal neurons integrate more complex information. Second, a study by Sherwood and colleagues compared the density of glial cells relative to neurons in subregions of the PFC across 18 primate species including humans, gorillas, and chimpanzees (Sherwood et al. 2006). They showed that glial cell density relative to neuron density in the human PFC is considerably higher than in other primates. However, in that study, they did not examine glial cell densities other than in the PFC. Therefore, it was not clear whether the increased density of glial cells is specific to prefrontal areas. The third unique characteristic observed in the human PFC is a large number of bipolar spindle-shaped neurons, called von Economo neurons (VENs). We discuss VENs in the next section.

16.2.5 Von Economo Neurons

The seminal work by von Economo and Koskinas described large bipolar neurons in layer V of the anterior cingulate cortex (von Economo and Koskinas 1925). In humans, VENs are located in the anterior part of the insular cortex, the anterior cingulate cortex, and the rostral part of the middle cingulate cortex (Allman et al. 2010). VENs are mostly observed in anthropoids, such as humans, chimpanzees, gorillas, and bonobos, and, more recently, the existence of VENs was also reported in the anterior insular cortex of macaque monkeys (Evrard et al. 2012). Although the neurons were also reported as large spindle-shape neurons in other reports, including those of Betz and Cajal (Betz 1881; Cajal 1904), von Economo comprehensively described the morphology and location of these neurons, and these neurons are apt to be called VENs. The morphological characteristics of VENs are unique. VENs are projection neurons characterized as being elongated, gradually tapering, with large-sized somas, and are considerably larger than neighboring pyramidal neurons (Nimchinsky et al. 1995, 1999). The functional role of VENs is still elusive. It has been speculated that VENs might be involved in the processing of emotionally and socially salient information relevant to human awareness (Allman et al. 2010) because VENs are observed only in anthropoids and macaque monkeys and are located in the rostral part of the ACC and insula, areas highly associated with social cognition (Frith and Frith 2012). Although most of these speculations are not supported by experimental evidence, Santos and colleagues have reported the atypical characteristics of VENs in autistic children, suggesting the association of VENs with the theory of mind (Santos et al. 2011). The study investigated the number of VENs in the frontoinsular cortex of healthy and autistic children. Autistic spectrum disorder is a developmental psychiatric disorder characterized by difficulty in social interactions and atypical processing (mostly hypersensitive) of sensory information (American Psychiatric Association 2013; Frith 1991). That study showed that children with autism had a higher ratio of VENs to pyramidal neurons in the frontoinsular cortex than in healthy children, but the density of VENs was comparable between the two groups, indicating that the numbers of both VENs and other neurons were increased. In addition, the study reported that some VENs of autistic children had atypically swollen somata. The results suggested the association of VENs with autism, but it is still elusive how VENs contribute to social interactions, which are impaired in autism. One possible consideration that may be derived from comparative approaches is behavioral differences among species with VENs can be associated with differences in distributions of VENs among those species. In summary, we overviewed the unique characteristics of the human PFC; that is, larger granular PFC and area 10, and higher amounts of white matter volume, spines, and VENs. All those characteristics are quantitatively different from other primate species but not qualitatively different. Human cognition can be partly understood by interpolation from other species, but the complexity of brain networks may add new cognitive processes that can be studied only in humans.

16.3 Executive Function in Human and Nonhuman Primates

16.3.1 Top-Down Attentional Control Mechanism

In the second half of this chapter, we discuss the common neural mechanism of executive function in human and nonhuman primates. To discuss how studies of human and nonhuman primates are integrated to understand executive function, we consider neural mechanisms of attention in humans and macaque monkeys because attention has been extensively studied in both species. Attention refers to a cognitive process to enhance and suppress the processing of external information (Corbetta and Shulman 2002; Desimone and Duncan 1995; Kastner and Ungerleider 2000; Reynolds and Chelazzi 2004). Selective processing is important from an evolutionary perspective. First, living organisms need to process information relevant to survival according to environmental demands. Second, the processing of irrelevant information is not efficient from the perspective of energy consumption. Therefore, the prioritization of information processing is important for fitness in different environments. In addition, information relevant to a behavioral goal depends on the environmental demands at hand, and the brain therefore needs to flexibly control the focus of attention in favor of the processing of currently relevant information.

In the following sections, I overview the neural mechanisms of top-down control of attention in humans and nonhuman primates and discuss the unified neural mechanisms of attentional systems.

16.3.2 Existence of Top-Down Signals

Seminal work by Moran and Desimone has demonstrated that attention to a visual stimulus enhances the selectivity of neuronal response to the stimulus when distracting visual stimuli are simultaneously presented (Moran and Desimone 1985). This study suggested that attention creates control signals that enhance the selective processing of sensory information. However, it was not clear how the control signals that enhance the selective processing of visual stimuli are generated. One possibility is that these control signals are driven by the external sensory stimulus itself (bottom-up driven account). Another explanation is that the control signals were generated in the brain according to the demands of the environment without actual sensory input (top-down account). To distinguish between top-down and bottom-up processing, researchers examined the modulation of brain activity in sensory areas without any sensory stimuli. Kastner and colleagues have shown activation in the extrastriate cortex in the absence of visual stimuli when human subjects covertly directed attention to a peripheral location where visual stimuli were expected to be presented (Kastner et al. 1999) (Fig. 16.3a). In addition, they also observed

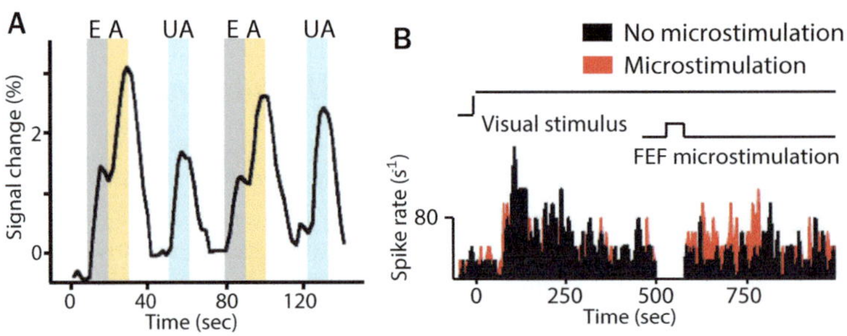

Fig. 16.3 Top-down signals during visual attention in humans and macaque monkeys. (**a**) Brain activity is measured by functional magnetic resonance imaging (MRI) in the human visual cortex. In one condition (*EA*), subjects were told to attend the forthcoming visual stimulus (*E* period, *gray*) and then an actual visual stimulus was presented (*A* period, *yellow*). In the other condition, the visual stimulus is presented, but subjects were not told to attend to the visual stimulus (*UA* period, *blue*). Note that the visual cortex was activated by the instruction to attend to the display without any visual stimulus (Figure is adapted and modified from Kastner et al. 1999). (**b**) Concurrent recording of visual neurons in area V4 and microstimulation of FEF. Neuronal firing of V4 neurons is recorded while presenting the visual stimulus to the receptive field of the V4 neurons. In the middle of the presentation of visual stimuli, microstimulation is briefly applied to FEF (*red*). The FEF microstimulation increased the firing rate of the V4 neurons after offset of microstimulation (Figure adapted from Moore and Armstrong 2003)

activation in the PFC during orienting attention without a visual stimulus. Tomita and colleagues also demonstrated that neurons in the inferior temporal (IT) cortex, which selectively respond to a specific visual stimulus, increase the firing rate when monkeys retrieve a visual item during a paired-association memory retrieval task (Tomita et al. 1999). In that task, monkeys were first trained to learn the association of two visual items, and then they were presented with one of the visual items (stim 1) and were instructed to retrieve the other visual item (stim 2). The firing rate of IT neurons selectively responding to stim2 was increased after the presentation of the first visual item but before the presentation of the paired item, indicating an increase in the firing rate of the neurons without a visual stimulus. In addition, they also slit the posterior half of the corpus callosum and again presented the first visual stimulus in the single visual hemifield. Then, they showed that the firing rate of the ipsilateral IT neurons were increased during retrieval. Because a visual stimulus presented in one visual hemifield is processed in the contralateral side of the visual cortex, no bottom-up visual information will be transmitted to the ipsilateral side of the visual cortex. Therefore, the neuronal firing in the ipsilateral side is supposed to be driven by signals via the PFC. Those results clearly showed the existence of top-down control signals originating from the PFC.

The studies left an open question of whether top-down signals originating from the PFC causally influence activity in the visual areas. Moore and colleagues addressed that question (Moore and Armstrong 2003) in a study in which they recorded neuronal activity from the V4 visual area while electronically stimulating the frontal eye field (FEF). The FEF is a subdivision of the PFC and is directly connected with V4 (Stanton et al. 1995). The FEF is also shown to have involvement in top-down attentional processes (Corbetta and Shulman 2002; Maunsell and Treue 2006). Electronic stimulation of the FEF increased the neuronal response of V4 neurons to visual stimuli (Fig. 16.3b), whereas the enhancement was not observed in the absence of visual stimulation. Another study by Ekstorm and colleagues reported that the electronic stimulation of the FEF increased the activation in visual areas only in the presence of visual stimuli (Ekstrom et al. 2008). The results suggested that the FEF causally enhances the activity of visual areas in the presence of visual stimuli. The results may sound contradictory to the initial report by Kaster et al., wherein there was activation in the extrastriate cortex in the absence of visual stimuli (Kastner et al. 1999). These results may be explained by the differences in underlying mechanisms between artificial electronic stimulation and endogenous top-down control. The other possible explanation is that both studies with monkeys did not instruct them to orient attention to a specific location during electronic stimulation in the absence of visual stimuli.

16.3.3 Flexible Control of Top-Down Signals

Studies with macaque monkeys have clearly shown that the top-down signals originating from the PFC can causally enhance the activity of visual areas in the presence of visual stimuli. However, to achieve adaptive behavior, the biased processing of sensory information needs to be flexibly controlled according to environmental demands at hand. This concept gives an assumption that top-down signals are flexibly controlled in favor of processing information relevant to behavioral goals. To address this question, our previous study measured neuronal signal transmission from the PFC to visual areas by concurrently using transcranical magnetic stimulation (TMS) and electroencephalography (EEG) (Morishima et al. 2009). The rationale of the TMS-EEG method is the following. A weak single-pulse TMS excites neurons in the stimulated area, and the impulse of stimulated neurons spreads along the anatomical connections. The spread of TMS-evoked potentials depends on the excitability of neuronal populations at the time of stimulation. In fact, other studies have shown that sleep state-dependent changes in neural transmission occur in cortical networks; neuronal signal transmission is broken down during the slow wave (deep stage) sleep (Massimini et al. 2005). We invented an experimental paradigm to study the flexible nature of top-down control signals during visual attention. Subjects were presented with visual stimuli comprising a face image and a moving stripe (Fig. 16.4a). The subjects were required to discriminate either the gender of the face image or the direction of the moving stripe according to a task instruction cue, which was presented at the beginning of each trial. Because the visual stimuli always comprised a face and moving stripe, the bottom-up information was equivalent in both tasks. In addition, we manipulated the level of preparation by changing the time interval between the task instruction cue and the target visual stimuli. Thus, the subject could fully prepare for the forthcoming target visual stimuli in long preparation trials, but they could not prepare in the short cue–target interval trials. We then applied TMS on the FEF (frontal eye field), shown as a source of top-down signals in primate studies (Ekstrom et al. 2008; Moore and Armstrong 2003) and examined TMS-evoked potentials in visual areas. We showed that TMS-evoked EEG potentials in the occipital area and neural transmission from the FEF to the visual cortex were changed depending on the task engaged, and the change was observed only when the subject had a long preparation time (long preparation trials) (Fig. 16.4b). We have also shown that TMS-evoked potentials spread to the middle temporal (MT) area, which is specialized to process visual motion, when performing a motion discrimination task, whereas TMS-evoked potentials spread to the fusiform face area, which is specialized to process face information, when performing the face discrimination task (Fig. 16.4c). In sum, the results suggested that the signal transmission from the FEF to visual areas is flexibly controlled according to the task demands at hand.

16.3.4 Prefrontal-Visual Interactions Through Neural Oscillations

In the previous two sections, I overviewed the causal evidence that top-down control signals originating from the PFC modulate selective processing in visual areas

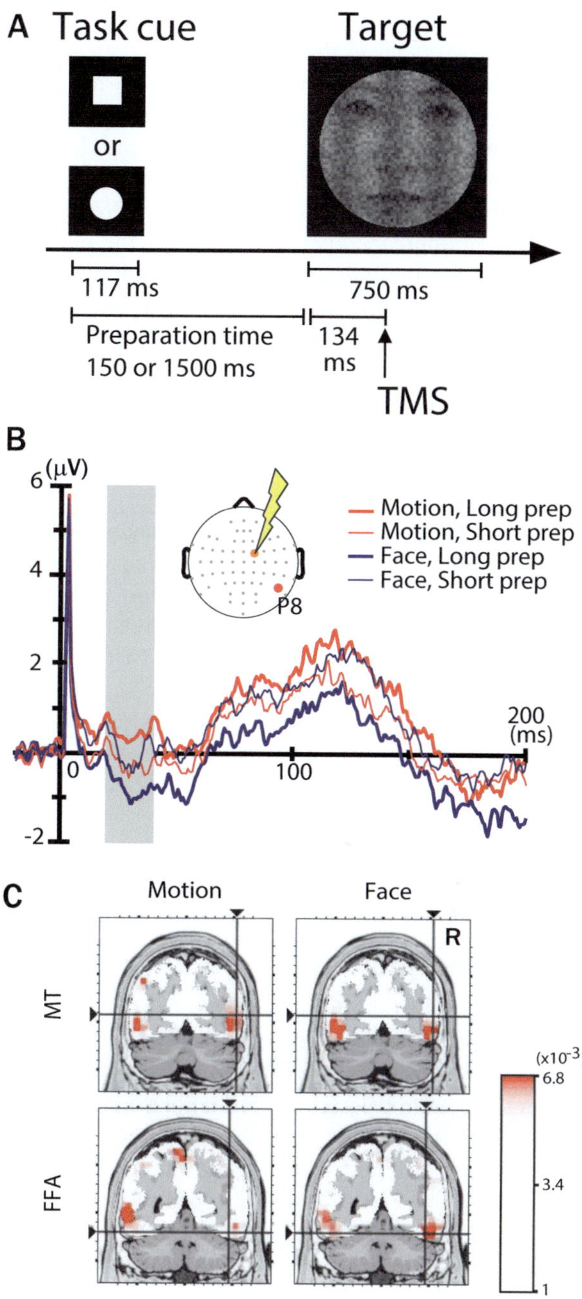

Fig. 16.4 Flexible prefrontal transmission to visual areas. (**a**) Tasks used in experiments. Subjects were asked to indicate either the direction of the grating motion or the gender of a face image according to a task cue. Between the task cue and target stimulus, there were two types of preparation, long and short. This allows for the manipulation of the level of preparation for the target stimulus. After 134 ms of target onset, a TMS pulse was delivered to the FEF. (**b**) TMS-evoked EEG potentials from the visual cortex. A TMS was delivered to the FEF (*orange dot* in the head map), and its response was recorded at occipital area P8 (*red dot* in the head map). TMS-evoked potentials are different between the face (*thick blue line*) and motion (*thick red line*) conditions in the long preparation trials, whereas the difference disappeared in the short preparation trials (*thin lines*). (**c**) TMS-evoked source brain activity was estimated in the motion processing MT area and face processing FFA area. TMS-evoked source activity in the MT area was enhanced in the motion task, whereas activity in the FFA was enhanced in the face task (Figures adapted and modified from Morishima et al. 2009)

during visual attention. The studies with macaque monkeys have shown that the FEF causally influences the activity of visual areas in the presence of visual stimuli. The studies with human subjects have shown that the FEF can flexibly control the prefrontal influence over visual areas according to task demands. The next question arising from these studies is how top-down signals are triggered and how interactions between the FEF and visual areas are established.

A study by Gregoriou and colleagues addressed this question (Gregoriou et al. 2009). They simultaneously recorded from the FEF and V4 visual area of macaque monkeys while the monkeys were performing a visual attention task. Then, they identified the receptive field of each recording site. When the receptive fields of both FEF and V4 recording sites were overlapped, neuronal coherence at the gamma frequency band (40–60 Hz) between the FEF and V4 was significantly increased. By contrast, the coherence was not modulated when the two receptive fields were not overlapped. The results are in fact consistent with the human EEG study that demonstrated occipital-frontal gamma-band coherence (Doesburg et al. 2008). Gregoriou and colleagues also examined the direction of influence between the FEF and V4 as a function of time by using Granger causality analysis. Granger causality analysis calculates time-lagged correlations, and it therefore provides the direction of influence from one time series to another. They found that the Granger causality from the FEF to V4 was significantly modulated by visual attention at 110 ms after the onset of the visual stimulus, whereas Granger causality from V4 to the FEF was modulated at 160 ms after the onset of the visual stimulus. The results suggest that attentional effects on top-down influence precede bottom-up influence. In summary, the PFC, in particular the FEF, is a major source of top-down signals that enhance the processing of relevant information at hand. Prefrontal influence followed by bottom-up influence at a gamma band frequency establishes the prefrontal-visual interactions.

16.4 Conclusion

In this chapter, I first showed that the primate brain is characterized by a considerably large volume of the PFC and overviewed the commonality and difference of prefrontal anatomical architecture among human and nonhuman primates. Those studies indicate that the human PFC is not qualitatively different but rather is quantitatively different from that of other primate species, which suggests that the human PFC would have the capacity to address more complex information compared to other primates, and the study of other primate species will contribute to the understanding of human cognition. In fact, as I discussed in the second part, human and primate experiments are successfully integrated to understand the neural mechanisms of visual attention. However, to understand cognitive processes unique to humans, we need to study the cognitive processes that involve the brain areas unique to humans, such as area 10 and area 44.

References

Allman JM, Tetreault NA, Hakeem AY, Manaye KF, Semendeferi K, Erwin JM, Park S, Goubert V, Hof PR (2010) The von Economo neurons in frontoinsular and anterior cingulate cortex in great apes and humans. Brain Struct Funct 214:495–517

Amunts K, Schleicher A, Bürgel U, Mohlberg H, Uylings HB, Zilles K (1999) Broca's region revisited: cytoarchitecture and intersubject variability. J Comp Neurol 412:319–341

American Psychiatric Association (2013) Diagnostic and Statistical Manual of Mental Disorders (DSM-5®). American Psychiatric Association, Washington, DC

Baddeley AD, Hitch GJ, et al (1974) Working memory. Psychol Learn Motiv 8:47–89

Badre D, D'Esposito M (2007) Functional magnetic resonance imaging evidence for a hierarchical organization of the prefrontal cortex. J Cogn Neurosci 19:2082–2099

Betz W (1881) Ueber die feinere Structur der Gehirnrinde des Menschen. Zentralbl Med Wiss 19:193–195

Botvinick MM (2008) Hierarchical models of behavior and prefrontal function. Trends Cogn Sci 12:201–208

Brodmann K (1909) Vergleichende Lokalisationslehre der Groshirnrinde. Barth, Leipzig

Burgess PW, Gilbert SJ, Dumontheil I (2007) Function and localization within rostral prefrontal cortex (area 10). Philos Trans R Soc B Biol Sci 362:887–899

Burman KJ, Rosa MG (2009) Architectural subdivisions of medial and orbital frontal cortices in the marmoset monkey (*Callithrix jacchus*). J Comp Neurol 514:11–29

Burman KJ, Palmer SM, Gamberini M, Rosa MG (2006) Cytoarchitectonic subdivisions of the dorsolateral frontal cortex of the marmoset monkey (*Callithrix jacchus*), and their projections to dorsal visual areas. J Comp Neurol 495:149–172

Butler AB, Hodos W (2005) Comparative vertebrate neuroanatomy: evolution and adaptation. Wiley, New York

Cajal R (1904) Textura del sistema nervioso del hombre y de los vertebrados. Alicante Instituto de Neurociencias, Madrid

Cantalupo C, Hopkins WD (2001) Asymmetric Broca's area in great apes. Nature 414:505–505. doi:10.1038/35107134

Cieslik EC, Zilles K, Caspers S, Roski C, Kellermann TS, Jakobs O, Langner R, Laird AR, Fox PT, Eickhoff SB (2012) Is there "one" DLPFC in cognitive action control? Evidence for heterogeneity from co-activation-based parcellation. Cereb Cortex 23(11):2677–2689

Corbetta M, Shulman GL (2002) Control of goal-directed and stimulus-driven attention in the brain. Nat Rev Neurosci 3:201–215

de Almeida J, Mengod G (2008) Serotonin 1A receptors in human and monkey prefrontal cortex are mainly expressed in pyramidal neurons and in a GABAergic interneuron subpopulation: implications for schizophrenia and its treatment. J Neurochem 107:488–496

De Schotten MT, Dell'Acqua F, Valabregue R, Catani M (2012) Monkey to human comparative anatomy of the frontal lobe association tracts. Cortex 48:82–96

Desimone R, Duncan J (1995) Neural mechanisms of selective visual attention. Annu Rev Neurosci 18:193–222

Doesburg SM, Roggeveen AB, Kitajo K, Ward LM (2008) Large-scale gamma-band phase synchronization and selective attention. Cereb Cortex 1991(18):386–396

Ekstrom LB, Roelfsema PR, Arsenault JT, Bonmassar G, Vanduffel W (2008) Bottom-up dependent gating of frontal signals in early visual cortex. Science 321:414–417

Elston GN (2003) Cortex, cognition and the cell: new insights into the pyramidal neuron and prefrontal function. Cereb Cortex 13:1124–1138

Elston GN, Benavides-Piccione R, Elston A, Zietsch B, Defelipe J, Manger P, Casagrande V, Kaas JH (2006) Specializations of the granular prefrontal cortex of primates: implications for cognitive processing. Anat Rec A Discov Mol Cell Evol Biol 288:26–35

Evrard HC, Forro T, Logothetis NK (2012) Von Economo neurons in the anterior insula of the macaque monkey. Neuron 74:482–489

Fleming SM, Weil RS, Nagy Z, Dolan RJ, Rees G (2010) Relating introspective accuracy to individual differences in brain structure. Science 329:1541–1543
Frith U (1991) Autism and asperger syndrome. Cambridge University Press, Cambridge/New York
Frith CD, Frith U (2012) Mechanisms of social cognition. Annu Rev Psychol 63:287–313
Fujii N, Hihara S, Nagasaka Y, Iriki A (2009) Social state representation in prefrontal cortex. Soc Neurosci 4:73–84
Fuster JM (1988) Prefrontal cortex. Springer, Berlin
Goldman-Rakic PS (1996) The prefrontal landscape: implications of functional architecture for understanding human mentation and the central executive. Philos Trans R Soc Lond B Biol Sci 351:1445–1453
Grabenhorst F, Rolls ET (2011) Value, pleasure and choice in the ventral prefrontal cortex. Trends Cogn Sci 15:56–67
Gregoriou GG, Gotts SJ, Zhou H, Desimone R (2009) High-frequency, long-range coupling between prefrontal and visual cortex during attention. Science 324:1207–1210
Iacoboni M, Lieberman MD, Knowlton BJ, Molnar-Szakacs I, Moritz M, Throop CJ, Fiske AP (2004) Watching social interactions produces dorsomedial prefrontal and medial parietal BOLD fMRI signal increases compared to a resting baseline. Neuroimage 21:1167–1173
Kane MJ, Engle RW (2002) The role of prefrontal cortex in working-memory capacity, executive attention, and general fluid intelligence: an individual-differences perspective. Psychon Bull Rev 9:637–671
Karafin MS, Tranel D, Adolphs R (2004) Dominance attributions following damage to the ventromedial prefrontal cortex. J Cogn Neurosci 16:1796–1804
Karim AA, Schneider M, Lotze M, Veit R, Sauseng P, Braun C, Birbaumer N (2010) The truth about lying: inhibition of the anterior prefrontal cortex improves deceptive behavior. Cereb Cortex 1991(20):205–213
Kastner S, Ungerleider LG (2000) Mechanisms of visual attention in the human cortex. Annu Rev Neurosci 23:315–341
Kastner S, Pinsk MA, De Weerd P, Desimone R, Ungerleider LG (1999) Increased activity in human visual cortex during directed attention in the absence of visual stimulation. Neuron 22:751–761
Koechlin E, Summerfield C (2007) An information theoretical approach to prefrontal executive function. Trends Cogn Sci 11:229–235
Mackey S, Petrides M (2010) Quantitative demonstration of comparable architectonic areas within the ventromedial and lateral orbital frontal cortex in the human and the macaque monkey brains. Eur J Neurosci 32:1940–1950
Margulies DS, Vincent JL, Kelly C, Lohmann G, Uddin LQ, Biswal BB, Villringer A, Castellanos FX, Milham MP, Petrides M (2009) Precuneus shares intrinsic functional architecture in humans and monkeys. Proc Natl Acad Sci USA 106:20069–20074
Marois R, Ivanoff J (2005) Capacity limits of information processing in the brain. Trends Cogn Sci 9:296–305
Massimini M, Ferrarelli F, Huber R, Esser SK, Singh H, Tononi G (2005) Breakdown of cortical effective connectivity during sleep. Science 309:2228–2232
Maunsell JH, Treue S (2006) Feature-based attention in visual cortex. Trends Neurosci 29:317–322
Miller EK, Cohen JD (2001) An integrative theory of prefrontal cortex function. Annu Rev Neurosci 24:167–202
Moore T, Armstrong KM (2003) Selective gating of visual signals by microstimulation of frontal cortex. Nature 421:370–373
Moran J, Desimone R (1985) Selective attention gates visual processing in the extrastriate cortex. Science 229:782–784
Morishima Y, Akaishi R, Yamada Y, Okuda J, Toma K, Sakai K (2009) Task-specific signal transmission from prefrontal cortex in visual selective attention. Nat Neurosci 12:85–91
Neubert F-X, Mars RB, Thomas AG, Sallet J, Rushworth MF (2014) Comparison of human ventral frontal cortex areas for cognitive control and language with areas in monkey frontal cortex. Neuron 81:700–713

Neubert F-X, Mars RB, Sallet J, Rushworth MFS (2015) Connectivity reveals relationship of brain areas for reward-guided learning and decision making in human and monkey frontal cortex. Proc Natl Acad Sci USA 112(20):E2695–E2704

Nimchinsky EA, Vogt BA, Morrison JH, Hof PR (1995) Spindle neurons of the human anterior cingulate cortex. J Comp Neurol 355:27–37

Nimchinsky EA, Gilissen E, Allman JM, Perl DP, Erwin JM, Hof PR (1999) A neuronal morphologic type unique to humans and great apes. Proc Natl Acad Sci USA 96:5268–5273

Öngür D, Price JL (2000) The organization of networks within the orbital and medial prefrontal cortex of rats, monkeys and humans. Cereb Cortex 10:206–219

Passingham R (2008) What is special about the human brain? Oxford University Press, Oxford/New York

Passingham RE, Wise SP (2012) The neurobiology of the prefrontal cortex: anatomy, evolution, and the origin of insight. Oxford University Press, Oxford

Petrides M, Pandya DN (1999) Dorsolateral prefrontal cortex: comparative cytoarchitectonic analysis in the human and the macaque brain and corticocortical connection patterns. Eur J Neurosci 11:1011–1036

Petrides M, Pandya DN (2002) Comparative cytoarchitectonic analysis of the human and the macaque ventrolateral prefrontal cortex and corticocortical connection patterns in the monkey. Eur J Neurosci 16:291–310

Petrides M, Pandya DN (2007) Efferent association pathways from the rostral prefrontal cortex in the macaque monkey. J Neurosci 27:11573–11586

Preuss T (1995) Do rats have prefrontal cortex? The Rose-Woolsey-Akert program reconsidered. J Cogn Neurosci 7:1–24

Ramnani N, Owen AM (2004) Anterior prefrontal cortex: insights into function from anatomy and neuroimaging. Nat Rev Neurosci 5:184–194

Ramnani N, Behrens TE, Johansen-Berg H, Richter MC, Pinsk MA, Andersson JL, Rudebeck P, Ciccarelli O, Richter W, Thompson AJ, et al (2006) The evolution of prefrontal inputs to the cortico-pontine system: diffusion imaging evidence from macaque monkeys and humans. Cereb Cortex 16:811–818

Real LA (1991) Animal choice behavior and the evolution of cognitive architecture. Science 253:980–986

Reynolds JH, Chelazzi L (2004) Attentional modulation of visual processing. Annu Rev Neurosci 27:611–647

Rilling JK, Glasser MF, Preuss TM, Ma X, Zhao T, Hu X, Behrens TEJ (2008) The evolution of the arcuate fasciculus revealed with comparative DTI. Nat Neurosci 11:426–428

Roth G, Dicke U (2005) Evolution of the brain and intelligence. Trends Cogn Sci 9:250–257

Rudebeck PH, Buckley MJ, Walton ME, Rushworth MF (2006) A role for the macaque anterior cingulate gyrus in social valuation. Science 313:1310–1312

Rushworth MF, Behrens TE, Rudebeck PH, Walton ME (2007a) Contrasting roles for cingulate and orbitofrontal cortex in decisions and social behaviour. Trends Cogn Sci 11:168–176

Rushworth MF, Buckley MJ, Behrens TE, Walton ME, Bannerman DM (2007b) Functional organization of the medial frontal cortex. Curr Opin Neurobiol 17:220–227

Sakai K (2008) Task set and prefrontal cortex. Annu Rev Neurosci 31:219–245

Sallet J, Mars RB, Noonan MP, Andersson JL, O'Reilly JX, Jbabdi S, Croxson PL, Jenkinson M, Miller KL, Rushworth MFS (2011) Social network size affects neural circuits in macaques. Science 334:697–700

Santos M, Uppal N, Butti C, Wicinski B, Schmeidler J, Giannakopoulos P, Heinsen H, Schmitz C, Hof PR (2011) von Economo neurons in autism: a stereologic study of the frontoinsular cortex in children. Brain Res 1380:206–217

Schoenemann PT, Sheehan MJ, Glotzer LD (2005) Prefrontal white matter volume is disproportionately larger in humans than in other primates. Nat Neurosci 8:242–252. doi:10.1038/nn1394

Semendeferi K, Armstrong E, Schleicher A, Zilles K, Van Hoesen GW (2001) Prefrontal cortex in humans and apes: a comparative study of area 10. Am J Phys Anthropol 114:224–241

Semendeferi K, Lu A, Schenker N, Damasio H (2002) Humans and great apes share a large frontal cortex. Nat Neurosci 5:272–276. doi:10.1038/nn814
Sherman SM (2007) The thalamus is more than just a relay. Curr Opin Neurobiol 17:417–422
Sherwood CC, Stimpson CD, Raghanti MA, Wildman DE, Uddin M, Grossman LI, Goodman M, Redmond JC, Bonar CJ, Erwin JM, et al (2006) Evolution of increased glia–neuron ratios in the human frontal cortex. Proc Natl Acad Sci USA 103:13606–13611
Squire RF, Noudoost B, Schafer RJ, Moore T (2013) Prefrontal contributions to visual selective attention. Annu Rev Neurosci 36:451–466
Stanton GB, Bruce CJ, Goldberg ME (1995) Topography of projections to posterior cortical areas from the macaque frontal eye fields. J Comp Neurol 353:291–305
Stuss DT, Knight RT (2013) Principles of frontal lobe function. Oxford University Press, Oxford/New York
Tomita H, Ohbayashi M, Nakahara K, Hasegawa I, Miyashita Y (1999) Top-down signal from prefrontal cortex in executive control of memory retrieval. Nature 401:699–703
Tsujimoto S, Genovesio A, Wise SP (2011) Frontal pole cortex: encoding ends at the end of the endbrain. Trends Cogn Sci 15:169–176
Uylings HB, van Eden CG, et al (1990) Qualitative and quantitative comparison of the prefrontal cortex in rat and in primates, including humans. Prog Brain Res 85:31–62
Uylings HB, Groenewegen HJ, Kolb B (2003) Do rats have a prefrontal cortex? Behav Brain Res 146:3–17
Van Dongen PAM (1998) Brain size in vertebrates. In: The central nervous system of vertebrates. Springer, Berlin, pp 2099–2134
von Economo CF, Koskinas GR (1925) Die Cytoarchitektonik der Hirnrinde der erwachsenen Menschen. Springer, Berlin
Walker AE (1940) A cytoarchitectural study of the prefrontal area of the macaque monkey. J Comp Neurol 73:59–86
Wilczynski W (2009) Evolution of the brain in amphibians. In: Encyclopedia of neuroscience. Springer, Berlin, pp 1301–1305

Part IV
Models and Designs

Chapter 17
Organisational Principles of Connectomes: Changes During Evolution and Development

Roman Bauer and Marcus Kaiser

Abstract The set of neural connections in an organism is now called the connectome. Using recent noninvasive techniques such as diffusion tensor imaging and traditional invasive techniques for tract tracing has uncovered a wide range of connectomes from *Caenorhabditis elegans* and *Drosophila melanogaster* to cat, mouse, rat, macaque, and human. We can therefore start to look at organisational changes during evolution. At the same time cell lineage information and measurements at different time steps allow us to observe network changes during individual, ontogenetic development. We find that the structure of a network is closely linked to its function, with distinct functional components first leading to network modules and, after the rise of further specialisation, to a hierarchical architecture with modules at different levels of network organisation. We first describe concepts that are used to characterize complex networks, then move on to briefly discuss computational models for development and evolution, before showing how network features change during the evolution and development of brain networks. We conclude with future challenges in the field of connectome development and evolution.

Keywords Complex networks • Connectome • Neuronal network • Network structure • Topology • Modelling • Development • Evolution

---

R. Bauer • M. Kaiser
Interdisciplinary Computing and Complex BioSystems (ICOS) research group, School of Computing Science, Newcastle University, Newcastle upon Tyne NE1 7RU, UK

Institute of Neuroscience, Newcastle University, Newcastle upon Tyne NE1 7RU, UK
e-mail: marcus.kaiser@ncl.ac.uk

S. Shigeno et al. (eds.), *Brain Evolution by Design*, Diversity and Commonality in Animals, DOI 10.1007/978-4-431-56469-0_17

17.1 Representing Brain Connectivity as a Network

The nodes of a neural network can be neurons, populations of neurons, or brain regions, depending on the scale under examination. Synaptic connections between such nodes can be of chemical or electrical nature. Neuronal activity is transmitted in only one direction by chemical synapses (A–>B), whereas electrical synapses allow for bidirectional communication (A<–>B). These networks or graphs can be represented in an adjacency matrix (Fig. 17.1), based on which various measures can be computed. Also, the network structure can be in weighted or binarized form, depending on the knowledge of connection strength (e.g., the number of chemical synapses between two neurons).

The synaptic connectivity reflects the structure of the neural network and shapes its function. Also this functional aspect can be captured using the network formalism, by establishing links between nodes that show similar activity patterns. Such similarity could, for example, be measured in the correlation of the activity patterns between two brain regions or two neurons. Again, the link could be a continuous value of correlation strength, or could be binarized in that connection weights are set to one if the corresponding correlation is above a certain threshold and zero otherwise. Importantly, a functional connection might indicate that two nodes are structurally connected, but it might also arise if both nodes are driven by common input. In this chapter, we focus on the structural connectivity, that is, the 'connectome' (Sporns 2013).

17.2 Properties of Complex Networks

Before we describe the organisation of biological neural networks, we first need to describe some concepts that are used to study complex networks. We only give a brief overview; a more complete list of network measures can be found in (Costa et al. 2007; Rubinov and Sporns 2010; Kaiser 2011).

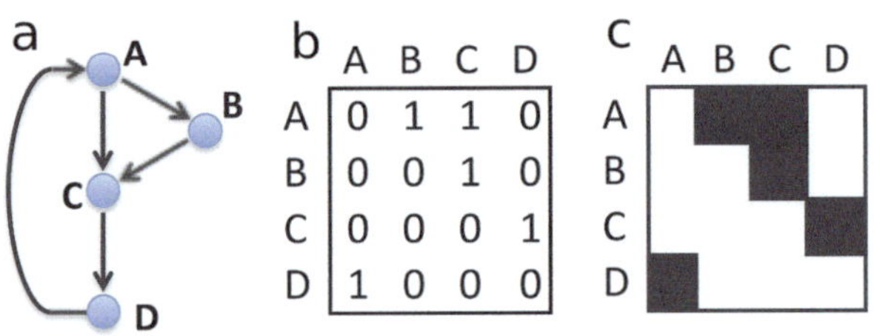

Fig. 17.1 Network representations of neural networks. (**a**) Network with four nodes and feedforward paths (A–>C, A–>B–>C) and feedback (A–>C–>D–>A) loops. (**b**) Representation in an adjacency matrix where '1' represents an existing connection and '0' stands for a connection that either has not been discovered yet or which is known to be absent. (**c**) For visualization, such binary matrices can be represented with *black squares* for existing and *white squares* for nonexisting connections

17.2.1 Modularity

Networks often show topological modules, also called clusters or communities. There is a relatively higher density of connections within modules than between modules. This difference allows rapid flow and integration of information within densely connected modules whereas information flow between modules can only use fewer links that form a potential bottleneck for passing information.

The measure of modularity Q is a reflection of the segregation within a network (Newman 2006), and serves as a tool in identifying the structural modules within. It quantifies how well a parcellation into nonoverlapping modules or communities represents the architecture of a network. Given two parcellations into distinct modules for the same network, the parcellation with the higher value of Q would be preferred.

From a biological aspect, modularity is an evolutionary beneficial network property because it allows for robustness and evolvability (Hintze and Adami 2008). Nonmodular network topologies entail strong interdependence among individual sub-networks, and so local changes can have detrimental effects on a more global level. It is therefore not surprising that modularity is a common feature of biological networks.

17.2.2 Hierarchy

A pervasive property of most complex networks is a hierarchical structure among nodes and/or modules. Usually, hierarchical networks are also modular, and the hierarchical composition can involve different functional levels or temporal orders. For example, a network might consist of several modules, where each module consists of several sub-modules, which again consist of several sub-sub-modules, and so on. A hierarchical structure has been shown to be a fundamental characteristic of many complex systems (Ravasz and Barabási 2003).

17.2.3 Small World (SW) Property

The small world phenomenon (Milgram 1967) refers to the property that two nodes in complex networks often are separated by much fewer edges than what one would expect. Small-world networks can be assessed using two network features (Watts and Strogatz 1998). First, the clustering coefficient describes how well neighbours of a node are connected where neighbours are all nodes that are directly connected to a node. For small-world networks, this proportion of links between neighbours is much higher than for randomly connected networks. Another more recent measure for this local connectivity is local efficiency (Latora and Marchiori 2001). Second, the characteristic path length describes the average number of connections one has to cross to go from one node to another node following the shortest possible path (the one with the lowest number of connections). This measure is only slightly higher than for a randomly organised network. Another more recent measure for this global feature is global efficiency (Latora and Marchiori 2001). For a small-world network, the clustering coefficient is thus much higher while the characteristic path length is comparable to that of a randomly connected network.

To ensure a comparable characteristic path length, small-world networks contain 'short-cuts' that directly link different parts of the network. Using these long-range connections, one can quickly reach different parts of the network over few intermediate links.

Most complex networks are also small-world networks. One main advantage of small-world networks is that they incorporate fast communication within functional modules (i.e., high clustering coefficient), and still allow for reliable and efficient signal propagation to nodes in different modules (i.e. short minimal path length). They also enable easier synchronisation of network activity (Masuda and Aihara 2004).

17.2.4 Scale-Freeness

Many complex networks have been shown to be scale-free or scale-invariant, a property of how the values for the number of connections of a node, its degree, are distributed. For randomly connected networks, the degree of a node will be close to the average degree of all nodes in the network, which means that the degree will be on the same characteristic scale: if the average degree is 10, all network degrees may be in the range of 0–99. On the other hand, scale-free networks do not show a characteristic scale: even if the average degree is 10, some nodes may have a degree of 100, 1000, or higher, thus reaching different orders of magnitude. For scale-free networks, the degree distribution follows the form $P(k) \sim k^{-\gamma}$, where $P(k)$ denotes the probability that a node is linked to k other nodes, and γ is the exponent of this power law. The seminal work of Barabási and Albert (Barabási and Albert 1999) has proposed an abstract model for the growth of such scale-free networks. Since then, many artificial networks and some biological ones have been demonstrated to be scale-free (Jeong et al. 2000). However, for structural neural networks usually only aggregate networks with connections between brain regions rather than between individual neurons have been reconstructed. The only organism for which the complete neuronal network structure is known is the roundworm *Caenorhabditis elegans* (White et al. 1986; Achacoso and Yamamoto 1992). However, a scale-free distribution is not supported in this case (Amaral et al. 2000). It therefore remains to be clarified whether whole-brain structural connectomes are scale-free or not.

17.2.5 Hubs

Scale-free and also other complex networks can have 'hubs,' nodes that participate in many more connections than one would expect. Because of their structural significance, hubs are usually also interesting from a functional point of view (Jeong et al. 2001; Goymer 2008). Studies show that such networks are very robust against random lesions, while being vulnerable towards removal or knockout of hubs (Newman 2003; Warren et al. 2014). This resilience is believed to be advantageous from an evolutionary point of view, which is in accordance with the finding that hubs have been observed in most biological networks.

17.2.6 Rich-Club Organization

Networks with hubs often incorporate rich-club organization, a bias for hubs to connect with one another, rather than with other nodes. It has been suggested that evolution favours both (hubs and rich-club organization) properties because they increase the robustness of networks to random breakdowns (McAuley et al. 2007). Along these lines, rich-club organization supports versatile information processing, allows for the dynamic resource allocation in a context-dependent manner and the collaborative integration of multisensory information (Zamora-Lopez et al. 2010; Collin et al. 2013; Senden et al. 2014).

17.3 Developmental and Evolutionary Patterns

As for other aspects of biology, it is useful to look at connectomes in terms of their evolutionary origins and developmental trajectories. Indeed, evolutionary mechanisms have been linked to topological network properties (Ebbesson 1980, 1984), and a number of complex network growth models have been proposed (Barabási and Albert 1999; Ravasz and Barabási 2003; Louf et al. 2013). Such models are usually framed on a rather abstract level, and it is ongoing work to elucidate how certain complex network properties arise using growth mechanisms based on local information exchange only (Sporns et al. 2004). Along these lines, Kaiser and Hilgetag (2004a, b) and Nisbach and Kaiser (2007) propose a local, spatial growth rule for the self-organization of network topologies with similar clustering coefficients and characteristic path lengths as for structural brain connectivity.

Advances in computing performance have led to the generation of novel research tools (Stanley and Miikkulainen 2002; Torben-Nielsen and De Schutter 2014; Zubler and Douglas 2009; Koene et al. 2009), paving the way for detailed computational models of neural network evolution (Verbancsics and Stanley 2011;

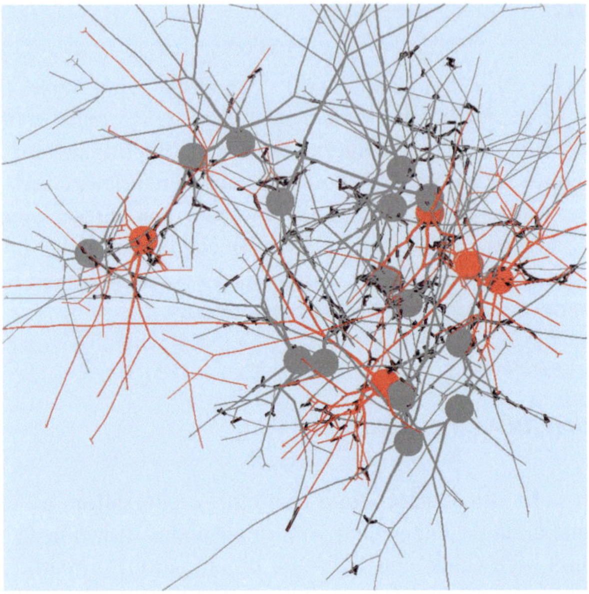

Fig. 17.2 Neuronal network morphology after simulated development (Bauer et al. 2014). The network is composed of differentiated neurons (*grey*). For better visualization, examples of neurons are coloured in *red*. Synapses (*black rectangles*) form at appositions between axons and dendrites

Gauci and Stanley 2010) and development (Bauer et al. 2012, 2014) (Fig. 17.2). In the future, such models will likely allow for a more extensive comparison to biological data across different spatial scales and developmental stages.

In the following, we give a short review of connectome patterns observed across different species and developmental stages.

One of the simplest species possessing a neural network is *Cnidaria*. These animals show a diffuse two-dimensional nerve net for the polyp stage, which, in terms of network science, is called a regular or lattice network (Fig. 17.3a). In such networks, neighbours are well connected (high clustering coefficient) but there are no long-distance connections. We therefore do *not* have a small-world network yet. Such lattice networks are an important part of neural systems such as the retina, as well as some cortical and subcortical layered structures.

For functionally specialized circuits, however, a regular organization is unsuitable. The connectomes of evolutionary higher progressed species therefore have modular topology (Kaiser 2015). Starting with the formation of sensory organs and motor units, neurons aggregate in ganglia. Such ganglia are often not only spatially clustered but also are modular in terms of connectivity (Fig. 17.3b). In this way, ganglia can process one modality without interference from neurons processing different kinds of information. A well-studied example of a modular network is the neuronal network of *C. elegans* (White et al. 1986; Achacoso and Yamamoto 1992), the first organism in which the complete set of neural connections or 'connectome'

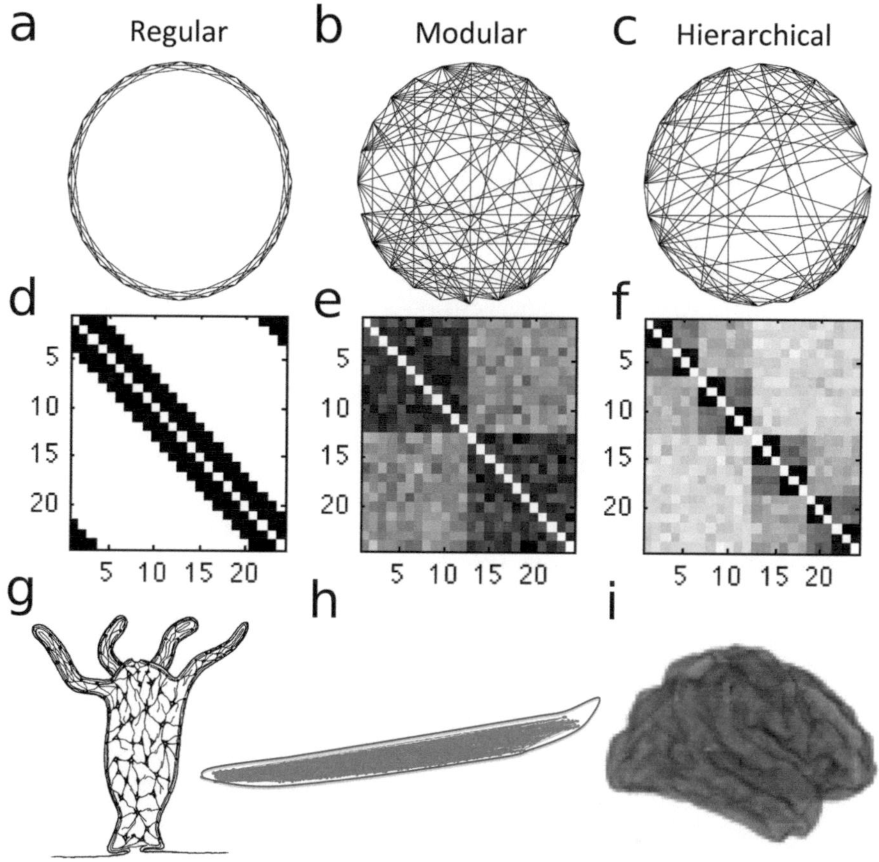

Fig. 17.3 Examples of different types of neural networks (From Kaiser and Varier 2011). (**a**) Regular or lattice network. (**b**) Modular network with two modules. (**c**) Hierarchical network with two modules consisting of two sub-modules each. (**d–f**) Matrices represent the circular network topologies. (**g–i**) Species possessing the afore detailed network architectures. (**g**) Polyp stage of *Hydra* (phylum Cnidaria) shows a nerve net. (**h**) Nematode *C. elegans* shows a modular network. (**i**) Global human neural network traced by diffusion tensor imaging

(Sporns et al. 2005) is known. In addition, the connectome of the fruit fly *Drosophila melanogaster* has been investigated in this respect (Cardona et al. 2010; Ito et al. 2013). Indeed, a high modularity in terms of both spatial proximity as well as topology are observed. However, with increasing complexity of neural processing, a single module for one modality or function is not sufficient; an example is the visual system in the rhesus monkey (macaque) where the visual module consists of two network components: the dorsal pathway for processing object movement and the ventral pathway for processing object features such as colour and form (Young 1992). These networks are hierarchical, because smaller sub-modules are nested within modules (Fig. 17.3c). A hierarchical structure has been observed in most modular brain connectomes (Chatterjee and Sinha 2008; Bassett et al. 2008; Felleman and van Essen 1991; Hilgetag et al. 2000). It has been argued that hierarchical topology embeds a rich dynamic and functional repertoire based on an economical wiring diagram (Meunier et al. 2010; Kaiser et al. 2010; Hilgetag and Hutt 2014).

From a developmental perspective, it is notable that certain poly-sensory and high-order association areas of cortex, which are the most complex areas in terms of their laminar architecture, also exhibit the most complex developmental trajectories (Shaw et al. 2008). Hence, structural and functional hierarchy is reflected also developmentally during brain ontogenesis.

Although there is a trend for spatial neighbours to be in the same module, it is not necessarily always true (da Fontoura et al. 2007). For the visual cortex in primates, for example, the frontal eye field is most closely linked to the topological module related to vision while being in the frontal lobe it is spatially distant from the other visual regions that are part of the occipital lobe.

On a smaller scale, a connectivity pattern composed of modules is the superficial patch system or daisy architecture, a patchy motive of clustered axonal projections in the superficial layers of cortex (Rockland and Lund 1983, 1982; Gilbert and Wiesel 1983). Interestingly, this connectivity pattern has been observed in mammalian species except rodents (Van Hooser et al. 2006). Different hypotheses exist for how it arises during development (Mitchison and Crick 1982; Buzas et al. 2006; Bauer et al. 2012). On the macroscale, modularity has been shown to develop early on during human development (van den Heuvel et al. 2014).

Modular systems (Fig. 17.4) usually have, in addition to the strong intramodular connectivity, sparse links between modules. These intermodular connections can

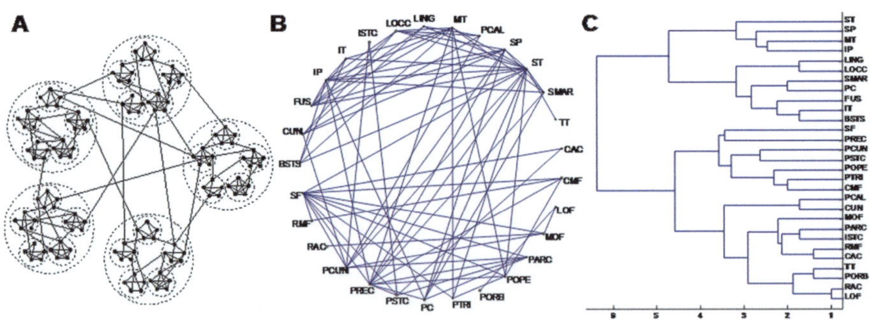

Fig. 17.4 Modules. (a) Schematic hierarchical modular network with modules at two levels, cortical regions (*large circles*) including columns (*smaller circles*) that include individual neurons. For modules, there are more connections within the same module than to other modules. (b) Modular organization of human corticocortical connectivity (Hagmann et al. 2008). Cortical areas were arranged around a circle by evolutionary optimization so that highly interlinked areas were placed close to each other. Note that nodes in the same cluster, having a high structural similarity, also have a similar function. (c) Dendrogram of the same network using hierarchical clustering. A dendrogram running from the root to the leaves (here, from *left* to *right*) consists of branches connecting objects in the tree. The distance of the branching point on the x-axis is the rescaled distance when clusters are combined

serve as shortcuts, hence rendering the average path length between any two to be short. A short path length supports global brain functions, as the distributed entities can efficiently be integrated (Sporns and Zwi 2004). Commonly, it has been shown that many neural networks possess a small-world organization, as, for example, *C. elegans* (Watts and Strogatz 1998), *Drosophila* (Ito et al. 2013), the fibre tract networks between brain regions in the cat (Scannell et al. 1995), the macaque (Hilgetag and Kaiser 2004; Sporns and Zwi 2004), and human brain (He et al. 2007; Hagmann et al. 2008). Recent work using injections of an anterograde tracer yielded the mouse connectome at the mesoscale resolution (between single-neuron and whole-brain imaging resolution) (Oh et al. 2014). Also in this case, a high clustering coefficient and the presence of hubs indicate small-world topology (Sporns and Bullmore 2014). The incorporation of the small-world property across many different species underlines its significance in promoting efficient and fast communication between any two nodes, while keeping the total wiring length comparably small (Karbowski 2001). However, these shortcuts come at rather high metabolic costs, as they require the development of (spatial) long-range connections. Interestingly, (Varier and Kaiser 2011) found that in *C. elegans* the majority of nodes connected via long-range connections are born around the same time. This finding suggests that developmental trajectories could allow for the efficient establishment of neuronal connections, by forming these long-range projections early during development, without the need for energetically expensive guidance cues. Related to this, a recent study on the *C. elegans* and human connectome found that the characteristic path length is longer than what one would expect based on the modularity alone (Kim and Kaiser 2014). Entropy-based considerations indicate that this discrepancy originates from evolutionary pressure towards efficient encoding of developmental processes.

Overall, in modular networks there is a multidimensional trade-off between saving axons, communication costs, and genetic efficacy. As for modularity, small-world organization has been shown to arise early during human brain development (van den Heuvel et al. 2014), and remain stable during brain maturation (Lim et al. 2013).

A further common hallmark of brain networks is the presence of hubs. For mammals such as macaques, subcortical regions such as the hippocampus and amygdala are the most highly connected nodes (Kaiser et al. 2007). The structural centrality of these nodes goes usually hand in hand with functional significance. Additionally, computational studies demonstrate that networks with hubs are more resilient towards random node removal or knockout (Kaiser et al. 2007; Newman 2003; Warren et al. 2014). It is therefore not surprising that many brain diseases usually involve malfunction of hub brain regions (Crossley et al. 2014). Interestingly, hubs are usually in the centre of the brain, forming early during development (Hwang et al. 2012; Varier and Kaiser 2011), and presumably originating earlier during evolution. It has been suggested that the time that is available for connection establishment, from node formation to brain maturation, has a crucial role in the developmentally efficient establishment of hubs in vertebrates as well as in *C. elegans* (Varier and Kaiser 2011).

Interestingly, most brain networks with hubs have been shown to exhibit a rich-club connectivity, for example, in the *C. elegans* (Towlson et al. 2013), cat (de Reus and van den Heuvel 2013), macaque (Harriger et al. 2012), and human brain (van den Heuvel and Sporns 2011). As for hubs, rich-club organization has been shown to arise early during development (Ball et al. 2014; van den Heuvel et al. 2014). Such a developmental priority points towards this connectivity pattern to serve as a developmental scaffold, and to confer several advantages to the network as a whole (Collin et al. 2013; van den Heuvel et al. 2012). This central role in the network is in accordance with pathological rich-club organization observed in neurodevelopmental and other brain diseases (Grayson et al. 2014; Ray et al. 2014; Daianu et al. 2013).

17.4 Conclusion

In summary, complex neural networks become less homogeneous during evolution in line with their increasingly varied functional tasks. Neural systems in species above a certain evolutionary stage show a modular, hierarchical and typically small-world topology with rich-club organization. This shift in structural complexity goes hand in hand with the (functional) specialization of the tasks that the organism performs. This relationship between structure and function is reflected in evolution (Sherwood et al. 2008; Semendeferi et al. 2011), as well as development (Hill et al. 2010). In addition to this functional perspective, certain network features emerged as a consequence of the network topology itself: as brain networks evolved to become more complex, there was the inherent pressure for greater resilience in the face of injury. For example, although hubs and rich-club organization entail the formation of additional axons, they are evolutionarily beneficial as they support such improved resilience towards lesions. Simpler, regular networks seen in primitive life forms have a higher degree of redundancy and are therefore less sensitive (Kaiser and Varier 2011).

Multiple studies have shown that the topology of biological neural networks satisfies a nontrivial 'fitness function,' that is, a combination of multiple natural requirements. Aspects such as wiring economy, fast information flow, richness of dynamics, functional specialization, integrative communication, robustness, and developmental efficiency (Bullmore and Sporns 2012; Kim and Kaiser 2014) influence connectome topologies. Hence, network science serves as a way of understanding the structure and function of neural networks in light of evolutionary pressure. Knowledge of how such multidimensional trade-offs can be satisfied will also likely help in the improved design and planning of many artificial networks.

The early (temporal) formation of many complex network properties underlines their significance and points to a genetically encoded blueprint. Possibly, these initial properties support the reliable unfolding of the developmental process. Interestingly, such characteristic network features are often disrupted in neurodevelopmental and neurodegenerative brain diseases, suggesting a better understanding of the connectome to be valuable from a clinical perspective (Stam 2014; Collin and van den Heuvel 2013). State-of-the-art computational models have been proposed to account for many real-world network characteristics (Barabási and Albert 1999; Ravasz and Barabási 2003). However, these models are usually phrased on a rather abstract level, and not directly relatable to biological mechanisms. The detailed modelling of connectome development will have a major part in the better understanding of the connectomes themselves.

Finally, elucidating the link between topological characteristics and functional processing (e.g., does consciousness structurally correlate with the top level of a hierarchical neural network and where is this 'top' level?) remains one of the main challenges of the field. Because the structure and function of neural networks are mutually influencing each other, insights into their dynamic interaction will constitute a crucial part of this endeavour.

Acknowledgments This work was supported by the Engineering and Physical Sciences Research Council of the United Kingdom (EP/K026992/1) as part of the Human Brain Development Project (http://www.greenbrainproject.org). R.B. was also supported by the Medical Research Council of the United Kingdom (MR/N015037/1).

References

Achacoso TB, Yamamoto WS (1992) AY's neuroanatomy of *C. elegans* for computation. CRC Press, Boca Raton

Amaral LAN, Scala A, Barthélémy M, Stanley HE (2000) Classes of small-world networks. Proc Natl Acad Sci USA 97(21):11149–11152

Ball G, Aljabar P, Zebari S, Tusor N, Arichi T, Merchant N, Robinson EC, Ogundipe E, Rueckert D, Edwards AD, Counsell SJ (2014) Rich-club organization of the newborn human brain. Proc Natl Acad Sci USA 111(20):7456–7461. doi:10.1073/pnas.1324118111

Barabási A-L, Albert R (1999) Emergence of scaling in random networks. Science 286:509–512

Bassett DS, Bullmore E, Verchinski BA, Mattay VS, Weinberger DR, Meyer-Lindenberg A (2008) Hierarchical organization of human cortical networks in health and schizophrenia. J Neurosci 28(37):9239–9248

Bauer R, Zubler F, Hauri A, Muir DR, Douglas RJ (2012) Developmental origin of patchy axonal connectivity in the neocortex: a computational model. Cereb Cortex. doi:10.1093/cercor/bhs327

Bauer R, Zubler F, Pfister S, Hauri A, Pfeiffer M, Muir DR, Douglas RJ (2014) Developmental self-construction and -configuration of functional neocortical neuronal networks. PLoS Comput Biol 10(12):e1003994. doi:10.1371/journal.pcbi.1003994

Bullmore E, Sporns O (2012) The economy of brain network organization. Nat Rev Neurosci 13(5):336–349

Buzas P, Kovacs K, Ferecsko AS, Budd JM, Eysel UT, Kisvarday ZF (2006) Model-based analysis of excitatory lateral connections in the visual cortex. J Comp Neurol 499(6):861–881. doi:10.1002/cne.21134

Cardona A, Saalfeld S, Preibisch S, Schmid B, Cheng A, Pulokas J, Tomancak P, Hartenstein V (2010) An integrated micro- and macroarchitectural analysis of the *Drosophila* brain by computer-assisted serial section electron microscopy. PLoS Biol 8(10):e1000502

Chatterjee N, Sinha S (2008) Understanding the mind of a worm: hierarchical network structure underlying nervous system function in *C. elegans*. Prog Brain Res 168:145–153. doi:10.1016/S0079-6123(07)68012-1

Collin G, van den Heuvel MP (2013) The ontogeny of the human connectome: development and dynamic changes of brain connectivity across the life span. Neuroscientist. doi:10.1177/1073858413503712

Collin G, Sporns O, Mandl RC, van den Heuvel MP (2013) Structural and functional aspects relating to cost and benefit of rich club organization in the human cerebral cortex. Cereb Cortex. doi:10.1093/cercor/bht064

Costa LF, Rodrigues FA, Travieso G, Villas Boas PR (2007) Characterization of complex networks: a survey of measurements. Adv Phys 56(1):167–242

Crossley NA, Mechelli A, Scott J, Carletti F, Fox PT, McGuire P, Bullmore ET (2014) The hubs of the human connectome are generally implicated in the anatomy of brain disorders. Brain Journal Neurol 137(8):2382–2395. doi:10.1093/brain/awu132

da Costa Fontoura L, Kaiser M, Hilgetag CC (2007) Predicting the connectivity of primate cortical networks from topological and spatial node properties. BMC Syst Biol 1:16

Daianu M, Dennis EL, Jahanshad N, Nir TM, Toga AW, Jack CR, Weiner MW, Thompson PM, Initia ADN (2013) Alzheimer's disease disrupts rich club organization in brain connectivity networks. I S Biomed Imaging 2013:266–269

de Reus MA, van den Heuvel MP (2013) Rich club organization and intermodule communication in the cat connectome. J Neurosci 33(32):12929–12939. doi:10.1523/JNEUROSCI.1448-13.2013

Ebbesson SOE (1980) The parcellation theory and its relation to interspecific variability in brain organization, evolutionary and ontogenetic development, and neuronal plasticity. Cell Tissue Res 213:179–212

Ebbesson SOE (1984) Evolution and ontogeny of neural circuits. Behav Brain Sci 7(3):321–331

Felleman DJ, van Essen DC (1991) Distributed hierarchical processing in the primate cerebral cortex. Cereb Cortex 1:1–47

Gauci J, Stanley KO (2010) Autonomous evolution of topographic regularities in artificial neural networks. Neural Comput 22(7):1860–1898. doi:10.1162/neco.2010.06-09-1042

Gilbert CD, Wiesel TN (1983) Clustered intrinsic connections in cat visual cortex. J Neurosci 3(5):1116–1133

Goymer P (2008) Network biology: why do we need hubs? Nat Rev Genet 9(9):650

Grayson DS, Ray S, Carpenter S, Iyer S, Dias TG, Stevens C, Nigg JT, Fair DA (2014) Structural and functional rich club organization of the brain in children and adults. PLoS One 9(2):e88297. doi:10.1371/journal.pone.0088297

Hagmann P, Cammoun L, Gigandet X, Meuli R, Honey CJ, Wedeen VJ, Sporns O (2008) Mapping the structural core of human cerebral cortex. PLoS Biol 6(7):e159. doi:10.1371/journal.pbio.0060159

Harriger L, van den Heuvel MP, Sporns O (2012) Rich club organization of macaque cerebral cortex and its role in network communication. PLoS One 7(9):e46497. doi:10.1371/journal.pone.0046497

He Y, Chen ZJ, Evans AC (2007) Small-world anatomical networks in the human brain revealed by cortical thickness from MRI. Cereb Cortex 17(10):2407–2419. doi:10.1093/cercor/bhl149

Hilgetag CC, Hutt MT (2014) Hierarchical modular brain connectivity is a stretch for criticality. Trends Cogn Sci 18(3):114–115. doi:10.1016/j.tics.2013.10.016

Hilgetag CC, Kaiser M (2004) Clustered organization of cortical connectivity. Neuroinformatics 2(3):353–360

Hilgetag CC, O'Neill MA, Young MP (2000) Hierarchical organization of macaque and cat cortical sensory systems explored with a novel network processor. Philos Trans R Soc Lond Ser B 355:71–89

Hill J, Inder T, Neil J, Dierker D, Harwell J, Van Essen D (2010) Similar patterns of cortical expansion during human development and evolution. Proc Natl Acad Sci USA 107(29):13135–13140. doi:10.1073/pnas.1001229107

Hintze A, Adami C (2008) Evolution of complex modular biological networks. PLoS Comput Biol 4(2):e23. doi:10.1371/journal.pcbi.0040023

Hwang K, Hallquist MN, Luna B (2012) The development of hub architecture in the human functional brain network. Cereb Cortex. doi:10.1093/cercor/bhs227

Ito M, Masuda N, Shinomiya K, Endo K, Ito K (2013) Systematic analysis of neural projections reveals clonal composition of the *Drosophila* brain. Curr Biol 23(8):644–655. doi:10.1016/j.cub.2013.03.015

Jeong H, Tombor B, Albert R, Oltwal ZN, Barabási A-L (2000) The large-scale organization of metabolic networks. Nature 407:651–654

Jeong H, Mason SP, Barabási A-L, Oltvai ZN (2001) Lethality and centrality in protein networks. Nature 411:41–42

Kaiser M (2011) A tutorial in connectome analysis: topological and spatial features of brain networks. Neuroimage 57(3):892–907

Kaiser M (2015) Neuroanatomy: connectome connects fly and Mammalian brain networks. Curr Biol 25(10):R416–R418

Kaiser M, Hilgetag CC (2004a) Modelling the development of cortical networks. Neurocomputing 58–60:297–302

Kaiser M, Hilgetag CC (2004b) Spatial growth of real-world networks. Phys Rev E Stat Nonlinear Soft Matter Phys 69(3):036103

Kaiser M, Varier S (2011) Evolution and development of brain networks: from *Caenorhabditis elegans* to *Homo sapiens*. Netw Comput Neural Syst 22:143–147

Kaiser M, Martin R, Andras P, Young MP (2007) Simulation of robustness against lesions of cortical networks. Eur J Neurosci 25(10):3185–3192. doi:10.1111/j.1460-9568.2007.05574.x

Kaiser M, Hilgetag CC, Kötter R (2010) Hierarchy and dynamics of neural networks. Front Neuroinform 4:112. doi:10.3389/fninf.2010.00112

Karbowski J (2001) Optimal wiring principle and plateaus in the degree of separation for cortical neurons. Phys Rev Lett 86(16):3674–3677. doi:10.1103/PhysRevLett.86.3674

Kim JS, Kaiser M (2014) From *Caenorhabditis elegans* to the human connectome: a specific modular organization increases metabolic, functional and developmental efficiency. Philos Trans R Soc Lond B Biol Sci 369:1653. doi:10.1098/rstb.2013.0529

Koene RA, Tijms B, van Hees P, Postma F, de Ridder A, Ramakers GJ, van Pelt J, van Ooyen A (2009) NETMORPH: a framework for the stochastic generation of large scale neuronal networks with realistic neuron morphologies. Neuroinformatics 7(3):195–210. doi:10.1007/s12021-009-9052-3

Latora V, Marchiori M (2001) Efficient behavior of small-world networks. Phys Rev Lett 87:198701

Lim S, Han CE, Uhlhaas PJ, Kaiser M (2013) Preferential detachment during human brain development: age- and sex-specific structural connectivity in diffusion tensor imaging (DTI). Cereb Cortex Adv Online. doi:10.1093/cercor/bht333

Louf R, Jensen P, Barthelemy M (2013) Emergence of hierarchy in cost-driven growth of spatial networks. Proc Natl Acad Sci USA. doi:10.1073/pnas.1222441110

Masuda N, Aihara K (2004) Global and local synchrony of coupled neurons in small-world networks. Biol Cybern 90(4):302–309. doi:10.1007/s00422-004-0471-9

McAuley JJ, Costa LDF, Caetano TS (2007) Rich-club phenomenon across complex network hierarchies. Appl Phys Lett 91(8). doi:10.1063/1.27723951

Meunier D, Lambiotte R, Bullmore ET (2010) Modular and hierarchically modular organization of brain networks. Front Neurosci 4:200. doi:10.3389/fnins.2010.00200

Milgram S (1967) The small-world problem. Psychol Today 1:60–67

Mitchison G, Crick F (1982) Long axons within the striate cortex: their distribution, orientation, and patterns of connection. Proc Natl Acad Sci USA 79(11):3661–3665

Newman MEJ (2003) The structure and function of complex networks. SIAM Rev 45(2):167–256

Newman ME (2006) Modularity and community structure in networks. Proc Natl Acad Sci USA 103(23):8577–8582. doi:10.1073/pnas.0601602103

Nisbach F, Kaiser M (2007) Developmental time windows for spatial growth generate multiple-cluster small-world networks. Eur Phys J B 58(2):185–191

Oh SW, Harris JA, Ng L, Winslow B, Cain N, Mihalas S, Wang Q, Lau C, Kuan L, Henry AM, Mortrud MT, Ouellette B, Nguyen TN, Sorensen SA, Slaughterbeck CR, Wakeman W, Li Y, Feng D, Ho A, Nicholas E, Hirokawa KE, Bohn P, Joines KM, Peng H, Hawrylycz MJ, Phillips JW, Hohmann JG, Wohnoutka P, Gerfen CR, Koch C, Bernard A, Dang C, Jones AR, Zeng H (2014) A mesoscale connectome of the mouse brain. Nature 508(7495):207–214. doi:10.1038/nature13186

Ravasz E, Barabási A-L (2003) Hierarchical organization in complex networks. Phys Rev E 67:026112

Ray S, Miller M, Karalunas S, Robertson C, Grayson DS, Cary RP, Hawkey E, Painter JG, Kriz D, Fombonne E, Nigg JT, Fair DA (2014) Structural and functional connectivity of the human brain in autism spectrum disorders and attention-deficit/hyperactivity disorder: a rich club-organization study. Hum Brain Mapp 35(12):6032–6048. doi:10.1002/hbm.22603

Rockland KS, Lund JS (1982) Widespread periodic intrinsic connections in the tree shrew visual cortex. Science 215(4539):1532–1534

Rockland KS, Lund JS (1983) Intrinsic laminar lattice connections in primate visual cortex. J Comp Neurol 216(3):303–318. doi:10.1002/cne.902160307

Rubinov M, Sporns O (2010) Complex network measures of brain connectivity: uses and interpretations. Neuroimage 52(3):1059–1069. doi:S1053-8119(09)01074-X [pii] 10.1016/j.neuroimage.2009.10.003

Scannell JW, Blakemore C, Young MP (1995) Analysis of connectivity in the cat cerebral cortex. J Neurosci 15(2):1463–1483

Semendeferi K, Teffer K, Buxhoeveden DP, Park MS, Bludau S, Amunts K, Travis K, Buckwalter J (2011) Spatial organization of neurons in the frontal pole sets humans apart from great apes. Cereb Cortex 21(7):1485–1497. doi:10.1093/cercor/bhq191

Senden M, Deco G, de Reus MA, Goebel R, van den Heuvel MP (2014) Rich club organization supports a diverse set of functional network configurations. NeuroImage 96:174–182. doi:10.1016/j.neuroimage.2014.03.066

Shaw P, Kabani NJ, Lerch JP, Eckstrand K, Lenroot R, Gogtay N, Greenstein D, Clasen L, Evans A, Rapoport JL (2008) Neurodevelopmental trajectories of the human cerebral cortex. J Neurosci 28(14):3586–3594

Sherwood CC, Subiaul F, Zawidzki TW (2008) A natural history of the human mind: tracing evolutionary changes in brain and cognition. J Anat 212(4):426–454. doi:10.1111/j.1469-7580.2008.00868.x

Sporns O (2013) The human connectome: origins and challenges. Neuroimage. doi:10.1016/j.neuroimage.2013.03.023

Sporns O, Bullmore ET (2014) From connections to function: the mouse brain connectome atlas. Cell 157(4):773–775. doi:10.1016/j.cell.2014.04.023

Sporns O, Zwi JD (2004) The small world of the cerebral cortex. Neuroinformatics 2(2):145–162. doi:10.1385/NI:2:2:145

Sporns O, Chialvo DR, Kaiser M, Hilgetag CC (2004) Organization, development and function of complex brain networks. Trends Cogn Sci 8(9):418–425

Sporns O, Tononi G, Kötter R (2005) The human connectome: a structural description of the human brain. PLoS Comput Biol 1(4):e42. doi:10.1371/journal.pcbi.0010042

Stam CJ (2014) Modern network science of neurological disorders. Nat Rev Neurosci 15(10):683–695. doi:10.1038/nrn3801

Stanley KO, Miikkulainen R (2002) Evolving neural networks through augmenting topologies. Evol Comput 10(2):99–127. doi:10.1162/106365602320169811

Torben-Nielsen B, De Schutter E (2014) Context-aware modeling of neuronal morphologies. Front Neuroanat 8:92. doi:10.3389/fnana.2014.00092

Towlson EK, Vertes PE, Ahnert SE, Schafer WR, Bullmore ET (2013) The rich club of the *C. elegans* neuronal connectome. J Neurosci 33(15):6380–6387. doi:10.1523/JNEUROSCI.3784-12.2013 van den Heuvel MP, Sporns O (2011) Rich-club organization of the human connectome. J Neurosci 31(44):15775–15786. doi:10.1523/jneurosci.3539-11.2011

van den Heuvel MP, Kahn RS, Goni J, Sporns O (2012) High-cost, high-capacity backbone for global brain communication. Proc Natl Acad Sci USA 109(28):11372–11377. doi:10.1073/pnas.1203593109

van den Heuvel MP, Kersbergen KJ, de Reus MA, Keunen K, Kahn RS, Groenendaal F, de Vries LS, Benders MJ (2014) The neonatal connectome during preterm brain development. Cereb Cortex. doi:10.1093/cercor/bhu095

Van Hooser SD, Heimel JA, Chung S, Nelson SB (2006) Lack of patchy horizontal connectivity in primary visual cortex of a mammal without orientation maps. J Neurosci 26(29):7680–7692. doi:10.1523/JNEUROSCI.0108-06.2006

Varier S, Kaiser M (2011) Neural development features: spatio-temporal development of the *Caenorhabditis elegans* neuronal network. PLoS Comput Biol 7:e1001044. doi:10.1371/journal.pcbi.1001044

Verbancsics P, Stanley KO (2011) Constraining connectivity to encourage modularity in hyperNEAT. Gecco-2011: proceedings of the 13th annual genetic and evolutionary computation conference, pp 1483–1490

Warren DE, Power JD, Bruss J, Denburg NL, Waldron EJ, Sun H, Petersen SE, Tranel D (2014) Network measures predict neuropsychological outcome after brain injury. Proc Natl Acad Sci USA 111(39):14247–14252. doi:10.1073/pnas.1322173111

Watts DJ, Strogatz SH (1998) Collective dynamics of 'small-world' networks. Nature 393:440–442

White JG, Southgate E, Thomson JN, Brenner S (1986) The structure of the nervous system of the nematode *Caenorhabditis elegans*. Philos Trans R Soc Lond B Biol Sci 314(1165):1–340

Young MP (1992) Objective analysis of the topological organization of the primate cortical visual system. Nature 358(6382):152–155

Zamora-Lopez G, Zhou C, Kurths J (2010) Cortical hubs form a module for multisensory integration on top of the hierarchy of cortical networks. Front Neuroinform 4:1. doi:10.3389/neuro.11.001.2010

Zubler F, Douglas R (2009) A framework for modeling the growth and development of neurons and networks. Front Comput Neurosci 3:25. doi:10.3389/neuro.10.025.2009

Chapter 18
Muscular-Hydrostat Computers: Physical Reservoir Computing for Octopus-Inspired Soft Robots

Kohei Nakajima

Abstract The octopus has been one of the major sources of inspiration for roboticists for many years. It can harness its complex body dynamics in a highly sophisticated manner despite the fact that its body does not contain any rigid components through the interaction between the characteristic organization of its nervous system and its specific body morphology and muscle characteristics, called muscular-hydrostats. Inspired by these biological findings, we investigated the potential information processing capacity of its soft body dynamics with the help of a machine learning approach called *reservoir computing*. We review a series of concepts and platforms, called *muscular-hydrostat computers* throughout our study and suggest that the diverse body dynamics of soft materials can be exploited as a computational resource. This approach could enable some controls to be embedded into the robot body.

Keywords Reservoir computing • Natural computing • Soft robotics • Bio-inspired robotics • Octopus • Soft materials • Embodiment

18.1 Introduction and Outlooks

Inspired by the soft material ubiquitous in all living creature's body structures, a new family of robots has been constructed with the aim of incorporating flexible body elements. The resulting machines, called *soft robots* (Trivedi et al. 2008; Kim et al. 2013), have significant advantages over traditionally articulated robots in terms of morphological deformability and interactional safety. These robots are able to adapt their morphology to unstructured environments and can manipulate fragile objects without damaging them, making them especially attractive for interactions with humans. Specific possibilities include care for the elderly, prosthetics, and wearable

---

K. Nakajima
The Hakubi Center for Advanced Research & Graduate School of Informatics, Kyoto University, Yoshida Honmachi, Sakyo-ku, Kyoto, 606-8501, Japan

JST PRESTO, 4-1-8 Honcho, Kawaguchi, Saitama, 332-0012, Japan
e-mail: k_nakajima@acs.i.kyoto-u.ac.jp

S. Shigeno et al. (eds.), *Brain Evolution by Design*, Diversity and Commonality in Animals, DOI 10.1007/978-4-431-56469-0_18

technology (Pfeifer et al. 2012). Soft robots can generate diverse behaviours with simple types of actuation by partially outsourcing control to the morphological and material properties of their soft bodies (Shepherd et al. 2011), which is made possible by the tight coupling between control, body, and the environment (Pfeifer et al. 2007; Pfeifer and Bongard 2006; Nakajima et al. 2012a). Furthermore, production costs for these types of robots are relatively low. Thus, they can be easily incorporated into a wide range of machines and technology with various purposes (Morin et al. 2012).

Concomitant with these benefits is a great challenge in controlling their body dynamics. In general, soft body dynamics exhibit multifold properties, such as high dimensionality, nonlinearity, and a certain amount of "sluggishness" as a result of their elastic nature (Nakajima et al. 2015a, 2012b, 2011). In addition, they have higher levels of freedom of movement than traditional robots, often with greater degrees of freedom than the number of actuators (i.e., under-actuated systems), making them notoriously difficult to control.

Here, we aim to reverse this scenario and view it in a positive light. We claim that it is these same nonlinear and elastic body dynamics that make soft robots feasible and simple to control. In nature, some animals have soft bodies and control them in a sophisticated manner by capitalizing on their body dynamics. The octopus serves as a prime example (Hochner 2012). It does not have any rigid components in its body, which can move with virtually infinite degrees of freedom. One of the characteristic structures observed in the octopus arm is the muscular-hydrostat (Smith and Kier 1989; Kier and Smith 1985; Taylor and Kier 2003; Feinstein et al. 2011; Li et al. 2012a). In such structures, the volume of the organ remains constant during all movements, enabling the muscles themselves to perform all of the functions usually performed by the skeleton. By developing a series of platforms called *muscular-hydrostat computers*, inspired by the octopus, we have demonstrated that these body dynamics can be positively exploited as a computational resource. For example, we have shown that their muscular organization provides sufficient nonlinearity and elasticity that it can be used to embed multiple control programs. In this paper, we aim to review bio-inspirations and our series of proposed platforms. We begin by discussing the inspiration stemming from octopus movement and then introduce the physical reservoir computing approach.

18.2 Octopus Inspirations: Embodiment, Muscular-Hydrostats, and Embedded Motor Programs

Biological systems have certain morphologies and material characteristics that improve their adaptivity and increase their probability of survival. In nature, these body structures and organizations evolve based on the animal's ecological niche, suggesting that the mechanism controlling animal behaviour is not only located in its brain, but that there is a tight coupling between the brain, body, and its environment, an idea that is usually termed *embodiment* (Pfeifer et al. 2007; Pfeifer and Bongard

2006). An octopus has hyper-redundant limbs with a virtually unlimited number of degrees of freedom; its movements are known to be highly sophisticated. Its specific body structure is specialized to permit survival in a complicated underwater environment (Hochner 2012). From a conventional control perspective, the octopus' method of controlling its movement and harnessing its nonlinear body dynamics is outstanding and instructive. In this context, the octopus has been a good source of inspiration for designing a control strategy for soft robots (Trivedi et al. 2008; Kim et al. 2013).

One of the most characteristic body structures observed in the octopus is its muscular organization, called *muscular-hydrostats* (Smith and Kier 1989; Kier and Smith 1985; Taylor and Kier 2003; Feinstein et al. 2011; Li et al. 2012a). It acts to keep the volume of the organ constant during all movements, so that if the diameter of the muscular-hydrostats decreases, then its length increases, and vice versa. For example, in the octopus arm, muscles are organized into transverse, longitudinal, and obliquely oriented groups, which form the muscular-hydrostatic structure. Such flexible structures show major advantages over articulated arms that have a rigid skeleton and joints. Accordingly, in robotics, there have been several attempts to characterize the role of the muscular-hydrostat system in terms of its anatomical structure and morphology.

The nervous system of the octopus is characteristically highly distributed throughout the entire body. It has a relatively small central nervous system (CNS), which controls the large peripheral nervous system (PNS) of the arms. The CNS integrates information from the visual system, and then issues commands to the lower motor centres that control the elaborate neuromuscular system of the arms. A typical example showing the effectiveness of this nervous system distribution is the animal's reaching behaviour. Reaching behaviour consists of a *bend propagation* along the arm toward the tip in a highly stereotypical and invariant way. Sumbre *et al.* showed that the arm extensions can be evoked in arms whose connection with the brain has been severed (Sumbre et al. 2001). Because the evoked motions in denervated octopus arms were qualitatively and kinematically identical to natural bend propagations, an underlying motor program appears to be embedded in the neuromuscular system of the arm, which does not require continuous central control.

Based on these biological findings, we have been exploring a novel view of the property of the muscular-hydrostat system combined with the PNS in terms of its functionality. In this chapter, we introduce a series of platforms based on a machine learning technique called *reservoir computing*.

18.3 Physical Reservoir Computing

Reservoir computing has been proposed as a framework to train recurrent neural networks (Jaeger and Haas 2004; Maass et al. 2002; Verstraeten et al. 2007). In this framework, low-dimensional inputs are mapped to a high-dimensional nonlinear dynamical system, typically referred to as a *reservoir*, and the corresponding outputs

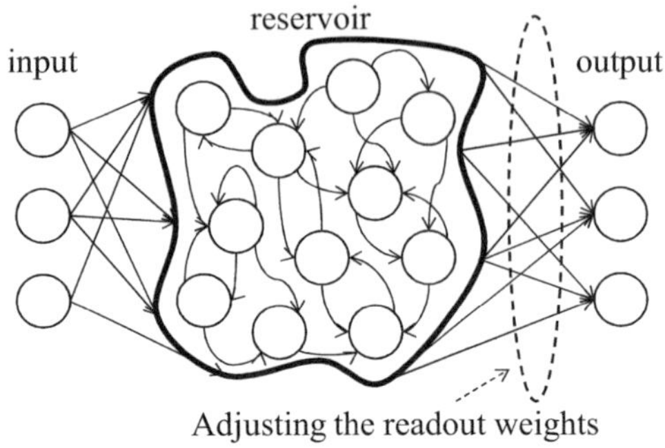

Fig. 18.1 Information processing in the reservoir computing scheme. Low-dimensional inputs are injected to a high-dimensional dynamic system called a reservoir. Outputs are calculated as a weighted sum of the states of the reservoir nodes. Network training only requires the adjustment of the linear readout weights

are expressed as weighted sums of the reservoir states (Fig. 18.1). To possess a computational capability, a reservoir should have both input separability and fading memory. Input separability is usually achieved through a mapping of the low-dimensional input to a high-dimensional state space. Fading memory is a property of a system that retains the influence of a recent input sequence within the system and permits the integration of stimulus information over time, enabling reproducible computation for which the recent history of the signal is significant. A remarkable property of the approach is that if the reservoir involves enough nonlinearity and memory, it can be shown that it is sufficient to add a simple linear and static readout from the high-dimensional state space to emulate nonlinear and complex computations (Fig. 18.1). The approach is extensively studied in neuroscience (Sussillo and Abbott 2009), and many applications can be found in the latest robotic applications (Li et al. 2013a, 2012b,c, 2013b; Kuwabara et al. 2012).

Because of its generic nature, reservoir computing is not limited to digital simulations of neural networks. Any high-dimensional dynamic system can serve as a reservoir if it has the appropriate properties. More specifically, if the reservoir consists of a physical system, we term the approach *physical reservoir computing* (e.g., see summary on robotics-related implementations in Hauser et al. 2014). A number of physical implementations for reservoirs have been proposed, such as the surface of water (Fernando and Sojakka 2003), and nonlinear mass spring systems (Hauser et al. 2011). More recently, electronic and optical implementations have also been reported (Appeltant et al. 2011; Woods and Naughton 2012).

In this chapter, we introduce a series of platforms that exploit the dynamics of soft structures as a reservoir. The first platform is a dynamic model of a muscular-hydrostat system that is inspired by the muscle organization of the octopus (Nakajima et al. 2013a). The second is a dynamic model of an entire octopus arm equipped with a muscular-hydrostat system (Nakajima et al. 2013b). It has been shown that this model can also be used as a model for certain soft robotic arms by tuning its mechanical parameters (Zheng et al. 2012). The final platform is a soft robotic arm made of silicone (Nakajima et al. 2014, 2015b, 2013c). In each platform, we demonstrate that the body dynamics of the system can be used as a successful reservoir; this means that, by simply attaching a static linear readout from the high-dimensional nonlinear dynamics, one can emulate complex nonlinear computations without altering the physical system itself. That is, we employ the physical body as part of a computational device. We call this series of platforms *muscular-hydrostat computers* (Nakajima et al. 2013d), and introduce them in the next section.

18.4 Muscular-Hydrostat Computers

In this section, we briefly introduce our physical reservoir computing platforms, called muscular-hydrostat computers. Note that the term "muscular-hydrostat" is not used in a strict sense, because the soft silicone arm that we introduce later does not have any muscular-hydrostat properties. However, we use this term because our arm shares the same operational characteristics as the octopus' arm.

18.4.1 Muscular-Hydrostat System

Several models emulating the muscular-hydrostat system of the octopus have been proposed (e.g., see Yekutieli et al. 2005a,b; Kang et al. 2012). The model used in this study is originally proposed in Kang et al. (2012). The overall structure of our muscular-hydrostat system is shown in Fig. 18.2a. We use a mass-spring-damper system in the shape of a truncated cone, consisting of a base plane, a ceiling plane composed of four transverse springs, a central strut, and four longitudinal springs, which emulate the anatomical structure of the muscle alignment in an actual octopus arm. The longitudinal springs control the position and orientation of the ceiling plane, while the transverse springs control the radius of the ceiling plane. The system has an isovolumetric structure, which provides forces constantly striving to maintain the arm volume, a property of the muscular-hydrostats. Thus, all the springs are assumed to be implicitly or explicitly coupled with each other. The values for all model parameters (e.g., spring coefficients, damping) are either inspired by the octopus or directly drawn from biological data. Because of these parameters, the overall model dynamics are highly nonlinear, even though the springs used in the model are linear.

By providing inputs as forces on the springs and adjusting the linear readout weights using the spring length dynamics as reservoir states, we demonstrate that the system can be directly used to accomplish complex nonlinear computations (Nakajima et al. 2013a). Together with the physical reservoir computing approach, the core concept of our technique was the introduction of different time scales between I/O signals and the system movement. In real-world applications, the

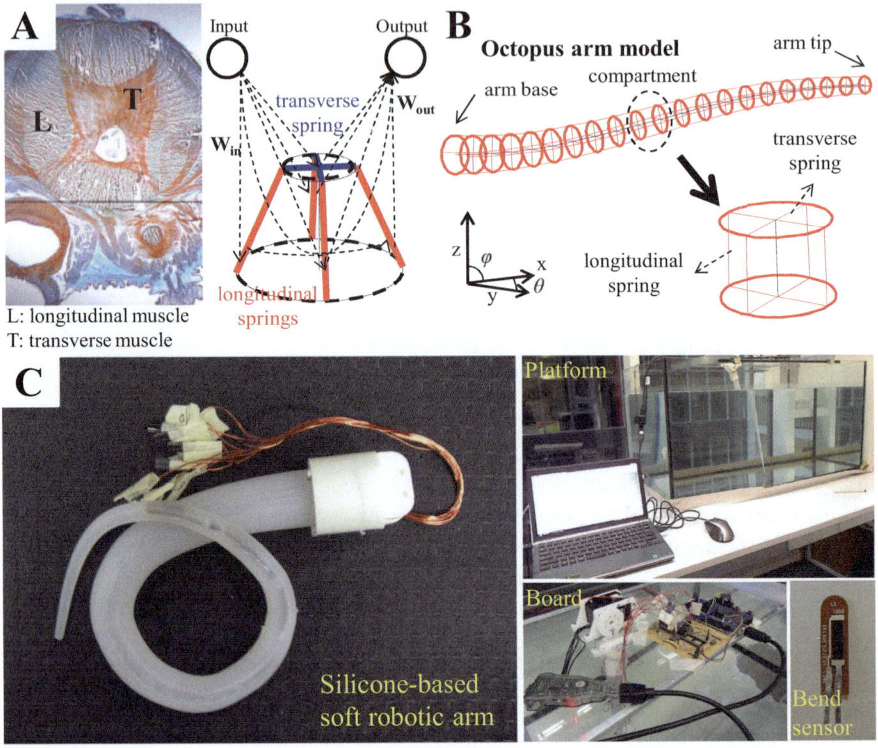

Fig. 18.2 A series of physical reservoir computing platforms called muscular-hydrostat computers. (**a**) Exploiting the muscular-hydrostat system as a computational resource, the mass-spring damper system emulates the muscle configuration of the octopus arm and the property of muscular-hydrostats, which maintain a constant system volume. Forces are provided as inputs to the springs and the corresponding dynamics of the transverse and longitudinal springs serve as a reservoir. By introducing a different time scale between I/O relations and the muscular-hydrostat system, we demonstrate that the system has sufficient computational capacity to emulate nonlinear dynamical systems. (**b**) A dynamic model of the octopus arm. The arm consists of 20 compartments, which are equipped with muscular-hydrostat systems (Fig. 18.2a). The mechanical parameters of the system are drawn from bio-inspirations. By controlling the base rotation of the model as inputs, the resulting passive dynamics of the springs are exploited as a reservoir. This system can be used to emulate nonlinear dynamical systems and embed robust closed-loop control. (**c**) The silicone-based soft robotic arm. The arm encloses 10 bend sensors, which monitor local bends during each time step. It is actuated at the base point and is able to rotate. It has been shown that the body dynamics (sensory time series) can be used as a successful reservoir. The behaviour of the arm successfully emulates functions that require memory and embeds a robust closed-loop control

assumption of this time scale difference is rather natural, usually caused by the sampling rate of the hardware device compared to the continuous physical dynamics of the robots. We have shown that this system can emulate desired nonlinear dynamical systems and reported that the system demonstrates a characteristic short-term memory profile. One possible application of this technique would be to embed a local control and outsource the computational load to the body part in order to make an external controller more efficient.

18.4.2 Dynamic Model of the Octopus Arm

The next platform we create is a 3D dynamic model inspired by the octopus arm. The overall structure of the entire arm is shown in Fig. 18.2b (Kang et al. 2012). The arm consists of 20 compartments; each compartment implements the same muscular-hydrostats construction explained in the previous section. It is assumed that the arm is to be immersed in an underwater environment in which water friction constants are approximated by computational fluid dynamics simulations (Kang et al. 2011; Kazakidi et al. 2012). As a result, the arm has highly nonlinear body dynamics when actuated.

By using this platform, we demonstrate that the body dynamics of the arm can be exploited as a computational resource: first, to emulate complex nonlinear dynamical systems; and second, to implement closed-loop control (Nakajima et al. 2013b). Usually, closed-loop control in robots is realized through a sensory-motor loop, where controllers take sensory values and calculate the corresponding motor commands. In our experimental setting, an external controller is absent, and the body dynamics themselves are exploited to control the next motion, meaning that the controller and the body part it is controlling are the same. We used several nonlinear limit cycles to see how they can be embedded directly into the soft robotic arm without any support for nonlinearity and memory from an external controller. For roboticists, our method will offer one way to quantitatively characterize which controls are efficient for a variety of body designs. In addition, it will also help outsourcing the control load to the body parts. In addition, through the emulation of nonlinear dynamical systems, we showed that each body part has a specific role according to its task type. These obvious and specific coherencies suggest that these properties originate from the intrinsic body structure.

18.4.3 Soft Robotic Arm Made of Silicone Materials

It is important to demonstrate whether the techniques illustrated in these models can actually work in a real physical platform. We have developed a simple but powerful physical platform implemented with silicone material (Fig. 18.2c) (Nakajima et al. 2014, 2015b, 2013c). The platform consists of a soft silicone arm, its sensing and actuation systems, data processing via a PC, and a water tank containing fresh water as an underwater environment. Note that in a precise sense, the silicone arm does not emulate the simulated octopus arm, which was introduced in the previous section. However, they share a similar operational property in terms of morphology and density. The arm has 10 embedded bend sensors within the silicone material. The base of the arm can rotate left and right through the actuation of a servo motor. Our system has one active degree of freedom, but there is a much higher number of passive degrees of freedom in the silicone arm. By rotating the base of the arm and generating body dynamics induced by the interaction between the underwater environment and the soft silicone material, we showed that the sensory time series reflecting body dynamics can be exploited as a part of a computational device. It could be used to emulate nonlinear dynamic systems and functions that require memory by embedding robust closed-loop control into the arm. The soft silicone arm presented here only partially mimics the properties, such as morphology, and the material of a real octopus arm. We expect that the real octopus arm has a greater computational capacity than the one we present here because of its diverse and dynamic nature.

18.5 Discussion and Future Directions

Many biological systems have soft bodies, which are able to adapt and behave effectively in their given ecological niche. The framework presented in this study may also shed light on the role of the body in biological systems. In this context, it would be interesting to investigate whether the soft arm could be used to embed more biologically plausible behaviours in future work. For example, as we mentioned earlier, it is well known that the octopus adopts a specific strategy for reaching, called *bend propagation* (Fig. 18.3a). In this specific motion, it is suggested that the CNS only initiates the motion and the muscle activations themselves are handled at the PNS level (Li et al. 2012a). Several researchers have investigated this behaviour by directly extracting the muscle contraction patterns from the real octopus and externally applying these patterns to the octopus arm models. The technique presented here may yield further insights into the overall scheme by including the role of the arm's body dynamics. Because the PNS does not have plasticity, it would be worth investigating how the arm's body dynamics, together with the PNS, modelled as a linear and static feedback loop in the arm, embeds the motor patterns of bend propagation according to the initiation command sent by the CNS (Fig. 18.3b, c). This line of experimentation can be investigated in future work.

In this study, we have introduced a series of platforms that exploit body dynamics as a computational resource. The technique can be potentially applied to a wide class of soft robots because the main component required is the soft body itself, which is already present in the soft robotic platform. For example, the model can be used to produce nonlinear limit cycles, which enable locomotion in robots. These systems can be embedded into the body in a closed-loop manner without additional nonlinearity and memory support from an external controller. Recently, this line of study has been initiated with the use of physical platforms (e.g., Caluwaerts et al. 2014; Zhao et al. 2013). Consequently, different types of robot morphology, which increases the computational capacity of the body, should be explored in the future.

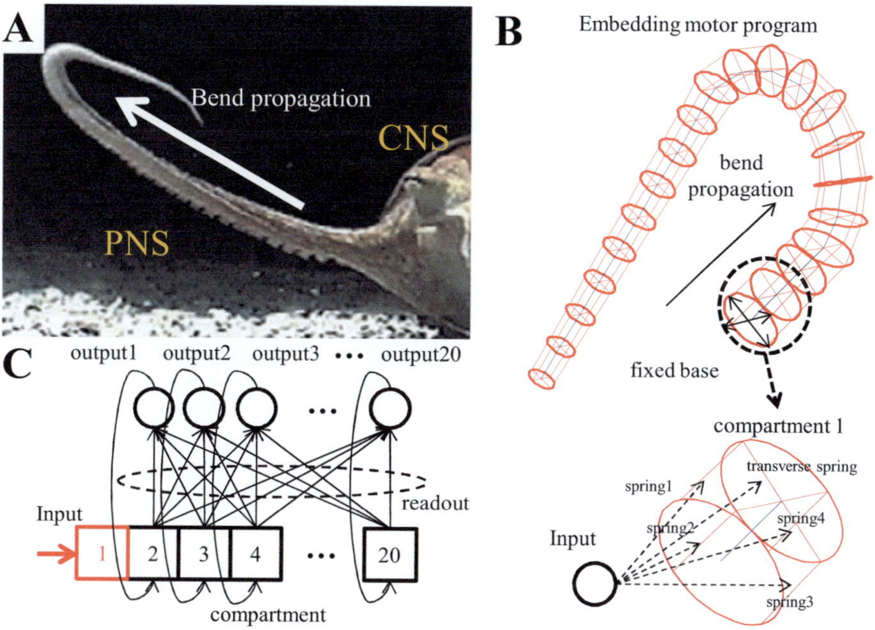

Fig. 18.3 Illustration of the scheme used to embed a motor program for the bend propagation using a physical reservoir computing approach. (**a**) Reaching behaviour of the octopus is based on stereotypical motion called bend propagation. The CNS only initiates the behaviour and does not handle the sequence of muscle controls, while the octopus arm together with the PNS controls the movement. (**b**) Schematics illustrating how to embed the bend propagation into the modelled arm. Bend propagation is achieved by contracting the springs of the compartments in a specific order from the base to the tip. (**c**) Information processing scheme used to embed the bend propagation using a physical reservoir computing approach. By exploiting the body dynamics of the modelled arm as a reservoir, it regulates the timing and strength of the local muscle contractions. The speed and degree of the bend propagation can be controlled by the initiation signal sent from the CNS as an input

In particular, recent advancements in biomaterial science suggest that numerous functionalities can be outsourced to the body materials. The way in which different types of materials regulate the computational capacity, from a reservoir computing point of view, is worth exploring. In addition, developments in new types of sensors, which can effectively monitor body dynamics, would make our technique applicable to further applications. Furthermore, with a novel technique that could be used to implement linear readouts as a device, it would be possible to control spatially distant points of the body through the use of body dynamics actuated on more local body parts. This approach would be especially useful when the actuation points are limited by the platform's physical constraints. To conclude, we believe that our approach will contribute to the use of soft materials in various engineering applications in multiple fields.

Acknowledgements The author expresses his gratitude towards all members of the ailab for the *CSC* discussions throughout his time in Zurich, Dr. Tao Li for engineering support during the OCTOPUS project, Dr. Helmut Hauser for his helpful suggestions, and Prof. Rolf Pfeifer for all the opportunities and valuable experiences. This work was supported by the European Commission in the ICT-FET OCTOPUS Integrating Project (EU project FP7-231608), the JSPS Postdoctoral Fellowships for Research Abroad, and the JST PRESTO program, KAKENHI No. 15K16076 and No. 26880010.

References

Appeltant L, Soriano MC, der Sande GV, Danckaert J, Massar S, Dambre J, Schrauwen B, Mirasso CR, Fischer I (2011) Information processing using a single dynamical node as complex system. Nat Commun 2:468

Caluwaerts K, Despraz J, Iscen A, Sabelhaus AP, Bruce J, Schrauwen B, SanSpiral V (2014) Design and control of compliant tensegrity robots through simulations and hardware validation. J R Soc Interface 11:20140520

Feinstein N, Nesher N, Hochner B (2011) Functional morphology of the neuromuscular system of the Octopus vulgaris arm. Vie Milieu 61(4):219–229

Fernando C, Sojakka S (2003) Pattern recognition in a bucket. Lecture notes in computer science, vol 2801. Springer, Berlin/New York, pp 588–597

Jaeger H, Haas H (2004) Harnessing nonlinearity: predicting chaotic systems and saving energy in wireless communication. Science 304:78–80

Hauser H, Ijspeert AJ, Fuchslin RM, Pfeifer R, Maass W (2011) Towards a theoretical foundation for morphological computation with compliant bodies. Biol Cybern 105:355–370

Hauser H, Fuchslin RM, Nakajima K (2014) Morphological computation – the body as a computational resource. In: Helmut H, Fuchslin RM, Pfeifer R (eds) E-book on opinions and outlooks on morphological computation, chapter 20, pp 226–244. ISBN:978-3033045156

Hochner B (2012) An embodied view of octopus neurobiology. Curr Biol 22:R887–R892

Kang R, Kazakidi A, Guglielmino E, Branson DT, Tsakiris DP, Ekaterinaris JA et al (2011) Dynamic model of a hyper-redundant octopus-like manipulator for underwater applications. In: Proceedings of 2011 IEEE/RSJ international conference on intelligent robots and systems (IROS), San Francisco, pp 4054–4059

Kang R, Branson DT, Guglielmino E, Caldwell DG (2012) Dynamic modeling and control of an octopus inspired multiple continuum arm robot. Comput Math Appl 64:1004–1016

Kier WM, Smith KK (1985) Tongues, tentacles and trunks: the biomechanics of movement in muscular-hydrostats. Zool J Linn Soc 83:307–324

Kim S, Laschi C, Trimmer B (2013) Soft robotics: a new perspective in robot evolution. Trends Biotechnol 31:287–294

Kazakidi A, Vavourakis V, Pateromichelakis N, Ekaterinaris JA, Tsakiris DP (2012) Hydrodynamic analysis of octopus-like robotic arms. In: Proceedings of 2012 IEEE international conference on robotics and automation (ICRA), St. Paul, pp 5295–5300

Kuwabara J, Nakajima K, Kang R, Branson DT, Guglielmino E, Caldwell DG, Pfeifer R (2012) Timing-based control via echo state network for soft robotic arm. In: Proceedings of the 2012 international joint conference on neural networks (IJCNN), Brisbane, pp 1–8

Li T, Nakajima K, Kuba MJ, Gutnick T, Hochner B, Pfeifer R (2012a) From the octopus to soft robots control: an octopus inspired behavior control architecture for soft robots. Vie Milieu 61:211–217

Li T, Nakajima K, Calisti M, Laschi C, Pfeifer R (2012b) Octopus-inspired sensorimotor control of a multi-arm soft robot. In: Proceedings of 2012 international conference on mechatronics and automation (ICMA), Chengdu, pp 948–955

Li T, Nakajima K, Cianchetti M, Laschi C, Pfeifer R (2012c) Behavior switching by using reservoir computing for a soft robotic arm. In: Proceedings of 2012 IEEE international conference on robotics and automation (ICRA), St. Paul, pp 4918–4924

Li T, Nakajima K, Pfeifer R (2013a) Online learning technique for behavior switching in a soft robotic arm. In: Proceedings of 2013 IEEE international conference on robotics and automation (ICRA), Karlsruhe, pp 1288–1294

Li T, Nakajima K, Pfeifer R (2013b) Learning from the octopus: sensorimotor control of octopus-inspired soft robots. In: Proceedings of 2013 international workshop on soft robotics and morphological computation (SoftRobot2013), Monte Verità, P-34

Maass W, Natschlager T, Markram H (2002) Real-time computing without stable states: a new framework for neural computation based on perturbations. Neural Comput 14:2531–2560

Morin SA, Shepherd RF, Kwok SW, Stokes AA, Nemiroski A, Whitesides GM (2012) Camouflage and display for soft machines. Science 337:828–832

Nakajima K, Li T, Sumioka H, Cianchetti M, Pfeifer R (2011) Information theoretic analysis on a soft robotic arm inspired by the octopus. In: Proceedings of 2011 IEEE international conference on robotics and biomimetics (ROBIO), Phucket, pp 110–117

Nakajima K, Ngouabeu AMT, Miyashita S, Goldi M, Fuchslin RM, Pfeifer R (2012a) Morphology-induced collective behaviors: dynamic pattern formation in water-floating elements. PLoS ONE 7(6):e37805

Nakajima K, Li T, Kang R, Guglielmino E, Caldwell DG, Pfeifer R (2012b) Local information transfer in soft robotic arm. In: Proceedings of 2012 IEEE international conference on robotics and biomimetics (ROBIO), Guangzhou, pp 1273–1280

Nakajima K, Hauser H, Kang R, Guglielmino E, Caldwell DG, Pfeifer R (2013a) Computing with a muscular-hydrostat system. In: Proceedings of 2013 IEEE international conference on robotics and automation (ICRA), Karlsruhe, pp 1496–1503

Nakajima K, Hauser H, Kang R, Guglielmino E, Caldwell DG, Pfeifer R (2013b) A soft body as a reservoir: case studies in a dynamic model of octopus-inspired soft robotic arm. Front Comput Neurosci 7:91

Nakajima K, Li T, Hauser H, Pfeifer R (2013c) Morphological computation in a silicone-based soft robotic arm. In: Proceedings of 2013 international workshop on soft robotics and morphological computation (SoftRobot2013), Monte Verità, P-49

Nakajima K, Li T, Hauser H, Pfeifer R (2013d) Muscular-hydrostat computers: toward a novel control scheme for soft robots. In: Proceedings of 2013 international workshop on soft robotics and morphological computation (SoftRobot2013), Monte Verità, O-29

Nakajima K, Li T, Hauser H, Pfeifer R (2014) Exploiting short-term memory in soft body dynamics as a computational resource. J R Soc Interface 11(100):20140437

Nakajima K, Schmidt N, Pfeifer R (2015a) Measuring information transfer in a soft robotic arm. Bioinspir Biomim 10(3):035007

Nakajima K, Hauser H, Li T, Pfeifer R (2015b) Information processing via physical soft body. Sci Rep 5:10487

Pfeifer R, Bongard J (2006) How the body shapes the way we think: a new view of intelligence. MIT Press, Cambridge

Pfeifer R, Lungarella M, Iida F (2007) Self-organization, embodiment, and biologically inspired robotics. Science 318:1088–1093

Pfeifer R, Lungarella M, Iida F (2012) The challenges ahead for bio-inspired 'soft' robotics. Commun ACM 55:76–87

Shepherd RF, Ilievski F, Choi W, Morin SA, Stokes AA, Mazzeo AD, Chen X, Wang M, Whitesides GM (2011) Multi-gait soft robot. Proc Natl Acad Sci USA 108:20400–20403

Smith KK, Kier WM (1989) Trunks, tongues, and tentacles: moving with skeletons of muscle. Am Sci 77:28–35

Sumbre G, Gutfreund Y, Fiorito G, Flash T, Hochner B (2001) Control of octopus arm extension by a peripheral motor program. Science 293:1845–1848

Sussillo D, Abbott LF (2009) Generating coherent patterns of activity from chaotic neural networks. Neuron 63:544–557

Taylor J, Kier W (2003) Switching skeletons: hydrostatic support in molting crabs. Science 301:209–210

Trivedi D, Rahn CD, Kier WM, Walker ID (2008) Soft robotics: biological inspiration, state of the art, and future research. Appl Bionics Biomech 5:99–117

Verstraeten D, Schrauwen B, D'Haene M, Stroobandt D (2007) An experimental unification of reservoir computing methods. Neural Netw 20:391–403

Woods D, Naughton TJ (2012) Photonic neural networks. Nat Phys 8:257

Yekutieli Y, Sagiv-Zohar R, Aharonov R, Engel Y, Hochner B, Flash T (2005a) Dynamic model of the octopus arm. I. Biomechanics of the octopus reaching movement. J Neurophys 94:1443–1458

Yekutieli Y, Sagiv-Zohar R, Hochner B, Flash T (2005b) Dynamic model of the octopus arm. II. Control of reaching movements. J Neurophys 94:1459–1468

Zhao Q, Nakajima K, Sumioka H, Hauser H, Pfeifer R (2013) Spine dynamics as a computational resource in spine-driven quadruped locomotion. In: Proceedings of 2013 IEEE/RSJ international conference on intelligent robots and systems (IROS), Tokyo, pp 1445–1451

Zheng T, Branson DT, Kang R, Cianchetti M, Guglielmino E, Follador M et al (2012) Dynamic continuum arm model for use with underwater robotic manipulators inspired by octopus vulgaris. In: Proceedings of 2012 IEEE international conference on robotics and automation (ICRA), St. Paul, pp 5289–5294

Chapter 19
Brain Evolution as an Information Flow Designer: The Ground Architecture for Biological and Artificial General Intelligence

Shuichi Shigeno

Abstract For centuries, neuroscientists have identified a number of neural systems involved in sensory, motor, state control, and cognitive functions. Modern comparative studies have proposed their diversity, origins, and basic functionality across animal phyla. Despite a number of attempts, however, a common functional plan of the complex brain remains controversial. For example, there is currently no prominent theory of how neural networks are structurally comparable between phylogenetically distant animals such as vertebrates, octopuses, worms, and insects, in which there are distinguishably different brain architectures. This chapter attempts to identify the types of information flow patterns that were specialized during brain evolution, when these patterns appeared as a prototype, and how the flow systems have been shaped based on the common morphological architecture. In a notable case, a number of sensory associative centers show comparable patterns in mammalian, insect, and octopus brains, representing a common input and output flow of information. One can speculate that a common underlying structure is shared between various animals because of common functionalities that produce highly effective learning, memory, and autonomous cognitive tasks. Such an underlying structure could help establish a large-scale framework for comparison between phylogenetically distant animal brains and perhaps even form the groundwork for artificial general intelligence.

Keywords Brain • Neuronal network • Brain diversity • Evolution

---

S. Shigeno (✉)
Department of Marine Biodiversity Research, Japan Agency for Marine-Earth Science and Technology, 2-15 Natsushima, Yokosuka 237-0061, Japan

Department of Biology and Evolution of Marine Organism, Stazione Zoologica Anton Dohrn, 80121 Naples, Italy
e-mail: sushigeno@gmail.com

S. Shigeno et al. (eds.), *Brain Evolution by Design*, Diversity and Commonality in Animals, DOI 10.1007/978-4-431-56469-0_19

19.1 Introduction

A number of neuroanatomical studies have elucidated a variety of brain structures (Cajal 1917; Hanström 1928; Bullock and Horridge 1965; Nieuwenhuys et al. 1998; Roth and Wullimann 2001; Butler and Hodos 2005; Striedter 2005; Kaas 2006, Strausfeld 2012; Schneider 2014; Schmidt-Rhaesa et al. 2016; Borden et al. 2016). These structures are complex, but they are often classified by common functional systems: sensory detection, motor generation, behavioral states, including reward and sleep, and cognitive systems (Swanson 2003, 2007). The origin of such neural systems can be traced back to ancestral forms in sponges, sea anemones, comb jellies, flatworms, nematodes, polychaetes, and other invertebrates (Roth 2013; Arendt et al. 2015; Holland 2016; Schmidt-Rhaesa et al. 2016; Wolff and Strausfeld 2016). Despite a well-established basic structure underlying small elementary nervous systems, there is no consensus about a basic underlying structure or unifying model in more complicated brains (Bullock 1993, 2002; Roth 2013). In fact, some invertebrates such as cephalopods and arthropods display a higher level psychological repertoire, with components such as cognition, emotion, planning, sleep, and consciousness, but a common neural basis between vertebrates remains in dispute (Hochner et al. 2006; Moroz 2009; Krubitzer 2009). A contemporary view posits that higher ordered brain centers, such as the cerebral cortex of mammals and the mushroom bodies of insects, share comparable organization, developmental control genes, neurotransmitter action, long-term memory consolidation, and in part, features of neuropsychiatric or behavioral disorders (Tomer et al. 2010; Strausfeld and Hirth 2013; Wolff and Strausfeld 2016).

For this reason, I was motivated to ask what basic architecture forms the dominant multisensory integration regions or binding sites, such as the cerebral cortex and the mushroom bodies, in neural evolution, when it appeared as a prototype in early animals, and how common neural systems have been shared among many phyla. The complex brains in vertebrates and cephalopods (octopuses and squids) that evolved independently are of particular interest. These two types of complex brains evolved independently from a common ancestor that may have had diffused neural nets or simple neural cords, yet the octopus or squid brains have similar sizes, functional regionalized units, and many similar neuronal and glial cell types in the sensory, motor, and cognitive systems (Nixon and Young 2003; Hochner et al. 2006; Darmaillacq et al. 2014; Shigeno et al. 2015).

19.2 The Classic and Modern Ladder-Like Models

How can we distinguish a common structure for the diverse neural centers spanning from invertebrates to advanced species such as mammals? The brains of phylogenetically distant animals often exhibit similarities despite anatomical differences, possibly the result of physiological and functional constraints (Farris 2008). Occurrence of the sensory, motor, and higher centers is summarized with the phyletic

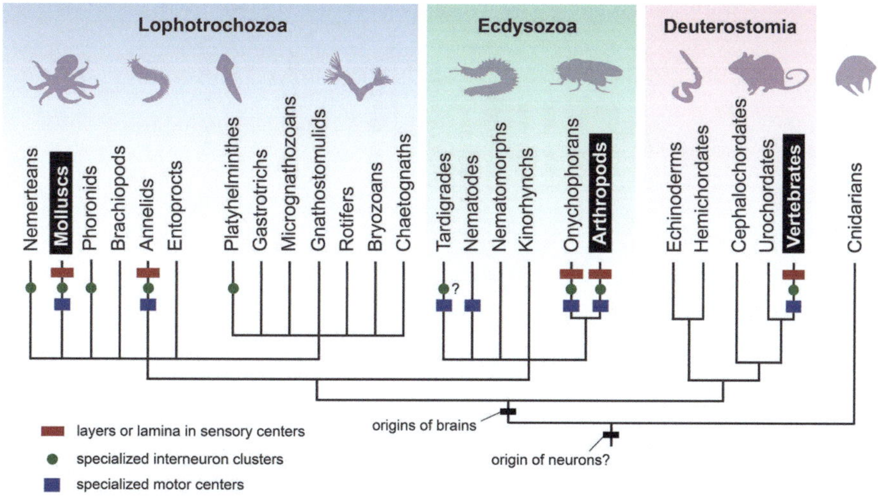

Fig. 19.1 Independent specialization of the higher brain centers. The common origins and homological relationships of higher brain centers are still in debate. The *red rectangles* indicate animals with a laminar or layer-like structure of the higher sensory processing centers, such as the cerebral cortex, the mushroom body, and the frontal-vertical lobe system found in cephalopods. The *green circles* indicate the presence of interneuron clusters in the anterior nervous system. The *blue rectangles* represent highly specialized motor centers (see Table X.1) (This phylogram was adapted from Hejnol et al. (2009; see also Philippe et al. 2009; Laumer et al. 2015) and the data were selected from Bullock and Horridge (1965) and Strausfeld (2012))

relationships of other phyla (Fig. 19.1). Such centers composed of a functional unit of neurons in early development can be arranged along the body axis that is usually defined with phylogenetically shared cell types, expression domains of the neurotransmitter, and developmental control genes (see Arendt 2008, 2015) (Fig. 19.2). The higher centers or associative centers for cognition are often defined by higher order and small intrinsic interneurons that usually appear at the anterior regions of the nervous systems (see Farris 2008) (Fig. 19.2; Table 19.1). The higher centers often form laminar patterns, although they exhibit a species-specific manner (Bullock and Horridge 1965; Butler and Hodos 2005; Strausfeld 2012). These centers are generally found in derived species, but not in basal lineages, suggesting that higher centers evolve independently (Hejnol and Martindale 2008; Farris 2008; Moroz 2009; for common origins and controversies, see Arendt and Nübler-Jung 1996; Reichert and Simeone 2001; Arendt 2008; Tomer et al. 2010).

If higher centers evolved independently, what structural and functional framework supports the commonality between higher order centers in different animals? In this section, the classic and modern models of brain structure are the focus

Table 19.1 The higher sensory, motor, and neurosecretory centers in the cerebral parts of selected animals

		Lophotrochozoa			Ecdysozoa	Deuterostomea
	Platyhelminthes	Molluscs		Annelids	Arthropods	Chordates
	Triclad	Gastropod	Cephalopod	Polychaete	Insect	Vertebrate
The cerebral cluster	Cerebral ganglia	Cerebral ganglia	**Supraesophageal mass**	Cerebral ganglia	Proto- and deutocerebrum	**Fore- and midbrain**
1. Interneuron-rich higher center	Globuli cell cluster	Procerebrum	**Frontal-vertical lobe system**	Mushroom body	Mushroom body	**Cerebral cortex, Pallium**
Specific cell types	*Globuli cells*	*Procerebral interneurons*	***Amacrine cells***	*Globuli cells*	*Kenyon cells*	***Granule cells***
2. Higher sensory-motor centers (basal area)	Spongy region?	Commissural area?	**Basal lobes**	Central body	Central complex	**Basal ganglia**
3. Cerebral neurosecretory center	n/a	Dorsal body	**Subpedunculate lobe, buccal lobe**	Neurosecretory center	PI	**Hypothalamus**

Analogical or homological relationships in some centers are still controversial. Some outgroups are included for comparison (data assembled from Bullock and Horridge 1965; Wells 1978; Butler and Hodos 2005; Hartenstein 2006; Strausfeld and Hirth 2013; and see text for more references). *PI* pars intercerebralis

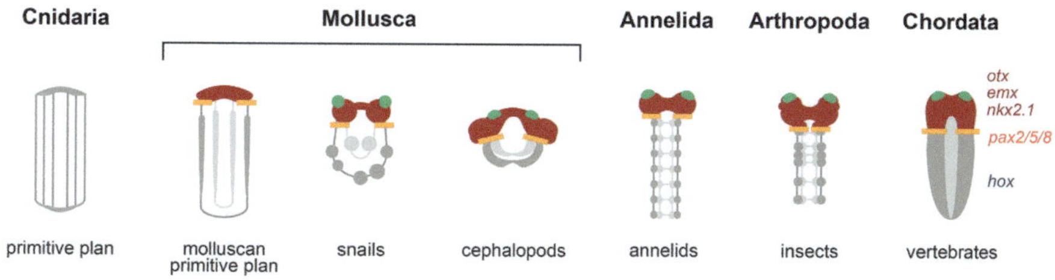

Fig. 19.2 Sketches of a shared plan of the central nervous systems in diverse animals. The figures indicate that two distinct evolutionary pathways can be identified across animal phyla: first, the discrete ganglion-type in snails and arthropods, and second, the centralized neural cords in cephalopods and vertebrates, both of which stem from the primitive plan found in species such as cnidarians. The discrete cords or ganglia are shown from a dorsal view. The bilaterian plan was adapted from Hejnol and Martindale (2008) (Data for the schematic are from Bullock and Horridge (1965), Rubenstein et al. (1998), Shigeno et al. (2010), Puelles and Rubenstein (2015) (see also Wollesen et al. (2015), Sugahara et al. (2016); Watanabe (Chap. 3, in this book) for more details). The expression patterns of some developmental regulatory genes are shared in many animals (Denes et al. 2007; Tessmar-Raible et al. 2007; Tomer et al. 2010))

(Fig. 19.3a, b). All neural systems have sensory, motor, and associative or cognitive subsystems, which has already been suggested by Cajal (1917) and a modern model by Swanson (2003, 2007). In Fig. 19.3a, b, we present a “ladder-like” network composed of two simple flows of information for sensory inputs (S) and motor outputs (M). We see the sensory, association areas, and motor areas, or the “S-A-M” axis, as the core structure. In the case of humans, the cerebral cortex can be seen as a large associative network capable of detailed cognitive processing. The state control networks support the sensory and motor pathways. Each pathway has feedback.

Notably, the S-A-M axis is an important basic architecture for the design of artificial intelligence in insect-like, worm-like, dog-like (Pfeifer et al. 2007; Pfeifer and Gomez 2009), and humanoid robots (Goerick 2009). For example, in the humanoid robot, Honda ASIMO, the central control system, ALIS (Autonomous Learning and Interacting System), has a ladder-like flow of sensory and motor information (Fig. 19.3c). ALIS is composed of a hierarchical control architecture called SYSTEMATICA, which represents a sensory system; the associative internal dynamics are organized in parallel for each sensory modality with interactions between top-down processing and long-term memory, which converge into motor commands for conflict resolution (Fig. 19.3c) (Goerick 2009). The similarity between biological and artificial systems is not surprising because artificial products were inspired by animal brain structures.

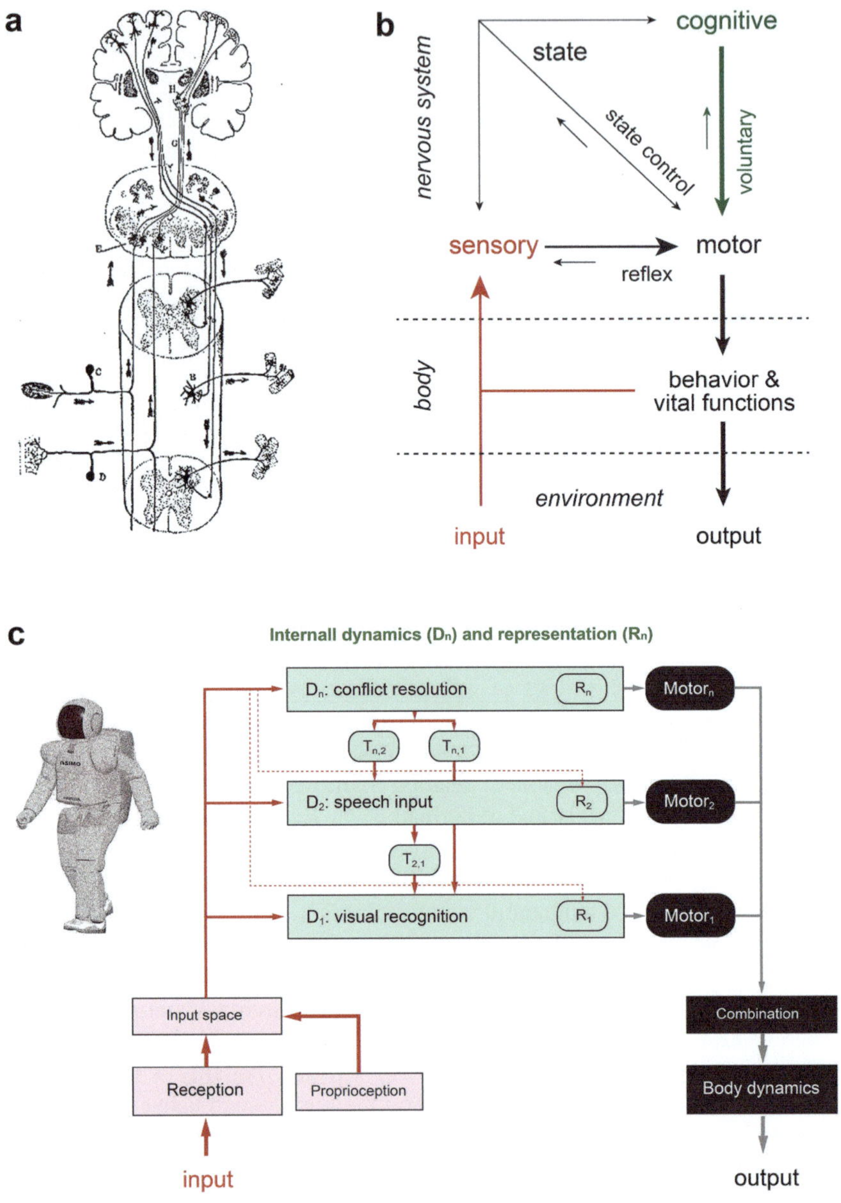

Fig. 19.3 Diagrams of the human brain and control architecture in a humanoid robot, Honda ASIMO, showing the common ladder-like information flow design. (**a**) In the classics, Cajal (1890) published the histological diagram based on the neuron doctrine to show how sensory information from the skin passes to the cerebral cortex and ends in motor output signals to the muscles. (**b**) The modern Swanson model shows the basic plan of the brain (Modified from Swanson 2003). The ladder-like model is composed of four systems: sensory, cognitive, state, and motor systems with reciprocal feedback. (**c**) Artificial autonomous learning and control architecture called ALIS in the humanoid robot Honda ASIMO (Simplified from Goerick 2009). *D* dynamics, *T* top-down information, *R* representations

19.3 The Cerebral Cortex Equivalents

Further evidence for the ladder-like S-A-M axis in different animal brains is the anatomically differentiable centers between the layered structure of the mammalian cerebral cortex and the nucleic nature of the pallium in birds. Karten (1997) originally emphasized the importance of both the developmental origins and the common plans of neural circuit connectivity. First, there is evidence that the cerebral cortex and pallium are both derived from the neurogenic region of the dorsal or rostral forebrain. Furthermore, in comparable regions, a conserved connective structure, the sensory input-interneuron-motor output axis, or S-A-M axis, is found (Karten 1997; see also Jarvis et al. 2005; Karten 2013; Puelles and Rubenstein; 2015). The S-A-M axis is often difficult to identify because of the many feedback loops, but it can be identified by its bundled, dominant fibers, which are often treated as "driver pathways" (Sherman and Guillery 1998, 2006). Based on the S-A-M structural framework, Dugas-Ford et al. (2012; see also Karten 2013) performed comprehensive molecular expression analyses for differentially organized nuclei in birds, cortical areas in reptiles, and six layered neocortical layers in mammals.

Roth also tried to identify a hierarchy of processing areas and the S-A-M axis in birds, octopuses, and honeybees (Roth 2013; see also Hochner 2010; Shomrat et al. 2011; Grasso 2014) (Fig. 19.4a, b). This comparative approach can be applied across phyla, including flatworms, ragworms, and cephalopods. In the schematic figure (Fig. 19.4b), local and feedback pathways are largely omitted for simplicity, and it is not obvious what function is being mapped for most of the pathways. This scheme, however, is important to help find comparable cell types and possible functional synaptic equivalents in complex centers derived from similar feed-forward information processing with cell types and layers. For instance, granule cells in the mammalian cerebral cortex, amacrine cells in octopus brains, and the Kenyon cells of insect mushroom bodies become the primary targets for comparison because of the small size and numerous numbers of interneuron types for information fine-tuning or memory pooling (Fig. 19.4a). Many levels of representation including primary and secondary, associative unimodal and multimodal, and global representations of one's physical body, the environment, and self-awareness may appear along such feed-forward information streams. Thus, the proposal of Roth (2013) and the theories of Cajal (1917) and Swanson (2007) are based on a "ladder-like" composition to represent the S-A-M axis as a whole.

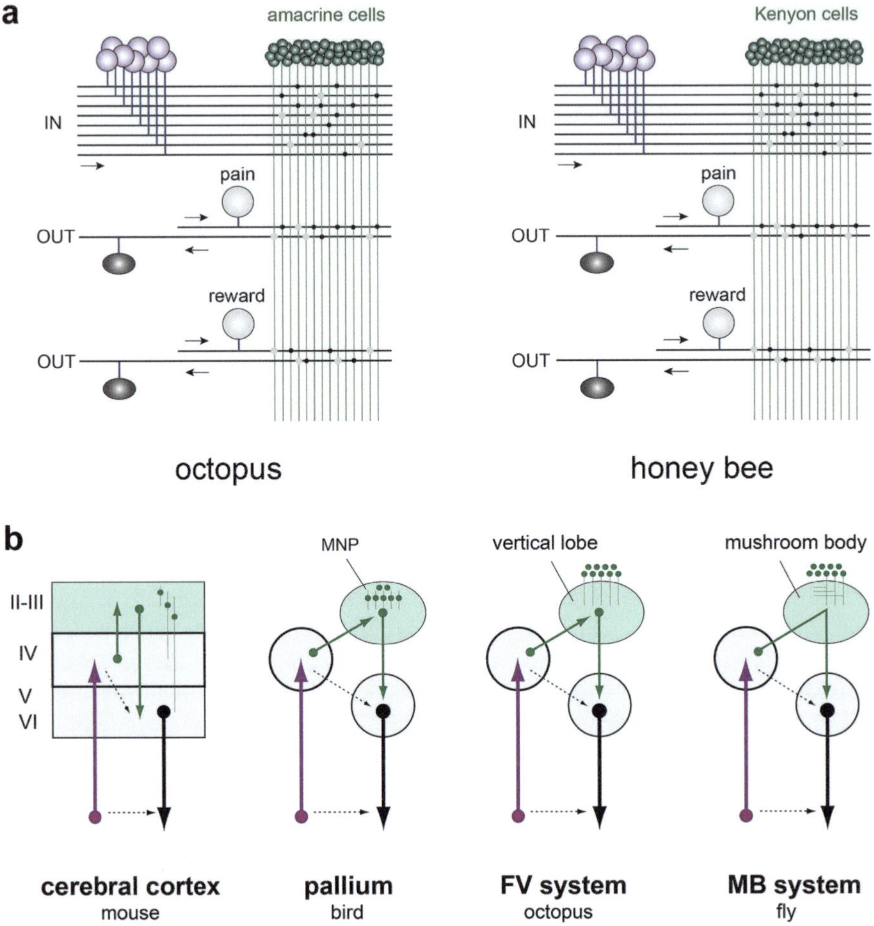

Fig. 19.4 Models for the learning systems and possible cerebral cortical equivalents. (**a**) The comparable models for learning systems in the frontal-vertical (*FV*) lobe system of *Octopus* (Wells 1978), the mushroom body (*MB*), and Kenyon cells in the honeybee brain (Heisenberg 2003). (**b**) A common connectivity structure, based on Karten's hypothesis (Karten 1997, 2013), between several species (see Roth 2013). The similar information flow composed of driver pathways represented by the input-interneuron-output networks (*arrows*) with some nested looping pathways (*dotted arrows*). The small interneurons are bipolar in vertebrates or monopolar in the octopus (The schematic was adapted from Karten (1997), Young (1971), and Strausfeld (2012)). *MNP* mesonidopallium

19.4 The Evolution of Information Flow Patterns

The ladder-like S-A-M axis, as just summarized, has traditionally been thought of as one of the most known structures for basic brain function. The structure consists of state controllers, such as the endocrine system, emotion for reward or reinforcement, and many reciprocal feedback loops. Here, a revised view is proposed. The brain is more than just a ladder and an input–output processing agent. This proposal extends the conventional view that information processing is based on a ladder-like organization. The approach proposed here was influenced by Adrian Bejan's design theory (Bejan and Lorente 2008, 2010), which explains and predicts the most efficient design for flow dynamics. In his theory, the pattern of convergence (many-to-one: signals tend to flow from area to point) or divergence (one-to-many: signals tend to flow from point to area) (Fig. 19.5a) of flow dynamics determines the distinct functional design and features in both animate and inanimate systems. Based on his theory, the networks of primitive and advanced brains can be classified into four representative types to establish the information flow configuration in evolution (Fig. 19.5a):

1. *The diffused type*: this is the typical type of nervous system in many sea anemones, comb jellies, and jellyfishes (Fig. 19.5a). Information processing occurs in simple elements called nodes, which are comparable, in part, with the single-layer perceptron (Rosenblatt 1962), parallel distributed processing (Rumelhart et al. 1986a, b), and the self-organizing map found in artificial networks (Kohonen 1995).
2. *The "many-to-one" type*: this is a typical central nervous system in primitive brains. The signals from sensory receptors of several body areas are localized to "one" point, for example, at the anterior of the body as a brain, but they still display a kind of diffuse arrangement. We see this type of network in the neural ring of cnidarians, the primitive brain of acoelomorphs, and primitive forms of the protostome and deuterostome brains (see Watanabe's chapter in this book).
3. *The "one-to-many" type*: this is defined by the appearance of higher order intrinsic neuronal clusters (Fig. 19.5a, b). Connectivity is organized in "one" centrally located brain, and then processed in "many" small, higher order interneurons, generally called a globuli cell cluster in protostome animals. The numerous small intrinsic neurons with short neuronal processes and synapses are organized in the cortex. The most frequently analyzed cases are the Kenyon or globuli cells and the mushroom bodies in insects and annelid brains. The signals from sensory receptors merge into one brain, and then these signals are processed at many synaptic sites in intrinsic neurons in cortically arealized domains. This type of organization is also seen in some protostomes, such as polyclad flatworms and land snails. Some species possess only a pair of small globuli cell clusters for chemical sensing, which usually lacks feedback loops between primary and higher order centers. Addition of such higher order association or internal representation units between the sensory and motor network axis in higher order animals is similar to a learning technique called "teacher node" or

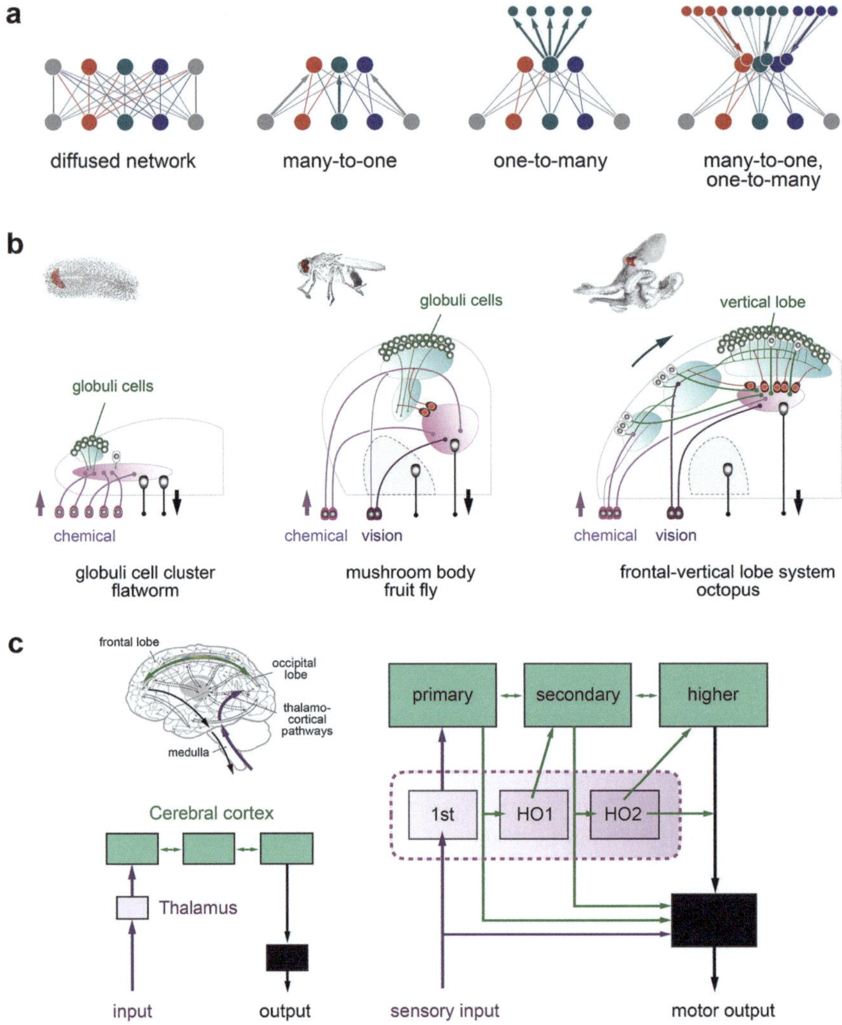

Fig. 19.5 Evolutionary transition of information flow design. (**a**) The four types of information flow design in brain evolution. (**b**) Transitions of major input–output pathways with small interneuron clusters from the brain of polyclad flatworm, fly, and octopus. (**c**) A model of human brains to propose comparable patterns. Many mammalian brains also have the identical systems. The *left* diagram shows simple input–output information flow, whereas a complex thalamocortical system is shown on the *right*. *HO* higher order (Modified from Bullock and Horridge (1965), Young (1971), Sherman and Guillery (2006), Strausfeld (2012))

back-propagation, which is used for training the multilayer perceptron in artificial networks (Rumelhart et al. 1986b).

4. *The "many-to-one, one-to-many" type*: this type is defined by the appearance of interactive association centers for cognitive function. The cortical centers are

composed of many small intrinsic neurons that send information to a relatively simple, or "one," subcortical center, and form the multisensory integration between hierarchically organized motor centers to perform simple behavioral tasks, such as attack or retreat. The subcortical centers project to multiple cortical centers to form reciprocal corticosubcortical loops or relay loops. Such patterns are commonly found in the octopus and vertebrate brains. The most developed loop is the thalamocortical relay loop found in mammals (Sherman and Guillery 2002, 2013; Sherman 2007; Fig. 19.5c).

In the traditional approach, sensory information, such as vision, enters the cortex via the thalamic pathway and is processed within the cortex. However, a recent model suggests that corticocortical information transfer is relayed via newly developed higher order thalamic nuclei, which form the characteristically ordered first-, second-, and higher order thalamocortical relay loops (Sherman and Guillery 2013; Fig. 19.5c). The functional significance is not known in detail, but such differentiable relay loops regulate brain functions, such as attention, by synchronizing sensory responses through burst or tonic modes of thalamic responses (Sherman and Guillery 2006). Models of artificial networks are discussed later, but many attempts have constructed internal representation using integration modules (Mareschal and Shultz 1996).

19.5 Diffused to Compact Form: Identifying a Transitory Pattern

The structure of information flow is important and is useful to provide models based on biophysical theories and to identify the best optimized structure, when we consider information flow along the neural tracts in function and evolution. It is obvious that the neural networks of animals initially have a diffuse structure rather than a ladder-like structure. There are many theories to explain the elaboration process of brain centers and circuitry in vertebrates and arthropods (as summarized in Butler and Hodos 2005; Strausfeld 2012). One of them, the parcellation theory by Ebbesson (1980), may provide a guideline for how the brain evolves. The theory suggests that the brain becomes more complex, not only by one system adding to another, but by a diffused "parcellation" process that involves the selective loss of diffused connections and aggregation of preexisting subsystems (Ebbesson 1980). This theory is based on the findings that (1) a diffuse, undifferentiated system appears first and (2) a range of segregated or "compact" patterns evolves from the diffused type. This process is also typical in early brain development, wherein the brain initially displays the diffused pattern and the segregated pattern emerges later. One principle underlying early growth, differentiation, and multiplication of circuits appears to be the process of parcellation (Ebbesson 1980).

Brain evolution in cephalopods may serve as a model organism. In cephalopods, more extensive subsystems have appeared in a primitive, diffused form. The cerebral cord of a polyplacophoran (e.g., the chiton) has nine transversely distinct zones with

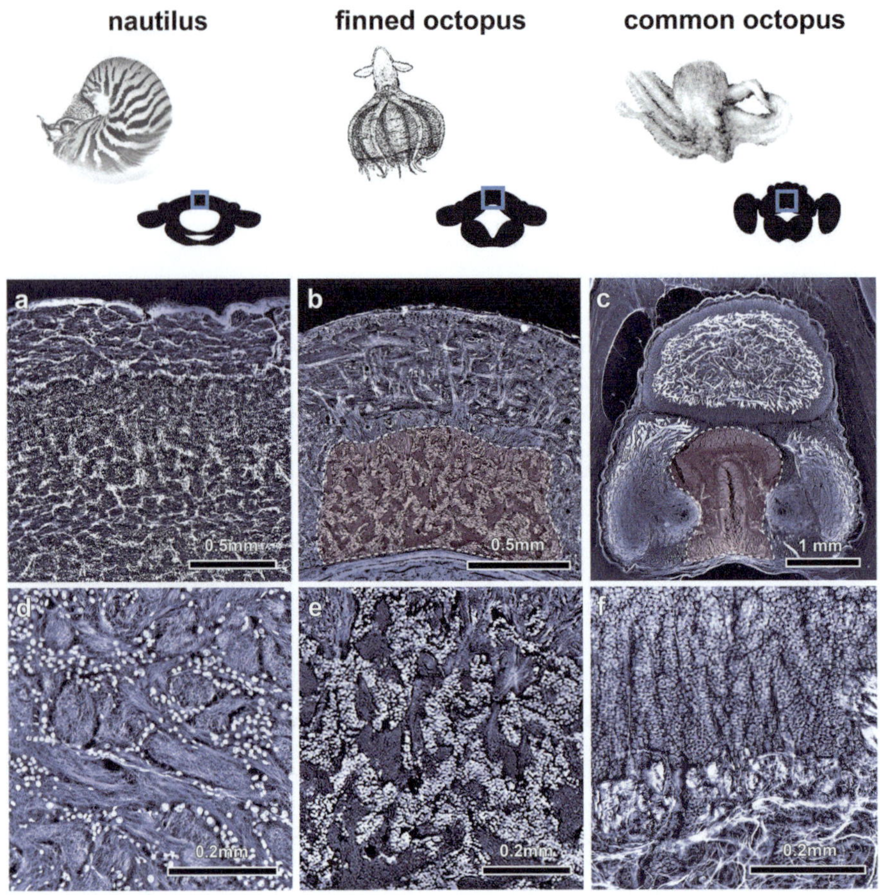

Fig. 19.6 Shift from the primitive diffused centers to the derived compact lobes in the cephalopod higher brain centers: the interlacing system of the nautilus higher centers. (**a**, **d**) The subfrontal lobe in the finned octopus (*red color*), *Stauroteuthis syrtensis* (**b**, **e**), and the subfrontal lobe (*red color*) or the vertical lobe of *Octopus vulgaris* (**c**, **f**), in which the small amacrine cells and neuropil layer are distinct. Many interweaving axonal bundles are still seen in the octopus. The subfrontal lobe is composed of numerous small interneurons with short axons. All photographs are negative images of Cajal staining from the Smithsonian JZ Young collection (kind courtesy of Dr. M. Vecchione, Smithsonian Institute; photographs by the author). The picture of the finned octopus, *Stauroteuthis syrtensis*, is modified from Roper and Sweeney (1984)

diffused neural tracts (Faller et al. 2012). Similarly, the origin of the cephalopod brain can be examined in the primitive nautiloid brain, where higher centers have numerous distributed cell bodies and intersected axonal fibers (Fig. 19.6a) (see also Young 1965). The brain of the deep-sea finned octopus, *Stauroteuthis syrtensis*, has an intermediate state between a distributed and centralized structure (Fig. 19.6b), but the most highly compact and centralized neuropils are in the derived coleoid cephalopods, such as the brain of *Octopus vulgaris* (Fig. 19.6c) (Young 1977; Nixon and Young 2003).

Interestingly, such "diffused to compact" systems have also been seen in the evolution of cognitive centers in arthropods. Although a trend may be difficult to identify because of the great diversity and parallelism, arthropod mushroom bodies display elaboration, acquisition, reduction, or loss of small intrinsic globuli cells, and neuropil compartments with variation in behavioral ecology (Bullock and Horridge 1965; Farris and Strausfeld 2001). One potential mechanism of brain evolution is that the precursors of globuli interneurons are initially distributed in the cerebral cord, resembling a band, with several olfactory inputs from receptors (Strausfeld 2012). In the next step of elaboration, the cell bodies and neuropils are clustered and the major bundled tracts are assembled. Some new functional modular tracts also appeared, but the basic plan has already established a complicated state in the basal lineages of arthropods such as Chelicerata and Crustacea, with consideration to Panarthtropoda, including Onycophora such as velvet worms (Strausfeld 2012).

19.6 Implications for Artificial Neural Networks

It is widely accepted that nervous systems function by physiological electronic signals activated via cellular networks. It is also known that many artificial neural networks are defined by electronically processed information flow, and their algorithms are often defined by sensor, motor, state control (reward and reinforcement), and an aspect detector, or planning system, similar to that seen in the brain (Braitenberg 1984; Sendhoff et al. 2009; Mnih et al. 2015). Even in the simplest cases, such as Von Neumann architecture, Turing machines, and neural Turing machines, artificial agents share a fundamental mechanism for information sorting, attentional processing, logical flow, associative recall, and working memory (Graves et al. 2014). Despite these similarities, the structure of the brain and artificial systems has been debated. Here, I use a comparative approach to describe novel artificial neural networks and identify a common basic organization. First, artificial neural networks have been historically created using a computational model inspired by biological perception and neural networks (Eames et al. 1990). Hebbian learning networks, where synaptic efficacy arises by repeated and persistent stimulation, is one example, but we focus on specific connectivity organization here.

The "perceptron" or a computational model of the "perceptron convergence procedure" was first invented by Rosenblatt (1957, 1962) as a linear classifier or pattern learning machine with simple feed-forward networks (Fig. 19.7a). Mapping sets of input data onto a set of appropriate output, an improved "multilayer perceptron" was constructed with three or more input layers of nodes and a feedback system, each of which in one layer connects to a certain information weight to every node in the next layer (see Minsky and Papert 1988; Rumelhart et al. 1986a). Effective learning occurs in the model by changing connective weights after each

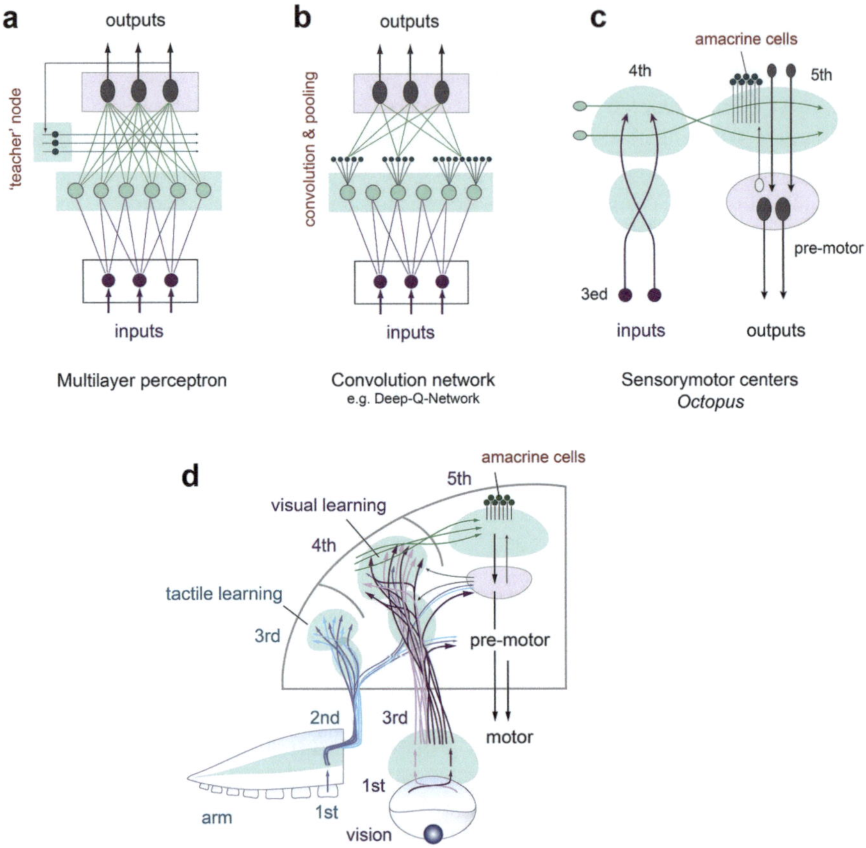

Fig. 19.7 A comparative model of the multilayer perceptron, convolution network (Deep-Q-network), and octopus brain. (**a–c**) Each network shows highly intersected and distributed pathways with hierarchically ordered nodes or centers along the input–output axis. Each network performs learning, recognition, and problem-solving tasks (The multilayer perceptron, convolution network, and octopus brain diagrams are modified from Minsky and Papert (1988), Mnih et al. (2015), and Young (1971, 1995), respectively). (**d**) A more detailed scheme of the octopus brain shows more locally distributed, but globally segregated sensory pathways, as in Fig. 19.6c (Modified from Young 1971)

datum is processed, based on the amount of error in the output compared to the expected output estimated by "teacher node" (Fig. 19.7a). Moreover, a large number of models have been created for a vast array of functions that represent multilevel, nonlinear, and parallel distributed processing (Rumelhart et al. 1986a). More recently, many improved neural networks have been created (Fukushima 1980; Kohonen 1995; Bengio 2009; LeCun et al. 2015), and notably, the brain-inspired agent called "Deep Q-Network" (DQN) has become skilled enough to beat a professional human player in various classic arcade games (Fig. 19.7b) (Mnih et al. 2015). The DQN can learn concepts and object categories directly via a sequence of observations, actions, and rewards. The architecture of the DQN has several layers of nodes that use progressively more abstract representation. Additionally, it has a deep convolutional network that is composed of three hierarchical tiled convolution filters that provide better image recognition using overwrapped visual fields. The basic idea of this design was inspired by the mammalian cerebral cortex to mimic the receptive fields of the visual cortex (Hubel and Wiesel 1963).

When we compare biological neural systems to the multilayer perceptron or deep neural networks, the similarity is obvious in the case of the octopus brain (Fig. 19.7c, d). Both biological and artificial systems are constructed by intersecting pathways and multiple layers, representing mainly feed-forward and parallel distributed networks. Differing from the perceptron or DQN, the octopus system has more nested pathways and a loop-like circuitry, which shape the lower and higher order hierarchy (e.g., lower order: from the sensory input to subvertical pathway; higher order: the input via the vertical lobe pathway) (Fig. 19.7d). Some octopus cell layers correspond to the perceptron or DQN layers. Amacrine cells are like teacher cells or small-scaled convolution processes. Again, recall that the perceptron and many other artificial networks were designed using the mammalian cerebral cortex and cerebellum as a template (Block 1962; Poon and Shah 1998; Voicu 2008) to create diverse applications such as speech, image recognition, and problem-solving abilities to obtain suitable solutions for extremely complex problems. For this reason, the identified similarities and basic organization between the biological and artificial systems might not to be surprising.

19.7 Cognitive Design from Classic to Modern Psychology

Even if the artificial neural networks display intelligence, such products cannot have animal-like cognitive states, such as emotion and consciousness. Creation of the DQN was inspired by the cerebral cortex, particularly the visual processing system, but its design did not draw from regions such as the thalamus, basal ganglia, hippocampus, amygdaloid complex, midbrain, or spinal cord. Complicated psychological phenomena have appeared in animal evolution, which modern artificial intelligence fails to mimic. In fact, the coordinated actions of the sensory, motor, and emotional systems with primary consciousness in vertebrates and cephalopods (Edelman et al. 2005; Edelman and Seth 2009; Mather 2008; Feinberg and Mallatt 2013) are never performed artificially. It is hard to mimic many neuropsychological functions produced by an entire system, including learning, sensation, working memory, sleeping, emotion, attention, spatial navigation, allocentric and egocentric awareness, and social interaction.

Modern neuroscientists have identified the major neuronal components of the S-A-M axis as a basic structure that is shared between phylogenetically distant animals (Fig. 19.3b) (Swanson 2007). This basic design was proposed by ancient Greek philosophers. Despite any empirical evidence, Plato proposed a preliminary concept that the major functions in humans are based on common sensory, motor, and cognitive cues, or a "sensory-cognitive-motor axis" (Plato, around 380 BCE). A more precise theory is attributed to Aristotle's comparative behavioral studies in which he described the functional subdivisions of the "mind" or "intellect" (Aristotle, around 350 BCE). These theories were largely speculative, but were influential until roughly the late seventeenth century, when French philosopher René Descartes formulated his theory of mind–brain dualism and the geometric rules of sensory integration (Descartes 1644).

More precise models of sensory perception and knowledge were proposed by Benedict de Spinoza (1677), John Locke (1690), David Hume (1748), and Immanuel Kant (1781). The most prominent studies were published by Kant (1781, 1788), who posed questions about the most basic and "pure" architecture of the complicated mental functions in the human. He theorized that mental functions are composed of a few major components such as concepts of creation and consciousness. The strategy of Kant in his *Critique of Pure Reason* was to propose the basic architecture for human minds, particularly sensory perception or aesthetic represented in space and time, categorical representation as analogies of experiences, cognition, judgment, and a rule for self-perception or consciousness. Although he never provided the neuroanatomical terms, he explained his theory as if an architect or comparative anatomist. For example, "Human reason is by nature architectonic, i.e., it considers all cognitions as belonging to a possible system . . . " (Kant 1781; p. 502, L23). "Systems seem, like certain worms, to be formed . . . " (p. 692, L33). In his *Critique*, the S-A-M axis is represented as the sensory (aesthetic), cognition (logic, reason, judgment), and motor (behavior, action) axis. "All our cognition starts from the senses, goes from there to the understanding, and ends with reason, beyond which there is nothing higher to be found in us . . . " (p. 387, L9).

One of the most important achievements of Kant was the identification of the "many-to-one, one-to-many" architecture as a crucial flow design for psychological processes (Fig. 19.8a). Many sensory faculties or centers for conceptual representation are integrated as "many-to-one." The concepts interact as "one-to-many." Kant found that the interaction between convergent and divergent information flow design is central to human mental processing. Such interactions are examples of "dialectic conflict" or "antinomy." This conflict is a compulsory trade-off: the centers for sensory representation tend to divide into "many" areal spaces whereas the many spaces tends to unite into "one." Many different types of sensory information have to be integrated to produce a simple "one" action. In Kant's *Critique of Pure Reason*, dialectic conflicts occupy about 220 of 700 pages to explain the mental functions of knowing, understanding, concepts, logic, decision making, illusion, estimation, and self-consciousness. Further theoretical debates have built upon Kant's theories, including those from the post-Kantian German idealism philosophers, such as Georg Hegel (1977), the phenomenologists, such as Edmund Husserl (1913), and the modern logicians, such as Bertrand Russell (1921).

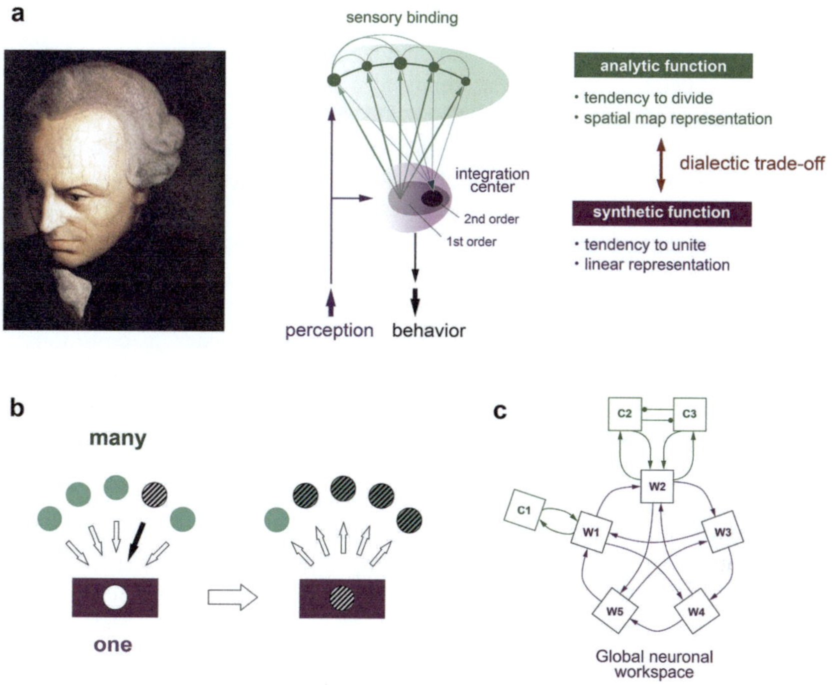

Fig. 19.8 Cognitive architectures. (**a**) A simplified interpretation for cognitive architecture in Kant's *Critiques of Pure Reason* (1781). There are two distinct faculties for cognition. First, the faculty of receptivity in which the sensory representations are cognized locally as concepts, sensory binding, or [elemental logic] (Kant 1781). Second, a unity of conceptions is demonstrated at the highest faculty for reason (Kant 1781), where the many categorized representations tend to be reduced to the smallest possible number by a total synthetic integration center. The functional differences of analytical (divisible) and synthetic (unified) functions give rise to a self-contradictory conflict (Kant 1781). The photograph from Public Domain. (**b**) A simplified diagram for the global workspace model for human consciousness by Baars (1988) and Baars et al. (2013) (Modified from Shanahan 2006). The many distinct and distributed contexts or information units bind to one massive center, and then the influenced contexts again change to distinct contexts along the linear timeline. (**c**) A network model and diagram simplified from Dehaene et al. (1998). The peripheral distinct processors connect to the central global workspace for long range and distributed activity

19.8 Current Frontiers on the Designs

The ideas of Kant have been incorporated into modern theories; for example, the concept-knowledge design theory (C-K theory) (Hatchuel and Weil 2009). This theory explains the logic that organizes the generation of unknown objects during problem solving. In the C-K theory, there are two distinct spaces: concept space and knowledge space. Concepts that depart from existing knowledge can be partitioned into "many" mutually interacting subconcepts in the concept space, and then derived concepts return to the simple "one" knowledge space for synthesis into a new single concept. This theory is based on a "many-to-one, one-to-many" information flow design, similar to those proposed by Kant (1781).

Interestingly, current neuropsychologists and computational neuroscientists have reached similar solutions. Even though their models are very different, they were placed on the strong influence (Baars 1988, 2002; Baars et al. 2013; Crick and Koch 1990; Damasio 2000; Edelman and Tononi 2000; Tononi 2004, 2008; Merker 2007; see Mudrik et al. 2014 for references). For example, Baars theorized about the simplest and the most complex computational designs in the brain. He focused on the parallel, sequential, and limited operations that occur in conscious and unconscious states in the corticothalamic pathways. Information from multiple sensory modalities is integrated into a space called "the global workspace," which forms and keeps a stream of rich experiences that happen one after another to form a conscious state. The content of a winning neuronal populations are broadcast to all other populations, forming a spatially integrated and continuous conscious state (Fig. 19.8b) (Shanahan 2006). Inspired by the Baars global workspace (GW) theory, several computational models and autonomous agents have been created in which the "sensory-GW-motor" or "S-GW-M" axis is the core design for information flow (Fig. 19.8c) (Dehaene et al. 1998; Shanahan 2006; Franklin 2003; Franklin et al. 2012). In these artificial models and in human GW theory, the "many-to-one, one-to-many" structure is the core morphological and functional structure in brain evolution.

One compelling interpretation of our perspective that the basic structure underlying the evolution of the brain is a whole interactive system represented by the "many-to-one, one-to-many" organization suggested by Kant, Hatchuel, and Baars (Fig. 19.7a, b), rather than the ladder-like axis that has been used for animals (Cajal 1917; Swanson 2007) and robots (Pfeifer et al. 2007; Goerick 2009). During animal brain evolution, this basic architecture may have evolved independently because of a stable framework that includes a dynamic trade-off of analytical (high resolution, but slow processing speed) and synthetic (low resolution, but fast processing speed) functions. Creating detailed spatial maps is essential to navigate one's environment to increase fitness, but information must be transported efficiently to enable effective behavior. The organization of the whole brain may provide a reasonable solution for such combinatorial optimization adaptive problems. The models described in the present paper are probably too global, large scale, and hard to study in neuroscience, and we do not yet have the methodology to gain access into the subjective realm in animal brains, except for recent advances in whole-brain activity imaging in vivo, such as in the zebrafish brain (Ahrens et al. 2012) and nematode nervous system (Nguyen et al. 2015), or fruit fly brains (Honegger et al. 2011).

19.9 Conclusion

Our current understanding of the evolution of the brain is too preliminary to permit us to propose a common functional layout in any simple way, despite many attempts in classic theories (Cajal 1917; Hanström 1928; Bullock and Horridge 1965; Reisinger 1972; Arendt and Nübler-Jung 1996) and in recent advanced studies and perspectives (Holland 2003; Arendt 2008; Hejnol and Martindale 2008; Moroz 2009; Pani et al. 2012; Northcutt 2012; Schmidt-Rhaesa et al. 2016; Borden et al. 2016). With such a long historical background, one may ask how a global information flow design has been shaped in brain evolution. I proposed a large-scale framework by which one can address questions about the functionality and performance shared between the brains of phylogenetically distant animals and even in artificial intelligence. A theory of information flow design is notably reminiscent to the theories proposed by the classic Greek philosophers Plato and Aristotle, which were further enhanced and structured by Immanuel Kant. Recently, theories have been proposed by Adrean Bejan (Bejan and Lorente 2010), who formed a concept-knowledge design theory (Hatchuel and Weil 2009), a global working space model for human consciousness was put forward by Baar (1988, 2002), and an information integration theory was presented by Tononi (2004, 2008). All these theories discussed information-processing patterns: the divergent "one-to-many," convergent "many-to-one," and their structure as a global underlying structure for the evolution of the brain. Despite its importance, such global patterns are not effectively adopted into recent evolutionary studies or the research for creating artificial intelligence. This proposal is based on a global connectivity design, but is neither a model nor a theory. It is a large-scale framework by which we can address questions about the structure and function shared between the brains of phylogenetically distant animals and even in the design of artificial general intelligence.

References

Ahrens M, Li MJ, Orger MB, Robson DN, Schier AF, Engert F, Portugues R (2012) Brain-wide neuronal dynamics during motor adaptation in zebrafish. Nature (Lond) 485:471–479

Arendt D (2008) The evolution of cell types in animals: emerging principles from molecular studies. Nat Rev Genet 9:868–882

Arendt D, Nübler-Jung K (1996) Common ground plans in early development in mice and flies. Bioessays 18:255–258

Arendt D, Tosches MA, Marlow H (2015) From nerve net to nerve ring, nerve cord and brain: evolution of the nervous system. Nat Rev Neurosci 17:61–72

Aristotle, De Anima [On the Soul] (350BCE) Delphi complete works of Aristotle, illustrated, Delphi ancient classics book 11, English edn. (2013) Amazon Ser Int, Inc

Baars BJ (1988) A cognitive theory of consciousness. Cambridge University Press, New York

Baars BJ (2002) The conscious access hypothesis: origins and recent evidence. Trends Cogn Sci 6:47–52

Baars BJ, Franklin S, Ramsoy TZ (2013) Global workspace dynamics: cortical "binding and propagation" enables conscious contents. Front Psychol 4:200
Bejan A, Lorente S (2008) Design with constructal theory. Wiley, Hoboken
Bejan A, Lorente S (2010) The constructal law of design and evolution in nature. Philos Trans R Soc Lond B 365:1335–1347
Bengio Y (2009) Learning deep architectures for AI. Found Trend Mach Learn 2:1–127
Block HD (1962) The perceptron: a model for brain functioning. Rev Mod Phys 34:123–135
Borden NM, Forseen SE, Cristian S (2016) Imaging anatomy of the human brain. A comprehensive atlas including adjacent structure. Demos Medical, New York
Braitenberg V (1984) Vehicles: experiments in synthetic psychology. Bradford/MIT Press, Cambridge
Bullock TH (1993) How are more complex brains different. Brain Behav Evol 41:88–96
Bullock TH (2002) Grades in neural complexity: how large is the span? Integr Comp Biol 42:757–761
Bullock TH, Horridge GA (1965) Structure and function in the nervous systems of invertebrates, vol I, II. Freeman, London
Butler AB, Hodos W (2005) Comparative vertebrate neuroanatomy: evolution and adaptation, 2nd edn. Wiley-Liss, New York
Cajal SR (1890) New ideas on the structure of the nervous system in man and vertebrates. English translation, Swanson N, Swanson LW (1990) MIT Press, Cambridge
Cajal SR (1917) Histologie du système nerveux de l'homme et des vertèbrès, 2 vols. Translated by Azoulay L, Maloine A, Paris. English translation, Swanson N, Swanson LW (1995) Histology of the nervous system of man and vertebrates, 2 vols. Oxford University Press, New York
Crick F, Koch C (1990) Towards a neurobiological theory of consciousness. Semin Neurosci 2:273–304
Damasio AR (2000) The feeling of what happens: body and emotion in the making of consciousness. Vintage Books, London
Darmaillacq AS, Dickel L, Mather J (2014) Cephalopod cognition. Cambridge University Press, Cambridge
Dehaene S, Kerszberg M, Changeux JP (1998) A neuronal model of a global workspace in effortful cognitive tasks. Proc Natl Acad Sci USA 95:14529–14534
Denes AS, Jekely G, Steinmetz PR, Raible F, Snyman H, Prud'homme B, Ferrier DE, Balavoine G, Arendt D (2007) Molecular architecture of annelid nerve cord supports common origin of nervous system centralization in Bilateria. Cell 129:277–288
Descartes R (1644) The principles of philosophy. Translation with explanatory notes. Rodger V, Miller RP, Reprint edition (1983). Reidel, Dordrecht
Dugas-Ford J, Rowell JJ, Ragsdale CW (2012) Cell-type homologies and the origins of the neocortex. Proc Natl Acad Sci USA 109:16974–16979
Eames C, Eames R, Fleck G, Cohen IB, Staples R (1990) A computer perspective: background to the computer age. Revised subject edition. Harvard University Press, Harvard
Ebbesson SOE (1980) The parcellation theory and its relation to interspecific variability in brain organization, evolutionary and ontogenetic development, and neuronal plasticity. Cell Tissue Res 213:179–212
Edelman DB, Seth AK (2009) Animal consciousness: a synthetic approach. Trends Neurosci 32:476–484
Edelman GM, Tononi G (2000) A universe of consciousness: how matter becomes imagination. Basic Books, New York
Edelman DB, Baars BJ, Seth AK (2005) Identifying hallmarks of consciousness in non-mammalian species. Conscious Cogn 14:169–187
Faller S, Rothe BH, Todt C, Schmidt-Rhaesa A, Loesel R (2012) Comparative neuroanatomy of Caudofoveata, Solenogastres, Polyplacophora, and Scaphopoda (Mollusca) and its phylogenetic implications. Zoomorphology 131:149–170
Farris SM (2008) Evolutionary convergence of higher brain centers spanning the protostome-deuterostome boundary. Brain Behav Evol 72:106–122

Farris SM, Strausfeld NJ (2001) Development of laminar organization in the mushroom bodies of the cockroach: Kenyon cell proliferation, outgrowth, and maturation. J Comp Neurol 439:331–351
Feinberg TE, Mallatt J (2013) The evolutionary and genetic origins of consciousness in the Cambrian period over 500 million years ago. Front Psychol 4:667
Franklin S (2003) IDA: a conscious artifact? In: Holland O (ed) Machine consciousness. Imprint Academic, Exeter
Franklin S, Strain S, Snaider J, McCall R, Faghihi U (2012) Global workspace theory, its LIDA model and the underlying neuroscience. Biol Insp Cogn Arch 1:32–43
Fukushima K (1980) Neocognitron: a self-organizing neural network model for a mechanism of pattern recognition unaffected by shift in position. Biol Cybern 36:193–202
Goerick C (2009) Towards cognitive robotics. In: Sendhoff B, Koerner E, Sporns O, Ritter H, Doya K (eds) Creating brain-like intelligence. From basic principles to complex intelligent systems. Springer, Heidelberg, pp 192–214
Grasso FW (2014) The octopus with two brains: how are distributed and central representations integrated in the octopus central nervous system? In: Darmaillacq A-S, Dickel L, Mather J (eds) Cephalopod cognition. Cambridge University, Cambridge, pp 94–122
Graves A, Wayne G, Danihelka I (2014) Neural turing machines. arXiv preprint arXiv:1410.5401
Hanström B (1928) Vergleichende Anatomie des Nervensystems der Wirbellosen Tiere unter Beruecksichtigung Seiner Funktion. Julius Springer, New York
Hartenstein V (2006) The neuroendocrine system of invertebrates: a developmental and evolutionary perspective. J Endocrinol 190:555–570
Hatchuel A, Weil B (2009) C-K design theory: an advanced formulation. Res Eng Des 19:181–192
Hegel GWF (1977) Phenomenology of spirit. Translation by Miller AV, Findlay JN. Clarendon Press, Oxford
Heisenberg M (2003) Mushroom body memoir: from maps to models. Nat Rev Neurosci 4:266–275
Hejnol A, Martindale MQ (2008) Acoel development supports a simple planula-like urbilaterian. Philos Trans R Soc Lond B 363:1493–1501
Hejnol A, Obst M, Stamatakis A, Ott M, Rouse GW, Edgecombe GD, Martinez P, Baguñà J, Bailly X, Jondelius U, Wiens M, Müller WG, Seaver E, Wheeler WC, Martindale MQ, Giribet G, Dunn CW (2009) Assessing the root of bilaterian animals with scalable phylogenomic methods. Proc R Soc Lond B 276:4261–4270
Hochner B (2010) Functional and comparative assessments of the octopus learning and memory system. Front Biosci 2:764–771
Hochner B, Shomrat T, Fiorito G (2006) The octopus: a model for a comparative analysis of the evolution of learning and memory mechanisms. Biol Bull 210:308–317
Holland ND (2003) Early central nervous system evolution: an era of skin brains? Nat Rev Neurosci 4:617–627
Holland ND (2016) Nervous systems and scenarios for the invertebrate-to-vertebrate transition. Philos Trans R Soc B 371:0. doi:10.1098/rstb.2015.0047
Honegger KS, Campbell RA, Turner GC (2011) Cellular-resolution population imaging reveals robust sparse coding in the *Drosophila* mushroom body. J Neurosci 31:11772–11785
Hubel DH, Wiesel TN (1963) Shape and arrangement of columns in cat's striate cortex. J Physiol 165:559–568
Hume D (1748) An enquiry concerning human understanding: a critical edition. The Clarendon edition of the works of David Hume (2006), Beauchamp TL, ed. Oxford University Press, Oxford
Husserl E (1913) Ideas pertaining to a pure phenomenology and to a phenomenological philosophy – first book: general introduction to a pure phenomenology, 1982. Kersten F, trans. Nijhoff Publisher, Hague
Jarvis ED, Güntürkün O, Bruce L, Csillag A, Karten H, Kuenzel W, Medina L, Paxinos G, Perkel DJ, Shimizu T, Avian Brain Nomenclature Consortium (2005) Avian brains and a new understanding of vertebrate brain evolution. Nat Rev Neurosci 6:151–159

Kaas JH (ed) (2006) Evolution of nervous systems. Vol. 1. Theories, development, invertebrates. Elsevier Academic Press, Oxford
Kant I (1781) Critique of pure reason. In: Complete works of Immanuel Kant. Translated into English by Meiklejohn JMD (1998). Cambridge University Press, Cambridge
Kant I (1788) Critique of practical reason. In: Complete works of Immanuel Kant. Translated into English by Meiklejohn JMD (1998). Cambridge University Press, Cambridge
Karten HJ (1997) Evolutionary developmental biology meets the brain: the origins of mammalian cortex. Proc Natl Acad Sci USA 94:2800–2804
Karten HJ (2013) Neocortical evolution: neuronal circuits arise independently of lamination. Curr Biol 7:12–15
Kohonen T (1995) Self-organizing maps. Springer, Heidelberg
Krubitzer L (2009) In search of a unifying theory of complex brain evolution. Ann NY Acad Sci 1156:44–67
Laumer CE, Bekkouche N, Kerbl A, Goetz F, Neves RC, Sørensen MV, Kristensen RM, Hejnol A, Dunn CW, Giribet G, Worsaae K (2015) Spiralian phylogeny informs the evolution of microscopic lineages. Curr Biol 25:2000–2006
LeCun Y, Bengio Y, Hinton G (2015) Deep learning. Nature (Lond) 521:436–444
Locke J (1690) An essay concerning human understanding. Penguin Classics, new edition (2008). Oxford University Press, Oxford
Mareschal D, Shultz TR (1996) Generative connectionist networks and constructivist cognitive development. Cogn Dev 11:571–603
Mather JA (2008) Cephalopod consciousness: behavioural evidence. Conscious Cogn 17:37–48
Merker B (2007) Consciousness without a cerebral cortex: a challenge for neuroscience and medicine. Behav Brain Sci 30:63–81, with discussion part pp 81–134
Minsky ML, Papert SA (1988) Perceptrons: an introduction to computational geometry, expanded edition. MIT Press, Cambridge
Mnih V, Kavukcuoglu K, Silver D, Rusu AA, Veness J, Bellemare MG, Graves A, Riedmiller M, Fidjeland AK, Ostrovski G, Petersen S, Beattie C, Sadik A, Antonoglou I, King H, Kumaran D, Wierstra D, Legg S, Hassabis D (2015) Human-level control through deep reinforcement learning. Nature (Lond) 518:529–533
Moroz LL (2009) On the independent origins of complex brains and neurons. Brain Behav Evol 74:177–190
Mudrik L, Faivre N, Koch C (2014) Information integration without awareness. Trends Cogn Sci 18:488–496
Nguyen JP, Frederick B, Shipleya FB, Linderc AN, Plummer GS, Liu M, Setru SU, Shaevit JW, Leifer AM (2015) Whole-brain calcium imaging with cellular resolution in freely behaving *Caenorhabditis elegans*. Proc Natl Acad Sci USA 113(8):E1074–E1081
Nieuwenhuys R, Ten Donkelaar HJ, Nicholson C (1998) The central nervous system of vertebrates, 3rd edn. Springer, Heidelberg
Nixon M, Young JZ (2003) The brains and lives of cephalopods. University Press, Oxford
Northcutt RG (2012) Evolution of centralized nervous systems: two schools of evolutionary thought. Proc Natl Acad Sci U S A 109:10626–10633
Pani AM, Mullarkey EE, Aronowicz J, Assimacopoulos S, Grove EA, Lowe CJ (2012) Ancient deuterostome origins of vertebrate brain signaling centres. Nature (Lond) 483:289–294
Pfeifer R, Gomez G (2009) Morphological computation: connecting brain, body, and environment. In: Sendhoff B et al (eds) Creating brain-like intelligence. Springer, Berlin, pp 66–83
Pfeifer R, Lungarella M, Iida F (2007) Self-organization, embodiment, and biological inspired robotics. Science 318:1088–1093
Philippe H, Derelle R, Lopez P, Pick K, Borchiellini C, Boury-Esnault N, Vacelet J, Renard E, Houliston E, Queinnec E, Silva CD, Wincker P, Guyader HL, Leys S, Jackson DJ, Schreiber F, Erpenbeck D, Morgenstern B, Worheide G, Manuel M (2009) Phylogenomics revives traditional views on deep animal relationships. Curr Biol 19:706–712
Plato (380BCE) Republic, Delphi complete works of Plato, illustrated, Delphi ancient classics book 5, English edition (2012) Amazon Serv Int, Inc

Poon CS, Shah JV (1998) Hebbian learning in parallel and modular memories. Biol Cybern 78:79–86

Puelles L, Rubenstein JL (2015) A new scenario of hypothalamic organization: rationale of new hypotheses introduced in the updated prosomeric model. Front Neuroanat 9:27

Reichert H, Simeone A (2001) Developmental genetic evidence for a monophyletic origin of the bilaterian brain. Philos Trans R Soc Lond B 356:1533–1544

Reisinger E (1972) Die Evolution des Orthogons der Spiralier und das Archicoelomatenproblem. Zeitsch Zool System Evol 10:1–43

Roper CFE, Sweeney MJ (1984) FAO species catalogue. Vol 3. Cephalopods of the world. An annotated and illustrated catalogue of species of interest to fisheries. FAO, Roma

Rosenblatt F (1957) The perceptron – a perceiving and recognizing automaton. Report 85-460-1, Cornell Aeronautical Lab, New York

Rosenblatt F (1962) Principles of neurodynamics. Spartan, New York

Roth G (2013) The long evolution of brains and minds. Springer, Heidelberg

Roth G, Wullimann MF (2001) Brain evolution and cognition. Wiley-Liss, Berlin

Rubenstein JL, Shimamura K, Martinez S, Puelles L (1998) Regionalization of the prosencephalic neural plate. Annu Rev Neurosci 21:445–478

Rumelhart DE, Hinton GE, Williams RJ (1986a) Learning internal representations by error propagation. In: Rumelhart DE, McClelland JL, PDP research group (eds) Parallel distributed processing: explorations in the microstructure of cognition. Volume 1: Foundations. MIT Press, Cambridge, pp 318–362

Rumelhart DE, McClelland JL, The PDP research group (eds) (1986b) Parallel distributed processing: explorations in the microstructure of cognition. Volume 1: Foundations. MIT Press, Cambridge

Russell B (1921) The analysis of mind. George Allen & Unwin, London

Schmidt-Rhaesa A, Harzsch S, Purschke G (2016) Structure and evolution of invertebrate nervous systems. Oxford University Press, Oxford

Schneider GE (2014) Brain structure and its origins: in development and in evolution of behavior and the mind. MIT Press, Cambridge

Sendhoff B, Koerner E, Sporns O, Ritter H, Doya K (eds) (2009) Creating brain-like intelligence. From basic principles to complex intelligent systems. Springer, Heidelberg

Shanahan M (2006) A cognitive architecture that combines internal simulation with a global workspace. Conscious Cogn 15:433–449

Sherman SM (2007) The thalamus is more than just a relay. Curr Opin Neurobiol 17:417–422

Sherman SM, Guillery RW (1998) On the actions that one nerve cell can have on another: distinguishing 'drivers' from 'modulators'. Proc Natl Acad Sci USA 95:7121–7126

Sherman SM, Guillery RW (2002) The role of the thalamus in the flow of information to the cortex. Philos Trans R Soc Lond B 357:1695–1708

Sherman SM, Guillery RW (2006) Exploring the thalamus, 2nd edn. Academic, San Diego

Sherman SM, Guillery RW (2013) Functional connections of cortical areas. A new view from the thalamus. MIT Press, Cambridge

Shigeno S, Sasaki T, Boletzky SV (2010) The origins of cephalopod body plans: a geometrical and developmental basis for the evolution of vertebrate-like organ systems. In: Tanabe K, Shigeta Y, Sasaki T, Hirano H (eds) Cephalopods—present and past. Tokai University Press, Tokyo, pp 23–34

Shigeno S, Parnaik R, Albertin C, Ragsdale C (2015) Evidence for a cordal, not ganglionic, pattern of cephalopod brain neurogenesis. Zool Lett 1:26

Shomrat T, Graindorge N, Bellanger C, Fiorito G, Loewenstein Y, Hochner B (2011) Alternative sites of synaptic plasticity in two homologous "fan-out fan-in" learning and memory networks. Curr Biol 21:1773–1782

Spinoza B de (1677). The ethics. Translated by GHR Parkinson. English new edition (1993). Tuttle Publishing, North Clarendon

Strausfeld NJ (2012) Arthropod brains: evolution, functional elegance and historical significance. Belknap Press of Harvard University Press, Cambridge

Strausfeld NJ, Hirth F (2013) Deep homology of arthropod central complex and vertebrate basal ganglia. Science 340:157–161
Striedter GF (2005) Principles of brain evolution. Sinauer Associates, Sunderland
Sugahara F, Pascual-Anaya J, Oisi Y, Kuraku S, Aota S, Adachi N, Takagi W, Sato N, Murakami Y, Kuratani S (2016) Evidence from cyclostomes for complex regionalization of the ancestral vertebrate brain. Nature (Lond) 10:1038
Swanson LW (2003). Brain architecture: understanding the basic plan. 1st Edition. New York: Oxford University Press, Japanese translation 2010
Swanson LW (2007) Quest for the basic plan of nervous system circuitry. Brain Res Rev 55:356–372
Tessmar-Raible K, Raible F, Christodoulou F, Guy K, Rembold M, Hausen H, Arendt D (2007) Conserved sensory-neurosecretory cell types in annelid and fish forebrain: insights into hypothalamus evolution. Cell 129:1389–1400
Tomer AS, Denes K, Tessmar-Raible K, Arendt D (2010) Profiling by image registration reveals common origin of annelid mushroom bodies and vertebrate pallium. Cell 142:800–809
Tononi G (2004) An information integration theory of consciousness. BMC Neurosci 5:42
Tononi G (2008) Consciousness as integrated information: a provisional manifesto. Biol Bull 215:216–242
Voicu H (2008) The cerebellum: an incomplete multilayer perceptron? Neurocomputing 72:592–599
Wells MJ (1978) *Octopus*: physiology and behaviour of an advanced invertebrate. Chapman & Hall, London
Wolff GH, Strausfeld NJ (2016) Genealogical correspondence of a forebrain centre implies an executive brain in the protostome–deuterostome bilaterian ancestor. Philos Trans R Soc B 371:20150055
Wollesen T, Rodríguez Monje SV, Todt C, Degnan BM, Wanninger A (2015) Ancestral role of Pax2/5/8 in molluscan brain and multimodal sensory system development. BMC Evol Biol 15:231
Young JZ (1965) The central nervous system of *Nautilus*. Philos Trans R Soc Lond B 249:1–25
Young JZ (1971) The anatomy of the nervous system of *Octopus vulgaris*. Oxford University Press, Oxford
Young JZ (1977) Brain, behaviour and evolution of cephalopods. Symp Zool Soc Lond 38:377–434
Young JZ (1995) Multiple matrices in the memory system of *Octopus*. In: Abbott JN, Williamson R, Maddock L (eds) Cephalopod neurobiology. Oxford University Press, Oxford, pp 431–443

MIX
Papier aus verantwortungsvollen Quellen
Paper from responsible sources
FSC® C105338
FSC
www.fsc.org

If you have any concerns about our products,
you can contact us on
ProductSafety@springernature.com

In case Publisher is established outside the EU,
the EU authorized representative is:
Springer Nature Customer Service Center GmbH
Europaplatz 3, 69115 Heidelberg, Germany

Printed by Libri Plureos GmbH
in Hamburg, Germany